Communications in Computer and Information Science 1492

More information about this series at https://link.springer.com/bookseries/7899

Yuqing Sun · Tun Lu · Buqing Cao ·
Hongfei Fan · Dongning Liu · Bowen Du ·
Liping Gao (Eds.)

Computer Supported Cooperative Work and Social Computing

16th CCF Conference, ChineseCSCW 2021
Xiangtan, China, November 26–28, 2021
Revised Selected Papers, Part II

Editors
Yuqing Sun
Shandong University
Jinan, China

Tun Lu
Fudan University
Shanghai, China

Buqing Cao
Hunan University of Science and Technology
Xiangtan, China

Hongfei Fan
Tongji University
Shanghai, China

Dongning Liu
Guangdong University of Technology
Guangzhou, China

Bowen Du
University of Warwick
Coventry, UK

Liping Gao
University of Shanghai for Science
and Technology
Shanghai, China

ISSN 1865-0929 ISSN 1865-0937 (electronic)
Communications in Computer and Information Science
ISBN 978-981-19-4548-9 ISBN 978-981-19-4549-6 (eBook)
https://doi.org/10.1007/978-981-19-4549-6

This Springer imprint is published by the registered company Springer Nature Singapore Pte Ltd.
The registered company address is: 152 Beach Road, #21-01/04 Gateway East, Singapore 189721, Singapore

Preface

Welcome to ChineseCSCW 2021, the 16th CCF Conference on Computer Supported Cooperative Work and Social Computing.

ChineseCSCW 2021 was organized by the China Computer Federation (CCF) and co-hosted by the CCF Technical Committee on Cooperative Computing (CCF TCCC) and the Hunan University of Science and Technology, in Xiangtan, Hunan, China, during November 26–28, 2021. The theme of the conference was "Human-Centered Collaborative Intelligence", which reflects the emerging trend of the combination of artificial intelligence, human-system collaboration, and AI-empowered applications.

ChineseCSCW (initially named CCSCW) is a highly reputable conference series on computer supported cooperative work (CSCW) and social computing in China with a long history. It aims at connecting Chinese and overseas CSCW researchers, practitioners, and educators, with a particular focus on innovative models, theories, techniques, algorithms, and methods, as well as domain-specific applications and systems, covering both technical and social aspects in CSCW and social computing. The conference was initially held biennially since 1998, and has been held annually since 2014.

This year, the conference received 242 submissions, and after a rigorous double-blind peer review process, only 65 were eventually accepted as full papers to be orally presented, resulting in an acceptance rate of 27%. The program also included 22 short papers, which were presented as posters. In addition, the conference featured six keynote speeches, six high-level technical seminars, the 2nd Hengdian Cup ChineseCSCW Big Data Challenge, an award ceremony for senior TCCC members, and a forum titled "Mobile Computing and Social Computing" jointly hosted by CCF Changsha and CCF YOCSEF Changsha. We are grateful to the distinguished keynote speakers, Tianruo Yang from Hainan University, Bin Hu from the Beijing Institute of Technology, Jiade Luo from Tsinghua University, Jing Liu from the Guangzhou Institute of Technology, Xidian University, Peng Lv from Central South University, and Tao Jia from Southwest University.

We hope that you enjoyed ChineseCSCW 2021.

November 2021

Yong Tang
Jianxun Liu
Yuqing Sun
Dongning Liu
Buqing Cao

Organization

The 16th CCF Conference on Computer Supported Cooperative Work and Social Computing (ChineseCSCW 2021) was organized by the China Computer Federation (CCF) and co-hosted by the CCF Technical Committee on Cooperative Computing (CCF TCCC) and the Hunan University of Science and Technology.

Steering Committee

Yong Tang	South China Normal University, China
Weiqing Tang	China Computer Federation, China
Ning Gu	Fudan University, China
Shaozi Li	Xiamen University, China
Bin Hu	Lanzhou University, China
Yuqing Sun	Shandong University, China
Xiaoping Liu	Hefei University of Technology, China
Zhiwen Yu	Northwestern University of Technology, China
Xiangwei Zheng	Shandong Normal University, China
Tun Lu	Fudan University, China

General Chairs

Yong Tang	South China Normal University, China
Jianxun Liu	Hunan University of Science and Technology, China

Program Committee Chairs

Yuqing Sun	Shandong University, China
Dongning Liu	Guangdong University of Technology, China
Buqing Cao	Hunan University of Science and Technology, China

Organization Committee Chairs

Xiaoping Liu	Hefei University of Technology, China
Zhiwen Yu	Northwestern University of Technology, China
Tun Lu	Fudan University, China
Jianyong Yu	Hunan University of Science and Technology, China
Yiping Wen	Hunan University of Science and Technology, China

Publicity Chairs

Xiangwei Zheng	Shandong Normal University, China
Jianguo Li	South China Normal University, China

Publication Chairs

Bin Hu	Lanzhou University, China
Hailong Sun	Beihang University, China

Finance Chairs

Pei Li	Hunan University of Science and Technology, China
Yijiang Zhao	Hunan University of Science and Technology, China

Paper Award Chairs

Shaozi Li	Xiamen University, China
Yichuan Jiang	Southeast University, China

Program Committee

Zhan Bu	Nanjing University of Finance and Economics, China
Tie Bao	Jilin University, China
Hongming Cai	Shanghai Jiao Tong University, China
Xinye Cai	Nanjing University of Aeronautics and Astronautics, China
Yuanzheng Cai	Minjiang University, China
Zhicheng Cai	Nanjing University of Science and Technology, China
Buqing Cao	Hunan University of Science and Technology, China
Donglin Cao	Xiamen University, China
Jian Cao	Shanghai Jiao Tong University, China
Chao Chen	Chongqing University, China
Jianhui Chen	Beijing University of Technology, China
Long Chen	Southeast University, China
Longbiao Chen	Xiamen University, China
Liangyin Chen	Sichuan University, China
Qingkui Chen	University of Shanghai for Science and Technology, China

Ningjiang Chen	Guangxi University, China
Weineng Chen	South China University of Technology, China
Yang Chen	Fudan University, China
Shiwei Cheng	Zhejiang University of Technology, China
Xiaohui Cheng	Guilin University of Technology, China
Yuan Cheng	Wuhan University, China
Lizhen Cui	Shandong University, China
Weihui Dai	Fudan University, China
Xianghua Ding	Fudan University, China
Wanchun Dou	Nanjing University, China
Bowen Du	University of Warwick, UK
Hongfei Fan	Tongji University, China
Yili Fang	Zhejiang Gongshang University, China
Shanshan Feng	Shandong Normal University, China
Jing Gao	Guangdong Hengdian Information Technology Co., Ltd., China
Yunjun Gao	Zhejiang University, China
Liping Gao	University of Shanghai for Science and Technology, China
Ning Gu	Fudan University, China
Bin Guo	Northwestern Polytechnical University, China
Kun Guo	Fuzhou University, China
Wei Guo	Shandong University, China
Yinzhang Guo	Taiyuan University of Science and Technology, China
Tao Han	Zhejiang Gongshang University, China
Fei Hao	Shanxi Normal University, China
Chaobo He	Zhongkai University of Agriculture and Engineering, China
Fazhi He	Wuhan University, China
Haiwu He	Chinese Academy of Sciences, China
Bin Hu	Lanzhou University, China
Daning Hu	Southern University of Science and Technology, China
Wenting Hu	Jiangsu Open University, China
Yanmei Hu	Chengdu University of Technology, China
Changqin Huang	South China Normal University, China
Bo Jiang	Zhejiang Gongshang University, China
Bin Jiang	Hunan University, China
Jiuchuan Jiang	Nanjing University of Finance and Economics, China
Weijin Jiang	Xiangtan University, China

Contents – Part II

Social Media and Online Communities

Contents – Part I

Crowd Intelligence and Crowd Cooperative Computing

Locally Linear Embedding Discriminant Feature Learning Model

Chensu Wang[1,2], Luqing Wang[1,2], Hongjun Wang[1,2(✉)], Bo Peng[1,2], and Tianrui Li[1,2]

[1] School of Computing and Artificial Intelligence, Southwest Jiaotong University, Chengdu 611756, China
wanghongjun@swjtu.edu.cn

[2] National Engineering Laboratory of Integrated Transportation Big Data Application Technology, Southwest Jiaotong University, Chengdu 611756, China

Abstract. Feature learning is one of the important research trends among researchers in machine learning and other fields, which can select compact representations as feature information from high-dimensional data as well as multi-label data. Discriminative feature learning strengthens discrimination between sample features. Therefore, the feature information of samples can be better discriminated against in algorithms. In this paper, we propose a new unsupervised discriminative feature learning model called UD-LLE (Unsupervised Discriminative Locally Linear Embedding) by the improvement on standard Locally Linear Embedding, which not only maintains the manifold structure of mapping from high-dimensional space to low-dimensional space but also increases the discriminative of features. Specifically, we propose the restructure cost function as an objective function by adding constraint conditions about discrimination to standard function, which is solved by using stochastic gradient descent and momentum gradient descent algorithms combined with standard LLE.

Keywords: Locally Linear Embedding · Feature learning · Manifold learning · Unsupervised learning · Discriminative learning

1 Introduction

With the development of information science and storage technology, most of the relevant data used by researchers have the characteristics of high-dimensional, multi-feature or multi-label. When analyzing and processing high-dimensional data directly, the insufficient number of samples is easy to lead to sparse sample distribution, which reduces the efficiency and accuracy of algorithm and model analysis. The purpose of feature learning [1–3] is to express the feature information in the original data with more lightweight and fewer indicators, to reduce the possibility of suffering the disaster of dimensionality.

Dimension reduction technique is one of the main methods to realize feature learning by transforming high-dimensional data into low-dimensional space

Y. Sun et al. (Eds.): ChineseCSCW 2021, CCIS 1492, pp. 3–14, 2022.
https://doi.org/10.1007/978-981-19-4549-6_1

served as the feature information [4]. According to the data space structure it can be simply divided into linear dimensionality reduction [5,6] and nonlinear dimensionality reduction [7–9]. In the feature learning algorithm of nonlinear data space, it is an important way to use manifolds dimensionality reduction based on the principle of manifold learning [10]. Manifolds can be divided into topological manifolds and differential manifolds. In this paper, topological manifolds are discussed. Compared with the linear dimensionality reduction, the manifold learning algorithm can retain the original high-dimensional spatial structure information better when learning the feature of nonlinear. According to whether there is supervision information it can be divided into supervised learning [11], semi-supervised learning [12], unsupervised learning [2] and self-supervised learning [13].

Locally Linear Embedding (LLE) is a widely used manifold learning method that is an unsupervised dimension reduction learning algorithm proposed by Roweis [8,9] in the early stage. It is used to calculate the corresponding manifolds in the low dimensional space based on neighborhood linear relationship preserving [8]. LLE plays an important role in feature learning in image [14–17], facial expression recognition [18], multi-view learning [19,20], machinery fault diagnosis and signal [21,22].

Researchers have done a lot of optimization based on the standard LLE. Donoho [23] proposed LLE based on Hessian matrix to extend the application range of LLE algorithm to differential manifold. He et al. [24] expanded LLE on quantum devices and proposes QLLE. Zhang et al. [25] proposed MLLE algorithm which is more stable. Ziegelmeier et al. [26] established a sparser matrix to improve the speed of matrix decomposition. Wang [27] made a further improvement based on RLLE [28] that considers noise information. Wen et al. [29] proposed the VK-HLLE algorithm to choose the parameter k adaptively. Zhang et al. [30] proposed an algorithm to obtain more discriminative feature selection by adding supervision information. To improve the poor performance of the standard LLE algorithm in the learning of complex manifolds, Hettiarachchi et al. [31], Liu et al. [32] make a secondary adjustment to the weight matrix.

The purpose of this paper is to improve LLE by considering discriminant information [33] while having feature learning. The model we proposed in this paper named UD-LLE (Unsupervised Discriminative Locally Linear Embedding) can make feature information obtained after feature learning can realize the discriminant optimization of features on the premise of maintaining the original data structure.

The rest chapters of this paper are arranged as follows: Sect. 2 of this paper shows the related work; Sect. 3 gives a brief description of the model; Sect. 4 shows the results of comparative experiments and analysis of related experimental results; Sect. 5 makes a summary of the whole article.

2 Related Work

The standard LLE algorithm uses Euclidean distance to measure the similarity between the data points of the sample, and the k-nearest neighbor algorithm is used to select the adjacent points used in the reconstruction. Suppose that the sample set as $X = \{x_1, x_2, ..., x_N\}, x_i \in \mathbb{R}^L$, corresponding to the low dimensional embedding projection existing in the high-dimensional data of the sample to be solved which is regard as feature space is marked as $Y = \{y_1, y_2, ..., y_N\}, y_i \in \mathbb{R}^d$; $G = \{V, W\}$ is denoted as weighted undirected graph (where V is vertex set and W is data reconstruction weight matrix). The reconstruction error of LLE is defined as Eq. 1:

$$\varepsilon(W) = \sum_{i=1}^{N} ||x_i - \sum_{j=1}^{N} x_j w_{ij}||^2 \quad (1)$$

where w_{ij} is the weight of the sample data x_j in the model when reconstructing the sample data x_i. If the sample x_j does not belong to the reconstruction neighbor point of the data sample x_i, then the $w_{ij} = 0$; add constraints $\sum_{j=1}^{N} w_{ij} = 1$ to realize data standardization. According to the embedding cost function of the standard LLE algorithm as Eq. 2, using Lagrange multiplier combined with matrix decomposition to get the low dimensional space mapping after dimension reduction.

$$\begin{aligned} \arg\min_Y \Phi(Y) = \textstyle\sum_{i=1}^{N} ||y_i - \sum_{j=1}^{N} y_j w_{ij}||^2 \\ s.t. \textstyle\sum_{j=1}^{N} w_{ij} = 1, \frac{1}{N} \sum_{i=1}^{N} y_i^T y_i = I \end{aligned} \quad (2)$$

There are still two main parameters to be determined as a part of the input parameters: the number of preserved features d, and the number k in the neighborhood of sample data points. Among them, the manifold structure obtained by the LLE algorithm is sensitive to k which makes it necessary to go through complex parameter adjustment comparison when you have to obtain more appropriate manifold results.

Most of the existing improvements based on the LLE algorithm are the secondary reconstruction of the weight matrix by combining label information so that the high-dimensional space structure can be better preserved which is likely to lead to the loss of discriminant information or require some supervision information. A new unsupervised discriminant locally linear embedding feature learning model called UD-LLE is proposed in this paper which can enhance the similarity between the similar sample to make the data sets retained the discriminant information more complete when reducing the number of features and to make the features learned more conducive to the further data analysis.

3 UD-LLE Model

To realize the discrimination feature learning, UD-LLE adds constraints to the original foundation. The discrimination information between data sets points is increased with the quantitative comparison of similarity within the data set when the reconstruction error becomes lower. And the gradient descent algorithm is used to solve it.

3.1 Construction of UD-LLE

Suppose that the sample set X can be divided into K class clusters, and each class cluster is marked as C_k. Discriminant information is used to guide feature learning. And the new cost function of the model can be obtained as Eq. 3:

$$\begin{aligned} \min_Y L(Y) &= \tfrac{\alpha}{2}\textstyle\sum_{i=1}^{N}||y_i - \sum_{j\in Q(i)} y_j w_{ij}||^2 + \tfrac{\beta}{2}\sum_k \sum_{y_i\in C_k} ||y_i - c_k||^2 \\ &= \tfrac{\alpha}{2}L_1 + \tfrac{\beta}{2}L_2 \\ s.t. \quad & \textstyle\sum_{i=1}^{N} y_i = 0, \tfrac{1}{N}\sum_{i=1}^{N} y_i^T y_i = I \\ & 0 \le \alpha \le 1, 0 \le \beta \le 1, \alpha + \beta = 1 \\ & 1 \le i \le N, 1 \le j \le N, 1 \le k \le K \end{aligned} \tag{3}$$

where c_k represents the center of C_k, $Q(i)$ represents the set of reconstructed data points belonging to the nearest data point of ith data sample point in the data set. L_1 is the objective function of LLE which means preserve the local linear relationship; L_2 is the constraint condition to strengthen the discriminant information based on the similarity of the class and cluster of the data set. α, β are used to weigh the proportion of spatial structure of original spatial information and discriminative feature retention in the process of feature extraction.

Gradient descent with momentum can be used to solve the target Eq. 3 and to accelerate the convergence rate of the model which is shown as Eq. 4:

$$\begin{aligned} m^{(t-1)} &= \rho m^{(t-2)} + \tfrac{\partial L(y_i^{(t-1)})}{\partial y_i} \\ y_i^{(t)} &= y_i^{(t-1)} - \lambda m^{(t-1)} \end{aligned} \tag{4}$$

where $m^{(t-1)}$ represents the momentum when the $(t-1)$th update, $y_i^{(t)}$ represents y_i after tth update, ρ represents the discount factor which is used to adjust the influence of the historical gradient on the current gradient, λ is the learning rate which represents the step size of each descent gradient.

There are three kinds of variables in the model to be decided: y_i, c_k, w_{ij}. Since the second term is fixed, only the optimal solution of the first term needs to be calculated during initialization. Firstly, the initial value of the low dimensional feature space during gradient descent is required to be solved. The weight matrix is derived from the original high-dimensional data space using the least square method, and Eq. 5 can be obtained as follows:

$$\begin{aligned} \varepsilon(W) &= \textstyle\sum_{i=1}^{N}||x_i - \sum_{j=1}^{N} x_j w_{ij}||^2 \\ &= \textstyle\sum_{i=1}^{N} W_{i,Q}(x_i 1_k^T - x_{j\in Q(i)})^T (x_i 1_k^T - x_{j\in Q(i)}) W_{i,Q}^T \end{aligned} \tag{5}$$

where 1_k represents a full 1-column vector of $k \times 1$, $W_{i,Q}$ represents the reconstruction weight of neighbor of y_i, $W_{P,i}$ represents the weight of y_i to reconstruct y_k which contains y_i as its neighbor, $W_{i,}$ represents the ith row vector of the weight matrix, $W_{,i}$ represents the ith column vector of the weight matrix.

Mark $S_i = (x_i 1_k^T - x_{j\in Q(i)})^T (x_i 1_k^T - x_{j\in Q(i)})$, and then the Lagrange multiplier method is used to solve Eq. 5 and get Eq. 6 as following:

$$W_{i,Q}^T = \frac{S_i^{-1} 1_k}{1_k^T S_i^{-1} 1_k} \tag{6}$$

The expression of L_1 is transformed into a constrained optimization problem, and make further optimization transformation to get Eq. 7:

$$\begin{aligned} L_1(Y) &= \sum_{i=1}^{N}(y_i - YW_{i,}^T)^T(y_i - YW_{i,}^T) \\ &= tr(Y(I-W)^T(I-W)Y^T) \end{aligned} \tag{7}$$

Mark $M = (I-W)^T(I-W)$, and the dimensionality reduction mapping which minimizes the initial reconstruction error is equivalent to the vector space formed by the corresponding eigenvectors of the minimum eigenvalues M. Since the smallest eigenvalue is close to 0, the corresponding eigenvectors of the eigenvalues in $[1, d+1]$ are selected in the order from small to large as the initial low dimensional feature space mapping. Then, the overall objective formula is optimized by gradient descent. The objective function is composed of adding up L_1 and L_2. When solving the gradient of L_1, we get the expansion result as shown in Eq. 8 in where the nearest neighbor reconstruction data points in the data set include the set of data points of the ith data sample point is represents as $P(i)$:

$$L_1 = ||y_i - \sum_{j\in Q(i)} w_{ij}y_j||^2 + \sum_{k\neq i} ||y_k - \sum_{j\in Q(k), j\neq i} w_{kj}y_j - w_{ki}y_i||^2 \tag{8}$$

The partial derivative of Eq. 8 is solved as Eq. 9:

$$\frac{\partial L_1}{\partial y_i} = 2(y_i - \sum_{j\in Q(i)} w_{ij}y_i + \sum_{k\in P(i)} w_{ki}(\sum_{j\in Q(k)} w_{kj}y_j - y_k)) \tag{9}$$

There are two summations in L_2, and it is solved as Eq. 10:

$$\frac{\partial L_2}{\partial y_i} = 2(y_i - c_k) \tag{10}$$

In conclusion, we can get the partial derivation of L about y_i as Eq. 11:

$$\frac{\partial L}{\partial y_i} = \alpha(y_i - \sum_{j\in Q(i)} w_{ij}y_i + \sum_{k\in P(i)} w_{ki}(\sum_{j\in Q(k)} w_{kj}y_j - y_k)) + \beta(y_i - c_k) \tag{11}$$

By substituting Eq. 11 into Eq. 4, the update formula of parameter y_i can be obtained and it is rewritten into matrix format as shown in Eq. 12:

$$\begin{aligned} \frac{\partial L(y_i^{(t-1)})}{\partial y_i} &= \alpha(y_i - YW_{i,}^T + (YW^T - Y)W_{,i})) + \beta(y_i - c_k) \\ &= y_i - \alpha(YW_{i,}^T - YW^TW_{,i} + YW_{,i}) - \beta c_k \end{aligned} \tag{12}$$

3.2 Algorithm Based on UD-LLE Model

The above chapters introduce the objective function of UD-LLE model in detail, complete the reasoning of the model, and obtain the update formula of low dimensional feature space. The algorithm flow of UD-LLE model is given below.

Algorithm 1. UD-LLE Algorithm

Input: Data set $X = \{x_1, x_2, ..., x_N\}, x_i \in \mathbb{R}^L$, the number of reconstructing neighborhood of a data sample k, number of data class clusters K, number of features to be retained after feature learning d, adjustment factor α, β, maximum number of iterations T, iterative step λ, learning rate ρ
Output: Feature space $Y = \{y_1, y_2, ..., y_N\}, y_i \in \mathbb{R}^d$

1: Using KNN algorithm to obtain k adjacent data sample points of each data sample point
2: $W \leftarrow$ Reconstruction of weight matrix
3: $M \leftarrow (I - W)^T(I - W)$
4: $Y_0 \leftarrow$The $[1, d + 1]$ matrix eigenvectors of M
5: $c_k \leftarrow$Using clustering algorithm to get clustering center
6: **for** t in $1, 2, ..., T$ **do**
7: $\quad y_i \leftarrow$ Using Eq.12
8: $\quad c_k \leftarrow$Using clustering algorithm to get clustering center
9: $\quad L \leftarrow$ Using Eq.3
10: $\quad$ **if** L is minimum **then**
11: $\quad\quad$ **return** Y
12: **return** Y

4 Experimental Results

To verify the effect of the UD-LLE model, one evaluation index called clustering accuracy is used with three different clustering algorithms on the selected data sets for comparative experiments.

4.1 Evaluation Measures

To quantitatively describe the effect of feature learning and facilitate the comparison of experimental results, we use the clustering effect to measure the amount of discrimination information retained.

(1) *Accuracy* (ACC)
The calculation formula of clustering accuracy is as Eq. 13:

$$Acc = \frac{\sum_{i=1}^{N} match(map(l_i), t_i)}{N} \tag{13}$$

where N means the number of the data set, l_i is the label of ith sample when having clustering, and t_i is the real label, $map(l_i)$ is the mapping function to match two labels which can get the real label of label l_i. When $x = y$, $match(x, y) = 1$, or $match(x, y) = 0$.

(2) *Friedman test and Iman – Davenport*
The definition of Friedman statistics is shown as Eq. 14:

$$\chi_F^2 = \frac{12n}{a(a+1)} \left[\sum_i R_i^2 - \frac{a(a+1)^2}{4} \right] \tag{14}$$

Iman and Davenport put forward a more comprehensive index to evaluate the performance of the algorithm. The definition is as Eq. 15:

$$F_F = \frac{(n-1)\chi_F^2}{n(a-1)-\chi_F^2} \tag{15}$$

where n means the number of data sets, a means the number of the algorithms, R_i means the average rank of the algorithm, the F_F distribution with degrees of freedom$(a-1)$ and $(a-1)(n-1)$ is obeyed.

4.2 Experimental Settings

In this paper, the experiment is carried out on 8 real datasets, which are from the image datasets with high reliability in Microsoft Research Asia multimedia (MSRA). Table 1 lists the detailed information of the experimental data sets. Meanwhile, the comparison experiment is carried out in two parts: different feature spaces and parameter adjustment. K-means [34], DP [35,36] and AP [37] are used to cluster the data set after feature learning. Accuracy is used as external evaluation index to compare the clustering effect. The parameter adjustment experiment verifies the effectiveness of the model by comparing the average accuracy through the β in the [0, 1] interval.

The number of features solved by feature learning is not only the key parameter but also has the most significant effect on the results in this experiment. To compare the fairness, this paper compares the performance of each algorithm by reducing the same original data space to the same number of features. At the same time, each algorithm is run ten times under each dataset, and the average value is taken as the final result. The parameters of the UD-LLE model in this comparative experiment are set as follows: the adjustment factors α and β are set to 0.6 and 0.4, the learning rate is set to 0.002, and the discount factor of momentum gradient decline is set to 0.5, The maximum number of iterations is set to 2000. For the UD-LLE model and standard LLE model, the number of features selected and retained is set to one percent of the original number of

Table 1. Detail information of the data sets.

Idx	Data sets	Samples	Features	Classes
D01	Ambulances	930	892	3
D02	Anonovo	732	892	3
D03	Balloon	830	892	3
D04	Boot1	845	892	3
D05	Butterflytattoo	738	892	3
D06	CANE	1080	856	9
D07	Ufo11	881	899	3
D08	Vista	799	899	3

features in the dataset, and the number of nearest neighbors k is set to one-tenth of the original data items in the dataset.

4.3 Comparison Results

Model Comparison. Tables 2 and 3 show the clustering results of K-means, DP and AP clustering algorithms in the original data space, the data space after feature learning using standard LLE algorithm, and the data space after feature learning using UD-LLE feature learning model. The data results ranking first in statistics are shown in bold.

Accuracy. From Table 2, the accuracy of the clustering obtained by using the feature information used in K-means and DP clustering using UD-LLE model is the highest among the nine combination algorithms. Meanwhile, the combination of UD-LLE and DP algorithm has obtained the highest average accuracy rate of 56.69% in 8 data sets. Compared with the original data space and standard LLE feature learning algorithm, the accuracy of the three clustering algorithms is improved by 9.93% and 10.24% on average.

Table 2. UD-LLE and LLE use different clustering algorithms to compare the clustering accuracy of each data set.

Idx	Raw_Data			LLE			UD-LLE		
	Km	AP	DP	Km	AP	DP	Km	AP	DP
D01	0.4713	0.0548	0.5667	0.5612	0.0473	0.5204	**0.6398**	0.1075	**0.6398**
D02	0.4817	0.0792	0.4822	0.4577	0.0751	0.4303	**0.4918**	0.1708	**0.4918**
D03	0.4398	0.0687	0.4289	0.4570	0.0675	0.5169	**0.5542**	0.1566	**0.5542**
D04	0.4734	0.0746	0.4059	0.4697	0.0592	0.4604	**0.4911**	0.1621	**0.4911**
D05	0.5440	0.1125	0.5556	**0.6598**	0.0935	0.4566	0.6355	0.4404	0.6355
D06	0.5297	0.2213	0.3370	0.7093	0.2139	0.2444	**0.7278**	0.6019	**0.7278**
D07	0.4115	0.4257	0.3825	0.4217	0.0511	0.3587	**0.4427**	0.1669	**0.4427**
D08	0.4690	0.0914	0.4881	0.5064	0.0763	**0.6083**	0.5519	0.1039	0.5519
ave-Acc	0.4776	0.1410	0.4559	0.5304	0.0855	0.4495	**0.5669**	0.2388	**0.5669**

It can be seen from Table 3 that the statistical test index Friedman is used for hypothesis test and comprehensive evaluation of UD-LLE model. It can be seen from Table 3 that the UD-LLE model is the best one with a sequence value of 1.0000. By substituting Eq. (14), the Friedman statistic can get an approximate value as:

$$\chi_F^2 = 8 \times \left[2.5625^2 + 2.4375^2 + 1.0000^2 - \frac{3 \times 4^2}{4} \right] \approx 12.0625 \tag{16}$$

By substituting Eq. (15), the Iman Davenport statistic is obtained as:

$$\chi_F^2 = \frac{7 \times 12.0625}{7 \times 3 - 12.0625} \approx 9.4475 \tag{17}$$

It is known that the calculated ρ value of $F(2, 14)$ distribution is 0.0025, so the feature performance of UD-LLE model is better than other comparative models.

Table 3. UD-LLE and LLE use different clustering algorithms to compare the clustering average rank and average accuracy of each data set.

Idx	ave-Rank			ave-Acc		
	Raw_Data	LLE	UD-LLE	Raw_Data	LLE	UD-LLE
D01	5.6667 (2.0)	6.0000 (3.0)	**3.3333 (1.0)**	0.3643	0.3763	**0.4623**
D02	5.0000 (2.0)	6.6667 (3.0)	**3.3333 (1.0)**	0.3477	0.3210	**0.3848**
D03	6.3333 (3.0)	5.3333 (2.0)	**3.3333 (1.0)**	0.3125	0.3471	**0.4217**
D04	6.0000 (3.0)	5.6667 (2.0)	**3.3333 (1.0)**	0.3180	0.3298	**0.3814**
D05	5.6667 (3.0)	5.3333 (2.0)	**4.6667 (1.0)**	0.4040	0.4033	**0.5705**
D06	6.3333 (2.5)	6.3333 (2.5)	**2.3333 (1.0)**	0.3627	0.3892	**0.6858**
D07	4.6667 (2.0)	6.6667 (3.0)	**3.6667 (1.0)**	**0.4066**	0.2772	0.3508
D08	6.3333 (3.0)	4.6667 (2.0)	**4.0000 (1.0)**	0.3495	0.3970	**0.4026**
Ave	2.5625	2.4375	**1.0000**	0.3581	0.3551	**0.4575**

MNIST Visualization. As shown in Fig. 1 UD-LLE can make the feature between different clusters further than LLE, which can highly improve the discrimination of points on the boundary of class clusters. The two most discriminant features are visualized in the figure to show the discriminant improvement of features and prove the validity of the model.

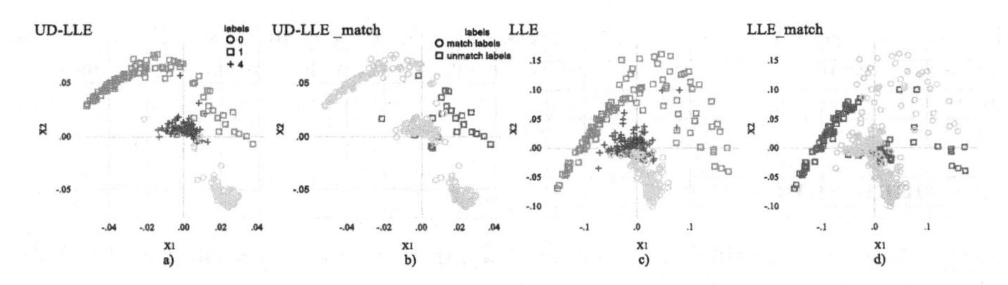

Fig. 1. The results of MNIST dataset in LLE model and UD-LLE model. Set $k = 30$, $d = 5$, α =0.6, $\rho = 0.002$, $\lambda = 0.5$, $T = 1000$. We choose 3 kinds of numbers in MNIST dataset with 100 data per number. a) and c) display the real tag corresponding to the feature, b) and d) point out the matching results between clustering results and real labels.

Parameter Adjustment. Figure 2 shows the average accuracy of the three clustering algorithms obtained by adjusting parameters on 8 experimental data sets. Through the analysis and comparison of the line chart in Fig. 2, we can get the following conclusions:

- The effect of the UD-LLE model is influenced by trade-off factors. When $\beta = 0$ means only the discriminant information is considered, the clustering accuracy is approximately equal to that of the standard LLE algorithm. When $\beta = 1$ compared with the standard LLE algorithm, the UD-LLE model can better retain the high-dimensional spatial structure of the data, to increase the discriminant information between the data and achieve a better clustering effect.
- UD-LLE model in the sample data set about the adjustment factor of the broken line fully proved that the UD-LLE model added based on clustering algorithm discriminant constraint bar can effectively improve the discriminability of learning features compared with the standard LLE algorithm and the original data space.

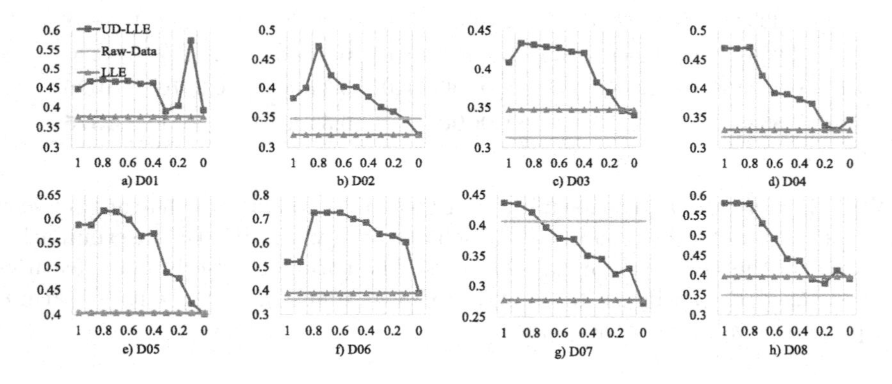

Fig. 2. Comparing the adjustment results of parameter β in UD-LLE model, the abscissa represents β, and the ordinate is the average accuracy in the three clustering algorithms

5 Conclusion

In this paper, a discriminant feature learning model based on locally linear embedding called UD-LLE is proposed by adding discriminant constraints in the process of feature learning to realize discriminant feature learning. The experiment shows that the model can carry out effective discriminant feature learning, which increases the accuracy of subsequent machine learning or data analysis. This model optimizes and improves the basic algorithm, and proposes a discriminant learning framework based on unsupervised information, which can be applied to other basic or improved algorithms.

Acknowledgments. This work is supported by the National Natural Science Foundation of China (No. 61806170), the Humanities and Social Sciences Fund of Ministry of Education (No. 18XJC72040001), and the National Key Research and Development Program of China (No. 2019YFB1706104).

References

1. Bengio, Y., Courville, A., Vincent, P.: Representation learning: a review and new perspectives. IEEE Trans. Pattern Anal. Mach. Intell. **35**(8), 1798–1828 (2013)
2. Wang, H., Zhang, Y., Zhang, J., Li, T., Peng, L.: A factor graph model for unsupervised feature selection. Inf. Sci. **480**, 144–159 (2019)
3. Li, J., et al.: Feature selection: a data perspective. ACM Comput. Surv. (CSUR) **50**(6), 1–45 (2017)
4. Zebari, R., Abdulazeez, A., Zeebaree, D., Zebari, D., Saeed, J.: A comprehensive review of dimensionality reduction techniques for feature selection and feature extraction. J. Appl. Sci. Technol. Trends **1**(2), 56–70 (2020)
5. Jolliffe, I.T., Cadima, J.: Principal component analysis: a review and recent developments. Philos. Trans. R. Soc. A Math. Phys. Eng. Sci. **374**(2065), 20150202 (2016)
6. Yang, F., Yang, W., Gao, R., Liao, Q.: Discriminative multidimensional scaling for low-resolution face recognition. IEEE Signal Process. Lett. **25**(3), 388–392 (2017)
7. Tenenbaum, J.B., De Silva, V., Langford, J.C.: A global geometric framework for nonlinear dimensionality reduction. Science **290**(5500), 2319–2323 (2000)
8. Roweis, S.T., Saul, L.K.: Nonlinear dimensionality reduction by locally linear embedding. Science **290**(5500), 2323–2326 (2000)
9. Saul, L.K., Roweis, S.T.: Think globally, fit locally: unsupervised learning of low dimensional manifolds. Departmental Papers (CIS), p. 12 (2003)
10. Lee, J.: Introduction to Topological Manifolds, vol. 202. Springer, Cham (2010). https://doi.org/10.1007/978-1-4419-7940-7
11. Conneau, A., Kiela, D., Schwenk, H., Barrault, L., Bordes, A.: Supervised learning of universal sentence representations from natural language inference data. arXiv preprint arXiv:1705.02364 (2017)
12. Van Engelen, J.E., Hoos, H.H.: A survey on semi-supervised learning. Mach. Learn. **109**(2), 373–440 (2019). https://doi.org/10.1007/s10994-019-05855-6
13. Grill, J.B., et al.: Bootstrap your own latent: a new approach to self-supervised learning. arXiv preprint arXiv:2006.07733 (2020)
14. Zhang, S., Lei, Y.K.: Modified locally linear discriminant embedding for plant leaf recognition. Neurocomputing **74**(14–15), 2284–2290 (2011)
15. Huang, M., Zhu, Q., Wang, B., Lu, R.: Analysis of hyperspectral scattering images using locally linear embedding algorithm for apple mealiness classification. Comput. Electron. Agric. **89**, 175–181 (2012)
16. Liang, D., Yang, J., Zheng, Z., Chang, Y.: A facial expression recognition system based on supervised locally linear embedding. Pattern Recogn. Lett. **26**(15), 2374–2389 (2005)
17. Luo, J., Xu, T., Pan, T., Sun, W.: An efficient method of hyperspectral image dimension reduction based on low rank representation and locally linear embedding. Integr. Ferroelectr. **208**(1), 206–214 (2020)

18. Yaddaden, Y., Adda, M., Bouzouane, A.: Facial expression recognition using locally linear embedding with LBP and HOG descriptors. In: 2020 2nd International Workshop on Human-Centric Smart Environments for Health and Well-being (IHSH), pp. 221–226. IEEE (2021)
19. Zhang, Y., Yang, Y., Li, T., Fujita, H.: A multitask multiview clustering algorithm in heterogeneous situations based on LLE and LE. Knowl.-Based Syst. **163**, 776–786 (2019)
20. Xie, Y., Jiang, D., Wang, X., Xu, R.: Robust transfer integrated locally kernel embedding for click-through rate prediction. Inf. Sci. **491**, 190–203 (2019)
21. Li, B., Zhang, Y.: Supervised locally linear embedding projection (SLLEP) for machinery fault diagnosis. Mech. Syst. Signal Process. **25**(8), 3125–3134 (2011)
22. Zhang, Y., Ye, D., Liu, Y.: Robust locally linear embedding algorithm for machinery fault diagnosis. Neurocomputing **273**, 323–332 (2018)
23. Donoho, D.L., Grimes, C.: Hessian eigenmaps: locally linear embedding techniques for high-dimensional data. Proc. Natl. Acad. Sci. **100**(10), 5591–5596 (2003)
24. He, X., Sun, L., Lyu, C., Wang, X.: Quantum locally linear embedding for nonlinear dimensionality reduction. Quantum Inf. Process. **19**(9), 1–21 (2020). https://doi.org/10.1007/s11128-020-02818-y
25. Zhang, Z., Wang, J.: MLLE: modified locally linear embedding using multiple weights. In: Advances in Neural Information Processing Systems, pp. 1593–1600. Citeseer (2007)
26. Ziegelmeier, L., Kirby, M., Peterson, C.: Sparse locally linear embedding. Procedia Comput. Sci. **108**, 635–644 (2017)
27. Wang, J., Wong, R.K., Lee, T.C.: Locally linear embedding with additive noise. Pattern Recogn. Lett. **123**, 47–52 (2019)
28. Chang, H., Yeung, D.Y.: Robust locally linear embedding. Pattern Recogn. **39**(6), 1053–1065 (2006)
29. Wen, G., Jiang, L., Wen, J.: Dynamically determining neighborhood parameter for locally linear embedding. J. Softw. **19**(7), 1666–1673 (2008)
30. Zhang, S.Q.: Enhanced supervised locally linear embedding. Pattern Recogn. Lett. **30**(13), 1208–1218 (2009)
31. Hettiarachchi, R., Peters, J.F.: Multi-manifold LLE learning in pattern recognition. Pattern Recogn. **48**(9), 2947–2960 (2015)
32. Liu, Y., Hu, Z., Zhang, Y.: Bearing feature extraction using multi-structure locally linear embedding. Neurocomputing **428**, 280–290 (2021)
33. Deng, P., Wang, H., Li, T., Horng, S.J., Zhu, X.: Linear discriminant analysis guided by unsupervised ensemble learning. Inf. Sci. **480**, 211–221 (2019)
34. Jain, A.K.: Data clustering: 50 years beyond K-means. Pattern Recogn. Lett. **31**(8), 651–666 (2010)
35. Chu, R., Wang, H., Yang, Y., Li, T.: Clustering ensemble based on density peaks. Acta Automatica Sinica **42**(9), 1401–1412 (2016)
36. Rodriguez, A., Laio, A.: Clustering by fast search and find of density peaks. Science **344**(6191), 1492–1496 (2014)
37. Frey, B.J., Dueck, D.: Clustering by passing messages between data points. Science **315**(5814), 972–976 (2007)

Cache Optimization Based on Linear Regression and Directed Acyclic Task Graph

Lei Wan[1], Bin Dai[1], Han Jiang[1](✉), Weixian Luan[2], Fan Ye[2], and Xianjun Zhu[3]

[1] ZheJiang Institute of Turbomachinery and Propulsion Systems, HuZhou 313219, Zhejiang, People's Republic of China
jiang078@163.com
[2] Unit 31102 of PLA, Nanjing 210000, People's Republic of China
[3] Jinling Hospital, Nanjing 210002, Jiangsu, People's Republic of China

Abstract. Based on the principle of locality, the current Cache hit rate has reached a high level. At the same time, its replacement algorithm has also encountered a bottleneck. The current algorithms show poor results on large-scale datasets. In this paper, machine learning technology is applied to the hit strategy of Cache. The calculation task is converted into a DAG graph, and then the DAG graph is represented by an adjacency matrix, and the multiple linear regression model is input to predict the number of the Cache block to be called. The final experiment shows that this method can greatly improve the accuracy of Cache block scheduling, and ultimately improve the efficiency of system operation.

Keywords: Cache · Replacement strategy · Machine learning

1 Introduction

With the development of computer hardware and software, some technologies have also ushered in major developments, such as big data technology [1] and deep learning technology [2]. At the same time, there has been a serious mismatch between the development of large-scale software and the computing power of current computers. And the industry often uses multiple devices and computing cores to solve this problem, especially in the cloud computing industry [3]. However, this kind of solution is destined to encounter its bottleneck, and how to improve the computing speed from the computer itself is also very important.

In recent years, more and more researches have focused on Cache hit strategy optimization. Due to the working mechanism of the Cache, the data in the Cache needs to be replaced. The most important problem to be solved is: how to allocate the limited space to the corresponding data in the memory, and minimize the scheduling. The current algorithms show unsatisfactory results when the calculation scale is large. In addition, due to the complexity of the system, the datasets accessed by different types of programs have different characteristics, but current algorithms do not provide such high-intensity locality. For any program executed on the CPU, the data that the algorithm needs to access is regarded as the same type by the Cache, so that the Cache will not

Y. Sun et al. (Eds.): ChineseCSCW 2021, CCIS 1492, pp. 15–24, 2022.
https://doi.org/10.1007/978-981-19-4549-6_2

treat different data differently. With the continuous development of various applications and dedicated hardware today, a single cache structure cannot meet the needs of data diversity. Therefore, solving this problem is the key to improving the current Cache hit rate bottleneck.

In response to this contradiction, more and more researchers have begun to pay attention to this problem. Qureshi et al. [4] proposed an LRU insertion strategy. When Cache searches for cache blocks, new cache blocks are inserted into the LRU side first, which makes LIP performance more stable and makes applications with relatively strong locality have an ideal hit rate. Guo F et al. [5] first proposed an analysis model for predicting the performance of a cache replacement strategy. The model takes a simple cycle sequence analysis of each application as an input, and uses the statistical properties of the application under different replacement strategies as an output, and the model uses Markov process to calculate the probability of cache miss. Ramo et al. [6] proposed a method that allows the block-based cache design interface to be exposed to middleware designers by designing a simple cache line-aware optimization interface.

The above methods have improved efficiency compared with traditional scheduling algorithms, but they also have the problem of large computational overhead and less improvement in Cache hit rate, and they do not perform well on large-scale datasets, so they still cannot meet the current Cache hit rate. Claim. Therefore, this article applies deep learning technology to the hit strategy of Cache, by converting the calculation task into a DAG graph, and then inputting the DAG graph and other system features into the multiple linear regression model to predict the block number that will be called out of the Cache, and then import the corresponding Memory block.

2 Prearrangement Knowledge

2.1 Multiple Linear Regression Algorithm

Linear regression algorithm is often used to analyze the correlation between independent variables and dependent variables. When there are multiple independent variables, this method is also called multiple linear regression [7]. The steps of multiple linear regression generally include modeling, solving unknown parameters, credibility testing, and prediction. Its model can be defined as:

$$f(x) = w_0 + w_1x_1 + w_2x_2 + \dots + w_nx_n \tag{1}$$

among them, x is a multivariate vector of independent variables, $f(x)$ is a dependent variable, and w_i represents the weight of each variable.

The existing datasets D contains data feature X and data label Y to train the regression model. Suppose the datasets has m data and j features. Assuming that the constant term is removed, the model requires a total of j weights.

$$X = \begin{bmatrix} x_1^{(1)} & x_2^{(1)} & \dots & x_j^{(1)} \\ x_1^{(2)} & x_2^{(2)} & \dots & x_j^{(2)} \\ \dots & \dots & \dots & \dots \\ x_1^{(m)} & x_2^{(m)} & \dots & x_j^{(m)} \end{bmatrix} \tag{2}$$

$$W = \begin{bmatrix} w_1 \\ w_2 \\ \dots \\ w_m \end{bmatrix} \tag{3}$$

$$Y = \begin{bmatrix} y^{(1)} \\ y^{(2)} \\ \dots \\ y^{(m)} \end{bmatrix} \tag{4}$$

The purpose of training is to find a series of parameters to make XW fit Y as much as possible, that is, make the Euclidean distance between XW and Y as small as possible. In order to achieve this goal, a reference indicator needs to be added to the model, namely the loss function:

$$J(w) = \frac{1}{m}(XW - Y)^T(XW - Y) = \frac{1}{m}\sum_{i=1}^{m}(f(x^{(i)}) - y^{(i)})^2 \tag{5}$$

Then the purpose of linear regression is transformed into minimizing $J(w)$ through iteration.

2.2 DAG Task Graph

From the perspective of the program running on the computer system, the program can be divided into code blocks with different functions. In these code blocks, if task A depends on task B, then the two tasks are connected by a directed edge Up, that is, $A \rightarrow B$, then the relationship of the entire task constitutes a directed graph, which is also called a task graph [8].

Task graphs can be used to convert computational tasks into graphs for analysis and research, so that discrete mathematics methods can be used to optimize the entire computational path. Because directed acyclic graphs are simple and easy to divide. Therefore, acyclic graphs are usually used in the research of task graphs.

3 Method

3.1 Problem Definition

Definition C is the collection of all Cache blocks in the system, T is the task graph information of the job to be calculated, S is the information of a series of computer systems, F is the the processed feature vector by using T and S. Define the Cache block call prediction function $y = CFun(F)$, where y represents the vector information of the Cache block to be called out, and F is the feature vector information during this scheduling. Define a function $r = CMRate(n, Y)$ that calculates the cache miss rate when accessing the Cache n times in the future, where Y represents a tuple of all scheduling schemes. The goal of the prediction function $y = CFun(F)$ is to make r reach the minimum value when the main memory is accessed n times in the future.

3.2 Feature Processing

The Cache scheduling situation in the computer is recorded by special hardware. Each Cache block number corresponds to a field, and the fields include the main memory block number, the Cache block number, the dirty bit, and the DAG subtask node number corresponding to the Cache block. Assuming that the number of Cache blocks in the computer is c, these features can be expressed as a matrix of $c \times 4$. Expand this matrix by column vector to the column vector of $4c \times 1$, and denote this vector as F_1. The DAG graph is a plane directed graph, which cannot be represented by one-dimensional data. Therefore, the method adopted in this paper is to express the DAG graph with an adjacency matrix, and then process it. The adjacency matrix is shown in formula (6):

$$D = \begin{bmatrix} a_1^1 \; a_1^2 \cdots a_1^n \\ a_2^1 \; a_2^2 \cdots a_2^n \\ \vdots \quad \vdots \quad \ddots \quad \vdots \\ a_n^1 \; a_n^2 \cdots a_n^n \end{bmatrix} \tag{6}$$

Assuming that the subtasks divided by the DAG graph correspond to n nodes respectively, then a_i^j represents the relationship coefficient between the ith node and the jth node. For the value of the relationship coefficient in the matrix, including the following Possible values:

When $a_i^j = 0$, it means that the ith node has no relationship with the jth node.
When $a_i^j = 1$, it means that the task of the ith node depends on the task of the jth node.
When $a_i^j = 2$, it means that the task of the ith node depends on the task of the jth node.
When $a_i^j = 3$, it means that the tasks of the ith node and the jth node has relationship with each other.

3.3 Model Training

Due to the particularity of the problem, it is difficult to find an optimal solution to the problem, but a better solution can be obtained through algorithms. Because this paper uses linear regression algorithm to solve the problem, and linear regression is a supervised algorithm, it is necessary to provide labels for the input of each datasets. In this regard, the strategy we adopt is to assist in providing an approximate solution of the label through other algorithms, thereby assisting the network to converge, and finally through the additional loss function to make the model obtain better performance. Therefore, when training the model, the y label of this schedule is selected by obtaining the solution of the conventional algorithm. Assuming that the set of algorithm functions is $Func = \{F_1(x), F_2(x), ..., F_t(x)\}$, then the final label is $y = Max(count(Func(x)))$. Among them, the function Max is used to obtain the maximum value, and the function $count$ is used to count the number of times each function in $Func$ takes the value of the input value x.

In the unsupervised algorithm, setting the correct reward and punishment mechanism can significantly improve the performance of the network. Therefore, this article uses this idea to improve the performance of the network, adding a reward and punishment loss function to the traditional loss function, which adjusts the parameters by analyzing the scheduling effect produced by the future algorithm. In order to evaluate the effect of the algorithm at a certain stage, it is necessary to count the number of cache misses in a period of time. Therefore, this article counts the number of Cache Misses in this stage during each s time period when the main memory address is accessed. In order for the algorithm to consider the timing characteristics, it is necessary to count the number of cache misses in different stages. The loss function is shown in formula (7):

$$J(\eta) = \sum_{i=1}^{n} \frac{Cm_i}{\eta - i + 1} \tag{7}$$

among them, η is the number of stages to be considered in the current stage of the algorithm, i indicates the number of stages, and Cm_i indicates the number of cache misses in the i stage. The selected stage includes the η stage from the current stage to the previous stage, and the η stage is the current stage. η should be a reasonable value, when η is too large, the convergence time of the algorithm will be prolonged, when η is too small, the algorithm will produce thrashing and it is not easy to converge. The summation of Cm_i adopts the strategy of linear decay with time, so that the algorithm can consider the timing feature.

Table 1 shows the notations and explanations in this study.

Table 1. Notations and explanations

Notation	Explanation
C	Set of all Cache blocks
T	Task graph information
S	Extra information of a series of computer systems
F	Processed feature vector
$CFun()$	Cache block call prediction function
y	Vector information of the Cache block to be clean
$CMRate()$	Cache miss rate compute function
r	Cache miss rate
Y	Tuple of all scheduling schemes
D	Adjacency matrix of DAG graph
c	The number of Cache blocks
F_1	Flattened matrix of Cache information
a_i^j	Value of the node of in row i, column j
$Max()$	The function of getting the max value in the array
$count()$	The function of counting the number of element in the array

(*continued*)

Table 1. (*continued*)

Notation	Explanation
η	The number of stages to be considered
i	The identifier of the stage
Cm_i	The number of cache misses in the i stage

3.4 Component Design

For all current Cache scheduling structures, it is impossible to implement a scheduling strategy based on multiple features. Therefore, this research needs to add new components to the cache structure to implement the above mechanism. Through analysis, the functions that the new components need to achieve include feature acquisition, data block division, frequency statistics of data block access, event detection, data collection and combination, and data block information recording.

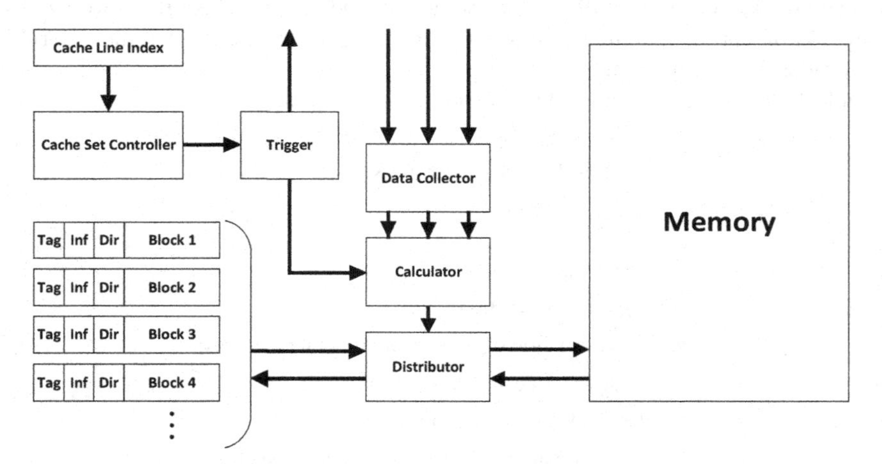

Fig. 1. Hardware structure

As shown in Fig. 1, the components added by the computer implementing the algorithm in this paper mainly include Trigger, Data Collector, Calculator, and Distributor.

Trigger: The trigger is responsible for monitoring the status of the Cache Set Controller. When a request arrives in the cache group, the Cache Set Controller needs to search for the cache block in the current cache group according to the physical address. If the corresponding row is not found according to the address, it sends a Cache Miss signal, otherwise it sends a Cache Hit signal. When the cache group is full, a Cache Full signal is sent, and a scheduling signal is sent to the Calculator. In addition, Trigger is also responsible for storing information from the Cache Set Controller and calculator in the cache.

Data Collector: The data collector is responsible for collecting information from caches and other information sources. And the information is processed and sent to the calculator in the form of a binary stream in the logical form of a vector.

Calculator: Mainly responsible for implementing the scheduling algorithm. It contains data input lines, data output lines, signal lines and calculation circuits. The input line is used to read the features required for calculation from the data collector. The output line is used to send the calculated scheduling result to the scheduler. The signal line is used to communicate with the trigger. The calculation circuit is implemented based on logical expressions, and its purpose is to implement scheduling algorithms.

Distributor: Whenever a message from the calculator is received, the scheduler will schedule it. The main functions it realizes include receiving and processing the scheduling signal sent from the calculator, the replacement of the Cache line and the main memory block. At the same time, it is also responsible for modifying the information bit (Inf) and dirty bit (Dir) of the Cache line, and writing the content of the block containing the dirty bit back to the main memory.

4 Experimental Analysis

In the current research field, the research on Cache is based on the form of system simulation to judge and improve the method. Using the form of system simulation to evaluate the algorithm mainly has the following advantages. (1) Simple deployment. By means of simulation, researchers can quickly conduct experiments and evaluate the proposed methods. The use of hardware for experiments requires a long preparation period. (2) Convenient for experiment. The effectiveness and deficiencies of the algorithm can be quickly verified by means of simulation. And can quickly iteratively improve the algorithm. (3) Powerful functions. The simulation software integrates a variety of experimental conditions, which can test the performance of the algorithm and explore the robustness of the algorithm in different environments.

Therefore, this article conducted a simulation experiment on the GEM5 platform [9], and the specific configuration of the machine is shown in Table 2.

Table 2. System configuration

Parameter	Configuration
Operating system	Ubuntu 16.04 LTS
GEM5	System simulation software
CPU	Single-core X86
TLB	Instruction TLB 16-Entry Data TLB: 16-Entry
Memory	1 GB

4.1 Evaluation Index

This study uses Cache Hit Rate to evaluate the efficiency of the method. The calculation method of the cache hit rate is shown in formula (9).

$$HR = \frac{N_{hit}}{N} \times 100\% \tag{9}$$

among them, HR represents the cache hit rate, N_{hit} represents the total number of Cache hits, and N represents the number of times the CPU accesses the main memory.

4.2 Experimental Results

In the experiment, we divide the memory into main memory blocks of equal size, and then perform random access to the main memory. For datasets of different sizes, we record the frequency of datasets reading and writing to the main memory.

Table 3. Statistics of the number of read and write of the datasets

Type	Size of dataset									
	2 KB	4 KB	8 KB	16 KB	32 KB	64 KB	128 KB	256 KB	512 KB	1024 KB
Read	16390	34035	67832	140343	269438	534938	1003842	2130290	4530304	9034249
Write	10034	24983	48340	94334	173293	363942	633948	1302399	2693039	5132902

It can be seen from Table 3 that as the datasets increases, the number of times the program reads and writes to the main memory also increases. And the increase shows a linear law. Due to the different mechanisms of the read and write strategy, this article will test and calculate read hits and write hits separately.

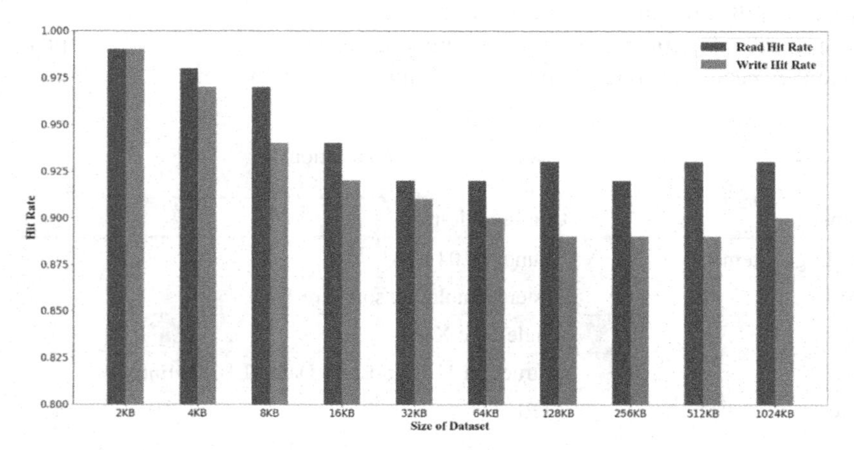

Fig. 2. The performance of the algorithm under different size datasets

It can be seen from Fig. 2 that as the datasets increases, the read hit rate and write hit rate of the model are gradually decreasing. This is because as the scale of the program involved in the calculation is gradually increasing, the scale of the variable is also increasing. This has caused the Cache to greatly increase the difficulty in recognizing the memory that needs to be transferred in or out.

In addition, in order to prove the superiority of the model in this paper on large-scale computing tasks, we used the algorithm in this paper and other algorithms to conduct comparative experiments on large-scale datasets of the same size. The algorithms for comparison include the LRU algorithm, the algorithm of Qureshi et al. [4], and the algorithm of Guo F et al. [5]. The size of the experimental datasets is 1024 KB.

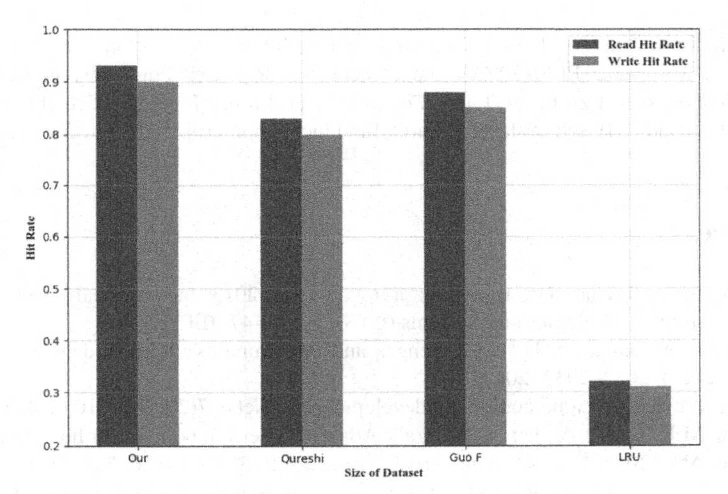

Fig. 3. Comparison of three algorithms

It can be seen from Fig. 3 that there are obvious differences in the read and write hit rate of the Cache between the algorithms. Among them, the algorithm effect of the LRU algorithm is extremely poor. This is because when the datasets is too large, because the total size of the Cache is less than the size of the data block required for the program to run, the LRU algorithm will call out the cache that has not been accessed for the longest time. This makes it possible for many memory blocks that have just been called out to be accessed again in the next time, which in turn makes the algorithm "Thrashing" [10]. At the same time, compared to the other two algorithms, our algorithm has a higher read hit rate or write hit rate. This is because the model in this article calculates DAG task graph features and other spatio-temporal features to make the algorithm effective to determine whether a Cache block will be accessed again in the next period of time, the model in this article also shows excellent results on larger datasets.

5 Summary

The current Cache replacement algorithm does not perform different types of processing for different types of datasets, which makes the system efficient. This paper uses machine learning technology for the hit strategy of Cache, converts the computing task into a DAG graph, and then inputs the DAG graph and other system features into the multiple linear regression model to predict the block number that will be called out of the Cache, which greatly improves the system Cache hit rate on large-scale datasets. In the future work, we should focus on the hardware structure design based on this type of algorithm and reduce the computational overhead of the algorithm, so that these algorithms can be put into practical application.

Acknowledgment. This work was supported in part by the Postdoctoral Science Foundation of China (No. 2020T130129ZX), the Postdoctoral Research Fund of Jiangsu Province (No. 2019K086), the High-Level Talent Foundation of Jinling Institute of Technology (No. JIT-B-201703), and 2019 school-level research fund incubation project (No. jit-fhxm-201912).

References

1. Sagiroglu, S., Sinanc, D.: Big data: a review. In: 2013 International Conference on Collaboration Technologies and Systems (CTS), pp. 42–47. IEEE (2013)
2. Labrinidis, A., Jagadish, H.V.: Challenges and opportunities with big data. Proc. VLDB Endow. **5**(12), 2032–2033 (2012)
3. Ma, S.: A review on cloud computing development. J. Netw. **7**(2), 305–310 (2012)
4. Qureshi, M.K., Jaleel, A., Patt, Y.N., et al.: Adaptive insertion policies for high performance caching. ACM SIGARCH Comput. Archit. News **35**(2), 381–391 (2017)
5. Guo, F., Solihin, Y.: An analytical model for cache replacement policy performance. In: Proceedings of the Joint International Conference on Measurement and Modeling of Computer Systems, pp. 228–239 (2006)
6. Ramos, S., Hoefler, T.: Cache line aware algorithm design for cache-coherent architectures. IEEE Trans. Parallel Distrib. Syst. **27**(10), 2824–2837 (2016)
7. Tranmer, M., Elliot, M.: Multiple linear regression. Cathie Marsh Centre Census Surv. Res. (CCSR) **5**(5), 1–5 (2008)
8. Li, Y., Cao, B., Peng, M., et al.: Direct acyclic graph based blockchain for internet of things: performance and security analysis. arXiv preprint arXiv:10925.2019 (2019)
9. Binkert, N., Beckmann, B., Black, G., et al.: The gem5 simulator. ACM SIGARCH Comput. Archit. News **39**(2), 1–7 (2011)
10. Denning, P.J.: Thrashing: its causes and prevention. In: Proceedings of the 9–11 December 1968, Fall Joint Computer Conference, Part I, pp. 915–922 (1968)

A University Portrait System Incorporating Academic Social Network

Yu Lai, Liantao Lan(✉), Rui Liang, Li Huang, Zihan Qiu, and Yong Tang

South China Normal University, Guangzhou 510631, Guangdong, China
{laiyu,lanlt,ruiliang,lhuang,hhhqzh,ytang}@m.scnu.edu.cn

Abstract. There are more than 2800 higher education institutions in China, all of which have a wealth of basic attributes and introductory information. However, by investigating common university and college information service platforms, we find a problem that users cannot quickly access key information. Inspired by user profile and corporate portraits, we propose a university portrait system incorporating academic social networks. We first collect two types of data, then utilize text mining techniques integrated with statistics-based methods and topic-based methods to extract features and generate tags of universities. Additionally, we incorporate data related to the universities on the academic social network SCHOLAT.COM including scholars, academic news, courses and academic organizations to enrich our university portraits.

Keywords: University portraits · Text mining · Information extraction · Academic social network · SCHOLAT

1 Introduction

The information explosion has led to difficulties in obtaining accurate and effective information and various solutions have been proposed for this problem, such as recommendation system, knowledge graph, data portrait, with data portrait technology being a successful example. User profile, also called user portrait, is a model to generate tag-based profile with users' classic information(name, age, etc.) or social attributes [10]. Furthermore, user portrait can be used in personalized recommendation, for the reason that it can reflect users' characteristic and interest [6]. Corporate portrait utilizes the technology similar to user profile and big data techniques to generate tags for corporations to describe their characteristics and reputation, which is helpful for the business management [4,25]. With the massive application of big data technology, user profile and enterprise portrait are widely researched and applied.

Wu et al. [23] proposed a probabilistic approach to fuse the underlying social theories for jointly users temporal behaviours in social networking services. Gu et al. [5] proposed a psychological modeling method based on computational linguistic features to analyse Big Five personality traits of Sina Weibo users

Y. Sun et al. (Eds.): ChineseCSCW 2021, CCIS 1492, pp. 25–36, 2022.
https://doi.org/10.1007/978-981-19-4549-6_3

and their relationship with social behaviour. Pan et al. [13] proposed a social tag expansion model to complete a user profile through exploiting the relations between tags to expand them and include a sufficient set of tags to alleviate the tag sparsity problem. TF-IDF is a classic and useful weighting scheme in tag-based user profile [10]. Tu et al. [22] integrated TF-IDF with TextRank to mine and analyze users' personal interests from Sina Weibo.

Information of higher education institutions contains a wealth of data, which are valuable for administrators, teachers, students and other related researchers. However, the information of higher education institutions in the Internet is often too cumbersome for users to access the knowledge that they really want to know. There is a wealth of information about universities on their official websites, encyclopaedic sites or other websites that publish information about them, including profiles, histories, faculties and so on. However, they are often similar and it is difficult to access key information in a short time. To the best of our knowledge, there are few studies on processing university information data and constructing university portraits. Therefore, it is necessary to integrate the data of universities and extract the appropriate tags based on their attributes and text of introduction so that the data can be used effectively.

In this paper, we propose a university portrait system incorporated with data on academic social network SCHOLAT [17]. We collect data on more than 2,800 universities in China in three categories, which are university basic information, university introduction and related data from SCHOLAT. With the data, we analyze and mine important information of universities such as attributes, features, projects and so on. In particular, we extract tags from the textual data of universities that are able to describe them effectively, concisely and painly. Moreover, we incorporate university-related data from SCHOLAT, including academics, academic news, courses and academic organizations to enrich the university portrait.

The reminder of this paper is organized as follows. The related research about profiling is reviewed in Sect. 2. Section 3 presents the design of university portrait system. Section 4 describes the methods of system implementation and the evaluation of the system is presented in Sect. 5. Finally, Sect. 6 draws the conclusion.

2 Preliminary

2.1 Text Mining Technique

In recent years, researchers have proposed several models on user profile or user analysis. However, we find few research on analyzing data of higher education institutions or university profile. Therefore, to construct university portraits, we employ the techniques of user profile which extract important information from textual data by utilizing keyword extraction methods such as TF-IDF, Word2Vec, LDA topic model, textRank and so on (see e.g. [11,12]).

This paper mainly uses TF-IDF as data processing model to extract tags of universities. TF-IDF is the abbreviation for term frequency(TF) and inverse document frequency(IDF). It is an important weighting techniques for information retrieval and data mining, which is statistical method used to evaluate the degree of importance for a word/term in a file from a corpus or file set.

Suppose there are N documents $D_1, D_2, ..., D_N$ and w is a keyword. The function TF_{w,D_i} is the frequency of the keyword w in the document D_i :

$$TF_{w,D_i} = \frac{\text{count}(w)}{|D_i|} \tag{1}$$

where *count*(w) is t he times of w appearing in D_i and $|D_i|$ is the number of all words in D_i. The function IDF_w is the logarithm of the ratio of N to the number of documents containing w:

$$IDF_w = \log \frac{N}{|\{i : w \in D_i\}|} \tag{2}$$

Clearly, IDF_w is larger if w is more prevalent; it is smaller otherwise. If the keyword w does not appear in any documents, the denominator in the formula (2) is 0, thus the following formula is usually used.

$$IDF_w = \log \frac{N}{1 + |\{i : w \in D_i\}|} \tag{3}$$

Then the function $TF\text{-}IDF_{w_i D_i}$ is defined as:

$$TF\text{-}IDF_{w,D_i} = TF_{w,D_i} \cdot IDF_w \tag{4}$$

From the definition of TF-IDF, it can be seen that:

- The higher the frequency of a word in a document and its freshness (i.e., low prevalence), the higher its value of TF-IDF.
- TF-IDF takes into account both frequency and freshness, filtering some common words and keeping important words that provide more information.

2.2 Scraping Techniques

Selenium, Requests and Beautiful Soup are common python libraries that can be used for data crawling. Selenium is a tool for web browser automation which provides extensions to emulate user interaction with browsers [19]. Selenium tests run directly in the browser, as if a real user were operating them so it can help address anti-crawling mechanism of some websites [8]. Using Requests can simulate browser Requests, and Beautiful Soup can parse the content of web pages.

2.3 SCHOLAT Data

SCHOLAT is an academic social network system specifically served for scholars, including teachers, researchers and students [1]. Millions of real data are available here. According to statistics, SCHOLAT currently has over 150,000 users, most of whom are scholars from universities. Lin et al. [7] proposed a scholar portrait system utilizing academic social network data in SCHOLAT to mine scholars' important information. There are over 200,000 university-related data in SCHOLAT, including but not limited to user, academic news, course, and academic organization. Thanks to the data from SCHOLAT, we can analyze a university more comprehensively, allowing for a more comprehensive and unique university portrait.

3 System Design

As mentioned above, we draw on the techniques of user profile to a great extent. The data needed to build user profile includes static data such as basic user information and dynamic data such as user behaviour data. Since university information data has no behaviour data, we mainly collect basic information data and introduction text data of them. We overview our university portrait system in Fig. 1. The system mainly consists of two components, namely data collection and tags generation. Before generating tags for university, there is a necessary step, data preprocessing.

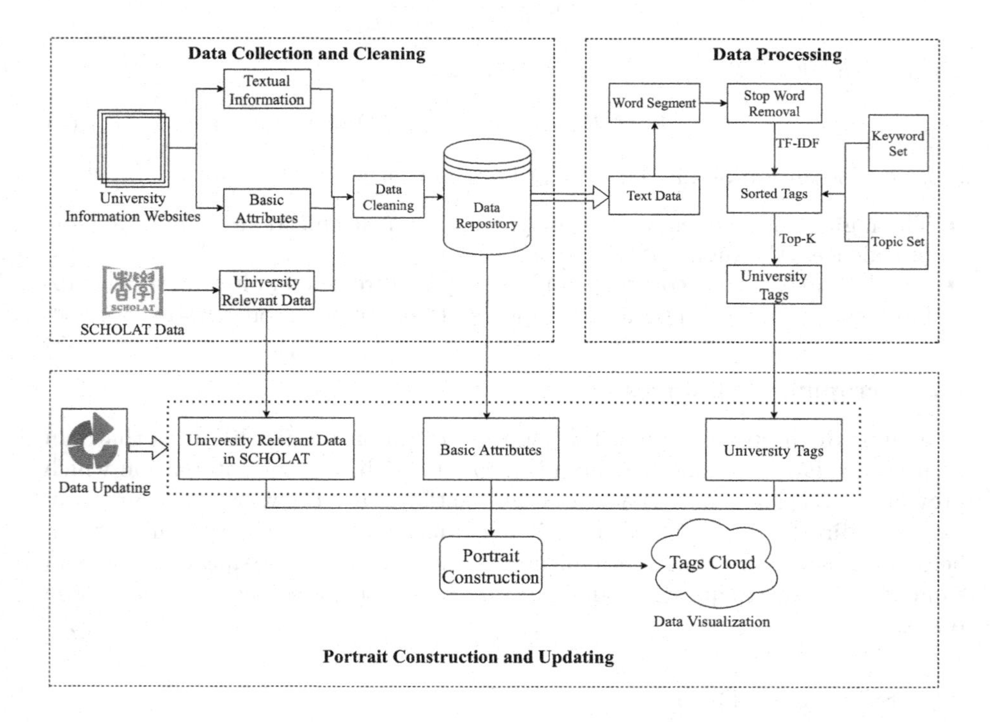

Fig. 1. Overview of university portrait.

3.1 Data Collection and Data Cleaning

University data play an important role in the university portrait because it depends on large amounts of university information data to generate tags of universities and construct university portrait. In order to obtain accurate, authoritative and sufficient data, we select four university information platforms in China, including China Higher Education Student Information and Career Center (CHSI) [21], Education Online of China (EOL) [2] and Baidu Encyclopedia [3]. The data types, source and their description are roughly shown in Table 1.

Table 1. Data types of university information.

Field	Source	Description
univCode	CHSI	The code assigned by Chinese Ministry of Education
univName	CHSI/Baidu Encyclopedia	The name of university in Chinese
univNameEn	Baidu Encyclopedia	The name of university in English
univUrl	CHSI/Baidu Encyclopedia	The official website of university
univAdmin	CHSI	The administrative department of the University
univType	CHSI	Discipline categories of university
univDegree	CHSI/ Baidu Encyclopedia	Education degree level of university
univAddress	Baidu Encyclopedia	Province and detailed address of university
univAttributes	CHSI/Baidu Encyclopedia	Attributes and features of university
univIntro	CHSI/EOL	The long text of university introduction
univSummary	Baidu Encyclopedia	The shorter, concise, summarized university introduction

After collecting the data and storing them in the database, we find that dirty data exist in both structured and unstructured text data. In particular, data of university attributes are synonymous, which will affect the effectiveness of tag generation. Therefore, we make a script to remove meaningless symbols from fields in database table and normalize words with the same meaning in university attributes into one word.

3.2 University Tags Extraction

With the preprocessed data, we use Jieba for word segmentation and stop words removal, which are essential procedure of keywords extraction technique [20]. There are many Chinese proper nouns in the text in Chinese that describe the attributes of the university. If we use default word segmentation method, it does not work well because the above-mentioned proper nouns cannot be identified. Liu et al. [9] built a Topical PageRank onword graph to measure word importance with respect to different topics and further calculate the ranking scores of words and extract the top ranked ones as keyphrases. Wu et al. [24] proposed a Keyphrase Identification Program (KIP), which extracts document keyphrases by using prior positive samples of human identified domain keyphrases to assign weights to the candidate keyphrases. Inspired by the above two methods, we count the term frequencies of all attribute words occurring and calculate their weights as indicators to describe their importance. We selected top-K words according to the ranking of the weights and construct a special dictionary of attribute words in Chinese universities. With this special dictionary, we are able to identify most of the university attribute words as keywords [15]. On the other hand, we make a set of words for university types as the topic set, because different types of university have their specific attribute words. We integrate both special dictionary and university type set when using TF-IDF for tags extraction [16].

4 Implementation

4.1 Data Collection

We collect university information data from the sources by using crawler libraries of python including Selenium, Requests and Beautiful Soup. When some sites allow us to crawl data, it will be more convenient to use Requests combined with Beautiful Soup.

By employing crawler tools above, we are able to access several types of data of university information and store into our MYSQL database. By the time this article was completed, we collect information of more than 2800 colleges and universities in China.

For collecting relevant data in SCHOLAT, we connect to the database and query the data we need by using JDBC in Java with the support of SCHOLAT. Currently, we mainly collect data on users, academic news, courses and academic organizations of universities on SCHOLAT, to enrich the university portrait through the statistics and presentation of the data. In addition, the data related to colleges and universities on SCHOLAT are constantly updating, since users on SCHOLAT are constantly generating new data. So our university portraits are updated as a result. We leave data analysis for our future work.

After collecting university information data above, we start the procedure of university tags extraction. First of all, we have to preprocess the data. Dirty data exist in the data we collected. Specificially, there are three problems as follows:

1. There are extra, meaningless symbols or numbers in several fields.
2. There are synonyms for separate university names, resulting in data mismatching up.
3. Some keywords in the university attributes field are synonymous, i.e. more than one keyword has the same meaning. For example, the phrase Institutions of Higher Learning Innovation Ability Enhancement is also call Project 2011 in Chinese while Pilot Project for Training Top-notch Students in Basic Subjects also known as Everest Project. There are many more such cases.

To address the first problem, we make a simple script in Python to remove the dirty data from the fields of the database table. For the second problem, we look for the mismatched data and correct the university names artificially. As for the third problem, we find out all the synonyms of attribute words and choose the shorter one to replace all of others, as we want the final generated keywords to be as concise as possible.

4.2 University Tagging

For the basic information of university such as name, type and address, we do not need to do secondary preprocessing because they are concise enough. What we have to focus on is the long text data of university, that is, summary and introduction mentioned in Table 1. According to Sect. 3.2, we define a special dictionary including most of the proper attribute words of universities by means of TF. We make a word set of university types matching to the university name, to be the topic set which can also guide the Chinese word segmentation. With loading our predefined special dictionary, we can avoid the problem of incorrectly separating the attribute words of university. At the same time, we perform stop word removal in the text with loading stop words dictionary to remove words that are meaningless for keyword extraction such as auxiliaries. At this point, we get the corpus of university introduction $Corpus_{intro}$, set of attribute words Set_{attri} and set of topics Set_{topic}.

We then use the corpus above $(Corpus_{intro}, Set_{attri}, Set_{topic})$ for keyword extraction. In this paper, we combine statistics-based methods and topic-based methods [14] and our approach is shown in Algorithm 1. First, we calculate the TF-IDF value of each term in $Corpus_{intro}$ and select those with a weight value greater than 0.2 for sorting. If the number of terms is small, we will utilize topic-based methods since good keywords should be relevant to the topics of the given text. We use the set of keywords Set_{attri} and a set of topics Set_{topic} to allocate keywords for different universities with different types by matching their topics to the attribute words. Finally, we merge the keywords obtained by the above two methods into the final set of keywords as the tags of universities.

Algorithm 1. University tagging with hybrid methods

Require:
  Input: $Corpus_{intro}$, Set_{attr}, Set_{topic};
Ensure:
  Output The set of keywords for university tagging, $Tags_{univ}$;
1: $univCode$ = getFirstUnivCode($Corpus_{intro}$); /*get the first university code*/
2: **for** item in $Corpus_{intro}$ **do**
3:   $initTags$ = calculateTFIDF(item); /*calculate each item of $Corpus_{intro}$*/
4:   **if** $initTags$ is not null **then**
5:     /*select the items whose weight value is greater than 0.2*/
6:     $sortedTags$ = sortByWeight(initTags)
7:   **end if**
8:   **if** number of $sortedTags$ <= k **then**
9:     $allocatedTags$ = allocateTagsByTopic($univCode$, Set_{topic}, Set_{attr})
10:   **end if**
11:   $Tags_{university}$ = mergeTags($sortedTags$, $allocatedTags$)
12:   $univCode$ = getNextUnivCode($Corpus_{intro}$);
13: **end for**

4.3 Portrait Construction and Updating

With basic attributes, generated tags and relevant data in SCHOLAT of universities, we select tags that can describe universities accurately and delineate the dimensions to construct the university portrait. Finally, we visualize the data in the system. In particular, we use tag cloud techniques to present keywords of university information.

In general, the information of universities do not change easily. We will update the name, address, level of education degree etc. in the university portrait based on the information published by the Chinese Ministry of Education to ensure that the information is accurate and authoritative. In addition, the keywords for university information will be updated by the introduction of universities.

5 Evaluation

5.1 Evaluation of University Portrait

Our university portrait includes not only basic attributes such as name, university code, type, but also tags that are able to describe the university more effectively and briefly. In addition, we integrate relevant data of the university from SCHOLAT to make the university portrait more informative and diverse. We visualize the information of the system. Taking Peking University as an example, we show the basic attributes module and tag cloud visualization module in Fig. 2.

We represent our final tags of university generated by our university portrait system in tag cloud. The tag cloud words we selected mainly include university name, attributes, features, development programs or projects which are extracted from introduction text data of the university. With basic attributes mentioned above and these tags, we can carry out further application or research on university recommendation in the future.

Due to the limitation of space in this paper, we do not show other modules of the system that are integrated with SCHOLAT data. We count the number of users, academic news (especially related to recruitment), the number of courses and the number of academic organizations created related to the university in SCHOLAT. We then display the top-K scholars, courses, recruitment-related news, courses and academic organizations of the university according to the number of hits. We leave the problem that analyzing the university-related data in SCHOLAT as future work.

5.2 Comparison

Here, we would like to compare the common university and college information service platforms in China with our university portrait system. We mainly focus on characteristics and insufficiencies of each platform. The university and college information service platforms includes official homepage of the university, CHSI,

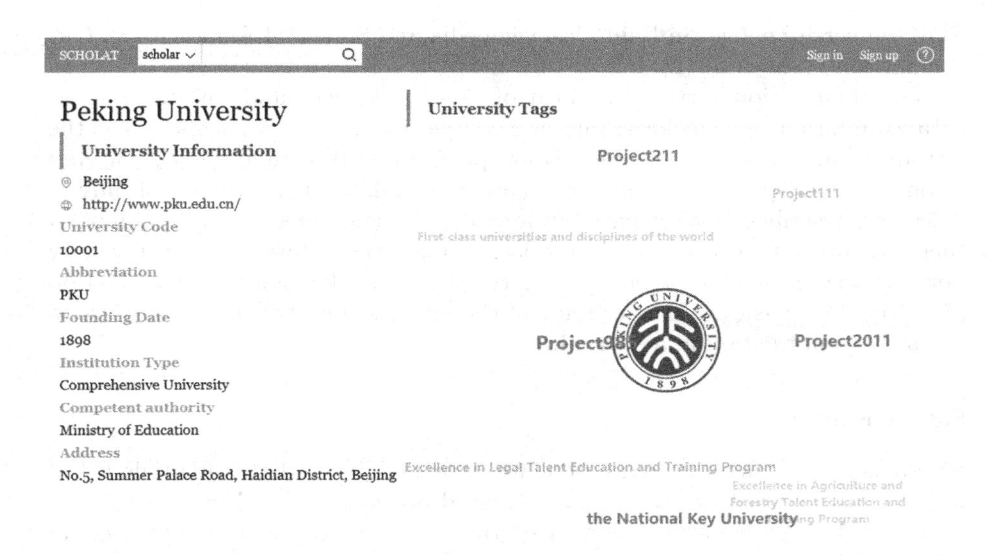

Fig. 2. The basic attributes module and tag cloud visualization module of university portrait system. On the left is the basic attributes module, including the university's name, location, official website, university code, abbreviation, founding date, institution type, competent authority and detailed address. On the right is the University Tags module, which mainly shows the tags of the university, which can accurately and effectively describe the university.

Table 2. Comparison of common university information service platforms in China and our college portrait

Platform	Characteristics	Insufficiencies
Official homepage of university	Most official and authoritative	Difficult to get key information because most of them are long text
CHSI	The information is also authoritative. Satisfaction survey data available	Some of the text information is not consistent with the official homepage
EOL	Learning index, life index, employment index and comprehensive score available. Rate of employment statistics available.	The evaluation indicators of the index are not open and transparent
University entries in Baidu Encyclopedia	Abundant information concentrated in one single page	Difficult to get key information in short time
Shanghai Ranking	Focusing on institution rankings, student quality rankings, and discipline rankings	Some of the evaluation indicators are debatable
Our university portrait	Extracting basic and key information. Combining SCHOLAT data to describe the university	Lack of analysis of SCHOLAT data

EOL, university entries in Baidu Encyclopedia and Shanghai Ranking [18]. Our comparison is shown in Table 2.

The information service platform of various colleges and universities has detailed information of colleges and universities, including but not limited to the introduction, statistical data, admission prospectus, discipline description and so on. Another type of platforms evaluate and rank the performance of universities and describe the university by rankings. As mentioned in Table 2, most of them are difficult to get the key information in a glance. However, our university portrait system extracts important information from the long text introduction of universities, ensuring the accuracy of the information while allowing users to access key information in a short time.

5.3 Discussion

Regarding university tag extraction, we have mainly used the TD-IDF based method for keyword extraction and achieved desired results. Next we will use deep learning methods for keyword extraction and combine the experimental results with the most appropriate method. With regard to the system visualisation module, we will conduct some quantitative evaluations, such as user's comments and feedback. In addition, teachers or students of the university can upload real data about the university, which can be used to enrich the university portrait.

6 Conclusion

In this paper, we propose a university portrait system incorporating academic social network SCHOLAT. The implementation mainly consists of two procedures, which are data collecting and university tagging. We collect university information data from multiples source and preprocess them by word segmentation and stop words removal techniques. Then we perform university tagging combining statistics-based methods and topic-based methods. Finally, we visualize our data and display the university-related data of SCHOLAT. Although text mining and information extraction techniques are used, university profiling is still a new domain-specific research topic. There is potential for improvement in the keyword extraction method used in this paper.

In summary, this paper has two contribution. Firstly, we proposed university portrait, which no one had proposed before. Secondly, we incorporated the university-related data on SCHOLAT to construct the university portrait, which are real and useful. We would like to perform keyword extraction directly on the original long text to obtain more useful information.

However, the university portrait still has some shortcomings that need to be improve in our future work. Although we obtain the expected results by using the current TF-IDF based method in keyword extraction, this is based on the extensive data pre-processing work we have done. We would like to perform keyword extraction directly on the original long text to obtain more useful information. In addition, the content of the extracted tags represent the existing information. In the future we will focus on some deep learning models that can cope with the challenges that can generate new tags about the university based on existing data. In addition, we will carry out further research on university recommendation system with our university portrait.

Acknowledgement. This work was supported by the National Natural Science Foundation of China under Grant U1811263, Grant 61772211 (Y. Tang), NSFC under Grant 11901210 and China Postdoctoral Science Foundation under Grant 2019M652924 (L. Lan).

References

1. Akram, A., Fu, C., Tang, Y., Jiang, Y., Lin, X.: Exposing the hidden to the eyes: analysis of SCHOLAT E-learning data. In: 2016 IEEE 20th International Conference on Computer Supported Cooperative Work in Design (CSCWD), pp. 693–698. IEEE (2016)
2. Application of college entrance examination service platform. https://gkcx.eol.cn
3. Baidu encyclopedia: the world's leading Chinese encyclopedia. https://baike.baidu.com
4. Davies, G., Chun, R., da Silva, R.V., Roper, S.: A corporate character scale to assess employee and customer views of organization reputation. Corp. Reput. Rev. **7**(2), 125–146 (2004)
5. Gu, H., Wang, J., Wang, Z., Zhuang, B., Su, F.: Modeling of user portrait through social media. In: 2018 IEEE International Conference on Multimedia and Expo (ICME), pp. 1–6. IEEE (2018)
6. Lee, W.J., Oh, K.J., Lim, C.G., Choi, H.J.: User profile extraction from Twitter for personalized news recommendation. In: 16th International Conference on Advanced Communication Technology, pp. 779–783. IEEE (2014)
7. Lin, R., Mao, C., Mao, C., Zhang, R., Liu, H., Tang, Y.: SCHONA: a scholar persona system based on academic social network. In: Milošević, D., Tang, Y., Zu, Q. (eds.) HCC 2019. LNCS, vol. 11956, pp. 223–232. Springer, Cham (2019). https://doi.org/10.1007/978-3-030-37429-7_22
8. Liu, Y., Yang, Z., Xiu, J., Liu, C.: Research on an anti-crawling mechanism and key algorithm based on sliding time window. In: 2016 4th International Conference on Cloud Computing and Intelligence Systems (CCIS), pp. 220–223. IEEE (2016)
9. Liu, Z., Huang, W., Zheng, Y., Sun, M.: Automatic keyphrase extraction via topic decomposition. In: Proceedings of the 2010 Conference on Empirical Methods in Natural Language Processing, pp. 366–376 (2010)
10. Mezghani, M., Zayani, C.A., Amous, I., Gargouri, F.: A user profile modelling using social annotations: a survey. In: Proceedings of the 21st International Conference on World Wide Web, pp. 969–976 (2012)
11. Nasar, Z., Jaffry, S.W., Malik, M.K.: Textual keyword extraction and summarization: state-of-the-art. Inf. Process. Manag. **56**(6), 102088 (2019). https://doi.org/10.1016/j.ipm.2019.102088
12. Onan, A., Korukoğlu, S., Bulut, H.: Ensemble of keyword extraction methods and classifiers in text classification. Expert Syst. Appl. **57**, 232–247 (2016)

13. Pan, Y., Huo, Y., Tang, J., Zeng, Y., Chen, B.: Exploiting relational tag expansion for dynamic user profile in a tag-aware ranking recommender system. Inf. Sci. **545**, 448–464 (2021)
14. Papagiannopoulou, E., Tsoumakas, G.: A review of keyphrase extraction. Wiley Interdisc. Rev. Data Mining Knowl. Discov. **10**(2), e1339 (2020)
15. Puigcerver, J., Toselli, A.H., Vidal, E.: Querying out-of-vocabulary words in lexicon-based keyword spotting. Neural Comput. Appl. **28**(9), 2373–2382 (2016). https://doi.org/10.1007/s00521-016-2197-8
16. Qaiser, S., Ali, R.: Text mining: use of TF-IDF to examine the relevance of words to documents. Int. J. Comput. Appl. **181**(1), 25–29 (2018)
17. Scholat. https://www.scholat.com
18. Shanghairanking-leading brand in higher education evaluation. https://www.shanghairanking.cn
19. Simon, S., Alexei, B.A.: A browser automation framework and ecosystem, January 2021. https://github.com/SeleniumHQ/selenium
20. Sun, J.: Jieba Chinese word segmentation tool, January 2020. https://github.com/fxsjy/jieba
21. Sunshine college entrance examination_the designated platform of the ministry of education's sunshine project for college entrance examination. https://gaokao.chsi.com.cn
22. Tu, S., Minlie, H.: Mining microblog user interests based on TextRank with TF-IDF factor. J. China Univ. Posts Telecommun. **23**(5), 40–46 (2016)
23. Wu, L., Ge, Y., Liu, Q., Chen, E., Long, B., Huang, Z.: Modeling users' preferences and social links in social networking services: a joint-evolving perspective. In: Proceedings of the AAAI Conference on Artificial Intelligence, vol. 30, pp. 279–286. AAAI Press (2016)
24. Wu, Y.F.B., Li, Q., Bot, R.S., Chen, X.: Domain-specific keyphrase extraction. In: Proceedings of the 14th ACM International Conference on Information and Knowledge Management, pp. 283–284 (2005)
25. Zhang, X., Yu, Z., Li, C., Zhai, R., Ma, H., Liu, L.: Construction of portrait system of listed companies based on big data. In: 2019 6th International Conference on Information Science and Control Engineering (ICISCE), pp. 210–214. IEEE (2019)

Multi-objective Optimization of Ticket Assignment Problem in Large Data Centers

Tariq Ali Arain, Xiangjie Huang, Zhicheng Cai(✉), and Jian Xu

School of Computer Science and Engineering,
Nanjing University of Science and Technology, Nanjing, China
caizhicheng@njust.edu.cn

Abstract. Software or hardware problems in large data centers are usually packaged to be tickets which are assigned to different experts to solve. It is very crucial to design multi-objective ticket scheduling algorithms to maximize the total matching degree and minimize the total flowtime. However, most of existing methods for assignment problems only consider single objective, while some methods optimizing multi-objectives are not for the same objectives of this paper. Meanwhile, exploring effectiveness of existing meta-heuristics for multi-objective optimization could be improved further. In this paper, a multi-objective heuristic algorithm called (GAMOA*) is proposed for ticket scheduling which is the combination of a genetic algorithm (GA) and a multi-objective A* (MOA*). In GAMOA*, ticket scheduling orders are evaluated and improved by GA, while MOA* is applied to find a Pareto set of solutions given an order of tickets effectively and efficiently. Experimental results illustrate that our approach obtains better results than state-of-art algorithms.

Keywords: Genetic algorithm · Ticket scheduling · Multi-objective A star · Cloud Computing · Routing problem

1 Introduction

End-users and application developers using Cloud Computing facilities may meet software or hardware problems inevitably. User problems are usually recorded to be tickets (contain problem descriptions) and sent to back-end experts to solve. From the perspective of Cloud Computing providers, solving these problems efficiently and effectively is beneficial to improving user experiences. Experts have diverse technique background and problem solving experience, then have different matching degrees for a specific ticket which determine the quality of solving the tickets. Meanwhile, experts also have different estimated processing times for the same ticket which have great impact on the response time to users. Therefore, it is crucial to develop ticket scheduling algorithms in terms of matching degrees and processing times.

The multi-objective scheduling of tickets to experts is an NP-hard problem. In this paper, the total matching degree and total flowtime of overall tickets

Y. Sun et al. (Eds.): ChineseCSCW 2021, CCIS 1492, pp. 37–51, 2022.
https://doi.org/10.1007/978-981-19-4549-6_4

are optimized simultaneously where the flowtime of a ticket is the finish time minus the arrival time of it. This ticket scheduling problem can be modeled to be a multi-objective assignment problem which is a well known NP-hard problem [1]. Most of existing works for assignment problems address the optimization of a single objective function [2,3]. The multi-objective optimization can not be solved by these methods directly [4]. Meanwhile, there are only a few existing algorithms for solving multi-objective assignment problems using genetic algorithms, simulated annealing and particle swarm optimization algorithms [1,5,6]. However, the exploring ability of these multi-objective meta-heuristic algorithms can be enhanced by some exact algorithms like A star algorithm.

In this article, a hybrid of genetic algorithm and multi-objective A* (called GAMOA*) is proposed for the ticket scheduling problem by taking advantages of genetic algorithms and A* simultaneously. The genetic algorithm first sorts out all tickets randomly to produce different ticket scheduling orders called genetic chromosomes. Afterward, scheduling ordered tickets to experts is transformed into a routing problem which can be solved by multi-objective A* effectively. Next, solutions found by the multi-objective A* (MOA*) is used to calculate the fitness of each chromosome, based on which better chromosomes are generated by crossover and mutations. Finally, new chromosomes are evaluated by MOA* again and the above process iterates until the stopping criteria are met. The main contributions include (1) a novel multi-objective heuristic algorithm called (GAMOA*) is proposed which is the hybrid of genetic algorithm and multi-objective A*. (2) a Pareto-set based fitness calculation method is developed for the individual with multiple solutions and a transforming method is proposed to transform the ticket scheduling problem into a routing problem which is the basis of using multi-objective A* as a local search method.

The rest of the article is organized as follows: Sect. 2 is related work. Section 3 demonstrates the problem formulation. The proposed heuristic is described in Sect. 4. Experimental setup and results are presented in Sect. 5. Finally, Sect. 6 concludes the paper.

2 Related Work

Ticket routing and ticket scheduling are two important problems about ticket solving. Ticket routing focus on finding the right order of candidate experts to assign each ticket. Tickets are transferred to next expert on the path when they can not be solved by the current experts. On the contrary, ticket scheduling is in charge of finding an appropriate assignment considering both the total matching degree (between tickets and experts) and the total flowtime. Ticket routing techniques could be used to enhance ticket scheduling.

2.1 Ticket Routing

When a client's request reaches the help desk, the operator has to view it, categorize it, and forward it to the responsible expert to investigate and fix the requested problem. The procedure is termed ticket routing. The ticket scheduling to the experts has a resemblance with the ticket routing. Ticket content-mining [7] and historical sequence [8,9] based methods have been developed for the ticket routing. Shao et al. [8] developed a route recommender system based on different orders of Markov models established from historical processing sequence data. In content-based methods [7], tickets are usually classified to be different categories by analysing text descriptions based on diverse machine learning techniques. Content-based methods are usually only better on the tickets with rich text descriptions. However, for general tickets with not too much descriptions, it has been proved that content-based methods are usually outperformed by sequence based methods [9]. There are also some hybrid methods based on Markov model and content-mining [9]. However, most of existing methods about ticket solving primarily focus on ticket routing ignoring the optimization of workload distribution among experts.

2.2 Ticket Scheduling

Many methods have been developed for different kinds of assignment problems and most of them are designed for single objective. Discrete particle swarm optimization (DPSO) based methods are extensively used in multi-objective optimization of assignment problems such as task scheduling in Grid [10], load balancing in heterogeneous computing systems [11,12] and task scheduling in a distributed computing system [13,14] in terms of makespan, flowtime, reliability and cost objectives, etc. Genetic algorithms are also widely used to solve multi-objective optimization of assignment problems. Subhashini et al. [15,16] developed a hybrid method of Elitist Non-dominated Sorting Genetic Algorithm (NSGA-II) and the Non-dominated Sorting Particle Swarm Optimization Algorithm (NSPSO) to schedule tasks in a heterogeneous environment for minimizing the makespan and flowtime. Khodadadi et al. [17] proposed an Genetic Algorithm with Variable Neighborhood Search (GA-VNS) for independent task scheduling in a grid environment to reduce the total cost and makespan. Many other nature-inspired meta-heuristics are also applied for assignment problems such as Simulated Annealing (SA) [18], Ant Colony Optimization (ACO) [18], and Cuckoo and Gravitational Search (CGSA) [19]. However, there is no algorithm designed for the ticket scheduling of this paper.

In a word, existing algorithms are usually designed for single objectives or not for the multi-objective ticket scheduling. Meanwhile, existing meta-heuristics could be improved by more efficient and effective local search methods because the convergency speed of traditional meta-heuristic algorithms are very slow for large scale ticket scheduling problems.

3 Problem Description

$E = \{e_i, i = 1, ..., N\}$ is the set of back-end experts in the Cloud Computing Data Centers. It is assumed that tickets are scheduled to experts periodically. For example, tickets are collected to be a batch and scheduled to experts every six hours. It is assumed that all experts have finished the processing of the previous batch and are idle currently. $T = \{t_j, j = 1, ..., K\}$ is the set tickets in the current batch. p_{ij} and m_{ij} denotes the processing time and matching degree of ticket t_i on expert e_j respectively. Multiple tickets can be assigned to the same expert. However, an expert cannot start a new ticket before finishing previous tickets. Let $d_{i,j}$ be the index of ticket i on expert j. If ticket i is not assigned on expert j, $d_{i,j} = 0$. For example, when tickets t_1, t_2 and t_3 are assigned on expert e_1 one by one, $d_{1,1} = 1$, $d_{2,1} = 2$ and $d_{3,1} = 3$. The flowtime $f_{i,j}$ of a ticket t_i on expert e_j is its own processing time plus the waiting time (used to process previous tickets). The objective of this article is to minimize the total flowtime and maximize the total matching degree. The mathematical model of this multi-objective tickets schedule is as follow.

$$\max \sum_{i=1}^{K} \sum_{j=1}^{N} m_{i,j} \cdot (d_{i,j} > 0) \tag{1}$$

$$\min \sum_{i=1}^{K} \sum_{j=1}^{N} f_{i,j} * (d_{i,j} > 0) \tag{2}$$

$$\sum_{j=1}^{N} (d_{i,j} > 0) = 1, i = 1, ..., K \tag{3}$$

$$f_k = \sum_{i=1}^{K} \sum_{j=1}^{N} p_{i,j}(d_{i,j} > 0) * (d_{i,j} \leq d_{k,j}), k = 1, ..., K \tag{4}$$

Equation (1) and (2) maximize the total matching degree and minimize the total flowtime. Equation (3) represents that one ticket is assigned to only one expert. Equation (4) illustrates that the flowtime of ticket t_k is the processing time of t_k plus processing times of tickets before t_k on the same expert.

4 Proposed GAMOA*

In this paper, a genetic algorithm-based multiple-objective A star algorithm (GAMOA*) is presented for the multi-objective ticket scheduling problem. Genetic algorithms (GA) have been widely used to solve multiple-objective scheduling problems and obtain good performance. The main reason is that GA has the ability to exploring the search space efficiently. However, the traditional GA lacks the ability to explore local optimal areas deeply. Therefore, some local search methods have been reported to aid GA. Nonetheless, an exhausted local search is very time-consuming. A star is an effective and efficient method for route searching. In this article, the ticket scheduling is transformed into a routing problem and solved by a multi-objective A star (MOA*) as an effective local search method of GA. Figure 1a explains the flow chart of the proposed GAMOA*. First, ticket orders are generated via a genetic algorithm. Second, the fittest orders are selected via the fitness function. Then ticket assignment on experts with the given ticket orders is transformed into a routing problem.

Next, the routing problem is solved by the MOA* to find assignments of tickets on experts in detail. Finally, a set of Pareto solutions is obtained for each order of tickets. The fitness of this ticket order is evaluated based on these solutions in terms of the total flowtime and matching degree. The above process iterates until the stopping criteria are met.

4.1 Transforming Ticket Scheduling to Routing

Traditional routing problems are usually modeled via a directed graph with weight on edges. The goal of routing algorithms is to find the route from source to sink nodes with the optimal objectives. However, the goal of ticket scheduling is to find an assignment of tickets to experts. Tickets can be assigned to experts in any order. The A-star algorithm cannot be used to solve the original ticket scheduling, which is very different from the routing problem. However, when the ticket scheduling order is given, the ticket scheduling can be transformed into a routing problem as follows: For example, in the left part of Fig. 1b, tickets are ordered to be $\{T_3, T_1, T2, T_4\}$ which means that they are scheduled from the top to bottom. On each row, nodes on the right of the ticket are candidate experts for that ticket. The same expert might appear on multiple rows, which means that multiple tickets can be assigned to the same expert. The calculating of a ticket's flowtime should consider already assigned tickets before this ticket. Experts of different levels are connected by up-down links, and a source and sink nodes are added to construct a directed graph. A route from the source node to the sink node on the graph means a schedule of tickets to experts. For each level, different selected nodes on the route mean assigning the corresponding tickets to different experts. For this routing problem, the A* algorithm can be used to solve it effectively. However, different orders of tickets have a significant impact on the A* for the above routing problem.

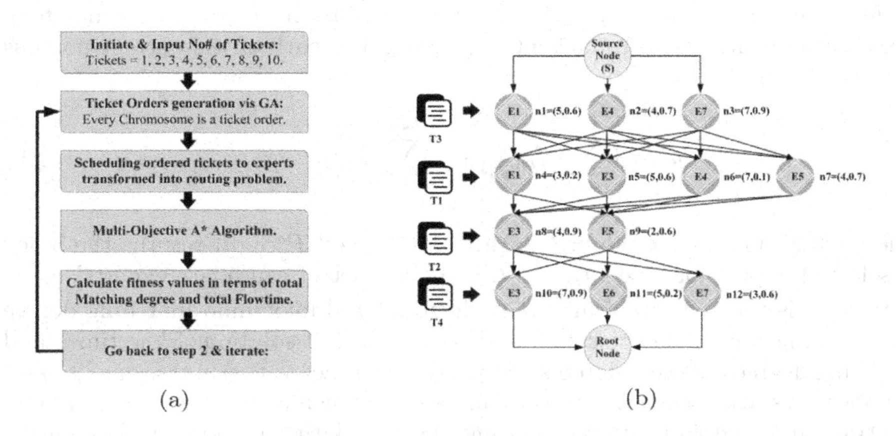

Fig. 1. An illustration of architecture of the proposed GAMOA* and transforming of assignment problems into routing problem

4.2 Multi-objective A* for Scheduling Ordered Tickets

Based on the proposed estimation function, the existing multi-objective A* (MOA*) [20] is applied to solve the above-mentioned routing problem. MOA* applies a best-first search technique which uses $\vec{F_n} = \vec{g_n} + \vec{h_n}$ as an evaluation method to select the next open node for expansion where $g(n)$ keeps the values (in tuples of multiple objectives) of the routes from the source node to the current node in the search graph and $h(n)$ estimates the path values (in tuples too) from the current node to the final destination node. The calculating methods of $h(n)$ determine the computation time of MOA* because $f(n)$ is used to cut searching branches. In this paper, an appropriate $g(n)$ function is found to reduce the time while maintaining search effectiveness.

Let $(t_{w_1}, t_{w_2}, t_{w_3}, ...t_{w_k}..., t_{w_K})$ be the ordered tickets and n is the node at the level of ticket t_{wk} in the routing graph. $S_p = (n, \vec{g_n}, F(n, \vec{g_n}))$ is a partial solution from s to node n in the graph of the routing problem, where $F(n, \vec{g_n})$ is the function to calculate $\vec{F_n}$ based on S_p. For each S_p, $\vec{g_n}$ and the path to obtain S_p is stored in $G_{op}(n)$ in the form of $\{(\vec{g_n}, P^s_{\vec{g_n}})\}$ where $P^s_{\vec{g_n}}$ is the set of paths obtaining $\vec{g_n}$, like $\{(n1, n4, ..., ni), (n2, n5, ..., ni)\}$. $P^s_{\vec{g_n}}$ will be updated when a new path with the same $\vec{g_n}$ was extended. For one successor node m, $P^s_{\vec{g_m}}$ will be updated by adding paths $(P^s_{\vec{g_n}}[w], n), w = 1, ..., |P^s_{\vec{g_n}}|$. In the partial solution S_p, the flowtime of ticket t_i on expert e_j is

$$f_{i,j} = \sum_{l=1}^{K} p_{l,j}(d_{l,j} > 0) * (d_{l,j} \le d_{i,j}) \tag{5}$$

where $d_{l,j}$ is determined by $P^s_{\vec{g_n}}$. For each node n, there might be multiple partial solutions, i.e., $G_{op}(n)$ may have more than one elements with different $\vec{g_n}$. Meanwhile, multiple paths might have the same $\vec{g_n}$ making $P^s_{\vec{g_n}}$ have multiple paths. The calculation of $g(n)$ is the sum up of total flowtime and matching degree from source node to current node along the path of the current node as follows.

$$\vec{g_n} = (\sum_{i=t_{w_1}}^{t_{w_k}} f_{i,P^s_{\vec{g_n}}[0][i]}, \sum_{i=t_{w_1}}^{t_{w_k}} m_{i,P^s_{\vec{g_n}}[0][i]}) \tag{6}$$

where $P^s_{\vec{g_n}}[0]$ means get the first path from $P^s_{\vec{g_n}}$ and $P^s_{\vec{g_n}}[0][i]$ returns the index of selected expert for ticket t_i. Let CE_i be the set of candidate experts' index of t_i. $\vec{h(n)}$ is the sum-up of minimum flowtime and maximum matching degree of remaining unassigned tickets based on the partial solution. Flow times and matching degrees are calculated separately, which means that unassigned tickets are not necessarily scheduled to the same expert when calculating the minimum flowtime and matching degree. During the calculating of flowtime, execution times of tickets scheduled in the partial solution and tickets before the current ticket should be considered. Let $P^{sf}_{\vec{g_n}}$ and $P^{sm}_{\vec{g_n}}$ record partial paths to estimate flowtime and matching degree based on $\vec{g_n}$, respectively.

$$\vec{h_n} = (\sum_{i=t_{w_{k+1}}}^{t_{w_K}} \min_{j \in CE_i} \{f_{i,j}\}, \sum_{i=t_{w_{k+1}}}^{t_{w_K}} \max_{j \in CE_i} \{m_{i,j}\}) \tag{7}$$

where $f_{i,j}$ is determined by $P^{sf}_{\vec{g_n}}$.

Algorithm 1. Multi-objective A* (MOA).

Input: Ticket order T_o;

1: Initialize the source node (s), destination node (e), reached goal node set $GOALN$, non-dominated cost set $COSTS$, the sets of non-dominated cost vectors of paths reaching n that have or have not been explored $G_{cl}(n)$ and $G_{op}(n)$, further expanded path set $OPEN$ and add S to set $OPEN = (s, \vec{g_s}, F(s, \vec{g_s}))$;
2: **while** $OPEN$ is not empty **do**
3: Select non_dominated $(n, \vec{g_n}, F(s, \vec{g_n}))$ with smallest flowtime in $OPEN$
4: Move $\vec{g_n}$ from $G_{op}(n)$ and add $\vec{g_n}$ to $G_{cl}(n)$ if $\vec{g_n}$ not dominated by $G_{cl}(n)$;
5: **if** n is final node **then**
6: Include n in $GOALN$ and $\vec{g_n}$ in $COSTS$;
7: Filtering $F(x)$ in $OPEN$;
8: continue;
9: **else**
10: Get all successors nodes m and calculate $\vec{g_m} = \vec{g_n} + \vec{c}(n, m)$;
11: **if** m is new **then**
12: Calculate F_m filtering estimates dominated by $COSTS$;
13: **if** F_m is not empty **then**
14: Put $(m, \vec{g_m}, F(m, \vec{g_m}))$ in $OPEN$
15: $P^s_{\vec{g_m}} \leftarrow (P^s_{\vec{g_n}}[w], n), w = 1, ..., |P^s_{\vec{g_n}}|$. And add ($\vec{g_m}$,$P^s_{\vec{g_m}}$) to $G_{op}(m)$
16: continue;
17: **end if**
18: **else**
19: **if** $\vec{g_m} \in G_{op}(m) \bigcup G_{cl}(m)$ **then**
20: $P^s_{\vec{g_m}} \leftarrow (P^s_{\vec{g_n}}[w], n), w = 1, ..., |P^s_{\vec{g_n}}|$. And add ($\vec{g_m}$,$P^s_{\vec{g_m}}$) to $G_{op}(m)$
21: continue;
22: **else**
23: Calculate F_m filtering estimates dominated by $COSTS$;
24: **if** F_m is not empty **then**
25: Put $(m, \vec{g_m}, F(m, \vec{g_m}))$ in $OPEN$
26: $P^s_{\vec{g_m}} \leftarrow (P^s_{\vec{g_n}}[w], n), w = 1, ..., |P^s_{\vec{g_n}}|$. And add ($\vec{g_m}$,$P^s_{\vec{g_m}}$) to $G_{op}(m)$
27: continue;
28: **end if**
29: **end if**
30: **end if**
31: **end if**
32: **end while**
33: **return** Non_dominated results R;

In the multi-objective A* algorithm, the $OPEN$ set record the reachable path. If the $OPEN$ set is not empty, the partial solution $S_p = (n, \vec{g_n}, F(n, \vec{g_n}))$ with the minimum flowtime is taken from $OPEN$, and corresponding $(\vec{g_n}, P^s_{\vec{g_n}})$ is removed from the unexplored set G_{op} and added to the explored set G_{cl} in the form of $\{(\vec{g_n}, P^s_{\vec{g_n}})\}$. If the fetched node n is the destination node, it is added to the goal node-set GOALN and the $\vec{g_n}$ value is added to the $COSTS$ set. If it is not the destination node, then the reachable node is obtained. The extension node will be filtered if it is dominated by G_{op} and G_{cl}, and finally add the remaining extension nodes to $OPEN$. The formal description of multiple-objective A* (MOA*) algorithm is shown in Algorithm 1.

An Example of MOA*: The process of MOA* for scheduling ordered tickets in Fig. 2a is as follows. Triples of partial solutions and values of GOALN and COSTS are shown in Table 1.

- **Iteration-1**: G_{op}(S) $= (0, 0)$, and G_{cl}(S) $= \emptyset$. From the S node to the destination node, the flowtime and matching degree of T1, and T2 need to be calculated for calculating $h(S)$. For ticket T1, the earliest finish time is 4 on E2, and the maximum matching degree is 0.7 on E2. With T1 executed on E2, the earliest finish time of T2 is 3 on E1, and the maximum matching degree of T2 is 0.6 on E3. Therefore $h(S) = (7, 1.3)$. The only path in OPEN is selected, and its two extensions to nodes $n1$ and $n2$ are added to $OPEN$, as shown in Fig. 2a.
- **Iteration-2**: When calculating $h(n1)$, T1 is executed on E1. If T2 is executed on E1, the finish time of T2 is 8 including 5 min of waiting and 3 min of execution. The finish time of T2 is executed on E3 is 5. Therefore, the earliest finish time is 5 on E3. The maximum matching degree is 0.6 on E3. With T1 executed on E2, the earliest finish time of T2 is 3 on E1, and the maximum matching degree of T2 is 0.6 on E3. The value of $h(n1)$ and $h(n2)$ are (5,0.6) and (3,0.6) respectively. Update $OPEN = \{(n1, (5, 0.6), (10, 1.2)), (n2, (4, 0.7), (7, 1.3))\}$ and insert $((5, 0.6), (S))$ into G_{op}(n1) and $((4, 0.7), (S))$ into G_{op}(n2). The non-dominated n2 with lowest flowtime is selected.
- **Iteration-3**, update $OPEN = \{(n1, (5, 0.6), (10, 1.2)), (n3, (7, 0.9), (7, 0.9)), (n4, (9, 1.3), (9, 1.3))\}$. The reason why $h(n3) = h(n4) = (0, 0)$ is that the flowtime and matching degree from n3 and n4 to the destination node are both zero. Insert $((7, 0.9), (S, n2))$ into G_{op}(n3) and $((9, 1.3), (S, n2))$ into $G_{op}(n4)$. And $((4, 0.7), (S))$ will be moved to $G_{cl}(n2)$ from $G_{op}(n2)$. $n3$ is non-dominated with smallest flowtime in $OPEN$ and selected for the extension towards node "e". The node "e" represents a dummy node, and $n3$ is added to the $GOALN$ array. The cost vector of $n3$ is included in the COSTS array. The $n3$ is closed and $((7, 0.9), (S, n2))$ will be moved to $G_{cl}(n3)$ from $G_{op}(n3)$.
- **Iteration-4**, update $OPEN = \{(n1, (5, 0.6), (10, 1.2)), (n4, (9, 1.3), (9, 1.3))\}$. n4 is non-dominated with smallest flowtime in OPEN and selected for the extension towards node "e". Update $GOALN = \{n3, n4\}$ and $COSTS = \{(7, 0.9), (9, 1.3)\}$.

- **Iteration-5**, update $OPEN = \{(n1, (5, 0.6), (10, 1.2))\}$. The node is selected for an extension to proceeds towards the goal nodes E1 and E3. The node E1 is closed and inserted into the $G_{cl}(\text{n1})$.
- **Iteration-6**, the estimate $(n3, (13, 0.8), (13, 0.8)), (n4, (10, 1.2), (10, 1.2))$ will be filtered by $COSTS$ causing neither $n3$ nor $n4$ is added to $OPEN$. Update $OPEN = \emptyset$. All the alternate solutions are filtered. The $GOALN$ and $COSTS$ arrays are solution node $n3(7, 0.9), n4(9, 1.3)$.

Table 1. An example of MOA* OPEN (<<< = selected node).

Iteration	OPEN	GOALN	COSTS
1	(s, (0, 0), (7, 1.3))	Ø	Ø
2	(n1, (5, 0.6), (10, 1.2)), (n2, (4, 0.7), (7, 1.3)) <<< S	Ø	Ø
3	(n1, (5, 0.6), (10, 1.2)), (n3, (7, 0.9), (7, 0.9)), <<< n2 (n4, (9, 1.3), (9, 1.3))	(n3)	(7, 0.9)
4	(n1, (5, 0.6), (10, 1.2)), (n4, (9, 1.3), (9, 1.3)) <<< n2	(n3, n4)	(7, 0.9), (9, 1.3)
5	(n1, (5, 0.6), (10, 1.2)) <<< S	(n3, n4)	(7, 0.9), (9, 1.3)
6	Ø	(n3, n4)	(7, 0.9), (9, 1.3)

4.3 Genetic Algorithm for Ticket Ordering

The number of ticket orders is exponential to the number of tickets. It is very time-consuming to solve every ticket orders using A*. In this article, a multiple-objective GA (MGA) is employed to find appropriate ticket orders. MOA* is used to collect the set of Pareto solutions for each individual.

Chromosome Encoding: Each chromosome is expressed as an array with a length of K which represents a ticket order. Each chromosomes gene represents the index of ticket, and a chromosome is an order of tickets. Figure 2b is an example of a chromosome with $K = 10$.

Pareto Set Based Fitness Function: Selecting an appropriate fitness function is very crucial to genetic algorithms (GAs). For instance, the maximin [1] is widely used to achieve optimum performance. However, a set of pareto solutions are generated for each individual in the population of the genetic algorithm of this paper (GA). Therefore, a pareto set based maximin fitness function is presented. Given an objective function f with K objectives, f_k^i denotes $k\text{-}th$ objective value of the $i\text{-}th$ solution in the specific generation ($k \in \{0, 1\}$ in this paper). The formula of maximin Fitness Function [1] is

$$fitness_i = \max_{j \neq i, j \in X}(\min_{k \in \{0,1,\ldots K\}}(f_k^i - f_k^j)) \tag{8}$$

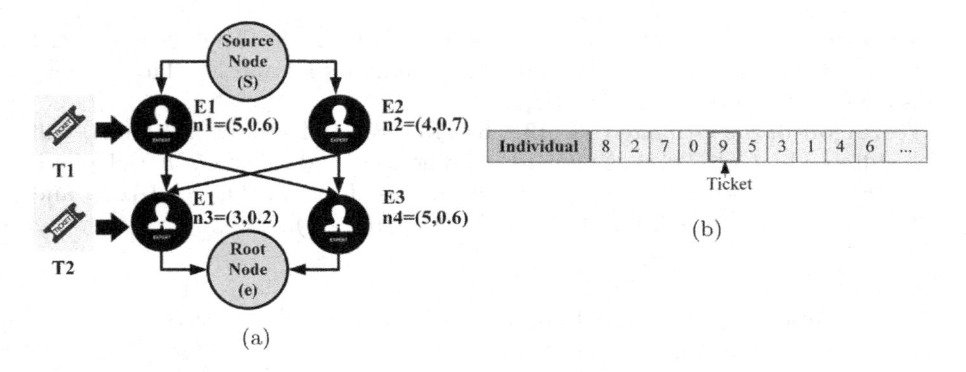

Fig. 2. Examples for MOA* and chromosome encoding

where X is the set of non-dominated solutions. All fitness values are normalized to be in the range of [0, 1] and negative label is employed to transform the maximum of matching degree to a minimum problem coherent with the minimizing of flowtime. The min is used to find the best objective of i compared with j and max_f is used to find the worst condition compared with different solution j. If a solution has a maximin fitness value > 0, which means that it is dominated by others. On the contrary, a maximin fitness value < 0 means the corresponding solution is non-dominated. In existing algorithms for multi-objective optimization, each individual has one corresponding solution. Therefore, the fitness value of each individual can be calculated using Eq. (8) directly. However, for each individual (an order of tickets) in this paper, a set of Pareto solution is obtained by MOA*. Therefore, the fitness of each individual is the average of fitness values of all Pareto solutions. In a population G, there are multiple individuals (chromosomes), and a set of Pareto solutions S_i is obtained by MOA* for each individual $i \in G$. The solutions of all individuals compose a large solution set $S^a = \cup_{i \in G} S_i$. For each solution, the fitness value is calculated in the range of S^a based on Eq. (8). Then, the fitness value of the individual i is obtained by averaging the fitness values of all solutions in S_i.

Selection and Crossover Operation: The Uniform crossover technique is applied in which parents are selected via the roulette wheel selection. For minimizing objectives, individuals with lower maximin fitness values have more chances of being chosen as a parent. Figure 3a demonstrates the process of crossover operation. Initially, two parents named P1 and P2 are selected for crossover as follows. P1 has the minimum fitness value and the roulette wheel method is applied to select P2 randomly. Then, P1's genes are copied to the offspring with a probability of 0.5, and unassigned positions of the offspring are set to be -1. Next, tickets not appeared in the current offspring are taken from P2 in order generating a temp GeneList. Finally, tickets of the temp Genelist replaced -1 in the offspring chromosome in order obtaining the final complete offspring.

Mutation Operation: For every individual, there is a certain probability to mutate for each gene. The mutation process iterates from the first gene to the last gene one by one. For each gene, there is a probability to select another gene randomly and swap it with the current gene. The Fig. 3b demonstrates an example of mutation process. Firstly, for the first gene, the random value generated by a random number generator is larger than the mutation probability which means the first gene need not to be swapped with others. For the second gene, the random value is generated again and is smaller than the mutation probability. Therefore, the second gene need to be swapped with another randomly selected gene (9 for this step). After genes 2 and 9 are swapped, the current gene changes to the third gene and the above process iterates until the last gene.

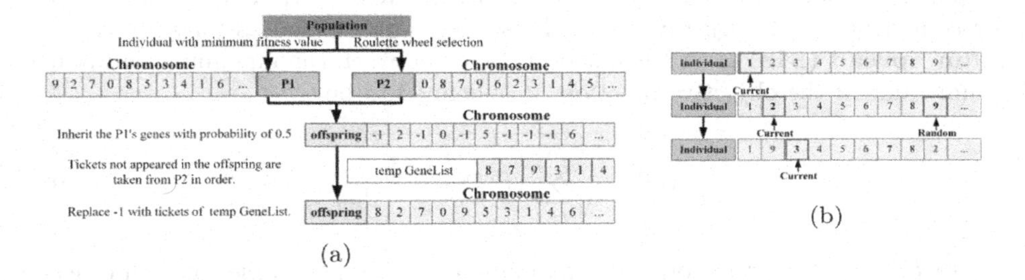

Fig. 3. An illustration of crossover and mutation

5 Performance Evaluation

Since there is no exact algorithms for the considered ticket scheduling problem in terms of flowtime and matching degree, our approach is compared with existing multi-objective NMCMO [1] and IHDPSO [13] algorithms for traditional assignment problems. Algorithms are coded using Java and executed on a Windows system with i5-8400 CPU and 16 GB memory. Algorithms are tested on a randomly generated set of tickets and experts to evaluate their performance. Test data set is divided into three groups named small (30–40 tickets and experts), medium (50–100 tickets and experts), and large (100–150 tickets and experts). Parameters of GAMOA*, NMCMO and IHDPSO are shown in Table 2 which are set to be the same as much as possible for fair comparison. Meanwhile, in order to evaluate the impact of the number of generations to compared algorithms, results of different generations are recorded. Algorithms are given the same computation times. For GAMOA*, results of the 10-*th* and 20-*th* generations are recorded because GAMOA* consumes more time to evaluate each individual. On the contrary, NMCMO and IHDPSO run more fast and results of 100-*th* to 30000-*th* generations are stored. Then, different generations' results of algorithms are compared pairwise.

Table 2. Parameter settings for GAMOA, NMCMO and IHDPSO

Parameters	Size
Population	100
Crossover probability	0.95
Mutation probability	0.175
GAMOA*'s number of generations (iteration times)	10\20
NMCMO's number of generations	1000\5000\10000\30000
IHDPSO's number of generations	100\1000\5000\20000

The maximum spread and the convergence of two sets [1] are popular metrics for comparing multi-objective algorithms which are defined as follows. **Maximum Spread (MS)** is an Euclidean distance between the maximum and minimum value of the objective function. It is defined to be

$$MS = \sqrt{\sum_{i=1}^{K} |f_i^{\max} - f_i^{\min}|} \tag{9}$$

where f_i^{max} and f_i^{min} denote the maximum and minimum values of each objective, respectively. The higher value of the maximum spread means the vast diversity in the search space, i.e., the maximum spread measures the diversity of a set of solutions. **Coverage of Two Sets (C)** compares the domination of two sets of populations which shows that how better population P' are dominated by population P. The formulation of coverage of two sets is defined to be

$$C(P, P') = \frac{|\{x' \in P'; \exists x \in P, x \; dominates \; x'\}|}{|P'|} \tag{10}$$

where x and x' are solutions. Larger $C(P, P')$ means more solutions in P' are dominated by solutions in P. For example, when $C(P, P') = 1$, all solutions in P' are dominated by some other solutions in P.

5.1 Experimental Results

Experimental results in terms of Maximum Spread is shown in Table 3 which denotes that MS of NMCMO are larger than those of GAMOA*. Meanwhile, MS increases as the generation size increases for NMCMO while the generation size has very little impact on MS of GAMOA*. NMCMO is a traditional genetic algorithm and the quality of solutions is diverse, while our approach uses multi-objective A* to find local optimal solutions every time. Therefore, the MS of NMCMO are larger than those of GAMOA*. For example, solutions with too large flowtime have been ignored by GAMOA*, but not by NMCMO. Meanwhile, as the generation size increases, NMCMO finds more different solutions by exploring the search space increasing MS in some degree. IHDPSO obtains

Table 3. Maximum spread of each instance

Group	Tickets-experts	NMCMO(GS)				IHDPSO(GS)				GAMOA*(GS)	
		1000	5000	10000	30000	100	1000	5000	20000	10	20
Small	30-30	11.9	14.3	14.4	15.0	4.1	2.7	3.1	3.8	5.1	5.1
	30-40	13.1	13.1	12.9	15.0	3.2	4.0	5.3	2.7	5.7	5.7
	40-30	15.9	15.9	14.3	15.3	0	17.0	11.9	7.5	6.0	6.0
	40-40	11.1	12.0	13.7	15.6	0	0	5.5	6.9	6.3	6.8
Medium	50-50	11.8	12.2	13.9	15.6	3.5	4.7	0	5.7	6.1	6.1
	50-100	9.6	7.4	13.1	14.0	2.1	2.6	2.8	2.7	4.0	2.0
	100-50	22.6	24.4	24.4	25.4	26.8	30.5	4.4	20.6	7.0	7.2
	100-100	19.8	18.6	19.9	19.1	0	3.5	0	0	6.1	5.1
Large	100-150	14.7	15.1	12.5	18.3	2.6	2.6	0	4.2	4.3	3.3
	150-100	14.2	14.9	15.2	17.5	0	16.4	26.6	14.2	5.2	5.2

Table 4. The coverage of compared algorithms with different generation sizes

Group	GAMOA* (GS)	NMCMO(GS)								IHDPSO(GS)							
		1000		5000		10000		30000		100		1000		5000		20000	
Small	**10**	**0.99**	0	**0.97**	0	**0.96**	0	**0.97**	0	**1**	0	**1**	0	**1**	0	**1**	0
	20	**0.99**	0	**0.97**	0	**0.96**	0	**0.97**	0	**1**	0	**1**	0	**1**	0	**1**	0
Medium	**10**	**0.88**	0	**0.88**	0	**0.89**	0	**0.90**	0	**0.5**	0.22	**0.56**	0.2	**0.5**	0.24	**0.5**	0.22
	20	**0.89**	0	**0.89**	0	**0.91**	0	**0.91**	0	**0.5**	0.2	**0.63**	0.17	**0.5**	0.22	**0.5**	0.2
Large	**10**	**1**	0	**1**	0	**1**	0	**0.97**	0	**0.25**	0.1	**0.38**	0.14	**0.42**	0.1	**0.38**	0.1
	20	**1**	0	**1**	0	**1**	0	**0.97**	0	**0.25**	0.06	**0.38**	0.13	**0.42**	0.06	**0.38**	0.08

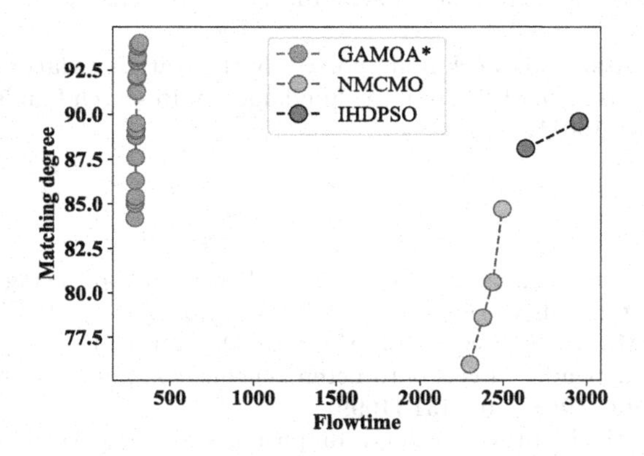

Fig. 4. Pareto-front comparisons on a large instance

similar MS with GAMOA*, but only gets one solution for many cases leading to a zero MS.

The Coverage $C(GAMOA^*, NMCMO)$ and $C(NMCMO, GAMOA^*)$ are shown in Table 4. $C(GAMOA^*, NMCMO) = 1$ and $C(NMCMO, GAMOA^*) = 0$ for all pairs of different generations which illustrate that solutions of NMCMO's 30000-*th* generation are even all dominated by solutions of GAMOA*'s 10-*th* generation, and solutions of GAMOA* are all non-dominated by those of NMCMO. In other words, our approach GAMOA* obtains better solutions at early generations taking advantages of multi-objective A*'s effective searching ability. On the contrary, the traditional NMCMO's convergence speed is very slow and it cannot find better solutions even at the 30000-*th* generation. For IHDPSO, $C(GAMOA^*, IHDPSO)$ is usually much larger than $C(IHDPSO, GAMOA^*)$ which means that more solutions of IHDPSO are dominated by those of GAMOA*. As shown in Fig. 4, solutions of GAMOA* dominate all NMCMO's and IHDPSO's solutions, i.e., solutions of our method GAMOA* are all on the Pareto front.

6 Conclusion

In this paper, GAMOA* is proposed to address the multi-objective ticket scheduling problem in large data centers which is a hybrid of the genetic algorithm and multi-objective A star. GAMOA* aims to deal with the multi-objective ticket scheduling to find a set of Pareto expert assignments regarding the total matching degree and the total flowtime. In GAMOA*, the searching effectiveness of traditional genetic algorithm is enhanced by a multi-objective A star algorithm by the aid of transforming ticket scheduling to routing problems. Experimental results illustrate that most solutions found by our approach dominate solutions of NMCMO in terms of the matching degree and the flowtime. Combing ticket routing techniques to enhance our ticket scheduling is a promising future work.

Acknowledgements. This work is supported by the National Natural Science Foundation of China (Grant No. 61972202), the Fundamental Research Funds for the Central Universities (No. 30919011235).

References

1. Alemzadeh, S., Dastghaibyfard, G.: Time and cost trade-off using multi-objective task scheduling in utility grids. In: ICCKE 2013, pp. 362–367. IEEE (2013)
2. Maheswaran, M., Ali, S., Siegel, H.J., Hensgen, D., Freund, R.F.: Dynamic mapping of a class of independent tasks onto heterogeneous computing systems. J. Parallel Distrib. Comput. **59**(2), 107–131 (1999)
3. Zhu, L., Li, Q., He, L.: Study on cloud computing resource scheduling strategy based on the ant colony optimization algorithm. Int. J. Comput. Sci. Issues (IJCSI) **9**(5), 54 (2012)
4. Gong, M., Jiao, L., Du, H., Bo, L.: Multiobjective immune algorithm with nondominated neighbor-based selection. Evol. Comput. **16**(2), 225–255 (2008)
5. Chakravarthy, K., Rajendran, C.: A heuristic for scheduling in a flowshop with the bicriteria of makespan and maximum tardiness minimization. Prod. Plan. Control **10**(7), 707–714 (1999)
6. Vidya, G., Sarathambekai, S., Umamaheswari, K., Yamunadevi, S.: Task scheduling using adaptive weighted particle swarm optimization with adaptive weighted sum. Procedia Eng. **38**, 3056–3063 (2012)

7. Agarwal, S., Sindhgatta, R., Sengupta, B.: SmartDispatch: enabling efficient ticket dispatch in an it service environment. In: Proceedings of the 18th ACM SIGKDD International Conference on Knowledge Discovery and Data Mining, pp. 1393–1401 (2012)
8. Shao, Q., Chen, Y., Tao, S., Yan, X., Anerousis, N.: EasyTicket: a ticket routing recommendation engine for enterprise problem resolution. Proc. VLDB Endow. **1**(2), 1436–1439 (2008)
9. Sun, P., Tao, S., Yan, X., Anerousis, N., Chen, Y.: Content-aware resolution sequence mining for ticket routing. In: International Conference on Business Process Management, pp. 243–259. Springer, Cham (2010). https://doi.org/10.1007/978-3-642-15618-2_18
10. Izakian, H., Ladani, B.T., Abraham, A., Snasel, V., et al.: A discrete particle swarm optimization approach for grid job scheduling. Int. J. Innov. Comput. Inf. Control **6**(9), 1–15 (2010)
11. Sarathambekai, S., Umamaheswari, K.: Task scheduling in distributed systems using heap intelligent discrete particle swarm optimization. Comput. Intell. **33**(4), 737–770 (2017)
12. Sarathambekai, S., Umamaheswari, K.: Intelligent discrete particle swarm optimization for multiprocessor task scheduling problem. J. Algorithms Comput. Technol. **11**(1), 58–67 (2017)
13. Sarathambekai, S., Umamaheswari, K.: Multi-objective optimization techniques for task scheduling problem in distributed systems. Comput. J. **61**(2), 248–263 (2017)
14. Karimi, M.: Hybrid discrete particle swarm optimization for task scheduling in grid computing. Int. J. Grid Distrib. Comput. **7**(4), 93–104 (2014)
15. Subashini, G., Bhuvaneswari, M.: Non-dominated particle swarm optimization for scheduling independent tasks on heterogeneous distributed environments. Int. J. Adv. Soft Comput. Appl. **3**(1), 1–17 (2011)
16. Subashini, G., Bhuvaneswari, M.C.: Comparison of multi-objective evolutionary approaches for task scheduling in distributed computing systems. Sadhana **37**(6), 675–694 (2012). https://doi.org/10.1007/s12046-012-0102-4
17. Kardani-Moghaddam, S., Khodadadi, F., Entezari-Maleki, R., Movaghar, A.: A hybrid genetic algorithm and variable neighborhood search for task scheduling problem in grid environment. Procedia Eng. **29**, 3808–3814 (2012)
18. Abraham, A., Liu, H., Grosan, C., Xhafa, F.: Nature inspired meta-heuristics for grid scheduling: single and multi-objective optimization approaches. In: Metaheuristics for Scheduling in Distributed Computing Environments, pp. 247–272. Springer, Cham (2008). https://doi.org/10.1007/978-3-540-69277-5_9
19. Pradeep, K., Jacob, T.P.: CGSA scheduler: a multi-objective-based hybrid approach for task scheduling in cloud environment. Inf. Secur. J. Glob. Perspect. **27**(2), 77–91 (2018)
20. Mandow, L., Pérez-de-la Cruz, J.-L.: A new approach to multiobjective A* search, pp. 218–223 (2005)

Joint Embedding Multiple Feature and Rule for Paper Recommendation

Wen Li[1], Yi Xie[1,2], and Yuqing Sun[1,3](✉)

[1] School of Software, Shandong University, Jinan, China
sun_yuqing@sdu.edu.cn
[2] School of Computer Science and Technology, Shandong University, Jinan, China
[3] Engineering Research Center of Digital Media Technology, Ministry of Education, Shandong University, Jinan, China

Abstract. It is a common way to represent paper properties as a heterogeneous academic network graph, such as authorships, citations, by which the latent features of paper can be learnt. To better integrate both text and structural features, we propose the joint embedding method for paper recommendation. We adopt a pre-trained language model to learn the paper semantic features from titles, and adopt a graph convolution network to extract the structural features from the constructed academic network graph. These two embeddings are combined together through the attention mechanism as a joint one. To clarify the *real* negative samples on uncited papers, we introduce some expert rules as the selection strategy on samples in model training, which can exclude the far-unrelated negative samples and potential positive samples. User interests are modeled by their historical publications and references and thus papers are recommended according to the relatedness between user interests and paper embeddings. We conduct experiments on the ACM academic paper dataset. The results show that our model outperforms baseline methods on personalized recommendation. We also analyze the influence of model structure and parameter setting. The results show that our sample strategy effectively improves the precision of recommendation, which illustrate that the strategy enhances the quality of training data.

Keywords: Paper recommendation · Multiple features · Rules

1 Introduction

There are a large number of papers published every year, which means it would take a lot of time for researchers on finding papers of interest. So personalized paper recommendation is of great significance in scientific research and academic development. And it's worth studying how to accurately capture user preferences to help them find interested paper.

Existing works always introduce various attributes of users and items besides rating matrix to build recommendation models. Since single feature cannot fully reflect user interests, it's necessary to joint these different features. In addition, in order to obtain

Y. Sun et al. (Eds.): ChineseCSCW 2021, CCIS 1492, pp. 52–65, 2022.
https://doi.org/10.1007/978-981-19-4549-6_5

the training data for paper recommendation, existing works often use the citation relationships between papers to construct positive and negative samples. For example, a user-paper pair with citation relationship is labeled positive, indicating that the user is interested in this paper, otherwise negative [1, 2]. In fact, when users look for interested papers, not only the citation relationship is considered, but also the paper classification, keywords, and so on. Therefore, we believe sampling based on a single perspective is limited.

To tackle the above challenges, we propose the joint multi-feature and rules paper embedding model for paper recommendation (JMPR). In order to capture user interests more comprehensively, we integrate both the semantic and structural features, corresponding to paper titles and academic network graph, respectively. And we introduce rules as the sample strategy on samples in the training process. The contribution of this paper resides on two aspects:

(1) We propose the joint embedding method for paper recommendation. We construct the academic network graph based on the academic paper dataset, in which the entities such as papers, authors, venues are as nodes and the relations between them are as edges. To mine the potential relationships between entities which are not directly connected and get the paper entity representations incorporating neighborhood information and user interests, we optimize a graph convolution network (GCN) to extract the structural features from the academic network graph. To get the representations of paper title incorporating with domain knowledge, we pre-train a language model on domain dataset to learn the semantic features from titles. Then, to get the joint paper embedding, we combine the above two representations through attention mechanism. Finally, the recommendation is based on the relatedness between user interests and paper embeddings. We calculate the similarity between user vector and paper vector as the probability that the user is interested in the paper.

(2) We propose the rule-based sample selection strategy to clarify the real negative samples on uncited papers, so as to exclude the far-unrelated negative samples and potential positive samples from being mistakenly selected as negative samples. To jointly model the correlation between users and papers, we define three rules based on paper classification, references, and keywords. And according to the rules, we select the positive and negative samples as training set.

We conduct experiments on the ACM academic dataset, and compare our method with the baselines. The results show our method outperforms others on the recommendation task. Then, we analyze the influence of model structure and parameter setting. The experimental results show that our sample strategy effectively improves the precision of recommendation, which illustrate that the strategy enhances the quality of training data.

The rest of this paper is organized as follows. In Sect. 2, we will introduce the related works. In Sect. 3, we will introduce our proposed method in detail. In Sect. 4, we will analyze the experiments. Finally in Sect. 5, we will make the conclusions.

2 Related Work

2.1 Recommendations Based on Content and Rating Features

Traditional recommendation methods include content-based and CF-based methods. Content-based methods usually make recommendations by discovering the relation between user profiles and paper features. The user profiles are mainly constructed by the interaction between users and papers. For example, Gautam and Kumar [3] proposed a tag-based method that uses paper tags which users are interested in to represent user profiles. In addition, there are many ways to represent paper features. For example, Jeong et al. use BERT [4] to obtain the paper sentence representations [5]. Tao et al. uses LDA topic model [6] to obtain the paper feature representations. The basic idea of CF-based methods is that similar users' favorite items are also similar. Compared with content-based methods, CF-based methods, such as SVD [7], are more independent of the content of items, making it applicable to a wider range of scenarios. However, they rely on the rating matrix, so suffer from data sparsity and cold start problems. Through Combining the CF-based and content-based methods, the above two problems can be solved to a certain extent. For example, Sugiyama et al. use content-based method to model the user preferences, then use CF-based method to discover papers that users are potentially interested in [8]. The data feature that recommendation models mentioned above rely on is relatively single, so that it's difficult to mine potential user preferences.

2.2 Recommendations Based on Structural Features

With the development of graph-based information retrieval and data mining technology, more and more graph-based recommendation models have been proposed. For example, the emergence of social network contributed to the research of trust-aware recommendation systems [9], which could help us to infer the user's preferences indirectly by summarizing the user's friends' preferences. In addition, the Knowledge Graph (KG) as the side information besides rating matrix is also increasingly used for recommendation, such as DKN [10], PER [11], RippleNet [12], KGCN [13], KGCN-LS [14], etc. KG could improve the precision, diversity, and interpretability of the recommendation system [15]. Among the above-mentioned KG-based methods, KGCN [13] performs GCN to mine the high-order hidden information on the KG, which proves work well. And our work in this paper is inspired by this work.

There is a natural network structure between academic papers, since they are not isolated but connected with each other by citation relations, co-authors and so on. In fact, many graph-based recommendation models for paper recommendation have been proposed. For example, Pan et al. [16] proposed a model based on the similarity learning of citation network and keyword network. Manju et al. [17] proposed a model based on social network. It is also possible to combine graph-based methods with traditional methods. For example, Kong et al. [18] proposed a method that combines graph-based and CF-based methods. They use Word2vec and Struc2vec to construct citation network with semantic information, then calculate the cosine similarity between user representation and paper representation. Since graphs are rich in information, it's necessary to find an appropriate way to make full use of them.

3 Joint Multi-feature and Rules Paper Embedding Method

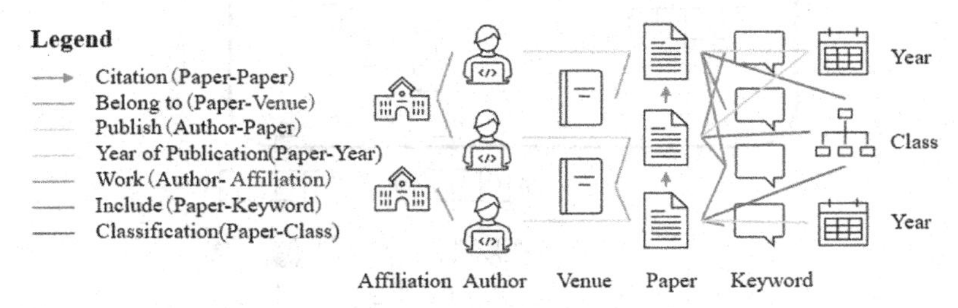

Fig. 1. Academic network graph based on ACM academic dataset.

3.1 Problem Definition

In this paper, the paper recommendation problem is formally defined as: Given a set of M users $U = \{u_1, u_2, ..., u_M\}$ and a set of N papers $V = \{v_1, v_2, ..., v_N\}$, we aim to learn a prediction function $\hat{y} = \mathrm{F}(u, v|\theta)$ that predicts whether user u has potential interest in paper v, $\hat{y}$ denotes the predicted probability that user u is interest in paper v, and θ denotes the parameters of function F.

In an academic dataset, each paper usually contains the title, keywords, authors, references, and some other attributes. As shown in Fig. 1, in order to capture inter-paper relatedness, we construct an academic network graph $G = (E, R)$, where E denotes the entity set and R denote the relation set. The types of entity include "Affiliation" "Author" "Venue" "Paper" "Keyword" "Year" and "Class". The types of relation include "Citation" "Belong to" "Publish" "Year of Publication" "Work" "Include" and "Classification". Each triple of entity-relation-entity is represented as the form (e_1, r, e_2), where $e_1, e_2 \in E$, and $r \in R$. It is a common way to represent these attributes as a heterogeneous academic network graph, such as authorships, citations [2]. We construct the academic network graph based on the ACM academic dataset, then we choose it as the structural feature.

3.2 Overall Framework

Generally, a paper title is the high-level summarization of the paper content, which are rich in semantic information. And in academic network graph, papers are not isolated but connected with each other through different relations. The entities involved in a publication and its relations together constitute an informative network. The above features reveal different aspects of papers, so we consider both semantic features and structural features on modeling paper. We propose the joint multi-feature and rules paper embedding model (JMPR), as shown in Fig. 2, which include three modules, i.e. the title embedding module, academic network GCN module, and joint recommendation module.

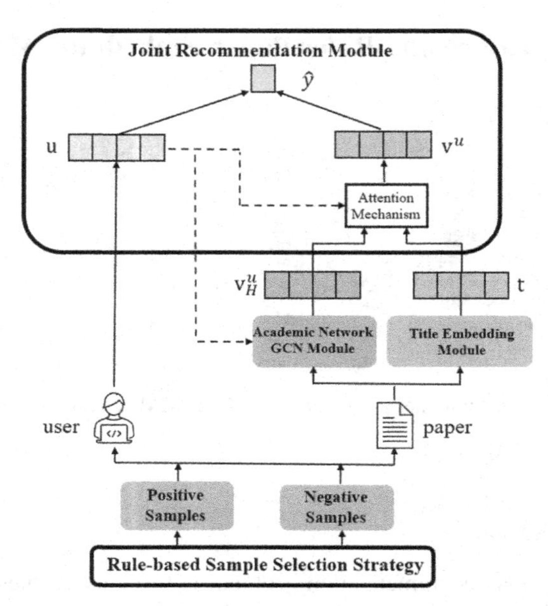

Fig. 2. JMPR model framework.

In the title embedding module, we pre-train a language model on domain dataset to get the semantic representation of paper title. In academic network GCN module, we optimize a graph convolution network (GCN) to capture the inter-paper relatedness and extract the high-order hidden information from academic network. And we obtain the representation of paper entity incorporating neighborhood information and user interests from this module. In joint recommendation module, we obtain the joint paper embedding through combining the title representation and paper entity representation mentioned above with attention mechanism. And we also obtain the user representation which incorporate the historical interest. Finally, we predict the user's potential interest in papers by calculating the vector similarity between user and paper. In addition, to handle with the sampling problem in the training process, we propose the rule-based sample selection strategy to clarify the real negative samples on uncited papers, so as to exclude the far-unrelated negative samples and potential positive samples from being mistakenly selected as negative samples.

3.3 Rule-Based Sample Selection Strategy

There are two forms in citation relation: cited and uncited, which are usually directly used as positive and negative samples to train models. We call it naive sample selection strategy. However, there are cases where users and papers are far-unrelated in uncited samples, such as a user in computer science and a paper in literary. Since the user and paper are from totally different domains, the naive sample selection strategy will lead to under-fitting problems. Any users and papers from the same domain will be predicted highly correlated by the recommendation model trained this way, which is obviously problematic. In addition, there may also be potential positive samples in uncited samples, because a user may have potential interest in the papers he didn't cite before.

To address the above problems, we propose the rule-based sample selection strategy to get user-paper pairs for training the paper recommendation model. Our strategy is user-centric. We select suitable papers for each user to construct positive samples and negative samples.

Consider user u, we take the papers that have citation relationship with u as positive samples about u and label them $y = 1$. To avoid labeling the potential positive samples and far-unrelated samples as negative samples, we introduce rules. According to the rules, we select the suitable ones from uncited papers as negative samples about u, and label them $y = 0$. By this way, the quality of training data and precision of the recommendation model are obviously improved.

Next, we will introduce three rules, and explain the sampling process in detail.

The three rules are the similarity of the research direction, Jaccard similarity of the references, and Jaccard similarity of the keywords, respectively.

Research Direction. The classification systems of the academic paper research direction are usually hierarchical structures, such as the ACM Computing Classification System (CCS) in computer science. We define two papers' research direction similarity as S_1, that is, the hierarchical logarithmic distance of paper v_1 and v_2 from the bottom paper nodes to their public parent node in classification system. R_1 and R_2 denote all nodes included in the path from the bottom paper node to the root node of paper v_1 and v_2, respectively. l_i denotes the level where the current node i is located.

$$S_1(v_1, v_2) = -\sum\nolimits_{i \in (R_1 \cup R_2 - R_1 \cap R_2)} \frac{1}{2^{l_i}} \tag{1}$$

References. We define the two papers' Jaccard similarity of references as S_2. Ref_1 and Ref_2 denote the reference sets of paper v_1 and v_2, respectively.

$$S_2(v_1, v_2) = \frac{|Ref_1 \cap Ref_2|}{|Ref_1 \cup Ref_2|} \tag{2}$$

Keywords. We define the two papers' Jaccard similarity of keywords as S_3. K_1 and K_2 denote the keyword set of paper v_1 and v_2, respectively.

$$S_3(v_1, v_2) = \frac{|K_1 \cap K_2|}{|K_1 \cup K_2|} \tag{3}$$

We synthesize the above three rules to calculate the correlation score $S(v_1, v_2)$ of two papers. In fact, the above rules are always calculated 0 between most paper pairs, due to the data sparsity problem. In this paper, we believe that the correlation between different paper pairs is reflected in different aspects, and the highly relevant paper pairs may only be highly relevant in just one aspect, so we choose the maximum value of the three rules as the final correlation score of paper v_1 and v_2.

$$S(v_1, v_2) = max(S_1(v_1, v_2), S_2(v_1, v_2), S_3(v_1, v_2)) \tag{4}$$

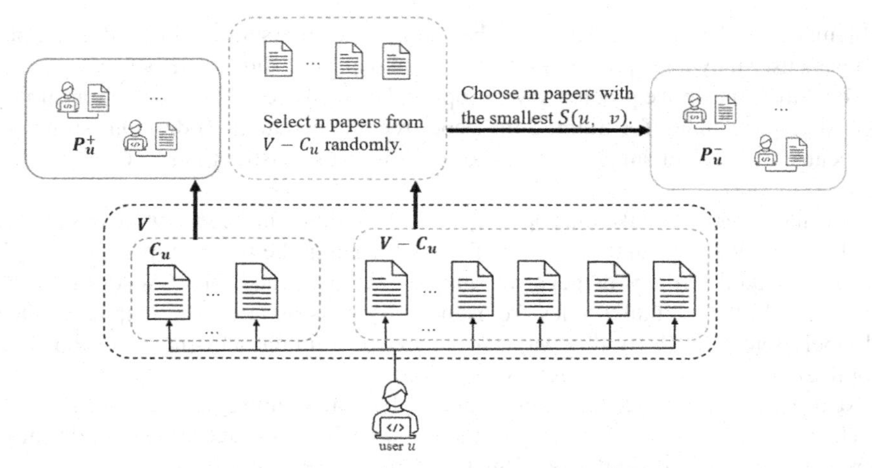

Fig. 3. Flow chart for sample selection strategy incorporating rules.

Then, we calculate the correlation score $S(u, v)$ of user u and paper v based on the correlation score $S(v_1, v_2)$ of two papers. P_u denotes the paper set published by u.

$$S(u, v) = max\{r(p_u, v)|p_u \in P_u\} \tag{5}$$

So far, we could label the positive and negative samples about user u according to $S(u, v)$. The process in detail is shown in Fig. 3. V denotes the set of all papers, C_u denotes the set of papers cited by u. If u has cited paper v, then $y = 1$, indicating a positive sample, and $P_u^+ = \{v|v \in C_u\}$ denotes the set of positive samples about u. Negative samples about u are selected from the paper set $V - C_u$ that u has not cited. Since the number of samples without citation relation is huge and the calculation of $S(u, v)$ is time-consuming and space-consuming, we only randomly select n papers from $V - C_u$ to calculate $S(u, v)$, $n \ll |V - C_u|$. We finally select m papers with the smallest $S(u, v)$ as the final negative sample set P_u^- about u, where $m < n$.

3.4 Joint Embedding Based Academic Paper Recommendation

3.4.1 Title Embedding Module

Inspired by SBERT [19]'s idea of fine-tuning the model parameters of pre-trained BERT through Siamese neural network on domain dataset to output sentence embeddings, we optimize a Siamese neural network on our constructed title dataset to output title embeddings.

The title dataset is constructed based on the ACM academic paper dataset. We define z as the true correlation score between t_1 and t_2. If there is a citation relationship between (t_1, t_2), we choose it as positive sample and label it as $z = 1$. Then we select negative samples from uncited ones based on rules. We randomly select ten times the number of positive samples from the uncited title pairs as candidate set. Then we calculate their $S(v_1, v_2)$ based on formula (4). Finally, we choose the samples with the smallest $S(v_1, v_2)$

from the candidate set as the negative samples, and we label them z = 0. The negative sample set is the same size as the positive one.

Our Siamese neural network includes three layers, which are Bert layer, mean pooling layer and the fully connected layer in sequence. The difference from SBERT is we add a fully connected layer to reduce the output vector dimension. We use the mean square error (MSE) function as the loss function. $\hat{z}$ denotes the predicted score.

$$loss = \left|\left|z - \hat{z}\right|\right|_2 = ||z - \cos(t_1, t_2)||_2 \tag{6}$$

3.4.2 Academic Network GCN Module

Inspired by KGCN [13]'s idea of implementing GCN on knowledge graph, we implement GCN on academic network graph.

Consider a user-paper pair (u, v), $u \in R^d$ and $v \in R^d$ are vector representations of u and v. r denotes the relations between entities on the academic network graph. And $r \in R^d$ denotes vector representations of r.

A user may be more interested in the papers from the same venue, another may be more interested in the papers of the same author. $r_{v,e}$ represents the relation between paper v and its neighbor entity$e$. We use $\phi^u_{r_{v,e}}$ to represent the score between user u and relation $r_{v,e}$. $\phi^u_{r_{v,e}}$ describes the importance of relation $r_{v,e}$ to u.

$$\phi^u_{r_{v,e}} = u \times r_{v,e} \tag{7}$$

We use E_v to denote the entity set directly connected to v in the academic network. In fact, the size of E_v may change a lot with different papers in the academic network graph. For computational convenience, we sample a fixed size of neighbors for each paper randomly instead of using all of them, which is defined as $E_v^{'}$. $\left|E_v^{'}\right| = K$ is a constant. $\tilde{\phi}^u_{r_{v,e}}$ is the normalized result of $\phi^u_{r_{v,e}}$.

$$\tilde{\phi}^u_{r_{v,e}} = \frac{exp\left(\phi^u_{r_{v,e}}\right)}{\sum_{e' \in E_v'} exp\left(\phi^u_{r_{v,e'}}\right)} \tag{8}$$

We compute the linear combination of the entities in $E_v^{'}$ to characterize the topological neighborhood structure of paper v. We use $v^u_{E_v'}$ to denote the vector representation of paper v's neighborhood. The score $\tilde{\phi}^u_{r_{v,e}}$ between u and $r_{v,e}$ plays an important role as the personalized filter. e is the vector representation of entity e.

$$v^u_{E_v'} = \sum\nolimits_{e \in E_v'} \tilde{\phi}^u_{r_{v,e}} e \tag{9}$$

$E_v^{'}$ can also be called as the single-layer receptive field of paper v. Then we aggregate paper $v's$ initial representation v and $v's$ neighborhood representation $v^u_{E_v'}$ into a single vector $v^u_1 \in R^d$ as $v's$ first-order representation.

$$v^u_1 = \sigma\left(W \cdot \left(v + v^u_{E_v'}\right) + b\right) \tag{10}$$

So far, we get the representation of entities in the academic network graph after single-layer GCN on academic network graph, which are also called first-order representation of entities. They only depend on itself and the neighborhood entities directly connected to them. We define the initial entity representation as zero-order representation, and the entity representation after single-layer GCN as first-order representation. In order to mine the long-distance interest of users, we extend the receptive field to multi hops which means the entities which are indirectly connected to the given entity are also selected to be its neighborhood entities. And we generate the neighborhood representation through implementing multi-layer GCN on the academic network graph. By this way, we get the high-order representations of entities. We use H to denote the maximum depth of the neighborhood. For a given user-paper pair (u, v), we compute the receptive field M of v iteratively, then generate the H-order representation $\mathrm{v}_H^u \in R^d$ of v through H times aggregation of entity representation and its neighborhood representation.

3.4.3 Joint Recommendation Module

We joint the paper title vector $\mathrm{t} \in R^m$ generated by title embedding module and paper entity vector v_H^u generated by academic network GCN module with attention mechanism. The dimension of t is bigger than v_H^u, so we add a fully connected layer to reduce the dimension of t.

$$\mathrm{t}' = \sigma(W_1 \cdot \mathrm{t} + b_1) \tag{11}$$

We use user vector u to calculate the attention weights α and β.

$$\alpha = \mathrm{u} \cdot \mathrm{t}' \tag{12}$$

$$\beta = \mathrm{u} \cdot \mathrm{v}_H^u \tag{13}$$

Then, we calculate the weighted sum of the t' and v_H^u to get the final paper representation v^u.

$$\mathrm{v}^u = \sigma\left(W_2 \cdot \left(\alpha \mathrm{t}' + \beta \mathrm{v}_H^u\right) + b_2\right) \tag{14}$$

$\hat{y}$ denotes the predicted probability that user u has potential interest in paper v, and is calculated by the inner product of paper vector v^u and user vector u.

$$\hat{y} = \mathrm{u} \times \mathrm{v}^u \tag{15}$$

We choose the cross-entropy loss function, and implement the rule-based sample selection strategy in the training process. P_u^+ and P_u^- are the positive sample set and negative sample set about user u, respectively, obtained by Sect. 3.3. The last term is the regularization term.

$$\sum\nolimits_{u \in U} \left(-\frac{1}{\left|P_u^+ \cup P_u^-\right|} \sum\nolimits_{v \in P_u^+ \cup P_u^-} y log \hat{y} \right) + \lambda \|\mathrm{F}\|_2^2 \tag{16}$$

4 Experiments

4.1 Dataset

We use the ACM dataset for experiments. It contains more than 40,000 academic papers in computer science. Since these papers are domain relevant, they are suitable for verifying the effectiveness of our rule-based sample selection strategy. In addition, the dataset contains the CCS classification labels which can be used to calculate the research direction similarity defined in Sect. 3.3. CCS refers to the ACM computing classification system, which is a standard classification system with hierarchical structure in computing science. The basic statistical information of this dataset is shown in Table 1.

Table 1. Statistics of ACM dataset

Statistics	Num
User	44953
Paper	31889
Sample	348856
Entity	148376
Relation	7
(e_1, r, e_2)	491679

4.2 Baselines and Experiment Setup

Baseline models are as follows.

SVD [7] is a traditional CF-based recommendation model, which needs rating matrix. The basic idea of SVD is to match the original data into a low-dimensional space, and calculate the predicted score of the unrated items, then recommend the items with high predicted scores to the user.

KGCN [13] implements GCN on knowledge graph. The basic idea is to aggregate the entities with their neighborhood in the knowledge graph to capture the inter-item relatedness and the potential interest of users.

KGCN-LS [14] introduces Label Propagation Algorithm (LPA) on the basis of KGCN, which is equivalent to introducing a regular term to prevent overfitting problem.

RippleNet [12] takes the items that users are interested in as seeds, and uses these seeds to spread out to other items on knowledge graph, refer to the idea of water wave propagation. This process is called preference propagation. RippleNet uses the method of spreading preferences in knowledge graph to discover the potential interests of users continuously and automatically, so that it achieves personalized recommendation.

Our model and its variants in this paper are as follows.

JMPR is the joint multi-feature and rules paper embedding model we proposed in this paper. JMPR includes three modules and implements the rule-based sample selection strategy.

ANGCN only uses academic network graph as feature, corresponding to academic network GCN module. And ANGCN implements naive sample selection strategy.

ANGCN-TE adds titles as features on the basis of ANGCN.

ANGCN-Neg replace ANGCN's naive sample selection strategy with our proposed rule-based sample selection strategy.

We use F1 score and AUC to evaluate the model performance on the task of judging whether users are interested in papers.

The experimental setup of JMPR is as follows. In the academic network GCN module, we use different activation functions as σ in formula (10) to aggregate the entity representation and its neighborhood representation. If it is not the last layer, we choose *ReLU*, else *tanh*. In the joint recommendation module, we feed the title embedding t to activation function *ReLU* and *tanh* in order to match the paper entity embedding v^u, then we choose *tanh* as σ in formula (14) to aggregate v^u and t.

4.3 Results

Table 2. Performance comparison of different models.

Model	AUC	F1
SVD	50.00	66.54
KGCN	86.61	79.53
KGCN-LS	86.62	79.01
RippleNet	90.46	82.78
JMPR	**95.50**	**87.87**

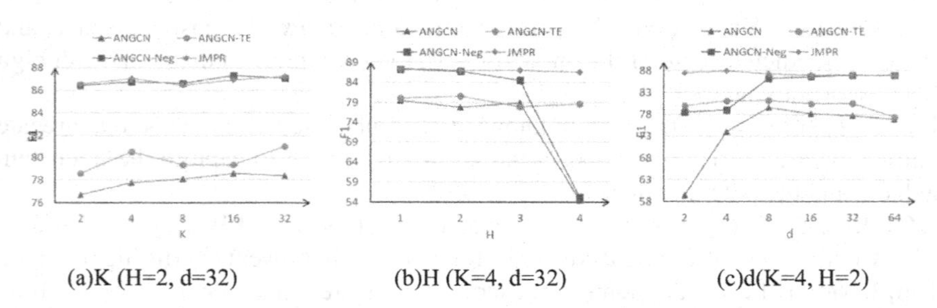

(a)K (H=2, d=32) (b)H (K=4, d=32) (c)d(K=4, H=2)

Fig. 4. Performance comparison of JMPR and its variants under different experimental setting.

Performance Analysis. Firstly, we compare our model with the baselines. As shown in Table 2, our model JMPR outperforms others. The reasons lie in two aspects: One is the rule-based sample selection strategy we proposed optimize the quality of training data and improve the precision of the model. The other is JMPR joints multiple features with attention mechanism, so that it could model the user preferences more accurately. What's more, GCN works well to mine high-order hidden information in academic network graph, and pre-trained paper title embedding is meaningful and rich in semantic information, which all contribute to the improvement of the model performance.

SVD performs worst. SVD only uses the rating matrix to train the recommendation model (in this article, the rating is 1 when it's a positive sample, otherwise 0), and doesn't introduce textual features or other structural features as additional information.

As an improved model of KGCN, KGCN-LS introduces LPA on the basis of KGCN, but the performance on ACM academic paper dataset is not significantly improved compared to KGCN.

RippleNet performs better than other baselines. Since it also uses a multi-layer neighborhood structure to capture the internal associations between neighborhood entities in the academic network graph just like KGCN and JMPR. It shows the importance of neighborhood information in academic network graph for recommendation.

Next, we will analyze the impact of model structure and parameter setting on model performance. The experimental results are shown in Fig. 4.

Model Structure. Title summarizes the main content of the paper, so it is rich in semantic information. Since the vocabulary of papers in different domains is quite different, the sentence embedding output by model trained in general domain is not suitable for the professional field. By pre-training the title embedding on the domain dataset, we can get the proper title representations which are meaningful and incorporated domain knowledge. In addition, the academic network graph is a heterogeneous graph which is rich in structural information. By implementing multi-layer GCN on it, we can capture the inter-paper relatedness and the long-distance interests of users. As shown in Fig. 4, ANGCN-TE outperforms ANGCN, since ANGCN-TE adds title as semantic feature on the basis of ANGCN. It indicates that the title embedding module and academic network GCN module complement each other and could model user preferences in all directions. What's more, compared with the naive sample selection strategy, our proposed rule-based sample selection strategy improved the performance of the model. ANGCN-Neg and JMPR replace the naïve sample selection strategy with the rule-based one based on ANGCN and ANGCN-TE, respectively. As shown in Fig. 4, ANGCN-Neg outperforms ANGCN, and JMPR outperforms ANGCN-TE. Because the rule-based strategy avoids labeling the potential positive samples and far-unrelated samples as negative samples from uncited samples. By this way, the quality of training data and the precision of the recommendation model are improved. However, it is worth mentioning that the correlation score $S(u, v)$ between user u and paper v in formula (5) is time-consuming and space-consuming, which led to a large increase in the workload.

Parameter Setting. Now we analyze the influence of neighborhood entity nodes' number K, the convolution depth H, and the embedding dimension d on the performance of the model, respectively. As shown in Fig. 4, we observe the ANGCN and ANGCN-Neg are sensitive to the setting of K, H, d. Since ANGCN and ANGCN-Neg only use academic network graph as feature, and he setting of K, H, d is mainly for academic network GCN module. And when K takes the value 16, H takes 1 and 2, and d takes middle value 8, the model performance is better. Because there may be over-fitting or under-fitting problem under other parameter setting conditions. Then, when the paper title feature is integrated with the academic network feature, corresponding to ANGCN-TE and JMPR, it is observed that the influence of the setting of K, H, d on the model performance is weakened. It indicates that the title embedding module makes up the academic network GCN module's lack of information, so that it improves the robustness of the recommendation model.

5 Conclusion

In this paper, we propose the joint multi-feature and rules paper embedding model for paper recommendation. We choose the academic network graph as structural feature and paper title as semantic feature to jointly model user interests. We pre-train the title embedding on domain dataset and implement GCN on the academic network graph, then aggregate them with attention mechanism. In addition, we propose the rule-based sample selection strategy to do with the sample problem in training process. The experimental results show that our method outperform the baselines in predicting whether users are interested in the paper.

References

1. Xie, Y., Wang, S., Pan, W., Tang, H., Sun, Y.: Embedding based personalized new paper recommendation. In: Sun, Y., Liu, D., Liao, H., Fan, H., Gao, L. (eds.) Chinese CSCW 2020. CCIS, vol. 1330, pp. 558–570. Springer, Singapore (2021). https://doi.org/10.1007/978-981-16-2540-4_40
2. Zhu, Y., Lin, Q., Lu, H.: Recommending scientific paper via heterogeneous knowledge embedding based attentive recurrent neural networks. Knowl. Based Syst. **215**, 106744 (2021)
3. Gautam, J., Kumar, E.: An improved framework for tag-based academic information sharing and recommendation system. In: Proceedings of the World Congress on Engineering, U.K., pp. 1–6 (2012)
4. Devlin, J., Chang, M.W., Lee, K., Toutanova, K.: BERT: pre-training of deep bidirectional transformers for language understanding. arXiv: Computation and Language (2018)
5. Jeong, C., Jang, S., Park, E., Choi, S.: A context-aware citation recommendation model with BERT and graph convolutional networks. Scientometrics **124**(3), 1907–1922 (2020). https://doi.org/10.1007/s11192-020-03561-y
6. Tao, M., Yang, X., Gu, G., Li, B.: Paper recommend based on LDA and PageRank. In: Sun, X., Wang, J., Bertino, E. (eds.) ICAIS 2020. CCIS, vol. 1254, pp. 571–584. Springer, Singapore (2020). https://doi.org/10.1007/978-981-15-8101-4_51
7. Koren, Y.: Factorization meets the neighborhood: a multifaceted collaborative filtering model. In: Proceedings of the 14th ACM SIGKDD International Conference on Knowledge Discovery and Data Mining, USA, pp. 426–434 (2008)
8. Sugiyama, K., Kan, M.Y.: Exploiting potential citation papers in scholarly paper recommendation. In: Proceedings of the 13th ACM/IEEE-CS Joint Conference on Digital Libraries, pp. 153–162 (2013)
9. Guo, G., Zhang, J., Yorke-Smith, N.: Leveraging multiviews of trust and similarity to enhance clustering-based recommender systems. Knowl. Based Syst. **74**, 14–27 (2015)

10. Wang, H., Zhang, F., Xie, X, Guo, M.: DKN: deep knowledge-aware network for news recommendation. In: Proceedings of the 2018 World Wide Web Conference, pp. 1835–1844 (2018)
11. Yu, X., Ren, X., Sun, Y., et al.: Personalized entity recommendation: a heterogeneous information network approach. In: Proceedings of the 7th ACM International Conference on Web Search and Data Mining, pp. 283–292 (2014)
12. Wang, H., Zhang, F., Wang, J., et al.: RippleNet: propagating user preferences on the knowledge graph for recommender systems. In: Proceedings of the 27th ACM International Conference on Information and Knowledge Management, Italy, pp. 417–426 (2018)
13. Wang, H., Zhao, M., Xie, X., et al.: Knowledge graph convolutional networks for recommender systems. In: Proceedings of the World Wide Web Conference, USA, pp. 3307–3313 (2019)
14. Wang, H., Zhang, F., Zhang, M., et al.: Knowledge-aware graph neural networks with label smoothness regularization for recommender systems. In: Proceedings of the 25th ACM SIGKDD International Conference on Knowledge Discovery & Data Mining, Anchorage, USA, pp. 968–977 (2019)
15. Wang, X., Wang, D., Xu, C., et al.: Explainable reasoning over knowledge graphs for recommendation. In: Proceedings of the AAAI Conference on Artificial Intelligence, vol. 33, no. 01, pp. 5329–5336 (2019)
16. Pan, L., Dai, X., Huang, S., Chen, J.: Academic paper recommendation based on heterogeneous graph. In: Sun, M., Liu, Z., Zhang, M., Liu, Y. (eds.) CCL/NLP-NABD -2015. LNCS (LNAI), vol. 9427, pp. 381–392. Springer, Cham (2015). https://doi.org/10.1007/978-3-319-25816-4_31
17. Manju, G., Abhinaya, P., Hemalatha, M.R., Manju, G.G.: Cold start problem alleviation in a research paper recommendation system using the random walk approach on a heterogeneous user-paper graph. Int. J. Intell. Inf. Technol. (IJIIT) **16**(2), 24–48 (2020)
18. Kong, X., Mao, M., Wang, W., et al.: VOPRec: vector representation learning of papers with text information and structural identity for recommendation. IEEE Trans. Emerg. Top. Comput. **9**, 226–237 (2018)
19. Reimers, N., Gurevych, I.: Sentence-BERT: sentence embeddings using siamese BERT-networks. In: Proceedings of the 2019 Conference on Empirical Methods in Natural Language Processing and the 9th International Joint Conference on Natural Language Processing (EMNLP-IJCNLP), pp. 3982–3992 (2019)

Predicting Drug-Target Interactions Binding Affinity by Using Dual Updating Multi-task Learning

Chengyu Shi, Shaofu Lin, Jianhui Chen(✉), Mengzhen Wang, and Qingcai Gao

Beijing University of Technology, Beijing 100020, China
chenjianhui@bjut.edu.cn

Abstract. The prediction of drug-target interactions binding affinity has received great attention in the field of drug discovery. The prediction models based on deep neural networks have shown the favorable performance. However, existing models mainly depend on large-scale labelled data and are unfit for the innovative drug discovery study because of local optimum on pre-training. This paper proposes a new deep learning model to predict the drug-target interaction binding affinity. By using multi-task learning, unsupervised pre-training tasks of drugs and proteins are combined with the drug-target prediction task for preventing local optimum on pre-training. And then the MAML based updating strategy is adopted to deal with the task gap problem in the traditional fine-tuning process. Experimental results show that the proposed model is superior to the existing methods on predicting the affinity between new drugs and new targets.

Keywords: Drug target interaction prediction · Multi-task learning · Pre-training · Graph embedding

1 Introduction

The discovery of new drugs is time consuming and very costly. But it is very inefficient by traditional wet laboratory experiments [1]. In the face of rapidly spreading diseases, such as COVID-19, drug development requires a shorter period of time. Drug target interaction prediction (DTI) with computer-aided design is an important way to accelerate drug discovery and drug repurposing. It not only has low cost but greatly reduces the experimental process [2]. Drug target interaction prediction binding affinity (DTA) [3] is a kind of specific DTI. Different from traditional DTI which neglects the interaction intensity between drugs and targets, DTA can quantitatively analyze the binding affinity between drugs and different types of targets, such as proteins, and thus has attracted widely research interesting in recent years.

With the development of deep learning, data-driven prediction has become an important solution for DTA. Because of ten millions of drugs and targets, it is difficult to effectively extract enough feature representations only depending on labelled drug-target data. In recent year, some work [4] utilized the pre-trained model based on large unlabelled

Y. Sun et al. (Eds.): ChineseCSCW 2021, CCIS 1492, pp. 66–76, 2022.
https://doi.org/10.1007/978-981-19-4549-6_6

data to obtain enhanced structural information of proteins. However, the traditional fine-tuning process could not consider the labelled data in the pre-training at all. The prediction model still often falls into overfitting on the limited labelled data and leads to low precision on predicting the affinity between new drugs and new targets.

Based on the above observations, this study proposes a DTA deep learning model based on dual updating multi-task learning to predict the drug-protein affinity. Main contributions are summarized as follows:

(1) In order to preventing local optimum on pre-training, this paper proposes a multi-task learning framework which combines the pre-training tasks of both drug compounds and proteins with the DTA task for preventing the representation learning falling into local optimum.
(2) In order to deal with the task gap problem in the traditional fine-tuning process, this paper proposes a dual adaptation mechanism based on MAML [5],which transfers prior knowledge from pre-training tasks to the downstream prediction task for preventing the fine-tuning process entirely driven by limited labelled data and solving the overfitting problem of DTA model training.
(3) A group of experiments are performed on the simulated innovative drug discovery dataset, which is constructed based on the drug database ZINC [6] and the protein databased PFAM [7]. Experimental results show that the proposed model is superior to the existing DTI models on the new drug-target affinity prediction.

2 Related Work

For DTA tasks, machine learning (ML) models can be formulated as an encoder-decoder architecture [4]. The encode model extracts feature representations from the input drug compound sequences or graphs and protein sequences. The decode model processes the feature representations to generate a binding affinity prediction. DeepDTA [8] is the first to introduce deep learning into the DTA task. It generated 1D representation of drug and protein sequences by using convolutional neural networks (CNN). GraphDTA [9] used RDKit, an open source chemical informatics software, to build the molecular graph of compound strings and learn the characteristic vectors of compounds by using the graph neural network. Inspired by attentional mechanism, Fan et al., [10] used an improved Transfomer to embed protein sequences for reinforcing structural information. These models have achieved good performers on the Davis [11] and Kiba [12] datasets.

However, the drug discovery is a fast changing disciplines. New drug compounds and proteins emerge in endlessly. The above models divided pre-training and prediction as serial tasks. The pre-training task was performed in advance on the existing large-scale compound and protein datasets, and neglected the inconsistency between pre-training and fine-tuning. Aiming at the affinity prediction between new drugs and new targets, the robustness of the model is doubted.

3 Materials and Methods

This paper proposes GeneralizedDTA, a new DTA model which is combined with pre-training of compounds and proteins by using dual updating multi-task learning. In the following sections, we will explain the dataset, the model architecture, the task process and the adaptation strategy.

3.1 Datasets

This paper uses Pfam [7] as the protein sequence pre-training dataset. Over 21M protein sequences are clustered into 16,479 families based on the similarity in protein sequences. For the drug pre-training, this paper uses the ZINC15 database [6] which has 2 million unlabelled compounds sampled for compound-level self-supervised pre-training. For DTI prediction, this paper uses Davis [11] dataset, which includes 30056 drug target pairs and is involved with 442 proteins, 68 compounds and their binding affinity indicated by the KD value.

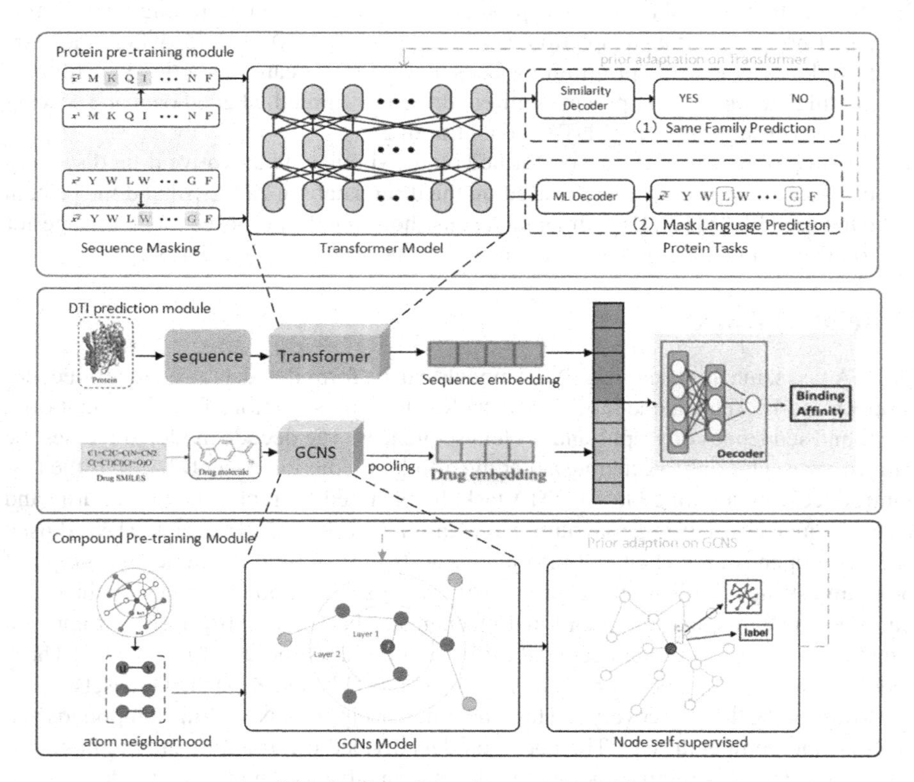

Fig. 1. Overview of proposed model

3.2 Model Architecture

Figure 1 gives the model architecture of GeneralizedDTA including three modules: the compound pre-training module, the protein pre-training module and the DTI pre-diction module.

Protein Pre-training Module: The protein pre-training module is to encode the protein sequence to a vector by the transformer model [13]. Inspired by BERT [13], this study adopts a transformer model with multi-head attention as encoder to receive protein sequence. Given a protein sequence $t = [t_1, \cdots, t_n]$ where $t_i \in \{$ 21 amino acid types $\}$, the transformer model [13] converts it into $z = [z_1, \cdots, z_n]$ as follows:

$$\begin{aligned} \mathrm{z} &= \mathrm{Transformer}\ (Q, K, V; t) = \mathrm{Concat}\ (\mathrm{head}_1, \ldots, \mathrm{head}_h)W^O \\ \mathrm{head}_i &= \mathrm{Attention}\ \left(QW_i^Q, KW_i^K, VW_i^V\right) \end{aligned} \tag{1}$$

where W^O is the weight of each head, Q, K, V are the parameters of attention, h is the number of heads, the parameter matrices $W^O \in R^{d_1 \times d_1}$, $W_i^V \in R^{d_1 \times d_2}$, $W_i^K \in R^{d_1 \times d_2}$, $W_i^Q \in R^{d_1 \times d_2}$ are the projections. It can be simplified as a parameterized function $Transformer(\cdot)$ with all parameters θ:

$$z = Transformer(\theta; t)\ \ \theta = \{\ Q, K, V, W^O\} \tag{2}$$

Compound Pre-training Module: Compound pre-training module is to encode the compound graph to a vector by the GCNs. This paper uses GCNs to excavate potential relationships from the compound graph structure of drug. Given a compound graph of drug $\mathcal{G} = (\mathcal{V}, \mathcal{E}, \mathcal{X}, \mathcal{Z})$ where $\mathcal{V}$ is the chemical atom set, $\mathcal{E}$ is chemical bond sets, $\mathcal{X} \in \mathbb{R}^{|\mathcal{V}| \times d_v}$ and $\mathcal{Z} \in \mathbb{R}^{|\mathcal{E}| \times d_e}$ are the atom and bond feature set. The GCN [14] involves two key computations "update" and "aggregate" for each atom v at every layer. They can be represented as one parameterized function $\Psi(\cdot)$ with parameters ψ.

$$\mathbf{h}_v^l = \Psi(\psi; \mathcal{A}, \mathcal{X}, \mathcal{Z})^l = \mathrm{UPDATE}\Big(\mathbf{h}_v^{l-1}, \mathrm{AGGREGATE}\Big(\Big\{\Big(\mathbf{h}_v^{l-1}, \mathbf{h}_u^{l-1}, \mathbf{z}_{uv}\Big) : u \in \mathcal{N}_v\Big\}\Big)\Big) \tag{3}$$

where z_{uv} is the feature vector of bond (u, v), $\mathrm{h}_v^0 = \mathrm{x}_v \in \mathcal{X}$ is the input of a GCN and represents the properties of this atom, A is the adjacency matrix of this compound, and $\mathcal{N}_v$ is the neighborhood of atom v.

In order to get a representation of the whole graph, the POOLING function at the last GCN layer is used to transform the whole chemical compound graph into a vector:

$$\mathbf{h}_{\mathcal{G}} = \mathrm{POOLING}\Big(\Big\{\mathbf{h}_v^l \Big| v \in \mathcal{V}\Big\}\Big) \quad \mathbf{h}_{\mathcal{G}} \in \mathbb{R}^{|x| \times 1} \tag{4}$$

where $h_{\mathcal{G}}$ is the representation of the whole compound graph $\mathcal{G}$, POOLING is a simple pooling function like max or mean-pooling [15, 16]. The GCNs model can be simplified as follows:

$$\mathbf{h}_{\mathcal{G}} = GCNs(\psi; \mathcal{G}) \tag{5}$$

DTA Prediction Module: The DTA prediction module is to associate the compound of drug with the protein for predicting their affinity. This paper adopts the fully connected neural network as decoder:

$$\hat{y} = FC(\gamma; \mathrm{Concat}(\mathrm{z}, h_{\mathcal{G}})) \tag{6}$$

where z is a protein sequence representation, h_G is a compound graph representation, FC is a fully-connected neural network with two hidden layer. The input is concatenated vectors of z and h_G, and the output is the binding affinity scores which represent interactivity strength between DT pairs.

3.3 Training Procedure

This section will explain the details of the training procedure in the proposed multi-task framework.

Protein Pre-training: For improving the representation learning ability on proteins, this study set two pre-training tasks which are Masked Language Modeling (MLM) [17] and Same Family Prediction (SFP) in a similar manner to Bert.

The MLM task is to screen some amino acids at random and predict the type of amino acids that has be screened. Given a masked protein sequence t, and masked amino acid set $m = \{m_1, m_1, \ldots, m_N\}$, the transformer produces sequence representations and the MLM decoder computes log probabilities for $\tilde{t}$ over 20 amino acid types:

$$z = Transformer(\theta; t) \tag{7}$$

$$m' = softmax(FC(\theta_1; z)) \tag{8}$$

where $FC(\cdot)$ is fully connected neural network with parameter θ_1 and m' is the probability distribution of masked amino acid. Then log-likelihood function is used as evaluation metrics for the MLM task:

$$\mathcal{L}^{\mathbf{MLM}}(\theta, \theta_1; m) = \sum_{i=1}^{N} -\ln p\big(m = \hat{m}_i | \theta, \theta_1\big) \tag{9}$$

By the MLM task, the transformer model of protein sequence could learn bidirectional contextual representation.

The SFP task enables the model to determine if two proteins were in the same family. In order to pre-train transformer with the SFP pre-training task, this paper samples two protein sequences t^1 and t^2 from the Pfam dataset [7], the probability of which coming from the same family and different families is the same.

Aiming at the sequence representation $z = [z_1^1, \cdots, z_{n_1}^1, z_1^2, \cdots, z_{n_2}^2]z \in \mathbb{R}^{|z| \times 1}$ of protein pair, the fully connected neural network with dropout is used to compute their similarity score $\hat{c}$:

$$\hat{c} = FC(\theta_2; z)$$

where $FC(\cdot)$ is full connection layer with parameters $\theta_2 \in \mathbb{R}^{|z| \times 2}$ and $\hat{c} \in \mathbb{R}^{2 \times 1}$ is the predicted similarity score, i.e., a probability that the pair belongs to a same protein family. The SFP task trains the model to minimize cross-entropy loss which is designed to deal with predicted errors on probabilities. This study employs the log-likelihood function for measuring SFP loss:

$$\mathcal{L}^{\mathbf{SFP}}(\theta, \theta_2; t) = -\ln p(n = n_i|\theta, \theta_2) \quad n_i \in [\text{same family}, \ \text{not same family}] \tag{11}$$

As the transformer is asked to produce higher similarity scores for proteins from the same family, the SFP task enables the transformer model to better assimilate global structural information.

Compound Pre-training: The purpose of the compound graph pre-training task is to improve the representation learning ability on drug.

This study uses GCNs to encode the compound graph as the vector representation. The aggregation is a key computation in each layer of GCNs. In compound-level aggregation, the compound neighborhoods aggregated their information in Eq. (3). For each compound $v \in \mathcal{G}$, GCN gets its representation h_v by $\Psi(\cdot)$. Then, given a random atom bond (u, v), the self-supervised loss function [19] is chosen to encourage similar embedding of neighboring nodes for the link between u and v as follows.

$$\mathcal{L}^{\text{atom}}\left(\psi; \mathcal{S}_{\mathcal{G}}\right) = \sum\nolimits_{(u,v)\in\mathcal{G}} - \ln\left(\sigma\left(\mathbf{h}_u^{\mathrm{T}}\mathbf{h}_v\right)\right) - \ln\left(\sigma\left(-\mathbf{h}_u^{\mathrm{T}}\mathbf{h}_{v'}\right)\right) \tag{12}$$

where v' is not really connected to u in the graph, ψ are the GCNs parameters of $\Psi(\cdot)$ and σ is the sigmoid function. The atom-level aggregation wants real bonds of a compound graph to get analogous representations, which makes GCNs leverage atom feature information for generalizing to unseen graphs.

DTA Prediction Training: For the DTA prediction task, both protein vectors and compound vectors are fed into the FC layers to get continuous binding scores. For instance, given a drug-protein pair (g, t) where $g = (v, e, x, z)$ is a chemical graph of compound and $t = [t_1, \cdots, t_n]$ is a protein sequence, the process of predicting their binding affinity values $\hat{y}$ is shown as follows.

$$l = Transformer(\theta; t) \tag{13}$$

$$\mathbf{h}_{\mathcal{G}} = GCNs(\psi; g) \tag{14}$$

$$\hat{y} = FC(\gamma; \text{Concat}(l, h_{\mathcal{G}})) \tag{15}$$

where $FC(\cdot)$ is full connection layers and $\gamma \in \mathbb{R}^{(|z|+|x|)\times 1}$ is the parameters of full connection layers.

The DTA prediction task trains the model to minimize the loss function. This study employs the mean squared error (MSE) as the loss function:

$$\mathcal{L}^{affinities}(\theta, \psi, \gamma; (t, g)) = \frac{1}{2}(\hat{y} - y)^2 \tag{16}$$

where $\hat{y}$ is the value predicted by the DTA model and y is true binding affinity values.

3.4 Adaptation Strategy

This study adopts multi-task learning to link the encoder, i.e. the protein pre-training task and compound pre-training task, and the decoder, i.e. the DTA prediction task, for prevent overfitting caused by local optimality under a relatively small variety of supervised samples. In order to make the overall framework more biased against the main task DTA prediction, this study adopts the updated strategy of MAML [5] as follows.

For the compound pre-training task and protein pre-training tasks, we adapt the compound-level aggregation prior parameter ψ, protein-level prior parameter θ with some steps for updating, by an compound-level learning rate α, protein-level learning rate β and η for MLM task and SFP task. It is shown as follows.

$$\psi' = \psi - \alpha \frac{\partial \mathcal{L}^{\text{atom}}\left(\psi; \mathcal{S}_{\mathcal{G}}\right)}{\partial \psi} \tag{17}$$

$$\theta' = \theta - \beta \frac{\partial \mathcal{L}^{\text{MLM}}\left(\theta, \theta_1; m\right)}{\partial \theta} - \eta \frac{\partial \mathcal{L}^{\text{SFP}}\left(\theta, \theta_2; n\right)}{\partial \theta} \tag{18}$$

where θ_1, θ_2 are task-specific parameters for the MLM task and SFP task respectively.

With the above-mentioned adaptations for protein and compound, the prior $\{\psi, \theta\}$ is adapted to $\{\psi', \theta'\}$. Then, the full connection layer parameters γ' will be updated through DTI loss function as follows.

$$\mathcal{L}^{affinities}\left(\psi', \theta', \gamma; (t, g)\right) \tag{19}$$

$$\gamma' = \gamma - \alpha \frac{\partial \mathcal{L}^{affinities}\left(\psi', \theta', \gamma; (t, g)\right)}{\partial \gamma} \tag{20}$$

After that, all the parameters are updated through the overall loss function of the multi-tasking learning framework. We define the overall loss function as followed:

$$\mathcal{L}^{all} = \lambda_{atom}\mathcal{L}^{\text{atom}} + \lambda_{MLM}\mathcal{L}^{\mathbf{MLM}} + \lambda_{SFP}\mathcal{L}^{\mathbf{SFP}} + \lambda_{affinities}\mathcal{L}^{affinities} \tag{21}$$

where λ_{atom} and λ_{MLM} and λ_{SFP} and $\lambda_{affinities}$ are the weight of the loss function for each subtask, that is set manually. This study updates all learnable parameters by gradient descent.

4 Experiments and Results

4.1 Experimental Setup

The discovery of new drugs means that the appearance of new drugs and proteins. In order to simulate this real scenario, this study constructed a simulated innovative drug discovery dataset. 20 compounds and 4 proteins were randomly selected from the Davis training dataset [11] as new drugs and targets. To ensure that at least one drug or target in the testset is innovative, we select all drug protein pairs which contain one of these compounds or proteins were extracted as test data of innovative drug discovery (innovative test data). Removing these drug-protein pairs, the remaining data in the Davis training dataset are used to construct the training data. In order to compare fully with existing models, this study also adopted the Davis test data set as traditional test data directly. The distribution of data is shown in Table 1.

Table 1. The data distribution in each dataset

Data set	Number of protein	Number of drugs	Number of correlations
All data	442	68	30056
Training data	424	64	22622
Traditional test data	442	68	5017
Innovative test data	442(20)	68(4)	2417

In the experiment, since the weight of the loss function of each subtask needs to be set manually, the training process is monitored through the training data. When the value of loss function is started rising and lasted 200 epochs, the model tends to be over fit, we terminate experiment.

4.2 Comparative Study

(1) DEEPDTA [8]: This model consists of a deep neural structure and a feedforward layer. The deep neural structure includes two independent CNN modules to learn features from Smiles strings and sequences respectively. The feedforward layer makes a complete connection between the drug and the protein representation for the DTA prediction.
(2) GraphDTA [9]: The model consists of a CNN module, a GAT module and a fully connected neural network. The CNN and GAT learn features from protein sequences and compound graphs, and the fully connected neural network realizes the DTA prediction. Different from other DAT models, GraphDTA learns the graph representation of compounds to be more consistent with the description of the drug.

The MSE value and Pearson's correlation coefficient(R) are used as quantitative evaluation indexes. 5-fold cross validation is performed for avoiding the overfitting problem.

4.3 Results

Table 2 gives experimental results. It can be seen that the MSE values of DeepDTA and GraphDTA on the training dataset are about 8 times that of the traditional test data, and 20 times that of the innovative test data. This means their models are overfitting. However, our model achieves the best performance on both the traditional test data and the innovative test data, indicating that our pre-training and secondary update strategy

Table 2. Experimental results

Data set	Method	MSE	R
Training data	DEEPDTA	0.0602	0.8531
	GraphDTA	0.0411	0.9755
	Proposed method	0.0599	0.9034
Traditional testing data	DEEPDTA	0.3263	0.7408
	GraphDTA	0.3073	0.7867
	Proposed method	0.2792	0.8439
Innovative test data	DEEPDTA	0.8747	0.3747
	GraphDTA	0.8150	0.4024
	Proposed method	0.6610	0.4802

can well reduce the over-fitting of the model. Hence, the proposed model can achieve better results on the DTA task for the real scenario of innovative drug discovery.

Figure 2 and Fig. 3 show scatter plots of the predicted and actual affinity value on the testing dataset and new type testing dataset. Each point is a drug-protein pair. The brightness value represents the absolute value of the true value minus predicted one, and the lighter the color, the more accurate the model's prediction. It can be found from the figure that the predicted value of the model has a high coincidence with the real value. Comparing Fig. 2 and Fig. 3, we can find that all spots still cluster on the main diagonal even if the test data changed from the traditional test data to the innovative test data. This shows that our model has good prediction ability and generalization ability for the drug-target pair which never appeared in training dataset.

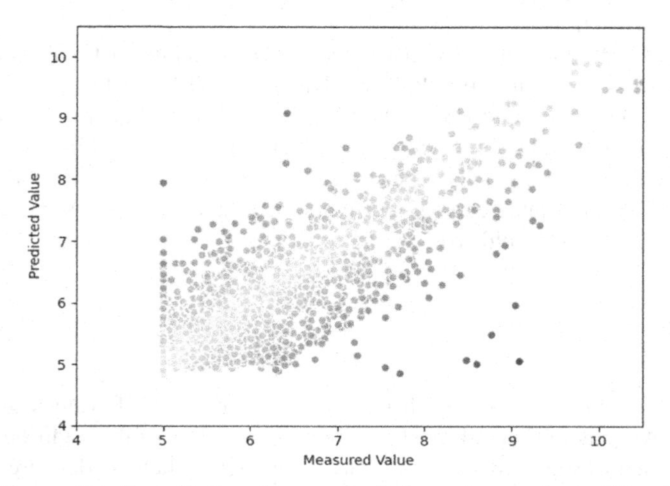

Fig. 2. Measured binding affinity values vs. prediction values on the traditional test data.

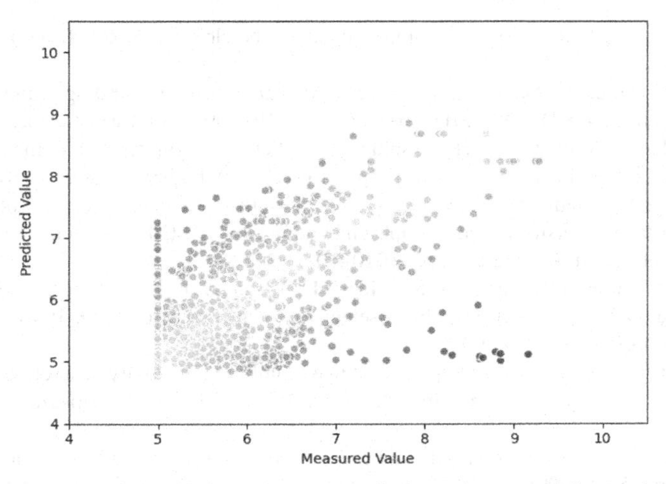

Fig. 3. Measured binding affinity values vs. prediction values on the innovative test data.

5 Conclusions and Discussion

This paper proposes GeneralizedDTA, a deep learning model for predicting drug-target binding affinity. Under a dual adaption strengthened multi-task framework, the GCN model and the transformer model are adopted to learn the representation of compound graphs and amino acid sequences, and the fully connected neural network is used to predict the value of affinity. Experimental results show that, our model can better predict drug-target affinity, especially the affinity of new drug-target pairs, than other existing models. It effectively improves the value of DTI in practical applications.

Acknowledgements. The work is supported by National Key Research and Development Program of China (Grant No. 2020YFB2104402).

References

1. Ezzat, A., Min, W., Li, X.-L., Kwoh, C.-K.: Computational prediction of drug–target interactions using chemogenomic approaches: an empirical survey. Brief. Bioinf. **20**(4), 1337–1357 (2018)
2. Martin, E.M., Jane, N., Louise, N.J.: Protein kinase inhibitors: insights into drug design from structure. Science **303**, 1800–1805 (2018). https://doi.org/10.1126/science.1095920
3. Cer, R.Z., Mudunuri, U., Stephens, R., Lebeda, F.J.: IC50-to-Ki: a web-based tool for converting IC50 to Ki values for inhibitors of enzyme activity and ligand binding. Nucleic Acids Res. **37**, W441–W445 (2009). https://doi.org/10.1093/nar/gkp253
4. Devlin, J., Chang, M.-W., Lee, K., Toutanova, K.: BERT: pre-training of deep bidirectional transformers for language understanding. arXiv preprint arXiv:1810.04805 (2018)
5. Finn, C., Abbeel, P., Levine, S.: Model-agnostic meta-learning for fast adaptation of deep networks. In: Proceedings of ICML, pp. 1126–1135 (2017)
6. Sterling, T., Irwin, J.J.: Zinc 15 – ligand discovery for everyone. J. Chem. Inf. Model. **55**(11), 2324–2337 (2015)

7. Finn, R.D., et al.: Pfam: the protein families database. Nucleic Acids Res. **42**(D1), D222–D230 (2014)
8. Ozturk, H., Ozgur, A., Ozkirimli, E.: DeepDTA: deep drug-target binding affinity prediction. Bioinformatics **34**, 821–829 (2018). https://doi.org/10.1093/bioinformatics/bty593
9. Nguyen, T.A.L., Venkatesh, S.H.: GraphDTA: prediction of drug-target binding affinity using graph convolutional networks. BioRxiv [preprint] (2019). https://doi.org/10.1101/684662
10. Hu, F., Hu, Y., Zhang, J., Wang, D., Yin, P.: Structure enhanced protein-drug interaction prediction using transformer and graph embedding. In: IEEE International Conference on Bioinformatics and Biomedicine, pp. 1010–1014 (2020)
11. Davis, M.I., Hunt, J.P., Herrgard, S., Ciceri, P., Wodicka, L.M., Pallares, G., et al.: Comprehensive analysis of kinase inhibitor selectivity. Nat. Biotechnol. **29**, 1046–1051 (2011). https://doi.org/10.1038/nbt.1990
12. Tang, J., et al.: Making sense of large-scale kinase inhibitor bioactivity data sets: a comparative and integrative analysis. J. Chem. Inf. Model. **54**(3), 735–743 (2014). https://doi.org/10.1021/ci400709d
13. Aswani, A.V., et al.: Attention is all you need. In: Advances in Neural Information Processing Systems, pp. 5998–6008 (2017)
14. Kipf, T.N., Welling, M.: Semi-supervised classification with graph convolutional networks. In: Proceedings of the International Conference on Learning Representations (ICLR) (2017)
15. Atwood, J., Towsley, D.: Diffusion-convolutional neural networks. In: Proceedings of NeurIPS, pp. 1993–2001 (2016)
16. Duvenaud, D., et al.: Convolutional networks on graphs for learning molecular fingerprints. In: Proceedings of NeurIPS, pp. 2224–2232 (2015)
17. Min, S., Park, S., et al.: Pre-training of deep bidirectional protein sequence representations with structural information. arXiv: arXiv:1912.05625 [cs, q-bio, stat] (February 2020)
18. Tang, J., Qu, M., Wang, M., Zhang, M., Yan, J., Mei, Q.: LINE: large-scale information network embedding. In: Proceedings of WWW, pp. 1067–1077 (2015)
19. Yang, K.K., Zachary, W., Bedbrook, C.N., Arnold, F.H.: Learned protein embeddings for machine learning. Bioinformatics **34**(15), 2642–2648 (2018)
20. Alley, E.C., Khimulya, G., Biswas, S., AlQuraishi, M., Church, G.M.: Unified rational protein engineering with sequence-based deep representation learning. Nat. Meth. **16**(12), 1315–1322 (2019)
21. Krause, B., Lu, L., Murray, I., Renals, S.: Multiplicative LSTM for sequence modelling. arXiv preprint arXiv:1609.07959 (2017)

GRE: A GAT-Based Relation Embedding Model of Knowledge Graph for Recommendation

Jihu Wang[1], Yuliang Shi[1,2(✉)], Lin Cheng[1], Kun Zhang[3], and Zhiyong Chen[1]

[1] School of Software, Shandong University, Jinan, China
jihu_wang@mail.sdu.edu.cn,
{shiyuliang,chenzy}@sdu.edu.cn,chenglin123_sdu@163.com
[2] Dareway Software Co., Ltd., Jinan, China
[3] School of Information Science and Engineering, University of Jinan, Jinan, China
kunzhangcs@126.com

Abstract. Compared with collaborative filtering, knowledge graph embedding based recommender systems greatly boost the information retrieval accuracy and solve the limitations of data sparsity and cold start of traditional collaborative filtering. In order to fully explore the relationship and structure information hidden in knowledge graphs, we propose the ***G****AT-based* ***R****elation****E****mbedding* (GRE) model. In our model, we propose a *Triple Set* to denote a set of knowledge graph triples whose head entities are linked by items in interaction records, and a *Triple Group* to denote a group of knowledge graph triples extracted from *Triple Set* according to different relations. The proposed GRE is a neural model that aims at enriching user preference representation in recommender systems by utilizing Graph Attention Network (GAT) to aggregate the embeddings of adjacent tail entities to head entity over *Triple Group* and embedding the representation of relation in the process of polymerization of *Triple Groups* in *Triple Set*. By embedding relation information into each *Triple Group* representation and concatenating *Triple Group* representations in *Triple Set*, this proposed novel relation embedding method addresses the problem that GAT-based models only consider aggregating the neighboring entities and ignore the effect of relations in triples. Through extensive experimental comparisons with the baselines, we show that GRE has gained state-of-the-art performance in the majority of the cases on two open-source datasets.

Keywords: Recommender systems · Knowledge graph · Graph attention networks

1 Introduction

In this age of information explosion, people are bombarded with more choices than they can effectively handle. Browsing too much irrelevant information will undoubtedly lead to the loss of consumers who are overwhelmed by information

Y. Sun et al. (Eds.): ChineseCSCW 2021, CCIS 1492, pp. 77–91, 2022.
https://doi.org/10.1007/978-981-19-4549-6_7

overload. Recommender Systems (RS) [22], proposed in 1997, aim to improve the efficiency of information acquisition for users. RS recommends personalized content to users according to their historical click or browse records. For example, content-based recommendation [18] intends to recommend to the user the goods similar to the other things that the user preferred previously. Collaborative filtering recommendation [14] aims to recommend to the user the items which other users with similar preference may like [19]. However, these traditional recommendation approaches are challenged with the limitations of data sparsity and cold start.

In 2012, Google built Knowledge Graph (KG) to optimize its search engine [7]. KG is a heterogeneous network consisting of various entities as well as different relations. Concretely, a triple (h, r, t) in the KG represents that there exists a head entity h and a tail entity t that are bridged by a relation r. KG, with its rich structural and semantic information, gradually becomes the important side information of various downstream tasks to strengthen the representation learning capacity of the model. As a widely used source of side information, open source KG has made significant progress. Many open source KGs, such as Wikidata [6], DBpedia [2], Freebase [3], and Microsoft Satori [1], provide rich structural and semantic side information for downstream tasks. The triples in these open source KGs contain the rich information of entities and relations. For example, (*Matt Damon, occupation, Actor*) and (*Matt Damon, occupation, Producer*) from Wikidata indicate that *Matt Damon* is not only a actor but also a film producer. KGs have been widely utilized in the researches of information retrieval, question-and-answer and personalized recommender systems [11,16,29]. Especially in the area of personalized recommendation, KG provides additional structural and semantic information for RS [15], and effectively solves the limitations of interaction sparsity and new user's cold start [21]. The application of KG as side information can be intuitively explained that the information of attributes, relations and entities can be embedded into the user's preference vector by linking items in interaction records to entities [10].

At present, Graph Attention Network (GAT) [23] is widely used in graph representation learning. GAT uses attention mechanisms to compute the weighted sum of the features of neighboring nodes. However, this method discards the rich information stored in the edge between two nodes, because it only takes two nodes into consideration and ignores the edge between the nodes.

To overcome the aforementioned limitations, we newly propose the ***G**AT-based **R**elation **E**mbedding* (GRE), which aims at enriching user preference representation in recommender systems by utilizing GAT to aggregate the information of adjacent tail entities to head entity over *Triple Group* and embedding the representation of relation in the process of polymerization of *Triple Groups* in *Triple Set*. By embedding relation information into each *Triple Group* representation and concatenating *Triple Group* representations in *Triple Set*, this proposed novel relation embedding method effectively addresses the problem that GAT-based approaches only consider aggregating the neighboring entities and ignore the effect of relations in triples. We conduct experiments to compare the proposed GRE with widely used recommendation baselines on two open-source datasets. The results imply that GRE has gained the best performance overall. Our contributions are as follows:

(1) We propose a novel KG-aware recommendation model, namely GRE. It aims at enriching the user final preference vector by combining GAT-based information propagation with relation embedding.
(2) We provide a new relation embedding method to embed relation information in the process of polymerization of *Triple Groups* in *Triple Set.*
(3) We prove the effectiveness of GRE and positive effect of the designed relation embedding method by comparative experiments with the baselines on two open-source datasets (i.e. MovieLens-1M and Last.FM) in Click-Through Rate (CTR) prediction and top-k recommendation tasks.

The remaining contents are conducted as follows. In Sect. 2, we retrospect relevant researches. Then, we elaborate details of GRE in Sect. 3. Experiments are conducted for GRE and baselines in Sect. 4. Finally, we provide a conclusion for this paper in Sect. 5.

2 Related Work

2.1 Graph Attention Networks

Graph Convolutional Network (GCN) [13] is a powerful neural framework for processing various graphs. It could capture the various dependency patterns in graphs via message propagation among the nodes [9,20]. Although GCN has achieved good results in many tasks, it faces limitations when dealing with inductive tasks in which the same graph structure is needed for model training and testing. With the proposal of GAT, these limitations have been solved, because GAT only considers the first-order neighborhood nodes when calculating the aggregation information of nodes. GAT is a network whose main idea is to boost the node representation via the value of each node's attention in the adjacent nodes. Unlike GCN, GAT does not make full use of the graph structure during training and testing.

Although GAT is very powerful in passing the information of neighboring tail entities to the head entity, it ignores the relations that store rich information between head entity and tail entity when GAT is introduced into KG-based RS [27]. This limits the effectiveness of GATs. For example, given a triple (*Tom Holland, ?, Avengers: Endgame*) without the relation *starred* in, we don't know if *Tom Holland* is an audience or an actor of *Avengers: Endgame.*

We introduce GAT for knowledge graph aware recommendation and solve the limitations by embedding relation information in the process of polymerization of *Triple Groups* in *Triple Set.* This method can take into account relations in triples in the process of calculating user's preference representation.

2.2 Knowledge Graph Aware Recommender Systems

As side information, KGs have made great progress in the research field of recommender system. For example, PER [30] introduces KGs as side information, and extracts meta-path-related compact vectors to generalize the relationships among different nodes along heterogeneous paths. KGAT [28] is a popular approach that incorporates representations of the multi-hop neighboring nodes to the central node based on the discriminative importance coefficients derived from an attention mechanism. It could effectively utilize the high-order nodes' information of KG to enrich the representation embeddings of users and items. KGCN [27] is a widely used approach that could effectively capture the inter-item correlation by exploring the associated entities or attributes in KG. Concretely, to iteratively utilize the high-order structural and semantic information, it randomly samples from the neighbors for every entity in the KG as their ego-centered network, in which the compact vector of a given entity is computed via aggregating the neighborhood information with bias. DKN [25] integrates the semantic- and knowledge-level information of news. At the same time, the attention mechanism is introduced to handle the variety of user's interests and integrate user's historical news records dynamically according to the current candidate news. MKR [26] builds cross and compress components, which mutually exchange implicit features and capture multi-hop interactivities between items in historical record and entities in KG. It has been demonstrated that the deployed cross and compress components have strong learning capacity. Besides, MKR is a more generalized framework than other knowledge graph-aware recommendation methods. RippleNet [24] draws on the strengths of embedding- and path-based recommendation methods. It progressively propagates users' latent preferences over the set of randomly sampled KG entities and seeks their multi-level interests along the paths extracted from the knowledge graph.

In conclusion, as side information, KG is used in RS to enrich the user's preference representation mainly through entity information propagation along extracted paths and semantic relation information embedding. In this article, we propose a new KG-based recommender system that efficiently handles entity information propagation via GAT and relation information via a new embedding method over *Triple Group* and *Triple Set*.

3 The Proposed Model

As illustrated in Fig. 1, GRE mainly includes the following three components: **(1) Triple Set and Triple Group Extraction.** In this part, we define *Triple Set* and *Triple Group* to reorganize triples linked by items in interaction records. **(2) GAT-based Entity Aggregation and Relation Embedding.** This part mainly describes the details of our GAT-based entity aggregation and relation embedding method. We use GAT for each *Triple Group* to propagate tail entities' information to the head entity to obtain the representation of the *Triple Group*, and then embed the relation information into each representation of the *Triple Group* by the product of the *Triple Group* representation vector and the

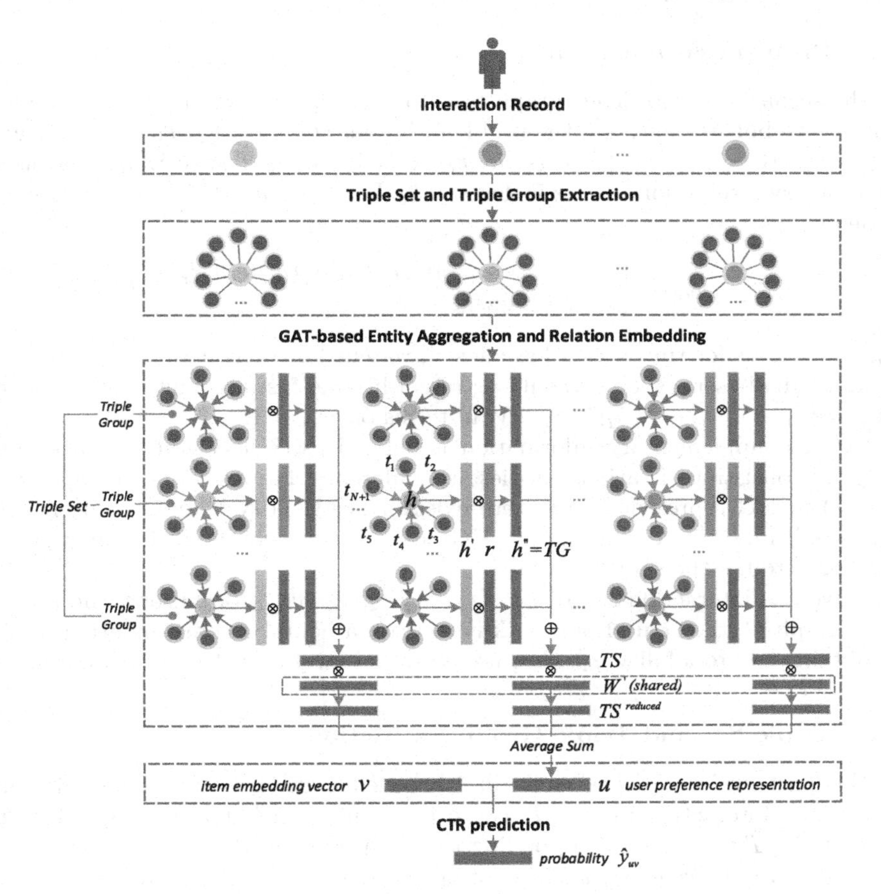

Fig. 1. The overall framework of GRE. Given an interaction record, we link items in record to knowledge graph to generate the *Triple Set* and the *Triple Group*. The user's preferences are calculated via GAT-based Entity Aggregation and Relation Embedding. Finally, GRE outputs the probability of CTR.

relation representation. The representation of an item in interaction records is obtained by reducing the dimension of concatenation of these different *Triple Group* representations in the *Triple Set*. Finally, we compute a user's preference vector by averaging all of his interaction item representations. The user's preference representation not only contains information from the tail entities, but also relation information. This is a novel relation embedding method for GAT-based knowledge graph aware model. **(3) Click-Through-Rate prediction.** Given a new item, we predict a user's final CTR using the inner product of his/her preference vector and item embedding vector. The details of GRE are provided in the following subsections.

3.1 Problem Formulation

In the scenario of knowledge graph based recommender system, we use the following symbols to illustrate the problem. We let $U = \{u_1, u_2, u_3, \cdots, u_{|U|}\}$ and $V = \{v_1, v_2, v_3, \cdots, v_{|V|}\}$ represent user and item set, respectively. The user-item binary interaction matrix is denoted by $Y = \{y_{uv}|u \in U, v \in V\}$, where y_{uv} is defined as

$$y_{uv} = \begin{cases} 1 \text{ , } if\ user\ u\ interacted\ with\ item\ v; \\ 0 \text{ , } otherwise. \end{cases} \tag{1}$$

$y_{uv} = 1$ stands for that user u has clicked, watched or bought item v. Therefore, for individual user $u \in U$, we could represent his/her historical interaction record as $I = \{v_1, \cdots, v_i, \cdots, v_{|I|}\}$, $v_i \in V$ in terms of $y_{uv_i} = 1$.

KG as supplementary information is defined as G. KG is a heterogeneous network consisting of various entities and different relations. Concretely, a triple (h, r, t) in KG represents that there exists a head entity h and a tail entity t that are bridged by a relation r. In KG, a head entity can point to different tail entities through the same relation.

Given an interaction record $I = \{v_1, \cdots, v_i, \cdots, v_{|I|}\}$ for a user u and knowledge graph G, the goal of our work is to train a neural recommendation model to obtain the probability $\widehat{y}_{uv}$ that user u would interact with a new item v.

3.2 Triple Set and Triple Group Extraction

In this part, we reorganize triples in a knowledge graph G by linking all items in a user's interaction history I to the head entities of knowledge graph G to extract the *Triple Sets* corresponding to each item, and then divide each *Triple Set* into several *Triple Groups* according to different relations.

Definition 1. *Triple Set (TS): A set of KG triples whose head entities are linked by items from interaction record* $I = \{v_1, \cdots, v_i, \cdots, v_{|I|}\}$.

$$TS_i = \{(h_i, r, t)|(h_i, r, t) \in G \text{ } and \text{ } h_i = v_i \in I\},\ i = 1, 2, \cdots, |I|. \tag{2}$$

From the above definition, triples in TS_i share the same head entity h_i. We assume all triples in TS_i have different relations $R = \{r_1, \cdots, r_j, \cdots, r_{|R|}\}$.

Definition 2. *Triple Group (TG): A group of KG triples extracted from TS according to different relations R.*

$$TG_i^j = \{(h_i, r_j, t)|(h_i, r_j, t) \in TS_i \text{ } and \text{ } r_j \in R\},\ j = 1, 2, \cdots, |R|. \tag{3}$$

In other words, TS_i can be formulated as

$$TS_i = \{TG_i^1, TG_i^2, \cdots, TG_i^{|R|}\}. \tag{4}$$

As a result, triples in TG_i^j own the identical head entity h_i as well as relation r_j, but disparate tail entities.

3.3 GAT-Based Entity Aggregation and Relation Embedding

We suppose that $TG_i^j = \{(h_i, r_j, t_1), \cdots, (h_i, r_j, t_k), \cdots, (h_i, r_j, t_N)\}$ is a triple group that consist of triples, where k and N stands for the k-th triple and the number of triples in triple group TG_i^j, respectively. As shown in Fig. 1, we use GAT to aggregate the information of tail entities to the head entity h_i along the same relation r_j over TG_i^j. The embedding vectors of the tail entities in TG_i^j is defined as

$$T = \{\mathbf{t}_1, \cdots, \mathbf{t}_k, \cdots, \mathbf{t}_{N+1}\},\ \mathbf{t}_k \in \mathbb{R}^d, \tag{5}$$

where $\mathbf{t}_k \in \mathbb{R}^d$ is the d-dimensional embedding vector of tail entity t_k. $\mathbf{h} \in \mathbb{R}^d$ and $\mathbf{r} \in \mathbb{R}^{d \times d'}$ denote the embedding vector of head entity and the embedding matrix of relation in TG_i^j, respectively, according to the embedding method (One-hot, TransR [17], TransE [8], etc.). In order to take into account head entity embedding in the process of tail entity propagation, we let $\mathbf{t}_{N+1}$ denote $\mathbf{h}$.

For a *Triple Group* TG_i^j, we obtain similarity coefficient e_k between head entity and its neighboring tail entities by

$$e_k = a([\mathbf{W}\mathbf{t}_k || \mathbf{W}\mathbf{h}]),\ k = 1, 2, \cdots, N+1, \tag{6}$$

where $\mathbf{W} \in \mathbb{R}^{d' \times d}$ stands for the shared learnable parameter matrix for feature augmenting. $[\cdot||\cdot]$ stands for the vector concatenation. We adopt a single-layer feedforward neural network denoted as $a(\cdot)$ to map a high-dimension feature vector to a similarity coefficient e_k.

Attention mechanism coefficient a_k between tail entity vector $\mathbf{t}_k$ and head entity vector $\mathbf{h}$ is obtained by normalizing the $LeakyReLU(e_k)$ via $softmax()$ function. The attention mechanism coefficient a_k is defined as

$$a_k = \frac{\exp(LeakyReLU(e_k))}{\sum_{k=1}^{N+1} \exp(LeakyReLU(e_k))},\ k = 1, 2, \cdots, N+1. \tag{7}$$

Finally, enriched head entity representation vector $\mathbf{h}'$ can be calculated as

$$\mathbf{h}' = \sum_{k=1}^{N+1} a_k \mathbf{W}\mathbf{t}_k. \tag{8}$$

The above calculations do not take into account the relation embedding information. To strength the robustness and expressiveness of the proposed GRE. We embed the relation representation vector via

$$\mathbf{h}'' = \mathbf{r}\mathbf{h}'. \tag{9}$$

Based on the above discussion, vector $\mathbf{h}''$ can be regarded as TG_i^j representation $\mathbf{TG_i^j}$. According to Eq. 4, each TS has several TGs, $TS_i = \{TG_i^1, TG_i^2, \cdots, TG_i^{|R|}\}$. We can obtain the representation $\mathbf{TS_i}$ of TS_i by concatenating $\mathbf{TG_i^j}$ in TS_i.

$$\mathbf{TS_i} = [\mathbf{TG_i^1} || \cdots || \mathbf{TG_i^{|R|}}] = [\mathbf{h''_i^1} || \cdots || \mathbf{h''_i^{|R|}}],\ i = 1, 2, \cdots, |I|. \tag{10}$$

We use linear transformation to reduce the dimension of $\mathbf{TS_i}$ because of the sparsity of high-dimension $\mathbf{TS_i}$.

$$\mathbf{TS_i^{reduced}} = \mathbf{TS_i} \cdot \mathbf{W}', \; i = 1, 2, \cdots, |I|, \tag{11}$$

where matrix $\mathbf{W}' \in \mathbb{R}^{(|R| \times d) \times d}$ is a shared coefficient matrix. Finally, according to Eq. 2 and Eq. 11, we compute the user final preference vector $\mathbf{u}$ by

$$\mathbf{u} = \frac{1}{|I|} \sum_{i=1}^{|I|} \mathbf{TS_i^{reduced}}, \tag{12}$$

where $|I|$ denotes the number of items in his/her interaction record.

3.4 Click-Through-Rate Prediction

Given a newly encountered item v, we can compute the probability $\widehat{y}_{uv}$ that user u would interact with this item v through

$$\widehat{y}_{uv} = \sigma(\mathbf{u}^{\mathrm{T}} \mathbf{v}), \tag{13}$$

where $\sigma(\cdot)$ stands for sigmoid function, $\mathbf{u} \in \mathbb{R}^d$ (see Eq. 12) and $\mathbf{v} \in \mathbb{R}^d$ denotes the user's final preference vector and the item embedding vector, respectively.

CTR prediction is a binary classification task, we opt for the cross-entropy loss as the optimization objective of GRE.

$$L = \frac{1}{M} \sum_{i=1}^{M} -[y_i \log(\widehat{y}_i) + (1 - y_i) \log(1 - \widehat{y}_i))], \tag{14}$$

where M represents the batch size. We optimize GRE through Adam [12] optimizer, where M is fixed at 1024 in our experiments.

4 Experiments

We first expatiate the datasets and baselines. Then, evaluation metrics and hyper-parameter settings are described in detail. Finally, we analyse the experiment results to evaluate the recommendation performance of GRE.

4.1 Datasets

We choose two open-source datasets to conduct experiments. The details of two datasets are shown as Table 1.

MovieLens-1M [5] is a famous open-source dataset for evaluating the performance of recommender system. It is a movie rating dataset provided by many users. This dataset include movie ratings, movie metadata (style and released year, etc.) and descriptive data for users.

Table 1. The details of datasets

Statistics	User number	Item number	Interaction number
MovieLens-1M	6036	2445	753730
Last.FM	1872	3846	42310

Last.FM [4] is also a popular recommendation dataset extracted from online music provider. It contains the listening records of many users and the metadata about listeners and musics.

Following [24], we use knowledge graph Microsoft Satori [1] as side information to extract triples used in our experiment corresponding to interaction items of MovieLens-1M and Last.FM, respectively.

4.2 Baselines

The following well-known and competitive approaches are chosen to compare with our proposed GRE:

- **MKR** [26] builds cross and compress components, which mutually exchange implicit features and capture multi-hop interactivities between items in historical record and entities in KG.
- **KGCN** [27] is a widely used approach that could effectively capture the inter-item correlation by exploring the associated entities or attributes in KG. Concretely, to iteratively utilize the high-order structural and semantic information, it randomly samples from the neighbors for every entity as their ego-centered network, in which the compact vector of a given entity is computed via aggregating the neighborhood information with bias.
- **RippleNet** [24] draws on the strengths of embedding- and path-based recommendation methods. It progressively propagates users' latent preferences over the set of randomly sampled KG entities and seeks their hierarchical interests along the paths extracted from the knowledge graph.
- **GRE-NR**. We add a GRE's variant named GRE-NR (GRE-NR means no relation embedding in GRE.) so as to evaluate the proposed relation embedding method.

For a fair comparison, the experiment settings of above methods refer to the default or recommended in the published papers.

4.3 Experiment Setup

In this part, we evaluate GRE over test datasets in the following two scenarios: (1) For CTR prediction task, we adopt ***AUC***, ***ACC*** and ***F1*** metrics to verify the recommendation performance of our proposed GRE. (2) For top-k recommendation task, we opt for ***Precision@K***, ***Recall@K*** and ***F1@K*** metrics to demonstrate GRE's ability of selecting k highest click probability items for each

Table 2. Hyper-parameters for MovieLens-1M and Last.FM

MovieLens-1M	$\|I\| = 8,\ \|R\| = 4,\ N = 8,\ d = 8,\ d' = 8,\ \alpha = 0.02$
Last.FM	$\|I\| = 6,\ \|R\| = 4,\ N = 8,\ d = 8,\ d' = 4,\ \alpha = 0.01$

user. The hyper-parameter settings of GRE are shown in Table 2 in this experiment. $|I|$ and $|R|$ stands for the number of TS and TG for user's interaction record, respectively. N means the number of triples in TG. Parameters d and d' are dimensions of initialized embedding representations for items and relations. We let α denote the learning rate to determine the convergence. We program GRE based on Tensorflow and run it on the NVIDIA RTX 3090 GPU.

4.4 Experiment Results

We feed the test datasets to the trained GRE model to obtain the test results. Table 3, Table 4, and Table 5 present the experiment results. Based on the observations, we state the validity of GRE and usefulness of proposed relation embedding method by answering the following two questions:

- **Q1**: How does GRE perform compared with the selected baselines?
- **Q2**: Does the proposed relation embedding method play a positive role in GRE compared with GRE-NR?

Table 3. Results of CTR prediction task.

Model	MovieLens-1M			Last.FM		
	AUC	***ACC***	***F1***	***AUC***	***ACC***	***F1***
MKR	0.9095	0.8325	0.8351	0.7932	**0.7544**	0.7256
KGCN	0.9007	0.8230	0.8260	0.7988	0.7246	0.7055
RippleNet	0.9202	0.8402	0.8427	0.8014	0.7363	0.7314
GRE-NR	0.9166	0.8399	0.8422	0.8011	0.7313	0.7337
GRE	**0.9206**	**0.8439**	**0.8454**	**0.8048**	0.7383	**0.7352**

(1) Comparison with the Baselines (Q1)

For CTR Prediction. According to Table 3, GRE has achieved outstanding performance in the majority of the cases in CTR prediction task. Specifically, compared with the benchmarks, GRE's performance has been improved by 0.04% to 1.99%, 0.37% to 2.09%, and 0.27% to 1.94% on ***AUC***, ***ACC***, and ***F1*** for MovieLens-1M, respectively. Similarly, GRE has improved the performance by 0.34% to 1.16%, 0.15% to 2.97% on ***AUC*** and ***F1*** for Last.FM. Despite MKR

Table 4. Results of top-k recommendation task on MovieLens-1M.

Model	***Precision@K***						
	K = 1	K = 2	K = 5	K = 10	K = 20	K = 50	K = 100
MKR	0.1400	0.1350	0.1280	0.1210	0.0965	0.0650	0.0493
KGCN	0.0900	0.0950	0.1180	0.1070	0.0930	0.0652	0.0518
RippleNet	0.1200	0.1400	0.1120	0.1080	0.0890	0.0710	0.0530
GRE-NR	0.2000	0.1350	0.1320	0.1230	0.1085	0.0738	0.0542
GRE	**0.2100**	**0.1400**	**0.1380**	**0.1270**	**0.1155**	**0.0770**	**0.0550**
Model	***Recall@K***						
	K = 1	K = 2	K = 5	K = 10	K = 20	K = 50	K = 100
MKR	0.0102	0.0220	0.0512	0.1118	0.1779	0.3074	0.4678
KGCN	0.0079	0.0162	0.0586	0.1116	0.1783	0.2904	0.4476
RippleNet	0.0121	**0.0290**	0.0673	**0.1305**	0.1921	0.3411	0.4981
GRE-NR	0.0237	0.0210	0.0621	0.1177	0.2103	0.3636	0.5144
GRE	**0.0256**	0.0254	**0.0684**	0.1229	**0.2247**	**0.3650**	**0.5144**
Model	***F1@K***						
	K = 1	K = 2	K = 5	K = 10	K = 20	K = 50	K = 100
MKR	0.0190	0.0379	0.0732	0.1162	0.1251	0.1073	0.0892
KGCN	0.0146	0.0277	0.0783	0.1093	0.1222	0.1065	0.0929
RippleNet	0.0220	0.0481	0.0841	0.1182	0.1216	0.1175	0.0958
GRE-NR	0.0423	**0.0492**	0.0845	0.1203	0.1431	0.1228	0.0981
GRE	**0.0456**	0.0430	**0.0915**	**0.1249**	**0.1526**	**0.1272**	**0.0994**

Table 5. Results of top-k recommendation task on Last.FM

Model	***Precision@K***						
	K = 1	K = 2	K = 5	K = 10	K = 20	K = 50	K = 100
MKR	0.0300	0.0380	0.0360	**0.0280**	0.0170	0.0114	0.0081
KGCN	0.0310	0.0350	0.0280	0.0260	0.0215	**0.0134**	0.0095
RippleNet	0.0200	0.0300	0.0260	0.0220	0.0170	0.0116	0.0082
GRE-NR	0.0400	0.0250	0.0220	0.0190	0.0145	0.0104	0.0071
GRE	**0.0420**	**0.0400**	**0.0410**	0.0270	**0.0250**	0.0130	**0.0113**
Model	***Recall@K***						
	K = 1	K = 2	K = 5	K = 10	K = 20	K = 50	K = 100
MKR	0.0125	0.0512	0.0873	0.1410	0.1643	0.2597	0.3436
KGCN	0.0165	0.0339	0.0606	0.1059	0.1616	0.2657	0.3655
RippleNet	0.0125	0.0325	0.0540	0.0798	0.1277	0.2278	0.3192
GRE-NR	0.0125	0.0158	0.0317	0.0808	0.1192	0.2482	0.3208
GRE	**0.0205**	**0.0358**	**0.0942**	**0.1558**	**0.1848**	**0.2728**	**0.3861**
Model	***F1@K***						
	K = 1	K = 2	K = 5	K = 10	K = 20	K = 50	K = 100
MKR	0.0176	0.0330	0.0410	0.0467	0.0308	0.0218	0.0158
KGCN	0.0215	0.0345	0.0383	0.0418	0.0380	0.0255	0.0185
RippleNet	0.0154	0.0312	0.0351	0.0345	0.0300	0.0221	0.0160
GRE-NR	0.0190	0.0294	0.0330	0.0308	0.0259	0.0200	0.0139
GRE	**0.0267**	**0.0394**	**0.0495**	**0.0484**	**0.0401**	**0.0278**	**0.0192**

has achieved best performance on ***ACC*** for Last.FM, it is worse than GRE on other metrics (***AUC***, ***F1***) over two datasets.

For Top-k Recommendation. As illustrated in Table 4 and Table 5, the results imply that GRE has also achieved superior performance in the majority of the cases in top-k recommendation task. For example, GRE outperforms the baselines on ***Precision@K*** for MovieLens-1M when k is greater than 1. For Last.FM, GRE outperforms the baselines on ***Recall@K*** and ***F1@K*** when k is greater than 1. In a word, experiment results indicates that GRE outperforms the benchmark methods in most cases of k over two open-source datasets.

From the above observations, we show that GRE has obtained state-of-the-art performance in the majority of cases in both two scenarios, compared with other all baselines.

(2) Comparison with GRE-NR (Q2)

For CTR Prediction. The results observed from Table 3 indicate that GRE is more superior than GRE-NR on all metrics for CTR prediction over two datasets. Specifically, the performance of GRE has been improved by 0.4%, 0.4%, and 0.32% on ***AUC***, ***ACC***, and ***F1*** for MovieLens-1M, respectively. For Last.FM, GRE increases the score of ***AUC***, ***ACC***, and ***F1*** by 0.37%, 0.7%, and 0.15%.

For Top-k Recommendation. As illustrated in Table 4 and Table 5, the results indicate that the proposed GRE has achieved better performance on all metrics for top-k recommendation over two datasets when k is not smaller than 2. We intuitively explain this observation with the reason that embedding relation information in GRE has an implicit positive effect on improving robustness and expressive ability of our model. That's why lager k comes better performance of GRE than GRE-NR.

From the above observations, the proposed relation embedding method contributes to GRE because of the embedded relation representation.

(3) Summary

Based on the above comparisons, we conclude that GRE has achieved outstanding performance in the majority cases of two benchmark datasets in two typical recommendation tasks. Additionally, we have proved the positive effect of the proposed novel relation embedding method for GRE through comparative experiments.

5 Conclusion

In this work, we propose GRE, a neural model that aims at enriching user's preference representation in recommender systems by utilizing GAT to perform aggregating the embeddings of adjacent entities to head entity linked by user's historical interaction records and embedding the representation of relation in the process of polymerization of *Triple Groups* in *Triple Set*. This model solves the problem that GAT-based methods only consider aggregating neighboring entities and ignore the embedding of relation information in triples by combining GAT-based information propagation with relation embedding. We show the superiority of GRE and positive effect of our proposed relation embedding method for GRE by experiments. Compared with the baselines, the experiment results indicate that the proposed GRE has gained better performance.

Acknowledgements. This work was supported by the National Key Research and Development Plan of China (No. 2018YFB1003804).

References

1. Bing Satori homepage. https://searchengineland.com/library/bing/bing-satori. Accessed 17 Feb 2020
2. DBpedia homepage. https://wiki.dbpedia.org/. Accessed 17 Feb 2020
3. Freebase homepage. http://www.freebase.be/. Accessed 17 Feb 2020
4. Last.fm homepage. https://grouplens.org/datasets/hetrec-2011/. Accessed 17 Feb 2020
5. MovieLens-1M homepage. https://grouplens.org/datasets/movielens/1m/. Accessed 4 Apr 2020
6. Wikidata homepage. https://www.wikipedia.org/. Accessed 17 Feb 2020
7. Acosta, M., Zaveri, A., Simperl, E., Kontokostas, D., Auer, S., Lehmann, J.: Crowdsourcing linked data quality assessment, pp. 260–276 (2013). https://doi.org/10.1007/978-3-642-41338-4_17
8. Bordes, A., Usunier, N., García-Durán, A., Weston, J., Yakhnenko, O.: Translating embeddings for modeling multi-relational data. In: Burges, C.J.C., Bottou, L., Ghahramani, Z., Weinberger, K.Q. (eds.) Advances in Neural Information Processing Systems 26: 27th Annual Conference on Neural Information Processing Systems 2013, Lake Tahoe, Nevada, United States, 5–8 December 2013, pp. 2787–2795 (2013). http://papers.nips.cc/paper/5071-translating-embeddings-for-modeling-multi-relational-data
9. Defferrard, M., Bresson, X., Vandergheynst, P.: Convolutional neural networks on graphs with fast localized spectral filtering. In: Lee, D.D., Sugiyama, M., von Luxburg, U., Guyon, I., Garnett, R. (eds.) Advances in Neural Information Processing Systems 29: Annual Conference on Neural Information Processing Systems 2016, Barcelona, Spain, 5–10 December 2016, pp. 3837–3845 (2016). http://papers.nips.cc/paper/6081-convolutional-neural-networks-on-graphs-with-fast-localized-spectral-filtering
10. Heitmann, B., Hayes, C.: Using linked data to build open, collaborative recommender systems. In: Linked Data Meets Artificial Intelligence, Papers from the 2010 AAAI Spring Symposium, Technical Report SS-10-07, Stanford, California, USA, 22–24 March 2010. AAAI (2010). http://www.aaai.org/ocs/index.php/SSS/SSS10/paper/view/1067
11. Jin, X., et al.: Explicit state tracking with semi-supervision for neural dialogue generation, pp. 1403–1412 (2018). https://doi.org/10.1145/3269206.3271683
12. Kingma, D.P., Ba, J.: Adam: a method for stochastic optimization. In: Bengio, Y., LeCun, Y. (eds.) 3rd International Conference on Learning Representations, ICLR 2015, Conference Track Proceedings, San Diego, CA, USA, 7–9 May 2015 (2015). http://arxiv.org/abs/1412.6980

13. Kipf, T.N., Welling, M.: Semi-supervised classification with graph convolutional networks. In: 5th International Conference on Learning Representations, ICLR 2017, Conference Track Proceedings, Toulon, France, 24–26 April 2017. OpenReview.net (2017). https://openreview.net/forum?id=SJU4ayYgl
14. Koren, Y., Bell, R.M., Volinsky, C.: Matrix factorization techniques for recommender systems. IEEE Comput. **42**(8), 30–37 (2009)
15. Krötzsch, M., Marx, M., Ozaki, A., Thost, V.: Attributed description logics: reasoning on knowledge graphs. In: Lang, J. (ed.) Proceedings of the 27th International Joint Conference on Artificial Intelligence, IJCAI 2018, Stockholm, Sweden, 13–19 July 2018, pp. 5309–5313. ijcai.org (2018). https://doi.org/10.24963/ijcai.2018/743
16. Lei, W., Jin, X., Kan, M., Ren, Z., He, X., Yin, D.: Sequicity: simplifying task-oriented dialogue systems with single sequence-to-sequence architectures, pp. 1437–1447 (2018). https://doi.org/10.18653/v1/P18-1133. https://www.aclweb.org/anthology/P18-1133/
17. Lin, Y., Liu, Z., Sun, M., Liu, Y., Zhu, X.: Learning entity and relation embeddings for knowledge graph completion. In: Bonet, B., Koenig, S. (eds.) Proceedings of the 29th AAAI Conference on Artificial Intelligence, Austin, Texas, USA, 25–30 January 2015, pp. 2181–2187. AAAI Press (2015). http://www.aaai.org/ocs/index.php/AAAI/AAAI15/paper/view/9571
18. Lops, P., Jannach, D., Musto, C., Bogers, T., Koolen, M.: Trends in content-based recommendation - preface to the special issue on recommender systems based on rich item descriptions. User Model. User Adapt. Interact. **29**(2), 239–249 (2019). https://doi.org/10.1007/s11257-019-09231-w
19. Niemann, K., Wolpers, M.: A new collaborative filtering approach for increasing the aggregate diversity of recommender systems. In: Dhillon, I.S., et al. (eds.) The 19th ACM SIGKDD International Conference on Knowledge Discovery and Data Mining, KDD 2013, Chicago, IL, USA, 11–14 August 2013, pp. 955–963. ACM (2013). https://doi.org/10.1145/2487575.2487656
20. Niepert, M., Ahmed, M., Kutzkov, K.: Learning convolutional neural networks for graphs. CoRR abs/1605.05273 (2016). http://arxiv.org/abs/1605.05273
21. Noia, T.D., Mirizzi, R., Ostuni, V.C., Romito, D., Zanker, M.: Linked open data to support content-based recommender systems. In: Presutti, V., Pinto, H.S. (eds.) 8th International Conference on Semantic Systems, I-SEMANTICS 2012, Graz, Austria, 5–7 September 2012, pp. 1–8. ACM (2012). https://doi.org/10.1145/2362499.2362501
22. Resnick, P., Varian, H.R.: Recommender systems - introduction to the special section. Commun. ACM **40**(3), 56–58 (1997). https://doi.org/10.1145/245108.245121
23. Velickovic, P., Cucurull, G., Casanova, A., Romero, A., Liò, P., Bengio, Y.: Graph attention networks. CoRR abs/1710.10903 (2017). http://arxiv.org/abs/1710.10903
24. Wang, H., et al: RippleNet: propagating user preferences on the knowledge graph for recommender systems. In: Cuzzocrea, A., et al. (eds.) Proceedings of the 27th ACM International Conference on Information and Knowledge Management, Torino, Italy, CIKM 2018, 22–26 October 2018, pp. 417–426. ACM (2018). https://doi.org/10.1145/3269206.3271739

25. Wang, H., Zhang, F., Xie, X., Guo, M.: DKN: deep knowledge-aware network for news recommendation. In: Champin, P., Gandon, F.L., Lalmas, M., Ipeirotis, P.G. (eds.) Proceedings of the 2018 World Wide Web Conference on World Wide Web, WWW 2018, Lyon, France, 23–27 April 2018, pp. 1835–1844. ACM (2018). https://doi.org/10.1145/3178876.3186175
26. Wang, H., Zhang, F., Zhao, M., Li, W., Xie, X., Guo, M.: Multi-task feature learning for knowledge graph enhanced recommendation. CoRR abs/1901.08907 (2019). http://arxiv.org/abs/1901.08907
27. Wang, H., Zhao, M., Xie, X., Li, W., Guo, M.: Knowledge graph convolutional networks for recommender systems. In: Liu, L., et al. (eds.) The World Wide Web Conference, WWW 2019, 13–17 May 2019, San Francisco, CA, USA, pp. 3307–3313. ACM (2019). https://doi.org/10.1145/3308558.3313417
28. Wang, X., He, X., Cao, Y., Liu, M., Chua, T.: KGAT: knowledge graph attention network for recommendation. In: Teredesai, A., Kumar, V., Li, Y., Rosales, R., Terzi, E., Karypis, G. (eds.) Proceedings of the 25th ACM SIGKDD International Conference on Knowledge Discovery & Data Mining, KDD 2019, Anchorage, AK, USA, 4–8 August 2019, pp. 950–958. ACM (2019). https://doi.org/10.1145/3292500.3330989
29. Wang, X., Wang, D., Xu, C., He, X., Cao, Y., Chua, T.: Explainable reasoning over knowledge graphs for recommendation, pp. 5329–5336 (2019). https://doi.org/10.1609/aaai.v33i01.33015329
30. Yu, X., et al.: Personalized entity recommendation: a heterogeneous information network approach. In: Carterette, B., Diaz, F., Castillo, C., Metzler, D. (eds.) 7th ACM International Conference on Web Search and Data Mining, WSDM 2014, New York, NY, USA, 24–28 February 2014, pp. 283–292. ACM (2014). https://doi.org/10.1145/2556195.2556259

Locating Hidden Sources in Evolutionary Games Based on Fuzzy Cognitive Map

Kai Wu[1], Xiangyi Teng[2](✉), and Jing Liu[2]

[1] School of Artificial Intelligence, Xidian University, Xi'an, China
kwu@xidian.edu.cn
[2] Guangzhou Institute of Technology, Xidian University, Guangzhou, China
tengxiangyi@xidian.edu.cn

Abstract. How to identify the enemy's hidden military power based on limited information is a great challenge in a military confrontation. A military confrontation environment can be naturally modeled as a complex system. Fuzzy cognitive map inherits the main characteristics of fuzzy logic and neural network. Thus, it is widely used to model complex systems and get a weighted directed network from existing data. In terms of great succuss of fuzzy cognitive map for modeling and analyzing complex systems, a hidden node localization strategy is proposed. This algorithm can measure the anomalies between fuzzy cognitive maps obtained from different data segments. The experimental results showed the our approach could effectively identify the enemy's hidden military power from the observed data. In several case studies, the influence of various parameters on the accuracy of positioning is analyzed through experiments. The framework for detecting hidden nodes is expected to be successfully applied in many fields.

Keywords: Fuzzy cognitive maps · Military game theory · Hidden source location · Time series

1 Introduction

The great magic of natural science is that it can predict the existence of things that experiments or observations cannot directly inform. We can formulate hidden sources prediction problem as follows. Consider a system whose topology is entirely unknown but whose nodes contain two parts: the first part of the nodes is accessible and the other part is inaccessible from the outside world. We can observe the accessible nodes whose states are available. Since the unreachable nodes are shielded from the outside, we cannot know their information. The question is, can we infer the existence and location of hidden nodes according to those accessible nodes? Solutions to this problem have potential applications in the field of military confrontation [19]. For example, it is a vital task in military confrontation to discover and destroy the enemy's important command sites or core killing weapons. However, these important nodes may be hidden where direct information about them cannot be obtained. We can understand that these nodes are likely to contact and operate through other members, and these members can often

Y. Sun et al. (Eds.): ChineseCSCW 2021, CCIS 1492, pp. 92–106, 2022.
https://doi.org/10.1007/978-981-19-4549-6_8

be detected or located by certain means. Therefore, this paper will focus more on how to identify the enemy's hidden military forces based on limited information.

There are many works in detecting hidden sources. Su et al. [1] proposed a compressive sensing-based method to detect hidden agents in evolutionary game (EG) models. However, this approach needs a large amount of prior knowledge of the EG model. To cope with a noisy environment, then, she developed a robust method [2] to locate hidden nodes. Moreover, the method to locate hidden nodes in geospatial networks was also proposed [3]. Shen et al. [4] proposed a method according to compressed sensing to locate hidden sources in propagation networks. These are also some works to locate sources with incomplete information [5, 6]. However, to achieve the high accuracy of the hidden source location, there are at least two problems to be solved. First, since hidden nodes are not directly observable, most of the time they act as some kind of "black box". Thus, how can we model this system to have insight into its internal mechanism, for example, in the form of networked structure? The above approaches are based on the assumption that there is prior knowledge of complex systems. For example, Ref. [1] assumes that each agent's strategy and the payoff are available. However, in real life, there are some difficulties in accessing the agents' strategies. Second, suppose that there is an effective tool to comprehend the system, namely, we can know the basic structure based on the network model. How can we locate the hidden nodes which are incomparable?

In this paper, we employ fuzzy cognitive maps (FCMs) [7, 9, 18, 20] to solve the first question. FCMs are a graph model that visualizes expert knowledge as weighted directed graphs. FCMs can act as a practical understanding tool for modeling and understanding complex systems. FCMs have been applied in predicting time series [8] due to their advantages in terms of abstraction, adaptability, flexibility, and fuzzy reasoning. Based on this fact, we employ this tool to model complex systems.

We then propose the following strategy to cope with the second question. The basic idea of detecting hidden nodes relies on missing information from the hidden node when trying to model complex systems. Since we do not have any information about the hidden source, the learned edge patterns of the neighboring nodes in FCMs will be inaccurate and anomalous. We may detect the neighborhood of the hidden nodes by discovering any abnormal link pattern in different data segments. The identified anomalies can then be employed to locate all nearest neighbors of the hidden source, which in turn imply the existence and position of the hidden source in complex systems.

The performances of our proposal have been validated in three real-life cases. The experimental results demonstrate that our proposal can locate hidden nodes with high accuracy. We also analyze the effect of various parameters on the performance of the proposed methods for locating hidden nodes.

2 Background

2.1 Fuzzy Cognitive Maps

Generally, an FCM is termed as a signed fuzzy graph with N concepts. The concepts represent real-world ideas, and weighted edges stand for the relations between concepts. The state values of these concepts are denoted as a vector

$$\boldsymbol{C} = [C_1, C_2, C_3, \ldots, C_N] \tag{1}$$

where C_i is in the range of [0, 1], $i = 1, 2, 3, \ldots, N$. We can establish the relationships among concepts by experts' knowledge or historical data. This paper employs a weight matrix W with the size of N dimension to define these relationships

$$W = \begin{bmatrix} w_{11} & w_{12} & \cdots & w_{1N} \\ w_{21} & w_{22} & \cdots & w_{2N} \\ \vdots & \vdots & \ddots & \vdots \\ w_{N1} & w_{N2} & \cdots & w_{NN} \end{bmatrix} \tag{2}$$

where $w_{ij} \in [-1, 1]$, and represents how much concept i affects concept j. Figure 1 shows a simple example FCM with four concepts.

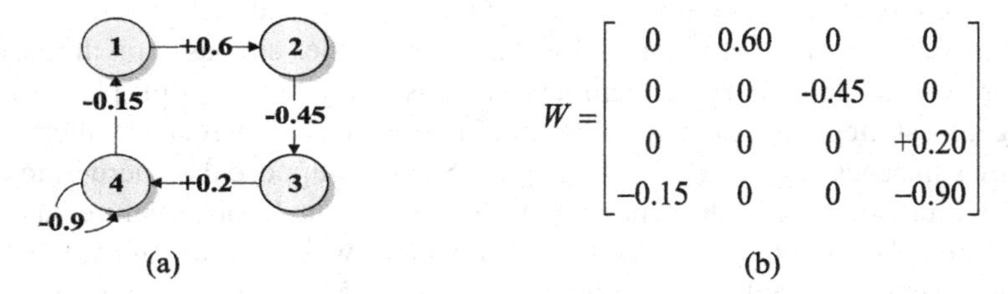

Fig. 1. A simple example of FCM.

The density of FCM with N nodes is defined as the ratio of the number of node connections and $N \times N$. The state value at the $t + 1$ iteration is influenced by the state values of connected concepts at the tth iteration and the weight matrix. To obtain the next iteration, we can employ the evolving Eq. (3),

$$C_i(t+1) = \psi\left(\sum_{j=1}^{N} w_{ji} C_j(t)\right) \tag{3}$$

where ψ is a transfer function that compresses the activation degree to the range of [0, 1] and $C_i(t)$ is the state value of concept i at the tth iteration. C is passed repeatedly through the FCM weight matrix W in order to evolve the complex system. We show an example of time series D with N nodes as follows,

$$D = \begin{bmatrix} C_1(1) & C_2(1) & \cdots & C_N(1) \\ C_1(2) & C_2(2) & \cdots & C_N(2) \\ \vdots & \vdots & \ddots & \vdots \\ C_1(M) & C_2(M) & \cdots & C_N(M) \end{bmatrix} \tag{4}$$

Various transfer functions can be employed in this paper. Based on the comparison study in [20], we find that the sigmoid function exceeds the others in general. Thus, this paper employs the sigmoid function,

$$\psi(x) = \frac{1}{1 + e^{-gx}} \tag{5}$$

where $g \in [0, 20]$ served as a parameter for characterizing how steep the function around zero.

2.2 EG Model

Evolutionary games [10, 16] model a common type of link in many complex, networked, social, and natural systems. In an EG, at any time, one agent has to choose one of the strategies (S): cooperation (C) or defection (D). 2×2 payoff matrices express the payoffs of the two agents in a game in agreement with the four possibilities. For example, in the prisoner's-dilemma game (PDG) [11], the payoff matrices are

$$P_{PDG} = \begin{pmatrix} R & S \\ Te & Pu \end{pmatrix} \tag{6}$$

where the agents obtain rewards $Pu(R)$ if both select to defect (cooperate) and the defector's and cooperator's payoff is Te (the temptation to defect) and S (sucker's payoff) if both choose different strategies. According to [12], $R = 1$, $Te = 1.2$ and $Pu = S = 0$.

We can characterize the links among agents in the network by an $N \times N$ adjacency matrix X with degree $< k >$. $x_{ij} = 1$ if agents i and j are linked, and $x_{ij} = 0$ otherwise. In tth round, all agents play the game with their neighbors and gain payoffs. For agent i, its payoff is

$$Y_i(t) = \sum_{l=1}^{N} x_{il} S_i^{\mathrm{T}}(t) P S_l(t) \tag{7}$$

where S_i represents the strategies of agent i at the time, the sum is over the neighbor-connection set Γ_i of agent i, and T stands for "transpose".

We describe the numerical simulation of EG as follows. Firstly, we set a fraction of agents to select the defection strategy and the remaining agents to select the cooperation strategy. We update nodal states in parallel. For agent i of degree $\langle k \rangle$, at round t, its payoff is calculated using Eq. (7). The strategy of agent i is updated using Eq. (8) in order to maximize its payoff. We term a Monte Carlo round t as the case where all the states at $t + 1$ were updated based on their states at t. After a round of the game, we use the Fermi rule [11] to update the strategy of agent i on the basis of its own and its neighbors' payoffs. We try to maximize the payoff of agent i in the next round. Fermi rule can be expressed as:

$$W\left(S_i \leftarrow S_j\right) = \frac{1}{1 + \exp\left[\left(Y_i - Y_j\right)/\kappa\right]} \tag{8}$$

where κ characterizes the stochastic uncertainties designed to allow unreasonable selection. In this paper, $\kappa = 0.1$.

3 Locating Hidden Agents in FCMs

3.1 Overall Design Process

The basic idea of our proposal is graphically illustrated in Fig. 2, which describes the procedure of detecting the existence and locations of hidden sources in one complex system. First, the time series of accessible nodes are obtained by simulating complex systems. The available data are normalized to the range $[-1, 1]$. Second, the time series is divided into different data segments, and each data segment is used to learn one FCM structure. Then, a strategy is employed to measure anomalies among these learned FCMs. Finally, we decide whether hidden nodes exist. The details of locating hidden nodes are summarized in the following subsections.

3.2 Learning FCMs from Time Series

This section establishes the following optimization objectives:

$$\min_{W_i}\left(\frac{1}{M}\|\Phi W_i - Y_i\|_2^2 + \lambda\|W_i\|_1\right) \tag{9}$$

where λ is a non-negative regularization parameter, $M = S \times (T - 1)$, Y_i, Φ, and W_i is shown as follows.

$$Y_i = \begin{bmatrix} \psi^{-1}(C_i(2)) \\ \psi^{-1}(C_i(3)) \\ \vdots \\ \psi^{-1}(C_i(M)) \end{bmatrix} \tag{10}$$

$$\Phi = \begin{bmatrix} C_1(1) & C_2(1) & \cdots & C_N(1) \\ C_1(2) & C_2(2) & \cdots & C_N(2) \\ \vdots & \vdots & \ddots & \vdots \\ C_1(M-1) & C_2(M-1) & \cdots & C_N(M-1) \end{bmatrix} \tag{11}$$

$$W_i = \begin{bmatrix} w_{1i} & w_{2i} & \cdots & w_{Ni} \end{bmatrix}^{\mathrm{T}} \tag{12}$$

With the increase of λ, the number of nonzero components in W_i decreases. The penalty of the form $\lambda\|W_i\|_1$ can be transformed into a condition where $\|W_i\|_1 \leq t$, in which $t \geq 0$ is a tuning parameter. $\|W_i\|_1$ in the lasso can ensure the sparsity of the solutions. Meanwhile, the least square term $\|Y_i - \Phi W_i\|_2{}^2$ makes the solution more robust against noise than L_1-norm-based optimization algorithms.

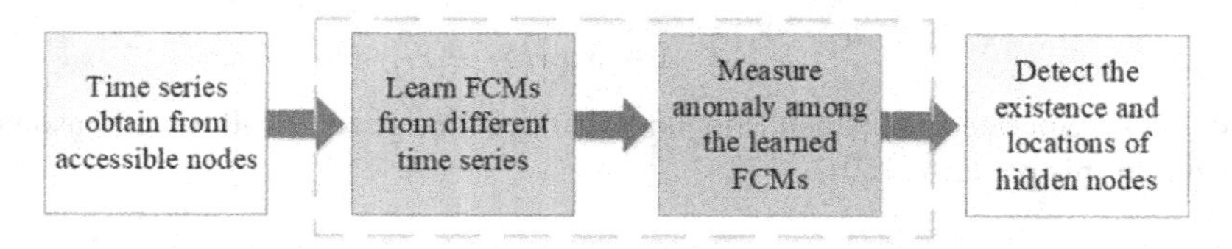

Fig. 2. Graphical illustration of our framework for LHN.

The problem of learning FCMs is decomposed into the task of learning local connections of nodes individually. Due to the sparseness of FCMs, we convert the problem of recovering local structures from time series into sparse signal reconstruction problems solved by the LASSO [13, 20]. We describe an example of learning FCMs with three nodes. First, for node i, a sub-network is constructed from the FCM. This sub-network contains node i and its neighbors. Therefore, we divide the FCMs learning problem into three sub-problems where each one is to construct one sub-network. Each sub-problem can be modeled as a signal reconstruction problem that involves both the differences between the original and predicted data and the sparse structure of FCM. The LASSO then optimizes each sub-problem. For example, we adopt the LASSO to learn the sub-network from nodes 1, 2, and 3 to node 3. We then obtain the link from node 2 to node 3 with $w_{23} = 0.28$. Finally, after learning all sub-networks, we compose the local connections into the whole FCM.

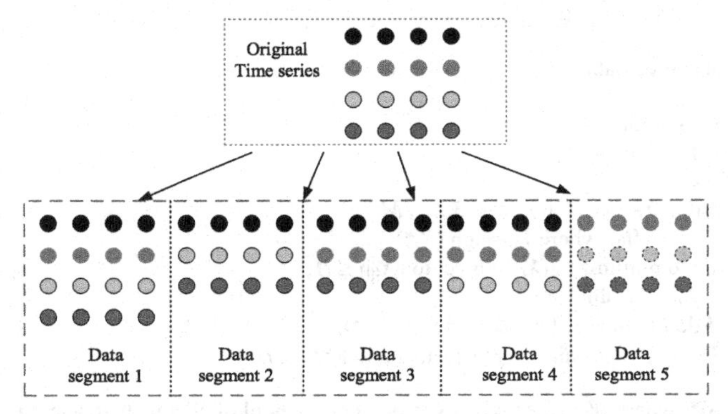

Fig. 3. Graphical illustration of assigning data segments for the original time series.

Let $g(X) = ||Y_i - \Phi W_i||_2$, and let $\eta_i,\ i = 1, 2,\ldots, 2^N$ be the N-tuples of the form $(\pm1, \pm1,\ldots, \pm1)$. The condition $||W_i||_1 \leq t$ is then equivalent to $\eta_i{}^T W_i \leq t$ for all i. For a given X_i, let $E = \{i\text{: } \eta_i{}^T W_i = t\}$ and $S = \{i\text{: } \eta_i{}^T W_i < t\}$. S is the slack set where those constraints for which equality does not hold. The set E is the equality set where those exactly met constraints. Denote by G_E the matrix whose rows are η_i for $i \in E$. Let **1** be a vector of ones of length equal to the number of rows of G_E. The algorithm below starts with $E = \{i_0\}$, where $\eta_{i0} = \text{sign}\ (W_i{*})$ with $X_i{*}$ being the overall least squares estimated. The LASSO optimizes the least-squares problem subject to $\eta_{i0}{}^T W_i \leq t$ and then checks whether $||W_i||_1 \leq t$. If the condition is met, the algorithm stops; if not, we add the violated constraint to the set E, and the process runs until $||W_i||_1 \leq t$. The details of the lasso are described in [13]. We choose the best value of λ by conducting cross-validation. We repeat the above process N times for different nodes. We summarize the process of FCM learning in Algorithm 1.

3.3 Model EG Using FCM

To provide necessary basic information for the abnormal measurement stage, this section divides Y into data, as shown in Fig. 3. The first step is to decompose the available data into $M + 1$ data segments based on the strategy in Fig. 3. Then, the FCM model is established for each data segment. After obtaining the FCMs, the outliers are calculated according to the formula described in Sect. 2.4. For example, we achieve FCM-1 for data segment 1. Then, {FCM-1, ..., FCM-k, ..., FCM-$M + 1$} are adopted to calculate the anomalies σ among them. Then, we can obtain the anomalies for each agent. We get $\sigma_1 = 0.4$ for agent 1, $\sigma_2 = 0.5$ for agent 2, and $\sigma_3 = \sigma_3 = 0$ for agents 3 and 4. The anomalies for agents 1 and 2 are much greater than for agents 3 and 4. Thus, we can consider that the hidden agent links to agents 1 and 2 and does not link to agents 3 and 4. Finally, we locate the position of the hidden agent 5 and its neighbors, agents 1 and 2. Moreover, if $\sigma_1 = \sigma_2 = \sigma_3 = \sigma_4 = 0$, we can ensure that none of agents connect hidden agent.

Algorithm 1 FCM Learning

Input:
  D: the observed data;
Output:
  W: the weight matrix;
$i \leftarrow 1$, process D to obtain Φ;
while ($i \leq N$) **do**
  For agent i, $Y_i \leftarrow \psi^{-1}(C_i(t))$, t=2, 3, ..., M;
  Start with E={i_0} where η_{i0}=sign (X_i*);
  Find X_i* to minimize $g(X_i)$ subject to $\|X_i\|_1 \leq t\mathbf{1}$;
  while ($\|X_i\|_1 > t$) **do**
    Add i to the set E where η_i=sign (X_i*);
    Find X_i* to minimize $g(X_i)$ subject to $\|X_i\|_1 \leq t\mathbf{1}$;
  end
  $W_i \leftarrow X_i$*, where W_i is the weight vector from the pool of all agents to agent i;
  $i \leftarrow i + 1$;
end

3.4 Measuring Anomaly

In this paper, we consider the case of one hidden agent. Owing to the lack of available data from the hidden agent, the right FCM could not be achieved for this agent. However, time series from other agents are available. The motivation for measuring anomaly is that if an agent does not connect with a hidden agent, the same FCM for such an agent can be obtained with different time series from itself and all its direct neighbors. If the results of inferring an agent's links vary significantly for different data segments, the agent is deemed to be connected with the hidden agent. The measurement of the anomaly can be accomplished by using different data segments. The following equation of the predicted results concerning different FCMs obtained from data segments can be used as a quantitative measure for the anomaly.

$$\sigma_i = \frac{1}{N}\sum_{i=1}^{N}\sqrt{\frac{1}{M+1}\sum_{k=1}^{M+1}\left(w_i^{(k)} - \langle W_i \rangle\right)^2} \tag{13}$$

where ${W_i}^{(k)}$ represents the weight vector from all agents to agent i inferred from the kth data segments, $\langle W_i \rangle = (1/(M+1)) \sum_{k=1}^{M+1} W_i^{(k)}$ is the mean value of ${W_i}^{(k)}$ and $M + 1$ is the number of data segments. Then, we prove the effectiveness of the proposed strategy, which is shown in Assumption 1 and Theorem 1.

Assumption 1. *Given enough time series D, the algorithm CS-FCM can fully learn the weighted matrix of FCM.*

Theorem 1. *For any neighbor agent of hidden agents, the value of* σ *is usually much larger than those of agents that are not in the immediate neighborhood of hidden agents.*
Proof. Case 1: The hidden agent does not link to agent i.

The weight vector ${W_i}^{(k)}$ needs to be learned from data segment k. According to Assumption 1, if each data segment is enough, each weight vector from different data segments is the same ($W_i^{(1)} = \cdots W_i^{(k)} = \cdots W_i^{(M+1)}$). Namely, $\sigma_i = 0$.

Case 2: The hidden agent links to agent i. We assume that $w_{Hi} \neq 0$, where w_{Hi} represents the weight from hidden agent to agent i inferred from the kth data segments. If we want to fully learn the weight matrix ${W_i}^{(1)}$ from data segment 1, the following equation needs to be solved by CS-FCM.

$$\begin{bmatrix} \psi^{-1}(C_i(2)) \\ \vdots \\ \psi^{-1}(C_i(M)) \end{bmatrix} = \begin{bmatrix} C_1(1) & \cdots & C_N(1) & C_H(1) \\ \vdots & \ddots & \vdots & \vdots \\ C_1(M-1) & \cdots & C_N(M-1) & C_H(M-1) \end{bmatrix} \times \begin{bmatrix} w_{1i} \\ \vdots \\ w_{Ni} \\ w_{Hi} \end{bmatrix} \tag{14}$$

where $C_H(t)$ is the activation degree of the hidden agent. However, in the real situation, $C_H(t)$ is unknown, and Eq. (6) is used to optimize ${W_i}^{(1)}$. Thus, it is difficult to obtain the exact ${W_i}^{(1)}$ due to $C_H(t)w_{Hi} \neq 0$. This case can be extended to other data segments. With different data segments, $W_i^{(1)} = \cdots W_i^{(k)} = \cdots W_i^{(M+1)}$ is difficult to fulfill, which leads to $\sigma_i > 0$. For any neighbor agent of hidden agents, the value of σ is usually much larger than those of agents that are not in the immediate neighborhood of hidden agents. The value of σ can thus be used to reliably identify the neighboring agents of hidden agents. Thus, Theorem 1 has been proved.

4 Experiment

4.1 Performance Measures

To locate a hidden agent, an important way is to detect its neighbors. In this paper, we use *SS_Mean* to measure the quality of the location of hidden agents. We calculate the measure *SS_Mean* as follows,

$$SS\_Mean = \frac{2 \times Specificity \times Sensitivity}{Specificity + Sensitivity} \tag{15}$$

$$Specificity = \frac{TP}{TP + FN} \tag{16}$$

$$Sensitivity = \frac{TN}{TN + FP} \tag{17}$$

where *TP*, *FN*, *TN*, and *FP* are defined in Table 1. *SS_Mean* $\in [0, 1]$, and the greater the value is, the better the algorithm's performance is.

Table 1. The definition of *TP*, *FP*, *FN*, and *TN*.

		Input FCM	
		0	1
Candidate FCM	0	*TP*	*FP*
	1	*FN*	*TN*

4.2 Case 1: Brazilian Amazon Example

The first case is about deforestation in the Brazilian Amazon [14]. The map mentions the 12 most concepts, and the influence graph is shown in Fig. 4(a). In the following, we discuss an example for locating a hidden agent. We first assume that concept C1 is a hidden agent due to its neighbors, and its information is unknown from the outside. However, the dynamic states of agents C2–C12 are known. The length of the time series of accessible agents is set to $M = 12$. The method in [14] is used to generate time series with various scales and properties. First, we randomly generate the initial state value of each agent from range [0, 1]. Then, based on Eq. (2), we obtain the response sequences. Based on the proposed strategy for locating hidden agents, we first divide the original time series into $M + 1$ data segments, as shown in Sect. 2.3. For data segment *i*, we predict the neighbors of all agents and construct one FCM. The predicted links of direct neighbors of hidden sources display anomalies. To evaluate the anomalies, we calculate the structural variance σ of each agent from different data segments using Eq. (13), which is demonstrated in Fig. 4(b).

In Fig. 4(b), we can observe that only the values of σ of agents C2, C6, C7, C8, and C10 are greater than other agents. According to Theorem 1, agents C2, C6, C7, C8, and C10 are found to link to hidden agent C1. From Fig. 5(a), agent C1 connects agents C2, C6, C7, C8, and C10. Thus, we can claim that we find the existence of hidden agent C1 and locate its neighbors, agents C2, C6, C7, C8, and C10. Then, we also assume that agent C2 is hidden and other agents are accessible. The results of the anomalies of the accessible agent are shown in Fig. 4(c). The two neighboring agents of agent C2 display much more significant values σ than those from the other agents, which provides the evidence that they might be the neighborhood of the hidden agent.

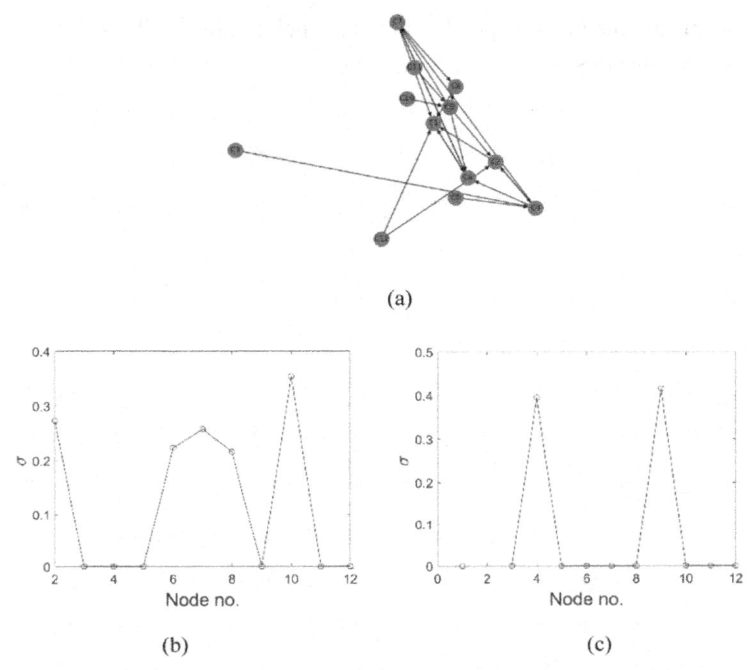

Fig. 4. Detecting a hidden agent. (a) The original FCM for dataset about deforestation in the Brazilian Amazon. (b) Structural variances of each agent. The time series of other agents except the source is assumed to be available. C1 is treated as a hidden agent. (c) C2 is treated as a hidden agent.

4.3 Case 2: Supervisory Control Systems

The second case is an FCM for supervisory control systems [15]. The concepts of an FCM are defined as features and states of this system, including the height of the liquid in each tank or the temperature. In the case, we employ eight concepts to form a model. The relationships among them are shown in Fig. 6(a). More details of supervisory control systems can be found in [15]. We first assume that agent C1 is a hidden agent, and we do not know any information about agent C1. However, the information about agents C2-C8 is known. The length of the time series of accessible agents is set to $M = 8$.

In Fig. 5(b), we find that only the values of σ of agents C2 and C3 are greater than zero, and that of other agents is equal to zero. Based on Theorem 1, agents C2 and C3 are found to link to hidden agent C1. From Fig. 5(a), agent C1 connects agents C2 and C3, which experimentally proves that the proposed strategy can exactly locate the position of hidden agent C1. Thus, we can claim that we ensure the existence of hidden agent C1 and locate its neighbors, agents C2 and C3. To enhance the persuasion of this case, we also assume that agent C4 is hidden and other agents are accessible. The results of the anomalies of an accessible agent are shown in Fig. 5(c). The two neighboring agents of agent C4 show much larger values than the other agents, which provides evidence that these two agents are the hidden agent's immediate neighborhood. This result also finds the existence of concept C4.

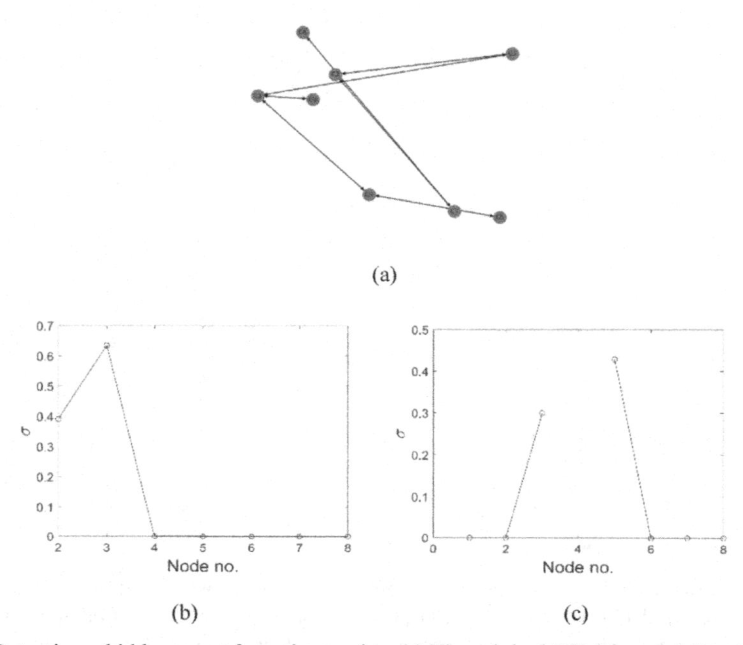

Fig. 5. Detecting a hidden agent from time series. (a) The original FCM for modeling intelligent supervisory control systems. (b) Structural variances of each agent. The time series of other agents except the source is assumed to be available. Agent C1 is treated as a hidden agent. (c) Agent C4 is treated as a hidden agent.

4.4 Case 3: Mobile Payment System Project

Finally, we study the case of the mobile payment system (MPS) project [17]. To increase the chances of the MPS project being perceived as successful by all the parties, FCMs are employed to identify the critical factors influencing that success at the outset of the project. The relationships among them are shown in Fig. 7(a) for this system. We first assume that agent C1 is a hidden agent, and we do not know any information about agent C1. However, the information about agents C2–C24 is known. The length for time series of accessible agents is set to $M = 24$.

In Fig. 6(b), we find that only the values of σ of agents C6 are greater than other agents. According to Theorem 1, agents C6 are found to link to hidden agent C1. From Fig. 6(a), agent C1 connects to agent C6. Thus, we can claim that we ensure the existence of hidden agent C1 and locate its neighbor, agent C6. Then, we also assume that agent C3 is hidden and other agents are accessible. The results of the anomalies of the accessible agent are shown in Fig. 6(c). The five neighboring agents of agent C3 display much more significant values σ than those of the other agents, which provides evidence that

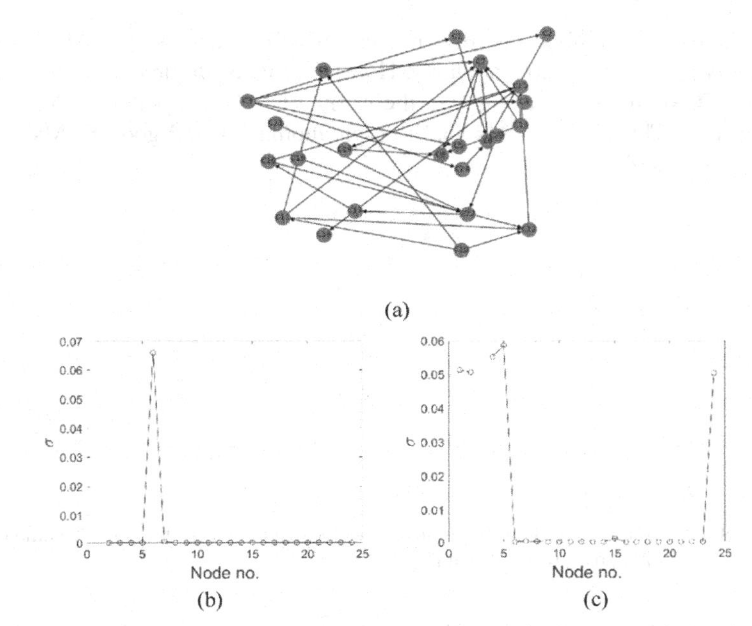

Fig. 6. Detecting a hidden agent from time series. (a) The original FCM for the MPS project. (b) Structural variances of each agent. The time series of other agents except the source is assumed to be available. C1 is treated as a hidden agent. (c) C3 is treated as a hidden agent.

these five agents are the immediate neighborhood of the hidden agent. We also find that σ_6 is slightly greater than 0, but the hidden agent does not connect to concept 15. This phenomenon appears due to the following factor: in many cases, it is difficult to make sure that Assumption 1 is right due to the ability of an optimization algorithm. Thus, we choose the agents whose value σ is far greater than the others. However, in terms of CS-FCM, the rationality of Assumption 1 can be ensured by employing enough time series in many cases.

4.5 Effect of N_M on *LHN*

Figure 7 shows the relationships among *SS_Mean*, f, and relative data length N_M for FCMs with different densities. Here, N is set to 20 and 40, and we set the density to 20% and 40%. We increase N_M from 0.1 to 1.0 in steps of 0.1. To obtain the high confidence of locating hidden agents and decrease the computation cost, the value of N_g can be set to 0.5 in this paper. The result of each point is obtained from ten dynamic realizations and ten configurations of the hidden node. Note that f can be achieved to maximize the value of *SS_Mean*. As seen from Fig. 7(a), as N_M is increased, the value of *SS_Mean* increases accordingly. Moreover, the *SS_Mean* of dense FCMs is worse than those of sparse FCMs. In Fig. 7(b), we discuss the effect of N_M on the best cut-off value of σ in terms of *SS_Mean*. We find that the curves show descending trend along with increasing N_M. At the same time, the value of f for FCMs with density = 20% is smaller than that for FCMs with density = 40%. This phenomenon appears due to the following factor:

If an agent does not neighbor any hidden agent, it will render valid FCM for such an agent, thereby resulting in a low anomaly. However, this is an ideal state which needs high-quality FCM. With increasing N_M, the learned FCMs can better model a certain complex system. Thus, poor FCMs lead to high anomalies, and good FCMs result in low anomalies (Fig. 8).

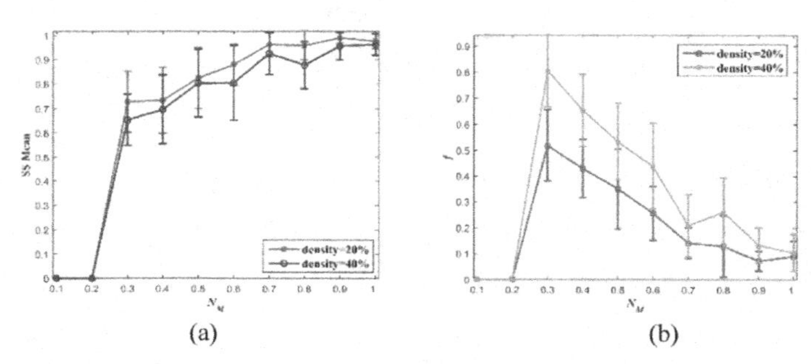

Fig. 7. The effect of the relative length of data on (a) *SS_Mean* and (b) the cut-off value of σ for FCMs size $N = 40$ and densities of 20% and 40%.

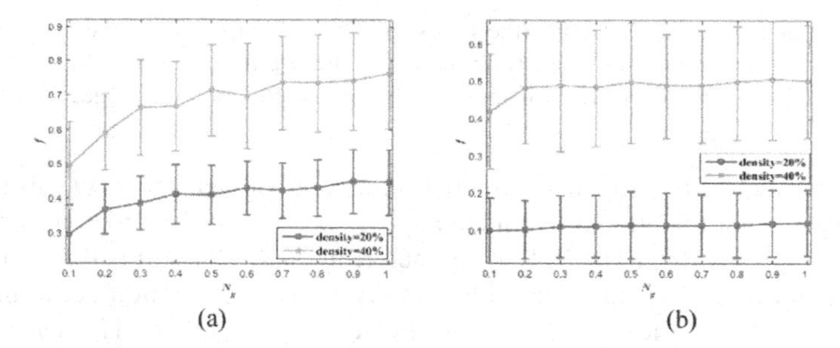

Fig. 8. The effect of g on the tradeoff of σ for FCMs size $N = 40$ and densities of 20% and 40%. (a) $N_M = 0.8$ and (b) $N_M = 1.6$.

5 Conclusions

The main contributions of this paper are as follows:

1) Aiming at the invisibility of complex evolutionary game systems, the FCM is used to model the complex military confrontation environment and generate a weighted directed network by inheriting the advantages of fuzzy logic and neural network.
2) According to the abnormal connection patterns of different data segments in the FCMs, we can identify all the nearest neighbor agents of the hidden game object so as to verify its existence and locate its location in the complex system. Experimental results proved the prososed algorithm could locate the hidden game object accurately.

3) Through several real case studies, the effects of various parameters on positioning accuracy are obtained, and the reasons are analyzed in detail.

Based on the known data of limited accessible agents, this paper effectively infers the existence and location of hidden agents by measuring the anomalies between the generated FCMs and reveals the internal law of hidden sources in complex systems. This strategy and method provide an important reference and application value for estimating enemy command sites and key killing weapons in the field of military confrontation. However, FCM has several limitations at the same time, and a large number of assumptions need to be satisfied to perfectly simulate complex systems. We intend to improve the FCM theory and apply it to more realistic scenes in the next work.

Acknowledgments. This work was supported in part by the Key Project of Science and Technology Innovation 2030, the Ministry of Science and Technology of China under Grant 2018AAA0101302, the General Program of the National Natural Science Foundation of China (NSFC) under Grant 61773300, and the Fundamental Research Funds for the Central Universities under Grant XJS211905.

References

1. Su, R.Q., Wang, W.X., Lai, Y.C.: Detecting hidden nodes in complex networks from time series. Phys. Rev. E **85**, 065201 (2012). (S2470-0045)
2. Su, R.Q., Lai, Y.C., Wang, X., et al.: Uncovering hidden nodes in complex networks in the presence of noise. Sci. Rep. **4**, 3944 (2014). S2045-2322
3. Shen, Z., Wang, W.X., Fan, Y., et al.: Reconstructing propagation networks with natural diversity and identifying hidden sources. Nat. Commun. **5**, 4324 (2014). S2041-1723
4. Su, R.Q., Wang, W.W., Wang, X., et al.: Data based reconstruction of complex geospatial networks, nodal positioning, and detection of hidden node. R. Soc. Open Sci. **1** (2015). S2054-5703
5. Shen, Z., Cao, S., Fan, Y., et al.: Locating the source of diffusion in complex networks by time-reversal backward spreading. Phys. Rev. E **93**, 032301 (2015). S2470-0045
6. Pinto, P.C., Thiran, P., Vetterli, M.: Locating the source of diffusion in large-scale networks. Phys. Rev. Lett. **109**, 068702 (2012). S0031-9007
7. Kosko, B.: Fuzzy cognitive map. Int. J. Man-Mach. Stud. **24**(1), 65–75 (1986). S0020-7373
8. Papageorgiou, E.I., Salmeron, J.L.: A review of fuzzy cognitive maps research during the last decade. IEEE Trans. Fuzzy Syst. **21**(1), 66–79 (2013). S1063–6706
9. Stach, W., Kurgan, L., Pedrycz, W., et al.: Genetic learning of fuzzy cognitive maps. Fuzzy Sets Syst. **153**(3), 371–401 (2005). S0165-0114
10. Nowak, M.A., May, R.M.: Evolutionary games and spatial chaos. Nature **359**, 826–829 (1992). S0028-0836
11. Szabó, G., Tőke, C.: Evolutionary prisoner's dilemma game on a square lattice. Phys. Rev. E **58**, 69 (1998). S2470-0045
12. Szabó, G., Fath, G.: Evolutionary games on graphs. Phys. Rep. **446**, 97–216 (2007). S0370-1573
13. Tibshirani, R.: Regression shrinkage and selection via the lasso. J. R. Stat. Soc. Ser. B (Methodol.), 267–288 (1996). S1369-7412
14. Kok, K.: The potential of fuzzy cognitive maps for semi-quantitative scenario development with an example from Brazil. Glob. Environ. Change **19**(1), 122–133 (2009). S0959-3780

15. Stylios, C.D., Groumpos, P.P.: Fuzzy cognitive maps: a model for intelligent supervisory control systems. Comput. Ind. **39**(3), 229–238 (1999). S0166-3615
16. Sun, Q., Lu, L., Yan, G., et al.: Asymptotic stability of evolutionary equilibrium under imperfect knowledge. Syst. Eng. Theory Pract. **23**(7), 11–16 (2003)
17. Rodriguez-Repiso, L., Setchi, R., Salmeron, J.L.: Modelling IT projects success with fuzzy cognitive maps. Expert Syst. Appl. **32**(2), 543–559 (2007). S0957-4174
18. Gao, X., Xie, W.: Development and applications of fuzzy clustering theory. Chin. Sci. Bull. **44**(021), 2241–2250 (1999)
19. Li, H., Liu, D., Liu, Y.: Architecture design research of military intelligent wargame system. Fire Control Command Control **45**(306), 118–123 (2020)
20. Wu, K.: Intelligent modeling algorithm for complex system and its application. Xidian University (2020)

Deep Bug Triage Model Based on Multi-head Self-attention Mechanism

Xu Yu[1], Fayang Wan[1], Bin Tang[2](✉), Dingjia Zhan[1], Qinglong Peng[1], Miao Yu[3], Zhaozhe Wang[1], and Shuang Cui[1]

[1] College of Information Science and Technology, Qingdao University of Science and Technology, Qingdao 266061, China
[2] College of Shipbuilding Engineering, Harbin Engineering University, Harbin 150001, China
tangbin@hrbeu.edu.cn
[3] College of Textiles and Clothing, Qingdao University, Qingdao 266071, China

Abstract. In the course of software maintenance and development, bugs are inevitable. At present, Bug tracking system uses bug reports to match bug with fixers. However, the previous bug triage model relies too much about the quality of the text of the bug report, introduces a lot of redundant information in natural language, and ignores the fixer community factor where the meta-field of the bug report, which makes the model performance worse. Aiming at the above problems, we propose a multi-head self-attention deep bug triage (MSDBT), which considers the text content of the bug report and generates a sequence of fixers with the same meta-field. The features of the input text and the fixer sequence are extracted by **Bi-directional Long Short-Term Memory network**. The multi-head self-attention mechanism is used to perform parallel attention calculation among the internal input elements. The model weakens the redundant information in the bug report, and further quantifies the influence of fixers with similar activities on bug triage through fixer sequence. We conducted texts on four open source software projects. We can get the MSDBT has clear strength over the previous model in recall index.

Keywords: Bug tracking system · Bug report · Bug triage · Deep learning · Fixer community · Multi-head self-attention

1 Introduction

Bugs are software errors that disrupt the normal operation of the program and make the system function invalid. With the increase in scale and complexity, inevitably, the majority of errors occur through the software development and maintenance process. Therefore, large open source software manages bugs through bug tracking systems (such as Bugzilla and JIRA), which not only realizes the recording, management and status update of bug reports, but also coordinates the work between fixers and records the resolution of bugs the way.

At present, most of the bug reports are processed manually. System managers check the bug reports one by one, and assign appropriate fixers manually according to historical

Y. Sun et al. (Eds.): ChineseCSCW 2021, CCIS 1492, pp. 107–119, 2022.
https://doi.org/10.1007/978-981-19-4549-6_9

experience. This is a labor-intensive assignment for large open source software projects. For example, platform receives approximately 100 new bug reports per day in Eclipse, and at least1800 fixers with different expertise. Faced with such the majority of bugs and fixers, it will consume a lot of time and human resources to assign suitable fixers manually. According to statistics, bugs in Eclipse take an average of 40.3 days from submission to allocation.

To make reasonable application for the human resources of the bug tracking system and enhance the software repair's efficiency, the literature [1] first an automatic error triage technique is proposed which automatically assigns bugs to a group of fixers with the most professional skills by learning historical data. Furthermore, the researchers proposed a series of automatic bug triage methods related to machine learning, information retrieval and deep learning. The machine learning method [2–4] regards the fixer as a category, and assigns the appropriate fixer label to the bug report by training the classifier. The information search method [5–7] takes the bug report as the query and the fixer as the return result, and ranks the fixers according to similarity between newly submitted bug reports and fixers deep learning method [8–10] uses the excellent feature learning ability of the neural network to characterize the data, thereby performing bug triage task.

In view of the current bug triage research, we found that there are still the following shortcomings. First, the previous models rely too much on the text quality of the bug report, and the quality directly affects the result of the bug triage. The bug report is a natural language description filled in by the reporter and records the key information of bug. The fixer needs to reproduce the bug based on the report in a short time and synchronize the reporter's ideas. However, as the user of the software project, the reporter does not have the professional bug recording ability. The redundant description in the bug report is inevitable, which will mislead the software fixer process. Figure 1 shows part of the summary of the bug report ID 485038. According to statistics, the report has 83 lines and 5307 characters. This description increases the workload of bug fixer. Secondly, the bug tracking system, as an open platform, contains multiple different fixer communities that composed of fixers with similar professional skills and development activities. However, the meta-fields (such as products, components, etc.) in the bug report as an important classification label for screening fixer communities are mostly ignored by previous models, and the influence of community factors on bug triage is not considered, resulting in a waste of resources. Figure 2 shows the repair history of fixers, joakim.erdfelt and jesse.mcconnell. We found that two fixers often participate in the bug of the product Jetty. They belong to a common fixer community.

Appears above the problems, the essay offers a deep bug triage method MSDBT based on the multi-head self-attention mechanism, which not only considers the text content of the bug report, generates a sequence of fixers with the same meta-field based on the product and component. The Bi-directional Long Short-Term Memory network is accustomed to excerpt the high-level features in the input text and the fixer series. The multi-head self-attention mechanism is used to perform parallel attention calculations between the internal input elements. MSDBT weakens the redundant information in the bug report, and further quantifies the influence of community factors on bug triage through the sequence of fixers.

```
EPP Error Reports        2015-12-31 04:53:25 EST                     Description
The following incident was reported via the automated error reporting:

    code:                  0
    plugin:                org.eclipse.oomph.gitbash_1.2.0.v20150923-0945
    message:               Git bash not found at null
    fingerprint:           4ebd42f0
    exception class:       java.lang.IllegalStateException
    exception message:     Git bash not found at null
    number of children:    0

    java.lang.IllegalStateException: Git bash not found at null
    at org.eclipse.oomph.gitbash.GitBash.openInputDialog(GitBash.java:133)
    at org.eclipse.oomph.gitbash.GitBash.getExecutable(GitBash.java:56)
    at org.eclipse.oomph.gitbash.GitBash.executeCommand(GitBash.java:74)
    at org.eclipse.oomph.gitbash.repository.GitStatusAction.run(GitStatusAction.java:27)
    at
org.eclipse.oomph.gitbash.repository.AbstractRepositoryAction.run(AbstractRepositoryAction.java:33)
```

Fig. 1. Part of the summary of the bug report ID 485038

ID	Product	Assignee	Status	Resolution	Changed ▲	ID	Product	Assignee	Status	Resolution	Changed ▲
271150	Jetty	jesse.mcconnell	RESO	FIXE	2009-04-10	283344	Jetty	joakim.erdfelt	RESO	FIXE	2009-07-17
271324	Jetty	jesse.mcconnell	RESO	FIXE	2009-04-10	283884	Jetty	joakim.erdfelt	RESO	FIXE	2009-07-17
271927	Jetty	jesse.mcconnell	RESO	FIXE	2009-04-10	284510	Jetty	joakim.erdfelt	RESO	FIXE	2009-07-23
270512	Jetty	jesse.mcconnell	RESO	FIXE	2009-05-20	289960	Jetty	joakim.erdfelt	RESO	FIXE	2009-09-30
271142	Jetty	jesse.mcconnell	RESO	FIXE	2009-05-20	309691	Jetty	joakim.erdfelt	RESO	FIXE	2010-05-12
277406	Jetty	jesse.mcconnell	RESO	FIXE	2009-05-21	313205	Jetty	joakim.erdfelt	RESO	FIXE	2010-05-24
277409	Jetty	jesse.mcconnell	RESO	FIXE	2009-05-21	314909	Jetty	joakim.erdfelt	RESO	FIXE	2010-05-25
282309	Jetty	jesse.mcconnell	RESO	FIXE	2009-07-02	315687	Jetty	joakim.erdfelt	RESO	FIXE	2010-06-07
277551	Jetty	jesse.mcconnell	RESO	FIXE	2009-07-02	271535	Jetty	joakim.erdfelt	RESO	FIXE	2010-10-14
282326	Jetty	jesse.mcconnell	RESO	FIXE	2009-07-02	318233	Jetty	joakim.erdfelt	RESO	FIXE	2011-06-29
283172	Jetty	jesse.mcconnell	RESO	FIXE	2009-07-10	316976	Jetty	joakim.erdfelt	RESO	FIXE	2011-08-22
281621	Jetty	jesse.mcconnell	RESO	FIXE	2009-08-08	293739	Jetty	joakim.erdfelt	RESO	FIXE	2011-08-30
288055	Jetty	jesse.mcconnell	RESO	FIXE	2009-08-30	357216	Jetty	joakim.erdfelt	RESO	FIXE	2011-09-09
288153	Jetty	jesse.mcconnell	RESO	FIXE	2009-08-31	353509	Jetty	joakim.erdfelt	RESO	FIXE	2011-09-15
309185	Jetty	jesse.mcconnell	RESO	FIXE	2010-05-26	357959	Jetty	joakim.erdfelt	RESO	FIXE	2011-09-19
320112	Jetty	jesse.mcconnell	RESO	FIXE	2010-07-16	361573	Jetty	joakim.erdfelt	RESO	FIXE	2011-10-20
320264	Jetty	jesse.mcconnell	RESO	FIXE	2010-07-19	358649	Jetty	joakim.erdfelt	RESO	FIXE	2011-10-20
323196	Jetty	jesse.mcconnell	RESO	FIXE	2010-08-23	362111	Jetty	joakim.erdfelt	RESO	FIXE	2011-10-26
325184	Jetty	jesse.mcconnell	RESO	FIXE	2010-09-14	362251	Jetty	joakim.erdfelt	RESO	FIXE	2011-10-27
320498	Jetty	jesse.mcconnell	RESO	FIXE	2010-10-04	362740	Jetty	joakim.erdfelt	RESO	FIXE	2011-11-03
297154	Jetty	jesse.mcconnell	RESO	FIXE	2010-10-06	362850	Jetty	joakim.erdfelt	RESO	FIXE	2011-11-03
294154	Jetty	jesse.mcconnell	RESO	FIXE	2010-10-13						

Fig. 2. The repair history of jesse.mcconnell and joakim.erdfelt

2 Related Work

For automatic bug triage from the perspective of machine learning, [1] treats fixers as labels for the first time, converts the text of bug reports into feature vectors, and predicts suitable fixers for new bug through the trained classification model. [2] extracted keywords from the heading and summary of the bug report, and use the bag-of-words model to train classifiers such as naive Bayes, support vector machines, and decision trees to assign fixers. [3] suggested a Bugzie method based on fuzzy sets. The membership degree of the fixer to the set is expressed by term relevance, and they are used to measure whether the fixer is suitable for new bug reports. [4] offered a semi-supervised bug triage method, which enhanced the naive Bayes classifier's ability to classify assigned and unassigned bugs through an expectation maximization algorithm. [5] extracted six features from bug reports and source files, including word meaning similarity, semantic similarity, class name similarity, collaborative filtering score, bug fixer progress and bug fixer efficiency to train weighted classification model.

In the method of information retrieval, the bug report gets used as a query, fixer is returned result, and a sorted list of the fixer's professional capabilities is returned by calculating the similarity between the bug report and fixers. [6] offered a bug triage method based on the learning to rank model. Combine the content information of the bug report and the location information of the source code to identify the appropriate fixer to handle the specific bug report. [7] suggested an IRRF method based on concept positioning, through the combination of information retrieval and relevance feedback mechanisms, to calculate the similarity between the bug report and the source file on the space vector model. In term of the theme of the bug report, [8] updated the part of every topic below the current meta-field by the fixer, and obtained a better topic distribution in a supervised manner.

Image classification and natural language have achieved remarkable success by deep learning, researchers have started to try to ask deep learning models to bug triage tasks to achieve the mapping from bug reports to fixers. [9] applied convolutional neural network to bug triage for the first time, converted text content into high-level features, and trained a predictive model to return the probability of each fixer being assigned. [10] offered a bug triage model with an attention mechanism by applying Recurrent Neural Network and combining bidirectional long- and short-term memory units to explore the sequence relationships existing in the text content. [11] used the text content of the bug report and the activity sequence of the fixer to find suitable fixers through the pooled recurrent neural network.

In addition to the above methods, increasingly researchers have begun to focus on the effect of the relationship network in the bug tracking system. [12] analyzed the social factors of network, such as out-of-degree, in-degree, PageRank, intermediate degree, and intimacy, to rank fixers for professional skills. [13] applied a first-order Markov chain to represent the transfer information of bugs, and combined construct a fixer redistribution graph model using the text content of bug reports. According to the job of Jeong, [14] incorporated a multi-feature ranking function into the redistribution graph and used a local weighted search algorithm to reduce the path. [15] offered a method of integrating heterogeneous networks into bug triage, searching for similar bug reports through the K-nearest neighbor algorithm, introducing a scoring mechanism and considering network link relationships to rank candidates.

3 MSDBT Model

To realize the automatic assignment of bugs, MSDBT model is divided into training phase and prediction phase. As this training stage, from the perspective of text content and meta-fields of historical bug reports, considering the impact of text context and fixer community on bug triage, construct a training set and input deep learning to generate feature vectors, and continuously optimize model through real fixer tags. In the prediction stage, a new bug report that has not been assigned is input. The model predicts the probability of each fixer participating in the bug report based on the fixer's historical activities to generate professional fixer list. The flowchart of the MSDBT model is displayed in the Fig. 3.

The framework of MSDBT consists of an input layer, a feature extraction layer, a multi-head self-attention layer, and an output layer. The model not only weakens the redundant information in the bug report, but also uses community factors to quantify the impact of fixers with similar development activities on bug triage. The input layer realizes the text preprocessing and feature embedding; the feature extraction layer realizes the feature calculation of the text content and the fixer community; the multi-head self-attention layer synthesizes the internal features to calculate the contribution of the features to the results, and the output layer calculates the probability of assigning fixers to the bug report.

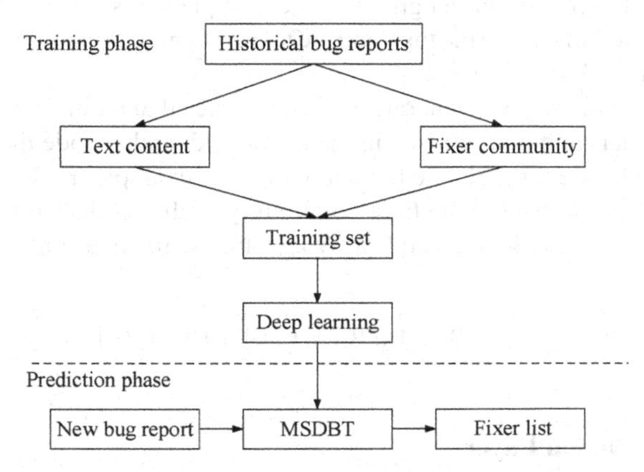

Fig. 3. The flowchart of the MSDBT model

3.1 Input Layer

Text Preprocessing. The text content of the bug report is composed of summary, description, and comments. The randomness of natural language makes the input have no fixed format and informal grammatical expression, which makes the characterization ability of the input information weak. Therefore, in order to obtain standardized input, Model uses natural language processing technology to analyze the text content, and the main processing is as follows:

- Word segmentation: divide the text content into a structure with words as units to generate a standardized term sequence stream.
- Stop word removal: Delete punctuation marks, special characters, numbers, and some common words, such as am, is, do, in, on, etc. Such data has no practical meaning for the classification results.
- Root restoration: Convert input words in reserved tense or singular and plural forms into stem or root form, for example, changes and changed into change.
- Abbreviation replacement: replace abbreviations of professional terms in the bug tracking system, such as replacing API with Application Programming Interface

Feature Embedding. Since the computer cannot directly recognize the sequence or text information, the input sequence is mapped into a low-dimensional numerical vector by feature embedding. Aiming at the text content of the bug report and the fixer community, we use Word2Vec and one-hot encoding to transform it into a real number vector.

For the text content, we apply Word2Vec to generate a k-dimensional word vector, and the semantic similarity of terms is characterized by the similarity of the vector. Its expression is:

$$E = [e1, e2, e3, e4, \ldots\ldots, el] \tag{1}$$

Among them, the maximum length of the text sequence is set to l, e_i is the feature representation of the i-th term in the text, and the text content of every bug report generates a word vector matrix E of size $l \times k$.

For the fixer community, we generate a fixer sequence that retains community factors based on the products and components in the meta-field, and encode the N fixers in the system with a valid code through one hot encoding. For example, Ed Merks, Dirk Fauth and Nathan Ridge participate in the fixer community of the product JDT. Assuming Ed Merks, Dirk Fauth, Nathan Ridge and Lars Vogel are the fixers of the entire system. Its expression is:

$$F = [[1, 0, 0, 0], [0, 1, 0, 0], [0, 0, 1, 0]] \tag{2}$$

3.2 Feature Extraction Layer

For the word vector matrix and community code submitted by the input layer, the feature extraction layer connects two to generate input matrix X, and uses Bi-LSTM to mine the features in the front and back directions, and more comprehensive feature extraction of the input content. The Bi-LSTM is a two-layer neural network composed of LSTM units, which determines the preservation degree of the cell state and the memory degree of the current input through the forget gate, input gate and output gate.

First step of LSTM is to determine the degree of discarding of the previous cell state C_{t-1} through the forget gate. Calculate the Sigmoid function of the hidden state h_{t-1} and the current input x_t. Among them, W_f represents the weight matrix corresponding to the forget gate, and b_f represents the constant vector.

$$f_t = \sigma(W_f \cdot [h_{t-1}, x_t] + b_f) \tag{3}$$

Second step determines the update degree of the information in the cell state. First, the Sigmoid function of the previous hidden state h_{t-1} and the current input x_t is calculated through the input gate to determine which information needs to be updated. Secondly, use the tanh function to generate the candidate vector $\tilde{C}t$. Among them, W_i represents the weight matrix corresponding to the input gate, and b_i represents the constant vector. W_c represents the weight matrix corresponding to the candidate vector, and b_c represents the constant vector.

$$i_t = \sigma(W_i \cdot [h_{t-1}, x_t] + b_i) \tag{4}$$

$$\tilde{C}_t = \tanh(W_C \cdot [h_{t-1}, x_t] + b_C) \tag{5}$$

Next, calculate the discarding and updating of information based on the output of the foregoing two measures to determine the current cell state C_t.

$$C_t = f_t * C_{t-1} + i_t * \tilde{C}_t \tag{6}$$

Finally, the output information is determined above on the cell state. First, the Sigmoid function of the previous hidden state h_{t-1} and the current input x_t is calculated through the output gate to determine which information needs to be output. Second, use the tanh function to normalize the cell state C_t and multiply it with the value of the output gate to determine the output h_t of the current cell. Among them, W_o represents the weight matrix corresponding to the input gate, and b_o represents the constant vector.

$$o_t = \sigma(W_o \cdot [h_{t-1}, x_t] + b_o) \tag{7}$$

$$h_t = o_t * \tanh(C_t) \tag{8}$$

The Bi-LSTM contains two hidden layers, which flow in two different directions, respectively. The formula of the recurrent neural network layer in the two directions is expressed as:

$$\overrightarrow{h_t} = \overrightarrow{LSTM}(x_i, \overrightarrow{h_{t-1}}) \tag{9}$$

$$\overleftarrow{h_t} = \overleftarrow{LSTM}(x_i, \overleftarrow{h_{t-1}}) \tag{10}$$

$\overrightarrow{h_t}$ represents the forward feature generated by combining the current input x_i and the result output $\overrightarrow{h_{t-1}}$ by the LSTM unit at the previous time in forward, $\overleftarrow{h_t}$ represents the reverse feature generated by combining the current input x_i and the output result $\overleftarrow{h_{t-1}}$ of the LSTM unit at the previous time in reverse, splicing and to generate feature s_t.

$$s_t = [\overrightarrow{h_t}, \overleftarrow{h_t}] \tag{11}$$

Finally, the input matrix X is synthesized to obtain the output sequence S of the feature extraction layer.

$$S = [s_1, s_2, s_3, s_4, \ldots\ldots, s_t] \tag{12}$$

3.3 Multi-head Self-attention Layer

Considering the redundant information contained in the text and the influence of the fixer community on the bug triage, we apply the multi-head self-attention mechanism to strengthen the key features of the output S of the feature extraction layer. The attention value at different s_i represents the contribution to the classification results, so as to further optimize the final output of the model. As shown in Fig. 4, multi-head uses multiple parallel queries to extract multiple groups of different subspaces from feature to obtain the relevant information, and self-attention feedbacks the internal dependence between the data, and captures the key information of the sequence from many aspects. The calculation method is as follows.

Firstly, feature extraction layer's output S is transformed linearly to generate query vector matrix Q, key vector matrix K and value vector matrix V. Among them, W_Q, W_K and W_V are weight matrices.

$$\begin{aligned} Q &= WQS \\ K &= WKS \\ V &= WVS \end{aligned} \tag{13}$$

We project Q, K and V into h different subspaces. W_{Q1}, W_{K1} and W_{V1} are the i-th weight matrices of Q, K and V respectively.

$$\begin{aligned} [Q_1, Q_2, \ldots, Q_h] &= [QW_{Q1}, QW_{Q2}, \ldots, QW_{Qh}] \\ [K_1, K_2, \ldots, K_h] &= [KW_{K1}, KW_{K2}, \ldots, KW_{Kh}] \\ [V_1, V_2, \ldots, V_h] &= [VW_{V1}, VW_{V2}, \ldots, VW_{Vh}] \end{aligned} \tag{14}$$

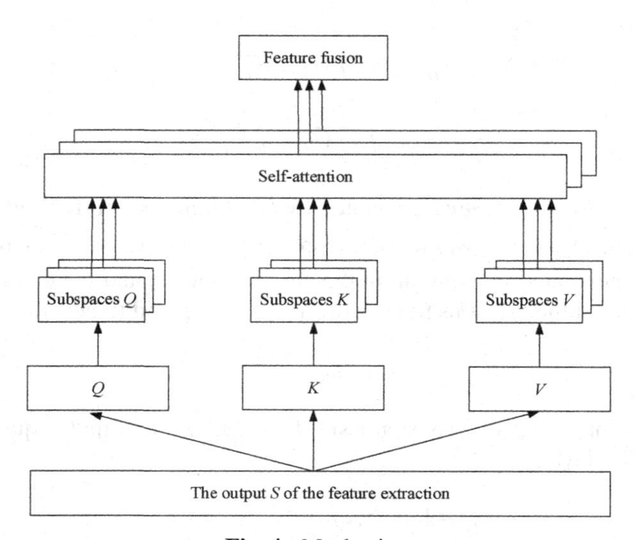

Fig. 4. Mechanism

The inner product of Q_i and K_i is calculated by scaling the point product attention, then normalized by softmax function, and multiplied by V_i to get the single head attention value $head_i$, which is calculated in h subspace in parallel. Among them, d is the scaling factor, which changes the inner product of Q and K into the standard normal distribution, making the result more stable.

$$head_i = \text{soft max}(\frac{Q_i, K_i^T}{\sqrt{d}})V_i \tag{15}$$

Finally, the attention values of all subspaces are fused. Among them, W_A is weight matrix.

$$A(Q, K, V) = Concat(head_1, \ldots, head_h)W_A \tag{16}$$

3.4 Output Layer

The output layer applies softmax classifier to calculate the probability of each fixer. Where W_d is the transformation matrix and b_d is the constant vector.

$$\hat{y} = \text{soft}\max(W_d A + b_d) \tag{17}$$

Finally, the loss function is set to minimize the cross entropy to train the prediction model. Where, y_i represents the real probability distribution, $\hat{y}_i$ represents the probability of each fixer predicted by the model, and moreover n is the total number of fixers.

$$C = -\sum_{i=1}^{i=n} y_i \log(\hat{y}_i) \tag{18}$$

4 Experimental

4.1 Data Set

The data set applied is a collection of bug reports from open source projects in the bug tracking system. According to the life cycle of the bug, the newly fixed bug is unstable, and it is prone to restart and fix again. Therefore, the historical bug with a relatively stable state is selected for analysis [3]. The submissions of bug reports are restricted to not less than 4 years before screening to obtain the maximum number of valid bug reports. Consistent with literature [3, 5], we collect reports that have been confirmed to be fixed. To decrease noise, remove fixers with less than 10 engagements and invalid fixers. The data statistics is in Table 1.

Table 1. Data information

Projects	Date	Fixers	Bug reports	Products	Components
Mozilla	2014.12.20–2016.1.5	679	14297	159	475
Eclipse	2014.12.20–2016.1.5	551	10514	57	404
Netbeans	2014.12.20–2016.1.5	234	11047	40	346
GCC	2014.12.20–2016.1.5	145	2287	2	40

4.2 Metrics

We use Recall@K in Eq. (19) as the metric.

$$\text{Recall}@K = \frac{1}{m}\sum_{j=1}^{m}\frac{|G_j \cap H_j|}{|H_j|} \tag{19}$$

$K = 3,5,10$, denotes size of the recommended results, m means the number of bug reports, G_j and H_j are the predicted and true fixers, respectively.

4.3 Comparative Models

SVM [2] applies support vector machines to finish bug report text classification.

MTM [8] makes an update of percentage of topics below the current meta-domain based on the topics reported in the bug, in order to get a better distribution of topics through monitoring.

DT [11] shows that the error report text is combined with the activity sequence of fixers by a recursive neural network to assign fixers.

To simulate the application of the model in actual situations, as shown in Fig. 5, we use ten-fold incremental experiments for verification, and divide the data set into 11 windows of the same size in chronological order to obtain 10 stacks of training and test data sets. For the first stack of data, the data in the first window used for the training set, another data in the second window as the test set. For the second stack of data, the data of first window and second window are merged as training set, then the data in the third window is used as the test set, and so on. For the 10th stack of data, the data of ten windows are merged as the training set, and then the data in the 11-th window is used as the test set. Finally, the model takes the average of ten tests as the final result.

For the parameter setting of the model in the essay, the length of the text content is set to 300, and the text content is adjusted by padding or truncation, and the vector dimension generated by Word2vec is installed to 256. The amount of hidden layer neurons in the BI-LSTM neural network is 300. The multi-head attention mechanism independently generates 12 linear subspaces, and each output dimension is 128 dimensions. The optimization method uses the gradient descent method. Moreover the learning rate is 0.01. The batch size is 64 and the number of iterations is 20.

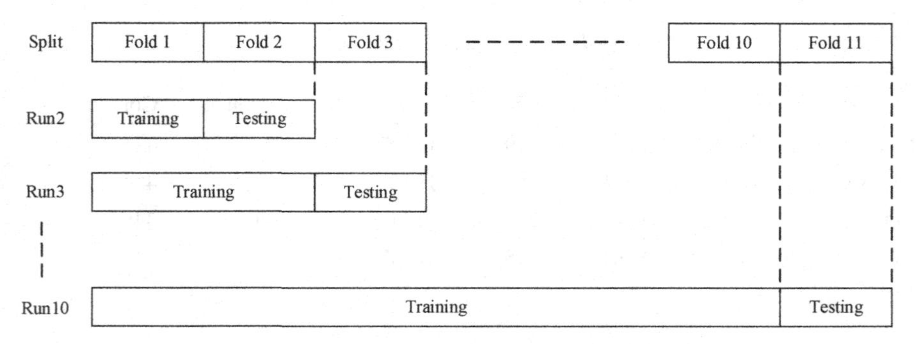

Fig. 5. Ten-fold incremental experiments

4.4 Experimental Result

To compare the achievement of bug triage between the MSDBT model and comparative models, as shown in Fig. 6, we calculate Recall@K on four large-scale open source software tasks, Eclipse, Mozilla, Netbeans, GCC. The average values of Recall@3, Recall@5, and Recall@10 on the four data sets of MSDBT are 0.5424, 0.6375, and 0.745, which are 11.21%, 8.97%, and 9.17% higher than SVM. Compared with MTM, it increased by 8.23%, 7.64%, and 8.41% respectively. Compared with DT, it increased by 4.97%, 3.76%, and 4.67% respectively.

To simulate the application of the model under actual conditions, MSBDT is verified in Recall@3, Recall@5, and Recall@10 in ten-fold increment. As shown in Fig. 7, in the chronological experiment, the recall rate did not increase significantly. Through the observation and understanding of the bug tracking system, we found that with the continuous increase of data, the fixers, products, and components in the system are also increasing, resulting in changes in the complexity of the data. Therefore, the recall rate of MSBDT is relatively stable.

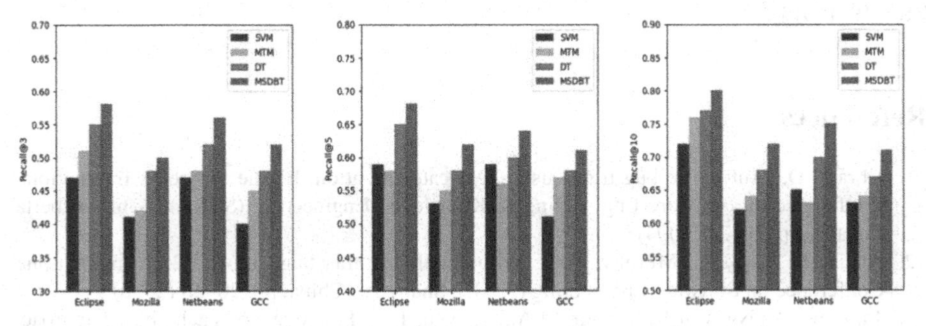

Fig. 6. Experimental results of MSBDT and comparative models

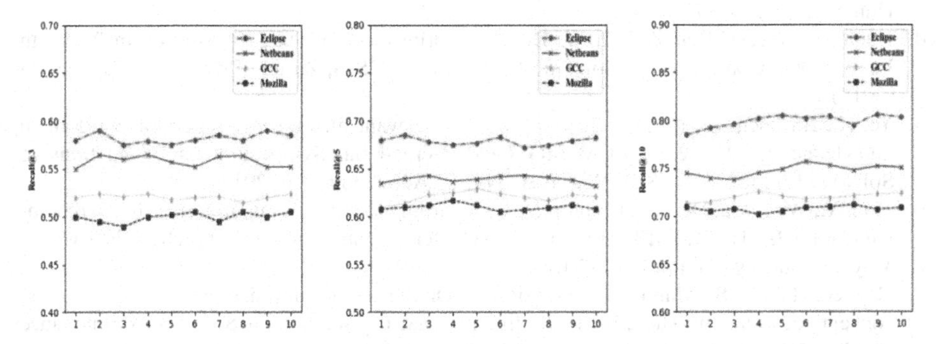

Fig. 7. MSBDT's ten-fold incremental experiments on different data set

5 Conclusion

Aiming at the problem of fixer recommendation, we propose a bug triage model MSDBT based on the multi-head self-attention mechanism. From the perspective of text description and meta-fields, train a neural network classification model. The model uses a Bi-LSTM to extract the features of the text context and fixer sequence, and uses a multi-headed self-attention mechanism to further compute parallel attention between internal elements. While weakening the redundant information in the text content, it increases the influence of the fixer community under the same product and the same component. Model is verified on the data sets of 4 large open source projects. We can know the MSDBT has clear profits over the classic bug triage method. In the long term, we will consider the impact of different data types (such as network, graph, source code, etc.), dynamic changes of the system, and cross-domain analysis between different data sets to help the achievement.

Acknowledgments. This work is jointly sponsored by National Natural Science Foundation of China (No. 6217072142), Natural Science Foundation of Shandong Province (No. ZR2019MF014).

References

1. Ubrani, D.: Automatic bug triage using text categorization. In: the Sixteenth International Conference on Software Engineering & Knowledge Engineering (SEKE), Banff, Alberta, Canada, pp. 20–24 (2004)
2. Anvik, J., Hiew, L., Murphy, G.C.: Who should fix this bug? In: the 28th International Conference on Software Engineering (ICSE), Shanghai, China, pp. 20–28 (2006)
3. Tamrawi, A., Nguyen, T.T., Jafar, M.A., Nguyen, T.N.: Fuzzy set and cache-based approach for bug triaging. In: the 13th European Software Engineering Conference and the ACM SIGSOFT Symposium on the Foundations of Software Engineering(ESEC/FSE), Szeged, Hungary, pp. 365–375 (2011)
4. Xuan, J., Jiang, H., Ren, Z., Zou, W.: Developer prioritization in bug repositories. In: The 34th International Conference on Software Engineering (ICSE), Zurich, Switzerland, pp. 25–35 (2012)
5. Ye, X., Bunescu, R., Liu, C.: Learning to rank relevant files for bug reports using domain knowledge. In: The 22nd ACM SIGSOFT International Symposium on Foundations of Software Engineering (FSE), New York, NY, USA, pp. 689–699 (2014)
6. Tian, Y., Wijedasa, D., Lo, D., Claire, L.G.: Learning to rank for bug report assignee recommendation. In: The 24th IEEE International Conference on Program Comprehension (ICPC), May, Piscataway, NJ, pp. 1–10 (2016)
7. Gay, G., Haiduc, S., Marcus, A., Menzies, T.: On the use of relevance feedback in IR-based concept location. In: The 25th IEEE International Conference on Software Maintenance (ICSM), Edmonton, AB, Canada, pp. 351–360 (2009)
8. Xia, X., Lo, D., Ding, Y., Al-kofahi, J., Nguyen, T., Wang, X.Y.: Improving automated bug triaging with specialized topic model. IEEE Trans. Softw. Eng. **43**, 272–297 (2017)
9. Lee, S.R., Heo, M.J., Lee, C.G., Kim, M., Jeong, G.: Applying deep learning based automatic bug triager to industrial projects. In: The 11th Joint Meeting on Foundations of Software Engineering (ESEC/FSE), New York, NY, USA, pp. 926–931 (2017)

10. Mani, S., Sankaran, A., Aralikatte, R.: DeepTriage: exploring the effectiveness of deep learning for bug triaging. In: The ACM India Joint International Conference on Data Science and Management of Data (CoDS-COMAD), New York, NY, USA, pp. 171–179 (2019)
11. Xi, S.Q., Yao, Y., Xu, F., Lv, J.: Bug triage approach based on recurrent neural networks. J. Softw. **29**, 2322–2335 (2018)
12. Wu, W., Zhang, W., Yang, Y., Wang, Q.: DREX: developer recommendation with k-nearest-neighbor search and expertise ranking. In: The 20th Asia-Pacific Software Engineering Conference (APSEC), Ho Chi Minh, Vietnam, pp. 389–396 (2011)
13. Jeong, G., Kim, S., Zimmermann, T.: Improving bug triage with bug tossing graphs. In: The 7th in the European Software Engineering Conference and the ACM SIGSOFT Symposium on the Foundations of Software Engineering (ESEC/FSE), New York, NY, USA, pp. 111–120 (2009)
14. Bhattacharya, P., Neamtiu, I.: Fine-grained incremental learning and multi-feature tossing graphs to improve bug triaging. In: The IEEE International Conference on Software Maintenance, Timisoara, Romania, pp. 1–10 (2010)
15. Zhang, W., Wang, S., Wang, Q.: KSAP: an approach to bug report assignment using KNN search and heterogeneous proximity. Inf. Softw. Technol. **70**, 68–84 (2016)

Taxi Pick-Up Area Recommendation via Integrating Spatio-Temporal Contexts into XDeepFM

Decheng Zhang[1,2], Yizhi Liu[1,2](✉), Xuesong Wang[1,2], Zhuhua Liao[1,2], and Yijiang Zhao[1,2]

[1] School of Computer Science and Engineering, Hunan University of Science and Technology, Xiangtan 411201, China
yizhi_liu@sina.cn

[2] Key Laboratory of Knowledge Processing and Networked Manufacturing in Hunan Province, Xiangtan 411201, China

Abstract. Taxi pick-up area recommendation based on GPS data can effectively improve efficiency and reduce fuel consumption. Most of the methods use the long-term GPS data, which makes recommendation accuracy low. Therefore, we propose a novel approach of integrating spatio-temporal contexts into the extreme Deep Factorization Machine (xDeepFM) for taxi pick-up area recommendation. In the training process, the urban area is divided into several grids of equal size, we extract pick-up points from the original GPS data. The pick-up points and points-of-interest (POIs) are mapped into the corresponding grids, we distil the spatio-temporal features from these grids to construct spatio-temporal contexts matrix. Then, the spatio-temporal contexts matrix is input into xDeepFM for training, and we get the taxi pick-up area recommendation model. xDeepFM not only can make feature interactions occur at the vector-wise in both implicit and explicit ways, but also learn both low-order and high-order feature interactions. xDeepFM can effectively enhance recommendation accuracy. Finally, the recommendation model is embedded in the system for testing. Evaluate on the public dataset of DiDi, we compare different recommendation methods. The experimental results show that our approach can effectively cope with the data sparseness problem, obtain excellent performance, and is superior to some state-of-the-art methods. The RMSE is only 0.8%, MAE is about 7%, and the explained variance score is over 98%.

Keywords: Trajectory mining · Location-based services (LBS) · Taxi pick-up area recommendation · Spatio-temporal contexts · Extreme Deep Factorization Machine (xDeepFM)

1 Introduction

Taxi is the GPS recorder of urban residents' mobility. The hidden patterns in these data are of great value for human travel, intelligent transportation, and urban planning,

Y. Sun et al. (Eds.): ChineseCSCW 2021, CCIS 1492, pp. 120–133, 2022.
https://doi.org/10.1007/978-981-19-4549-6_10

which promote many Location-Based Services (LBS). For example, the taxi pick-up area recommendation [1] can effectively improve driver's profit and reduce fuel consumption, etc. Compared with traditional recommendation systems, taxi pick-up area recommendation now is facing some new challenges.

Firstly, the challenge of GPS data update speed. Taxi produces a large amount of GPS data every day, if using long-term GPS data, such as, several years GPS data which will occupy a large amount of storage resources and generate a long computation time. In addition, the rapid urbanization makes the roads update quickly, and using long-term GPS data in the taxi pick-up area recommendation system will introduce more noise. For example, the original roads have been abolished or changed, and the city adds several roads. These noises may greatly reduce the recommendation accuracy. However, the using of recent short-term GPS data, such as, one month GPS data which faces the problem of data sparseness [2]. To cope with this problem, matrix factorization techniques [3] are widely used. But the matrix factorization technique lacks the effective use of contexts information. Therefore, researchers started to focus on models such as Factorization Machine (FM) [4].

Secondly, the regularity implied in GPS data needs to be further explored, for example, it contains many spatio-temporal contexts [6]. During the Taxi cruising process, the driver's choice of the pick-up area changes with the spatio-temporal information. For example, there is a large amount of travel demand in residential areas in the morning time; while there are often many passengers in recreational areas after 11 pm. Therefore, it is an important problem to integrate spatio-temporal contexts into taxi pick-up area recommendation.

To address the above challenges, we propose a novel approach of integrating spatio-temporal contexts into xDeepFM [5] for taxi pick-up area recommendation. Our contribution mainly lies in:

- We deeply mine the spatio-temporal contexts. We extract the pick-up points from the original GPS data, map the pick-up points and POIs into the corresponding grid. Then we extract the spatio-temporal features from these grids to construct the spatio-temporal contexts matrix, which not only can compensate the data sparseness, but also can improve the accuracy of recommendation.
- Spatio-temporal contexts are integrated into the xDeepFM model in the way of feature engineering. xDeepFM not only can make feature interactions occur at the vector-wise in both implicit and explicit ways, but also learn both low-order and high-order feature interactions. xDeepFM can effectively enhance the accuracy of recommendation.
- Evaluate on the public dataset of DiDi. The experimental results show that our approach can effectively cope with the data sparseness problem, and is superior to some state-of-the-art methods. The RMSE is only 0.8%, MAE is about 7%, and the explained variance score is over 98%.

2 Related Work

Taxi pick-up recommendation service can effectively improve the efficiency of Taxi driver. The main research directions focus on pick-up point recommendation, pick-up area recommendation and pick-up route recommendation.

Pick-Up Point Recommendation. Chen [7] et al. extracted the pick-up points from GPS data, calculated the economic benefits of each pick-up point, and then clustered the pick-up points by using the DBSCAN method to get the highest value pick-up points and recommended it to drivers. Song et al. [8] proposed a Markov-based pick-up point recommendation model, by calculating the travel time and distance parameters to provide the driver with a sequence of pick-up points with time and distance constraints. Agrawal et al. [9] proposed a Hotspot Recommendation Approach (HRA) that used a clustering approach on a large-scale Taxi dataset to identify hotspot. Wang et al. [10] predicted passenger demand points by using a large amount of GPS trajectory data. Phanhong et al. [11] used K-means to simulate Taxi stations and recommend the best stations to driver by estimating the time driver spend waiting for passengers at each station.

Pick-Up Area Recommendation. Yuan et al. [12] calculated the Moran 'I index to measure the spatial correlation between high-order areas and high-income areas to provide high-income areas for Taxi. Huang et al. [13] proposed the DBSCAN + algorithm and then sliced and cyclically clustered many pick-up points. Liao et al. [1] proposed a latent factor model combined with geographic information GeoLFM, the model integrated driver-related geographic information into the decomposition of the matrix, which compensated for the problem of data sparseness.

Pick-Up Route Recommendation. Li et al. [14] proposed an efficient driving route suggestion (DRS) algorithm based on inter-regional probability, which maximized the profit of Taxi drivers in a specified destination area. Hsieh et al. [15] proposed a multi-criteria route recommendation framework that considers real-time spatio-temporal forecasts and traffic network information to improve Taxi drivers' profits. Li et al. [16] proposed to evaluate the potential profit of driving routes by a profit objective function, then provided high profitability routes for Taxi drivers based on the current location of the Taxi. Lai et al. [17] proposed the concept of Coulomb's law for urban traffic, simulated the relationship between Taxi and passenger in the city and proposed a route recommendation scheme. Wang et al. [18] proposed a ranking-based extreme learning machine (ELM) model to evaluate the passenger-seeking potential of each road. Liu et al. [19] selected more location attributes from the historical pick-up points to obtain the spatial-temporal features, and the information entropy of the spatial-temporal features was integrated into the evaluation model, then the model was applied to obtain the next pick-up point and further recommend a series of sequential points which were constructed into a Taxi driver's cruising route.

The above-mentioned taxi pick-up recommendation methods have good recommendation results with large-scale GPS data, but does not consider solutions for short-term GPS data. Our work considers previous work. Mining the spatio-temporal features from GPS data and geographic information to construct the spatio-temporal contexts matrix. The constructed spatio-temporal contexts matrix is input to xDeepFM for training. The trained xDeepFM can generate driver-time slot-grid (DTSG) access probability matrix. According to the DTSG and the drivers' current spatio-temporal information, the grids with the top-N access probability will be recommended to the driver. Integrating spatio-temporal contexts into xDeepFM which effectively improves the performance of recommendation, and reduces the impact of data sparseness.

3 The Framework of Taxi Pick-Up Area Recommendation

3.1 Related Definition

Definition 1. Pick-Up Point. In this paper, The GPS data we use includes Taxi ID, GPS time, GPS longitude, GPS latitude and GPS status. The GPS status represents whether the Taxi has passenger. "0" represents that the taxi is no-load driving, "1" represents that the taxi is load driving. When GPS status changes from "0" to "1", it means that the taxi is carrying passenger. We record the current GPS point as the pick-up point.

Definition 2. POIs. POIs are the abbreviation of "Points of Interest". For example, companies, hospitals, schools, restaurants, etc. in a city can be called POIs.

Definition 3. Grid. In order to enhance the calculation efficiency. We divide the urban area into several grids of equal size. Then, we generate the grid set $Gridset = \{G1, G2, G2...Gn\}$.

Definition 4. Grid Attributes. The pick-up points and POIs are mapped into the corresponding grids. Then we count the pick-up points and POIs in each grid and we can get some important grid attributes. In this paper, we choose the geometric center location of the grid, the number of historical pick-up points, the average driving time, the average driving distance, the number of POIs, the ratio of each type POIs as grid attributes.

Definition 5. Driver-Time Slot-Grid (DTSG) Access Probability Matrix. We divide the day into 24 time slots. DTSG can be used to predict the driver's access probability to each grid in each time slot, for example, the access probability of driver 1 to grid 1 in the 6–7 time slot is 0.6.

3.2 Recommendation Framework

Figure 1 shows the framework of our approach, which is divided into three main parts. The data we use are Taxi GPS data and POI data.

The first part is to construct the spatio-temporal contexts matrix. Firstly, we extract the pick-up points from the original GPS data and the urban area is divided into several grids of equal size. Then we map the pick-up points and POIs into the corresponding grids according to their latitude and longitude. Finally, we distil the spatio-temporal features from these grids to construct spatio-temporal contexts matrix. The spatio-temporal contexts matrix consists of feature vector $\boldsymbol{X}$ and target Y. The feature vector $\boldsymbol{X}$ consists of driver feature D, grid feature G, time slot feature T, and grid attribute feature A. Target Y is obtained by normalizing the number of carrying passenger in the corresponding grid in the corresponding slot.

The second part is to integrate spatio-temporal contexts into xDeepFM. The constructed spatio-temporal contexts matrix is input to xDeepFM for training. xDeepFM can learn the low-order and high-order interactions between driver feature D, grid feature G, time slot feature T, and grid attribute feature A. The trained xDeepFM can generate driver-time slot-grid (DTSG) access probability matrix. DTSG can be used to predict the driver's access probability to each grid in each time slot.

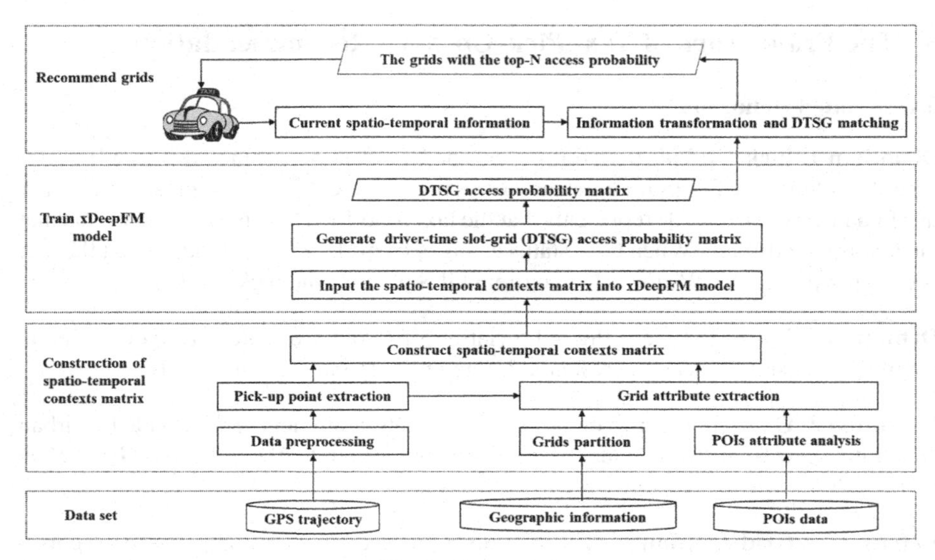

Fig. 1. Framework

The third part is to recommend grids to driver by using DTSG. According to the DTSG and the drivers' current spatio-temporal information, the grids with the top-N access probability will be recommended to the driver. The recommended grids not only have high access probability but also are close to the driver.

4 Methodology Overview

4.1 XDeepFM

The xDeepFM model was proposed by Lian et al. [5] in 2018. As shown in Fig. 2.

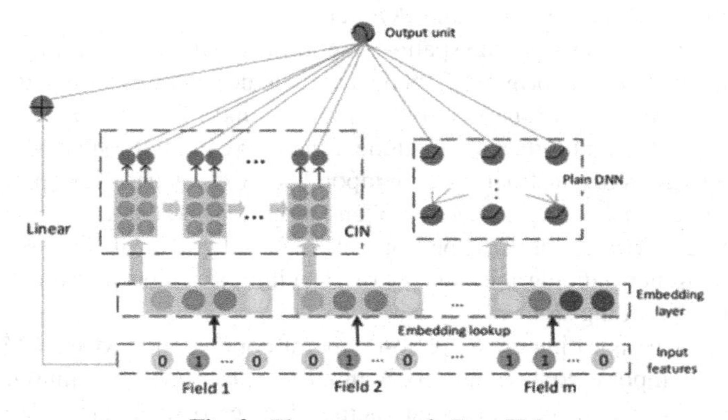

Fig. 2. The structure of xDeepFM

xDeepFM consists of a linear structure, a deep neural network (DNN), and a compressed interaction network (CIN). These three parts use the same input feature vector. The formula of xDeepFM is as follows:

$$\hat{y} = \sigma(y_{linear} + y_{cin} + y_{dnn}) \tag{1}$$

where $\hat{y}$ is the prediction result, y_{linear}, y_{cin}, and y_{dnn} are the output of the linear structure, CIN, and DNN. σ is the activation function.

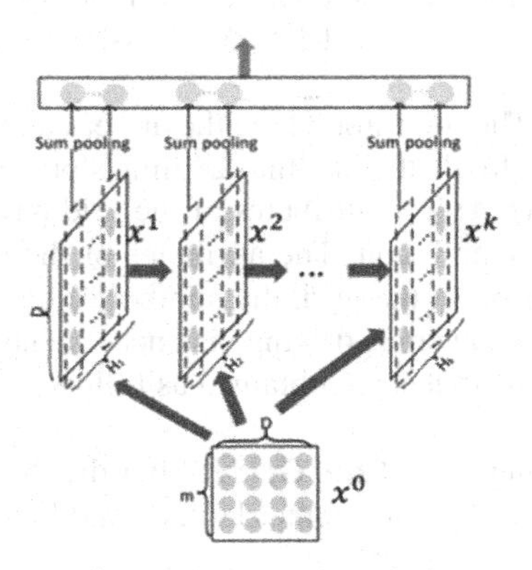

Fig. 3. The structure of CIN

The CIN in xDeepFM is designed to learn higher-order feature interactions explicitly and to make feature interactions occur at the vector-wise. The structure of CIN is shown in Fig. 3.

The input of CIN comes from the embedding layer. Assume that there are *m* fields and the embedding vector dimension of each field is *D*, so the input can be represented as a matrix $X^0 \in R^{m*D}$. Let $X^k \in R^{H_k*D}$ denote the output of the *kth* layer. H_k denotes the number of vectors in the *kth* layer, the dimension of the vector is always *D*. The formula of each vector in the *kth* layer is as follows:

$$X^k_{h,*} = \sum_{i=1}^{H_{k-1}} \sum_{j=1}^{m} W^{k,h}_{ij} \left(X^{k-1}_{i,*} \circ X^0_{j,*} \right) \in R^{1*D}, \quad where\ 1 \le h \le H_k \tag{2}$$

where $W^{k,h} \in R^{H_{k-1}*m}$ denotes the weight matrix of the *hth* vector in the *kth* layer. $\circ$ denotes the Hadamard, for example, $\langle a1, a2, a3 \rangle \circ \langle b1, b2, b3 \rangle = \langle a1b1, a2b2, a3b3 \rangle$.

As mentioned above, xDeepFM combines DNN and CIN. Two modules can make xDeepFM learn high-level feature interactions in both explicit and implicit ways, and can make feature interactions occur at the vector-wise.

4.2 Construction of Spatio-Temporal Contexts Matrix

The driver's pick-up behavior in different grid areas reflects the driver's preference for different grids. The demand of pick-up in different grids also varies in different time slots. The objective geographical information of the grid area and some historical pick-up information reflect the degree of hotspot of the grid area.

Therefore, when we construct the spatio-temporal contexts matrix, we select the following four features.

(1) **Driver feature *D*:** Drivers have preferences for different grids. Preferences may also be similar between drivers with similar grid pick-up experiences. The driver feature *D* is the driver-ID in this paper. Driver-ID is obtained by GPS data.
(2) **Grid feature *G*:** The grid is the recommended pick-up area for the drivers in this paper. The grid feature *D* is the grid-ID. The grid-ID is obtained when dividing the urban area.
(3) **Time slot feature *T*:** The time slot where the driver happens to pick up passengers in the grid region. We divide the day into 24 time slots.
(4) **Grid attribute feature *A*:** The attributes of the grid have a crucial impact on the accuracy of the recommendation. The attributes of the grid in this paper are: the geometric center location of the grid, the number of historical pick-up points, the average driving time, the average driving distance, the number of POIs, the ratio of each type POIs. Grid attributes are obtained as follows:

The Geometric Center Location of the Grid: After dividing the urban area into several grids of equal size, we can get the center latitude and longitude of each grid. Using O_{lat} and O_{lot} to represent.

The Number of Historical Pick-Up Points: After mapping the pick-up points into the Corresponding grid, we can count the number of historical pick-up points for each grid. It can reflect the historical pick-up demand for each grid. Using *His* to represent.

The Average Driving Time: It refers to the average of the driving time after carrying passengers in the grid. If this value is higher, it means that the driver can earn more profit after carrying passenger in the grid. The calculation formula is as follows:

$$T_i = \frac{1}{NUM} \sum_{j=1}^{NUM} (Time_j) \tag{3}$$

where T_i is the average driving time of the grid *Gi*, *NUM* is the number of orders in the grid *Gi*, and $Time_j$ is the driving time of the jth order in the grid *Gi*.

The Average Driving Distance: It refers to the average of the driving distance after carrying passengers in the grid. If this value is higher, it means that the driver can earn more profit after carrying passenger in the grid. The calculation formula is as follows:

$$D_i = \frac{1}{NUM} \sum_{j=1}^{NUM} (distance_j) \tag{4}$$

where D_i is the average driving distance of the grid *Gi*, *NUM* is the number of orders in the grid *Gi*, and $distance_j$ is the driving distance of the jth order in the grid *Gi*.

The Number of POIs: After mapping POIs into the Corresponding grid, we can count the number of POIs for each grid. The number of POIs in the grid can reflect the degree of hotspot for each grid. Using *POIs* to represent.

The Ratio of Each Type POIs: It refers to the ratio of each type POIs in the grid. For example, some grids have a high ratio of scenery POIs, some grids have a high ratio of residential POIs. The calculation formula is as follows:

$$type_{j(i)} = \left|POI_{type_{j(i)}}\right| \Big/ \left|\sum_{j=1}^{j=n} POI_{type_{j(i)}}\right| \tag{5}$$

where $type_{j(i)}$ is the ratio of the jth type POIs in the grid *Gi*, $|POI_{type_{j(i)}}|$ represents the number of the jth type POIs in the grid *Gi*, *n* represents the number of POIs type.

From the above, the feature vector ***X*** in this paper consists of four parts: driver feature *D*, grid feature *G*, time slot feature *T*, grid attribute feature *A*. Among them, time slot feature *T* and grid attribute feature *A* contain spatio-temporal contexts information, which is used to capture the driver's preference for the grid more accurately.

Target *Y* is obtained by normalizing the number of carrying passenger in the corresponding grid in the corresponding slot.

In this paper, we choose the normalization method as Min-Max scaling. The calculation formula is as follows:

$$X_{norm} = \frac{X - X_{min}}{X_{max} - X_{min}} \tag{6}$$

where X is the original data, X_{min} is the minimum value, X_{max} is the maximum value, and X_{norm} is the normalized value.

As shown in Fig. 4. Each row represents a feature vector $\boldsymbol{X}^{(i)}$ and its corresponding target $Y^{(i)}$. The first 4 columns represent the driver-ID. The next 6 columns represent the grid-ID. Then the next 4 columns represent the time slot. The last 6 columns represent the grid attribute. The rightmost column represents the number of carrying passenger in the corresponding grid in the corresponding slot. As seen in the Fig. 4, the driver-ID, grid-ID, and time slot are all one-hot encoded. The grid attribute and target *Y* are normalized.

Feature Vector X

	Driver-ID				Grid-ID						Time Slot				Grid Attribute						Target Y	
	D1	D2	D3	···	G1	G2	G3	G4	G5	···	T1	T2	T3	···	O_{lat}	O_{lot}	His	POIs	T	···		
$X^{(1)}$	1	0	0	···	0	0	1	0	0	···	1	0	0	···	0.62	0.54	0.55	0.71	0.32	···	0.54	$Y^{(1)}$
$X^{(2)}$	1	0	0	···	1	0	0	0	0	···	1	0	0	···	0.45	0.76	0.98	0.12	0.23	···	0.45	$Y^{(2)}$
$X^{(3)}$	0	1	0	···	0	1	0	0	0	···	0	1	0	···	0.12	0.13	0.45	0.56	0.74	···	0.67	$Y^{(3)}$
$X^{(4)}$	0	1	0	···	0	1	0	0	0	···	0	1	0	···	0.21	0.12	0.43	0.54	0.54	···	0.89	$Y^{(4)}$
$X^{(5)}$	0	1	0	···	0	1	0	0	0	···	0	1	0	···	0.14	0.50	0.60	0.30	0.10	···	0.34	$Y^{(5)}$
$X^{(6)}$	0	0	1	···	0	0	1	0	0	···	0	0	1	···	0.34	0.45	0.45	0.75	0.13	···	0.56	$Y^{(6)}$
$X^{(7)}$	0	0	1	···	0	0	0	0	1	···	0	0	1	···	0.12	0.30	0.80	0.90	0.2	···	0.45	$Y^{(7)}$

Fig. 4. Spatio-temporal contexts matrix

4.3 Pick-Up Area Recommendation

Taxi pick-up area recommendation is a location-based service, which necessarily has an important connection with the current location of the driver. In order to make driver obtain greater benefits, we make a restriction when recommending grids to driver. Get a circle with the driver's current position as the center and a radius of R (R as a hyperparameter). The grids inside the circle are used as candidate grids.

The constructed spatio-temporal contexts matrix is input into the xDeepFM for training, the trained xDeepFM generates driver-time slot-grid (DTSG) access probability matrix. Combine the driver's current time slot and DTSG to match the access probability of the candidate grids. Then, the candidate grids with the top-N access probability will be recommended to the driver. The recommended grids not only have high access probability but also are close to the driver. The recommendation process is as follows:

Recommendation Process:

1. The urban area is divided into several grids with side length of 100 m, we get the grid set *Gridset*. And we obtain the grid-ID of each grid, which is the grid feature *G*.
2. According to the driver unique identification in the Taxi GPS data, we obtain the driver-ID of each driver, which is the driver feature *D*.
3. Extracting the pick-up points from the GPS data. According to the latitude and longitude of the pick-up points and POIs, they are mapped into the corresponding grids. Counting the pick-up points and POIs in each grid to get the grid attribute feature *A*.
4. We divide the day into 24 time slots to obtain the time slot feature *T*. The extracted features are used to construct the spatio-temporal contexts matrix. The constructed spatio-temporal contexts matrix is input into the xDeepFM for training, the trained xDeepFM generates driver-time slot-grid (DTSG) access probability matrix.
5. Get a circle with the driver's current position as the center and a radius of R (R as a hyperparameter). The grids inside the circle are used as candidate grids.
6. Combine the driver's current time slot and DTSG to match the access probability of the candidate grids. Then, the candidate grids with the top-N access probability will be recommended to the driver.

5 Experiments

5.1 Data Set

The GPS data are derived from Didi GAIA open dataset. The data is the GPS data of a total of 13605 Taxi in Chengdu in August 2014. There are some abnormal data in the GPS data set, so some pre-processing is needed. For example, abnormal data elimination, trajectory compression, stopping point detection, and trajectory smoothing.

After preprocessing the GPS data, we choose Chengdu ([103.96, 30.59] - [104.17, 30.72]) as our research area. We divide the urban area into several grids with side length of 100 m, and we can get 33110 grids.

The POIs data set in this paper is used to construct the grid attribute feature, which is crawled from the Amap, including a total of 343,851 POI data points. Each piece of POI data includes POI name, POI type, POI address, POI latitude and longitude information. The POI data set contains 13 types of POI: transportation and accommodation, accommodation services, sports, companies, medical care, commercial housing, government, life services, science and education, culture, shopping, financial services, scenery, and catering. The number distribution of POI types is shown in Fig. 5, the top three are shopping, life services, and catering.

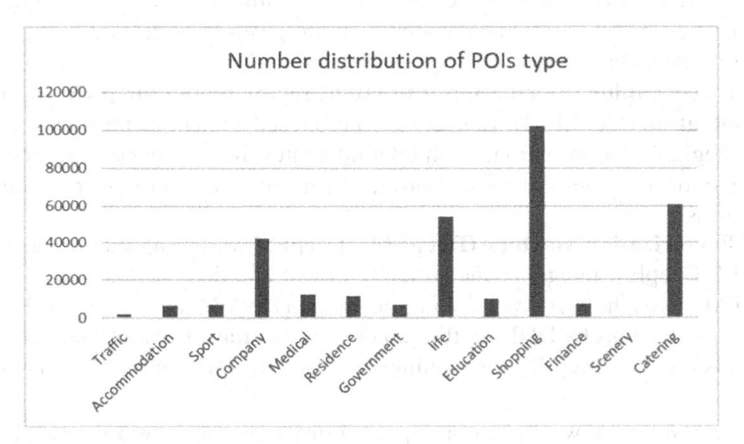

Fig. 5. Number distribution of POIs type

The GPS data are divided into two parts: weekdays (Monday to Friday) and weekends (Saturday and Sunday). Then the raw data is processed into a spatio-temporal contexts matrix. The division ratio between the training data and the testing data is 9:1.

Before training the xDeepFM model, we set some training parameters. We set the optimizer as Adam, the loss function as Mean square error, epochs = 10, batch size = 1024.

5.2 Performance Comparison with Other Recommendation Methods

Evaluation Metrics: The evaluation metrics in this paper use the ***mean absolute error (MAE)*** and ***root mean square error (RMSE)***, which are commonly used in recommendation systems. ***MAE*** is the average of the absolute error between the actual value and the predicted value. ***RMSE*** is the square root of the ratio of the square of the deviation between the predicted value and the true value to the number of observations n. The smaller the ***MAE*** and ***RMSE***, the better the performance of the recommendation. The calculation formulas for ***MAE*** and ***RMSE*** are as follows:

$$MAE = \frac{1}{m}\sum\nolimits_{i=1}^{m} |y_{di} - \hat{y}_{di}| \tag{7}$$

$$RMSE = \sqrt{\frac{1}{m}\sum\nolimits_{i=1}^{m}(y_{di} - \hat{y}_{di})^2} \quad (8)$$

where y_{di} is the actual access probability of driver d on the grid i, and $\hat{y}_{di}$ is the predicted access probability value. m is the number of test set.

We select the following methods for comparison with our method.

1. **User-based collaborative filtering (UCF):** In the user-based collaborative filtering, the similarity between drivers is calculated. The similarity is related to the grids that they have accessed. Grids recommendation are given to target driver based on the similarity between drivers.
2. **Fusing geographic information into latent factor model for pick-up area recommendation (GeoLFM):** Liao et al. [1] proposed a latent factor model combined with geographic information, the model integrates driver-related geographic information into the decomposition of matrix, which compensates for the problem of data sparseness.
3. **Deep Factorization Machine (DeepFM):** DeepFM was proposed by Guo et al. [20] in 2017. DeepFM integrates the structure of FM and deep neural network (DNN). DeepFM learns the low-order feature interactions by FM, and learns the high-order feature interactions by DNN. In this paper, the constructed spatio-temporal contexts matrix is input to DeepFM for training. It is used to compare with our method.

As shown in Fig. 6, we name our approach of integrating spatio-temporal contexts into xDeepFM for taxi pick-up area recommendation as STC_xDeepFM. The performance of STC_xDeepFM is the best. After analysis, we conclude that UCF has poor performance due to the low utilization of sparse data. GeoLFM has certain sparsity resistance, but it does not integrate spatio-temporal context well and does not consider the impact of feature interaction on recommendations, so its performance is not satisfactory.

We input spatio-temporal contexts matrix into DeepFM and xDeepFM respectively, as shown in Fig. 6, our approach using xDeepFM has better performance than DeepFM. After analysis, we believe that xDeepFM can better capture the higher-order interactions between driver feature D, grid feature G, time slot feature T, and grid attribute feature A, thus xDeepFM predict the driver's access probability to grid more accurately. However, xDeepFM takes the longest time to train.

5.3 Performance Evaluation of Spatio-Temporal Contexts

To verify the effect of spatio-temporal context on recommendation accuracy, we select time slot feature T and grid attribute feature A in combination with driver feature D and grid feature G to construct different input matrices. We choose weekdays data for the experiment.

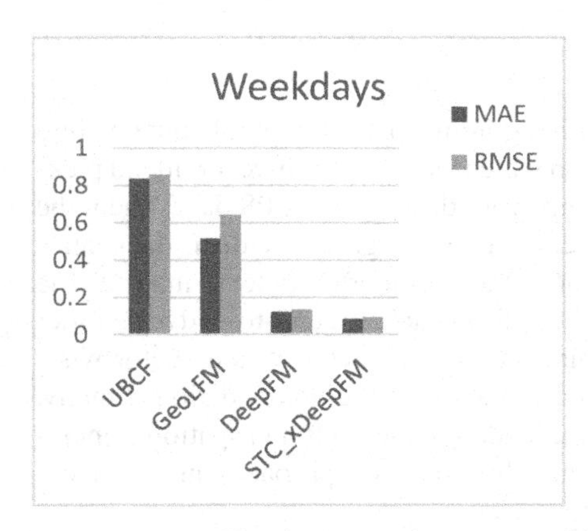

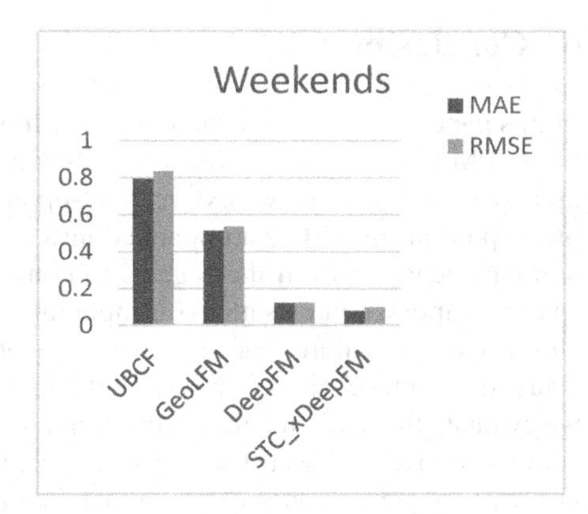

Fig. 6. Comparison among different recommendation methods

For this experiment, *explained variance score* is added as the rating index. This score is used to measure the ability of our model to interpret data set fluctuations. If the score is 1.0, it shows that our model is perfect and *Var* is variance. The formula is as follows:

$$explaine\_variance(y, \hat{y}) = 1 - \frac{Var\{y - \hat{y}\}}{Var\{y\}} \tag{9}$$

The performance of the feature combination is shown in Table 1.

Table 1. Performance of feature combination

	{D, G, Y}	{D, G, T, Y}	{D, G, A, Y}	{D, G, T, A, Y}
RMSE	7.2227%	7.1844%	7.1165%	7.0065%
MAE	1.0740%	1.0680%	0.9272%	0.8001%
explained_variance	97.9520%	97.9760%	97.9934%	98.0393%

As shown in Table 1. Adding time slot feature *T* to both {*D, G, Y*} and {*D, G, A, Y*} which can improve the performance of xDeepFM. Adding grid attribute feature *A* to both {*D, G, Y*} and {*D, G, T, Y*} which can also improve the performance of the xDeepFM. Meanwhile, xDeepFM performs best after the combination of {*D, G, T, A, Y*}. Therefore, it can be concluded that the time slot feature *T* and the grid attribute feature *A* can effectively improve the recommended performance.

The time slot feature *T* and grid attribute feature *A* represent the spatio-temporal contexts, so it can also be concluded that integrating the spatio-temporal [21] contexts into xDeepFM which can effectively improve the performance of the pick-up area recommendation. The *RMSE* is only 0.8%, *MAE* is about 7%, and *explained variance score* is over 98%.

6 Conclusion

In this paper, we propose a novel approach of integrating spatio-temporal contexts into xDeepFM for taxi pick-up area recommendation which can accurately recommend pick-up area to driver. First, we extract pick-up points from the original GPS data. Then, the pick-up points and POIs are mapped into the corresponding grids, we distil the spatio-temporal features from these grids to construct spatio-temporal contexts matrix. The spatio-temporal contexts matrix is input into xDeepFM to achieve end-to-end high-order and low-order feature learning, and the feature interactions occur at the vector-wise. Thus, our method can effectively enhance the accuracy of recommendation. Finally, we evaluate the performance of different recommendation methods and spatio-temporal contexts on a real dataset. The experimental results show that our approach can effectively improve the performance of recommendation.

In future work, we intend to mine more useful information from GPS data or try other advanced models to improve the performance of taxi pick-up area recommendation.

Acknowledgement. Vehicle trajectory data were acquired from the GAIA open data initiative (https://gaia.didichuxing.com) of Didi Chuxing.

Funding Statement. This project was funded by the National Natural Science Foundation of China (Grant No. 41871320), the Key Project of Hunan Provincial Education Department (19A172), Hunan Provincial Natural Science Foundation of China (2021JJ30276), A Project Supported by Scientific Research Fund of Hunan Provincial Education Department (No. 18K060).

Conflicts of Interest. The authors declare no conflicts of interest.

References

1. Liao, Z., Zhang, J., Liu, Y., Xiao, H., Zhao, Y., Yi, A.: Fusing geographic information into latent factor model for pick-up region recommendation. In: Proceedings of the 2019 IEEE International Conference on Multimedia & Expo Workshops, Piscataway, pp. 330–335. IEEE (2019)
2. Senthilselvan, N., Subramaniyaswamy, V., Sivaramakrishnan, N., Amir H.G.: Resolving data sparsity and cold start problem in collaborative filtering recommender system using Linked Open Data. Expert Syst. Appl. **149**, 113248 (2020)
3. Koren, Y., Bell, R., Volinsky, C.: Matrix factorization techniques for recommender systems. Computer **42**(8), 30–37 (2009)
4. Rendle, S.: Factorization machines. In: Proceedings of the International Conference on Data Mining, Piscataway, pp. 995–1000. IEEE (2010)
5. Lian, J., Zhou, X., Zhang, F., Chen, Z., Xie, X.: XDeepFM: combining explicit and implicit feature interactions for recommender systems. In: Proceedings of the 24th ACM SIGKDD Conference on Knowledge Discovery and Data Mining (KDD), pp. 1754–1763. ACM, New York (2018)
6. Pang, H., Wang, P., Gao, L., Tang, M., Huang, J., Sun, L.: Crowdsourced mobility prediction based on spatio-temporal contexts. In: Proceedings of the 2016 IEEE International Conference on Communications (ICC), Piscataway, pp. 1–6. IEEE (2016)

7. Chen, Y., Fu, Q., Zhu, J.: Finding next high-quality passenger based on spatio-temporal big data. In: Proceedings of the 5th International Conference on Cloud Computing and Big Data Analytics, Piscataway, pp. 447–452. IEEE (2020)
8. Song, M., Yuan, J.: Recommendation model of taxi passenger-finding locations based on Markov algorithms. Comput. Eng. Softw. **40**(6), 140–143 (2019)
9. Agrawal, A., Raychoudhury, V., Saxena, D., Kshemkalyani D, A.: Efficient taxi and passenger searching in smart city using distributed coordination. In: Proceedings of the 21st International Conference on Intelligent Transportation Systems, Piscataway, pp.1920–1927. IEEE (2018)
10. Wang, J., Zhang, L., Wang, L., Yan, K., Diao, Y., Zhao, Z.: The prediction analysis of passengers boarding and alighting points based on the big data of urban traffic. In: MacIntyre, J., Zhao, J., Ma, X. (eds.) SPIOT 2020. AISC, vol. 1282, pp. 17–24. Springer, Cham (2021). https://doi.org/10.1007/978-3-030-62743-0_3
11. Phanhong, M., Likitpanjamanon, P., Chareonwai, P., Srisurapanon, V., Phunchongharn, P.: A spot-recommendation system for taxi drivers using Monte Carlo optimization. In: Proceedings of the 1st International Conference on Big Data Analytics and Practices, Piscataway, pp. 1–5. IEEE (2020)
12. Yuan, C., Geng, X., Mao, X.: Taxi high-income region recommendation and spatial correlation analysis. IEEE Access **8**, 139529–139545 (2020)
13. Huang, Z., Gao, S., Cai, C., Zheng, H., Pan, Z., Li, W.: A rapid density method for taxi passengers hot spot recognition and visualization based on DBSCAN+. Sci. Rep. **11**(1), 1–13 (2021)
14. Li, X., et al.: A taxi recommender system based on inter-regional passenger mobility. In: Proceedings of the International Joint Conference on Neural Networks, Piscataway, pp. 1–8. IEEE (2019)
15. Hsieh, H., Lin, F.: Recommending taxi routes with an advance reservation–a multi-criteria route planner. Int. J. Urban Sci. **26**(1), 162–183 (2022)
16. Li, S., Su, F.: PTRA: a route recommendation system for idle taxi drivers. Comput. Technol. Dev. **31**(2), 33–37 (2021)
17. Lai, Y., Lv, Z., Li, K., Liao, M.: Urban traffic Coulomb's law: a new approach for taxi route recommendation. IEEE Trans. Intell. Transp. Syst. **20**(8), 3024–3037 (2019)
18. Wang, R., Chow, C., Lyu, Y.: TaxiRec: recommending road clusters to taxi drivers using ranking-based extreme learning machines. Trans. Knowl. Data Eng. **30**(3), 585–598 (2018)
19. Liu, Y., Wang, X., Liu, J., Liao, Z., Zhao, Y., Wang, J.: An entropy-based model for recommendation of taxis' cruising route. J. Artif. Intell. **2**(3), 137–148 (2020)
20. Guo, H., Tang, R., Ye, Y., Li, Z., He X.: DeepFM: a factorization-machine based neural network for CTR prediction. In: Proceedings of the 26th International Joint Conference on Artificial Intelligence, pp. 1725–1731. Morgan Kaufmann, San Francisco (2017)
21. Wu, L., et al.: Optimizing cruising routes for taxi drivers using a spatio-temporal trajectory model. ISPRS Int. J. Geo Inf. **6**(11), 373 (2017)

Learning When to Communicate Among Actors with the Centralized Critic for the Multi-agent System

Qingshuang Sun, Yuan Yao, Peng Yi, Xingshe Zhou, and Gang Yang(✉)

School of Computer Science, Northwestern Polytechnical University, Xi'an, China
sunqsh@mail.nwpu.edu.cn, yeungg@nwpu.edu.cn

Abstract. Centralized training and decentralized execution have become a basic setting for multi-agent reinforcement learning. As the number of agents increases, the performance of the actors that only use their own local observations with centralized critics is prone to bottlenecks in complex scenarios. Recent research has shown that agents learn when to communicate to share information efficiently, that agents communicate with each other in a right time during the execution phase to complete the cooperation task. Therefore, in this paper, we proposed a model that learn when to communicate under the centralized critic supporting, so that the agent is able to adaptive control communication under the centralized critic learned by global environmental information. Experiments in a cooperation scenario demonstrate the advantages of model. With our proposed cooperation model, agents are able to block communication at an appropriate time under the centralized critic setting and cooperation with each other at the task.

Keywords: Centralized critic · Communication · Multi-agent · Reinforcement learning · Cooperation

1 Introduction

Multi-agent systems (MAS) has been paid much attention in the past few decades, because a single agent with autonomy, perception, communication and computation is cannot cope with complex tasks in the dynamic and unpredictable nature of world [1]. Cooperation is fundamental characteristic of multi-agent systems where agents should achieve a global task to maximum the utility of overall systems [2]. For example, multiple unmanned aerial vehicles (UAVs) has been designed to cooperation with each other to accomplish in military tasks or natural disaster rescue tasks. The complexity of multi-agent cooperative tasks makes it difficult for agents' strategies to be designed with expert knowledge to adapt to dynamic environments. Due to the limitation of "pre-designed" methods, agents rely on self-learning by "learning-based" methods to find a better strategy to gradually improve the performance of the agent or the whole multi-agent system.

Y. Sun et al. (Eds.): ChineseCSCW 2021, CCIS 1492, pp. 134–146, 2022.
https://doi.org/10.1007/978-981-19-4549-6_11

Reinforcement Learning (RL) is an area of machine learning, mapping the environmental state to the action policy. However, Traditional reinforcement learning approach for MAS is difficult to handle high-dimensional continuous environment. Accompanied the developing of deep learning, the deep neural network (DNNs) is utilized as the function approximation of RL to develop an area of deep reinforcement learning (DRL). Recent years have witnessed successfully application of DRL in many sequential decision-making problems, such as game learning [3], multi-robot systems [4,5], and autonomous driving [6]. While DRL is effective handle high-dimensional data and learn how to interact with in a dynamic environment (i.e., taking actions). DRL provide a paradigm for learning of MAS, emerging area of multi-agent deep reinforcement learning (MDRL). Learning in multi-agent setting is fundamental more difficult than the single agent due to the non-stationary in MAS, where the learning of each agent need to considering the policy changing of other agents. MADDPG [7] is an extension of actor-critic method, which adopt the framework of centralized training with decentralized execution. The centralized critic allow to use extra information of all agents to keep the stationary of environment in the training phase, while the local actor only is allowed to use local information at execution phase. In other words, their is no communication in running process of MAS. However, in mostly MAS, the behavior of a single agent that only obtaining the local information is highly dependent on that of other members of the team. Therefore, communication between agents to share information effectively is an important way for the team to achieve common goals. Recently, there is an emerging subarea in MDRL: learning communication where agents can learn communication protocols to share information efficiently in cooperative tasks [8].

The DNNs structure of DRL is capable of solving the communication problem in MAS, and overcome the deficiency of communication protocols designed by expert knowledge. Recent studies [9,10] have shown that agents learn communication protocols based on DRL, which automatically learn communication protocols that are hard for human to design in advance. More importantly, the "learning-based" method effectively share information, which guides the agent decision-making more reasonably in a complex and dynamic environment. Generally, learning communication has at least three meanings. To be specify, firstly, restricted by the communication bandwidth, agents learn to extract valuable information from partially observable environments, which can be transported using a few packets at a timestep. Secondly, agents learn to determine whether to communicate with others at a specific timestep to avoid wasting the communication bandwidth. Finally, agents learn targeted communication. An individual agent can actively select other agents to send a message.

In this paper, we focus on when to communicate under the centralized critic supporting. Some studies enable agents to learn when to communicate. IC3Net [11] use gating mechanism to learn when to communicating. But this method uses REINFORCE [12] to train the actor, which only augments the extra information between agents' actors, not using the critic to estimate the wiser or foolish for actor's policy. In addition, it cannot generalize to a more complex environment with the continuous space environment. The centralized critic that adds extra information from all agents has become the basic setting of much multi-agent deep reinforcement learning, which keep the stationary of environment to a certain degree. However, the centralized critic with the local actor is difficult to scale to a large-scale multi-agent environment. To solve the issues mentioned above, we put forward the model that controlled communicating among actors with the centralized critic, called COCC, which explores learning when to communicate with the support of a centralized critic. It is noteworthy that the running processing of COCC is in a distributed manner. Experimental results show that the performance of our model better than the baselines in cooperation scenarios with continuous space. It is demonstrated that our method enables agents to learn effective communication under the centralized critic supporting.

Concretely, our contributions are as follow:

- we explore when to communicate under the centralized critic setting, which has good results in the continuous space environment.
- The process of the system run in a distributed manner. Each agent makes a decision and receives information independently.

2 Related Works

In real-world applications, agents that work together need to achieve a global task, but they only have partial observability of the environment. From the perspective of communication, the recent cooperation of MDRL works can be divided into two categories: learning cooperation with explicit communication and learning cooperation without explicit communication. Learning cooperation with explicit communication refers to agents learn communication protocols to solve cooperation tasks. The communication protocols of the agent can determine when to communicate, what kind of message to send, and who to send them to [13]. Learning cooperation without explicit communication refers to agents learn to cooperate using extract information such as actions and observations in the training phase, but only use the local observation in the execution phase.

In the works of learning cooperation with explicit communication, ATOC [14] first proposed an attentional communication model that designs an attention unit that enables agents to learn when to communicate. In addition, each agent according to their local state judge whether to connect with other agents. If communication, the agent, called initiator, select other agents to share information. Finally, it can output important information from received information by the LSTM unit for cooperating decision-making. However, the method is hand-tuned to choose neighbors to integrate information, which is manual. IC3NET using the gating mechanism enables agents to learn when to communicate. Each agent is trained with its individual reward, which can be applied to cooperation scenarios and competition scenarios. However, it uses REINFORCE to train its setup, only using other information for the actor without critic.

In the works of learning cooperation without explicit communication, MADDPG that extends DDPG [15] explore deep reinforcement learning methods for multi-agent domains, which applied it to multi-agent systems, and achieves good results. The framework is centralized training with decentralized execution. The centralized critic obtains extra information from all agents to ease training, keeping the environmental stationery. The actor only uses local observation in the test time without extra information from others. After training is completed, only the local actors are used at test time. However, When the number of agents gradually increases, only centralized critics and local actors will lose their effectiveness. MAAC [16] is an improvement of MADDPG, where also uses the centralized critic, but embedding attention mechanism which gauge the relevance between agents to selects relevant information for each agent. The main approach is to learn a centralized critic with an attention mechanism. However, its main disadvantage is the same as MADDPG, only exiting centralized critic is able to obtain other agents' information, lacking information exchange for actors who only use their own local information.

Centralized critic makes the agent perceive global environment in the training phase to keep the stationary of MAS, which not enough for large scale multi-agent systems. In the execution phase, like humans, communication is a necessary skill to share important information between agents, which is the basis for multi-agent systems cooperation. However, the fully-communicate not only consumes bandwidth but also generates redundant information. Therefore, it is a worthwhile exploration that the agent to learn when to communicate to partial communication with the support of centralized critic as the number of agents and the complexity of the environment increase.

3 Methods

COCC is an extension of the actor-critic model, which utilizes centralized training with decentralized execution. It means that critic is a centralized way, which is able to get the information of all agents in the environment in the training phase, and actors send messages to exchange information at a time step by learning when to communicate. COCC consists of independent critic for each agent, and a shared actor including environmental encoder, controlled module, information integration module, and policy network. We describe our framework from two parts: critic and actor as shown in Fig. 1.

Centralized Critic

MADDPG proposes the centralized critic that uses extra information for all agents during training, which can include additional state information if available. Here, we use the local observation and policy of the agent to represent additional state information. The critic takes the local observation $(o_1 \dots o_N)$ and the action policy of agent $(a_1 \dots a_N)$ as inputs to estimate the joint the action-value $Q_i^{\mu}(o_1 \dots o_N, a_1 \dots a_N)$ of agent i at every timestep. Each agent has its own estimator Q_i^{μ}. Since each Q_i^{μ} is learned separately. Agents can have a diverse reward structure, such as global reward and local reward. We treat the shared actor network as a whole network as μ, parameterized by θ. The

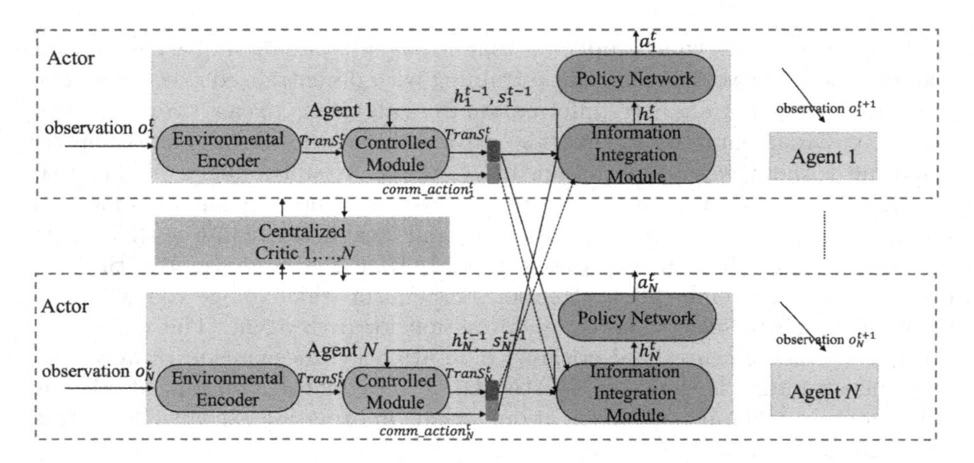

Fig. 1. An overview of COCC. Each agent has independent Critic and a shared Actor that includes four modules. For agent i, the local observation o_i^t is preprocessed by the environmental encoder to transform it to the $TranS_i^t$. The controlled module judges the $TranS_i^t$ whether to send to other agents in this time step. All received information for agent i is integrated by the information integration module. The policy network depend on integrated information to make a decision.

centralized critic is learned by temporal difference [17]. The gradient of the expected return $J(\theta) = \mathbb{E}[R_i]$ for agent i can be formulated as:

$$\begin{aligned} \nabla_\theta J(\theta) = \mathbb{E}\left[\nabla_\theta \log \mu_\theta\left(a_i^t \mid o_i^t\right)\right. \\ \left. Q_i^\mu\left(o_1^t, \ldots, o_N^t, a_t^1, \ldots, a_t^N\right)\right] \end{aligned} \tag{1}$$

Decentralized Actor

In structure of COCC, all agents share a whole actor network. They will run independently in the execution process, forming a distributed manner. The whole actor consists of four parts, including environmental encoder, controlled module, information integration module, and policy network.

Environmental Encoder. The agent initially perceives the original local observation information from the environment. Due to the different dimensions and sizes of the local observation information in different environments, it is necessary to use the environmental encoder to preprocess the original local observation information and compress it to a fixed dimension and length. This not only transmit a compress information by meeting limited bandwidth but also facilitates the generalization of the model to different scenarios. Firstly, the environmental encoder preprocesses the local observation o_i^t from the environment. The encoder function $e(.)$ is parameterized by a fully connected layer, which can transform the local observation o_i^t to a fixed dimension. Therefore, the dimension of o_i^t can

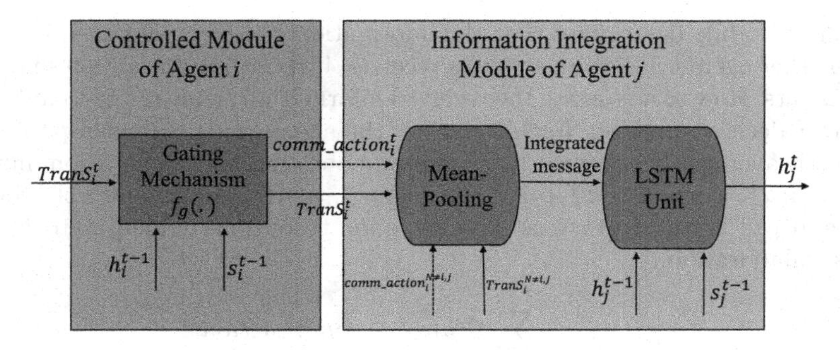

Fig. 2. The Controlled Module and Information Integration Module. In the sending process, the gating mechanism cohesion history experience h_i^{t-1} and current information to output signal of communicative action $comm\_action_i^t$ where judge $TranS_i^t$ whether to send to other agents. In the receiving process, received messages are integrated by mean-pooling. LSTM unit abstracts the hidden state h_i^t from the integrated message, which supports policy network decision-making.

be controlled so as to facilitate leveraged to the next module. The transformed o_i^t can be formulated as:

$$TranS_i^t = e\left(o_i^t\right) \tag{2}$$

Controlled Module. Hard attention is proposed in the caption generators task [18], where the model focuses attention on a special region of the image. If the region is selected, the value of the region is set to 1, otherwise, it is 0. This mechanism is suitable for focusing important information in MAS. In COCC, the controlled module is instantiated as gating function $f_g(.)$ with a fully connected neural network and seen as hard attention. The gating function $f_g(.)$ is also regarded as a binary classifier, inferring communicating action 1 or 0. If the output is 1, it means that the gating function opens the door and allows information to flow. If the output is 0, it means that the door is closed to block the flow of information. This method control the communication well, allowing the agent to learn when to communicate, so as to release the circulation of important information and block the circulation of non-important information. The controlled module takes $TranS_i^t$ and the saved hidden state h_i^{t-1} from the information aggregation module at time $t-1$ as input to output the actions binary communication actions, which determines whether the message of the agent needs to communicate with others. The $f_g(.)$ combining historical experience h_i^{t-1} and the local observation state $TranS_i^t$ of this time judge whether the information at this moment is worth sending as shown in Fig. 2.

$$comm\_action_i^t = f_g\left(TranS_i^t, h_i^{t-1}\right) \tag{3}$$

Information Integration Module. This module is composed of mean-pooling and $LSTM$ unit [19]. $LSTM$ unit retain part of the previous state and selectively output information. Therefore, we can extract more useful information by using $LSTM$. Figure 2 shows that when the agent acts as the sender, the controlled module determines that the information can flow, and then it will be sent to other agents. When an agent as receiver, it receives information sent from other agents. How to aggregate the received information is one of the foundations of correct decision-making. In this session, the agent first uses mean-pooling to arithmetic average the received information to obtain the integrated communication vector c_i^t. Then, $LSMT$ takes the $TranS_i^t$, c_i^t, the hidden state of previous h_i^{t-1}, and cell state of previous s_i^{t-1} as input to output hidden state h_i^t that guides collaboration.

$$c_i^t = \frac{1}{N-1} \sum_{j \neq i} TranS_j^t \odot comm\_actiont_j^t \tag{4}$$

$$h_i^t, s_i^t = LSTM\left(TranS_j^t, c_i^t, h_i^{t-1}, s_i^{t-1}\right) \tag{5}$$

Policy Network. The policy network based on the hidden state h_i^t predicts the action that acts on the environment.

$$a_i^t = \pi\left(h_i^t\right) \tag{6}$$

4 Experiments

We use the multi-agent particle environment [7] to validate the effectiveness of COCC, which consists of N agents and L landmarks exiting in a two-dimensional world with continuous space and discrete time. Experiments are performed on the cooperative navigation scenario that needs agents to collaborate with each other to finish the task. The training setup is an extensive DDPG. DDPG is a variant of Actor-Critic method where the policy and critic are approximated with deep neural networks. It has experience replay mechanism, which random sample minibatch data that are stored in the experience buffer during training. DDPG makes use of the online network and target network where the target network is updated by the soft updating way.

4.1 Setup

Hyperparameters. In the experiments, we use the Adam optimizer with the learning rate of 0.001. The discount factor of reward γ is 0.95. For the soft targeted network, we use τ is 0.001 updated. The critic network is parameterized by a two-layer MLP with 128 units per layer. The hidden layer of components of Actor is 128 units. We initialize all of parameters by the method of random normal. The capacity of the replay buffer is set 10^6, and every time we take of a minibatch of 1024. We run a thousand episodes to accumulate experience to the replay buffer before training.

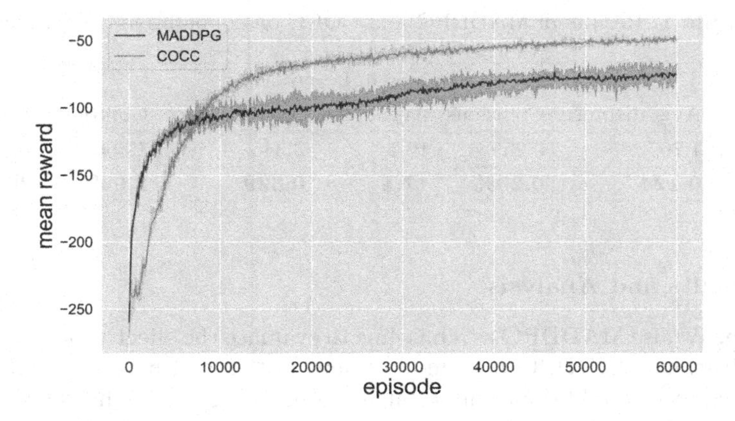

Fig. 3. Learning curves of MADDPG and COCC

Cooperative Navigation. There are N agents and L landmarks in the cooperative navigation environment. The goal of the team is to reach the position of landmarks through collaboration. If the number of agents is equal to the number of landmarks, N agents will occupy N landmarks respectively, that is, there is no overlap of multiple agents on the same landmark. Agents observe the relative positions of other agents and landmarks. In our experimental settings, we use a shared global reward or individual rewards as feedback to encourage multi-agent cooperation. $N = L = 3$ is designated with a shared global reward. $N = L = 6$ is designated with individual rewards, because individual reward converge better and scales better than global rewards [11]. The shared global reward $r_{textglobal}$ is the sum of the minimum distance of any landmark to each agent. The individual reward $r^i_{\text{inidivual}}$ is the sum of the minimum distance between agents and its nearest landmark and $r_{collision}$. Further, agents are penalized for $r_{\text{collision}} = -1$ when a collision occurs. The common goal of the cooperation is that agents cover all landmarks without collision. Finally, the shared global reward and individual reward can be formulated as:

$$r_{\text{global}} = \sum_{l=1}^{N} -d_l^a + \delta_{num} * r_{\text{collision}} \quad (7)$$

$$r^i_{\text{inidivual}} = -d_i^l + \sum_{l=1}^{N} -d_l^a + \delta_{num} * r_{\text{collision}} \quad (8)$$

where d_l^a denotes the distance between the landmark l and the nearest agent, d_i^l is the minimum distance between agent i and its nearest landmark, and δ_i is the total number of collisions between agents in a unit time step.

Table 1. Results of MADDPG and COCC on cooperative navigation

N = 3, L = 3				N = 6, L = 6		
Methods	Avg. distance	Collisions	Avg. steps	Avg. distance	Collisions	Avg. steps
MADDPG	1.767	0.209	19.7	3.345	1.366	–
COCC	**0.124**	**0.23%**	**17.1**	**0.329**	**2.64%**	**28.6**

4.2 Results and Analysis

Baseline. We use MADDPG as a baseline to evaluate the effectiveness of explicit communication. MADDPG uses independent critics and actors for all agents, only critics receive additional information from all agents in the training time. Actors only take local observation of themselves as input to output action policies in the execution time. In other words, the agent uses the trained model without communication at execution time.

The Efficiency of Explicit Communication. There is no communication between agents during the execution of MADDPG. The COCC keeps information sharing among agents at a certain time step when performing tasks. Therefore, compared with MADDPG, we can analyze the effectiveness of explicit communication to assist agents in completing cooperation tasks. Figure 3 shows learning curves of 60000 episodes in terms of mean rewards of episodes. Here, centralized critics of the two methods uses the same neural network structure and number of neural network units. It is obvious that our proposed method can faster converge to the better average reward value. COCC converge to a better policy using only 60,000 episodes, but MADDPG needs more episodes to reach convergence. In this experiment, we trained MADDPG for more than 150,000 episodes to make it convergence. We also found that the fluctuation of COCC training is smaller than that of MADDPG in the repetitive experiment. Our method is more stable in the training time. It is worth noting that MADDPG converges faster in the early stage of training, because MADDPG has a relatively simple structure compared to COCC so updating and iterating better model parameters faster. COCC requires more episodes to learn due to the complexity of the Actor structure, which accelerate the speed of learning and develop towards a better policy network after learning the initial knowledge.

We find out from here the agent has the ability to communicate during execution, which is able to learn a better strategy with fewer episodes. Table 1 shows that we explore two cases including $N, L = 3$ with shared global reward, and $N, L = 6$ with individual rewards. When the number of agents is few, such as $N = 3$, agents are able to converge to a policy that is able to guide agents complete the task. The centralized critic MADDPG help the agent perceive a stationary environment in a setting with a small number of agents. However, when the number of agents increases, MADDPG is difficult for the setting $N, L = 6$ to converge to smaller mean rewards, losing the ability to find landmarks in the environment. The actor of COCC with communication skills still accomplishes

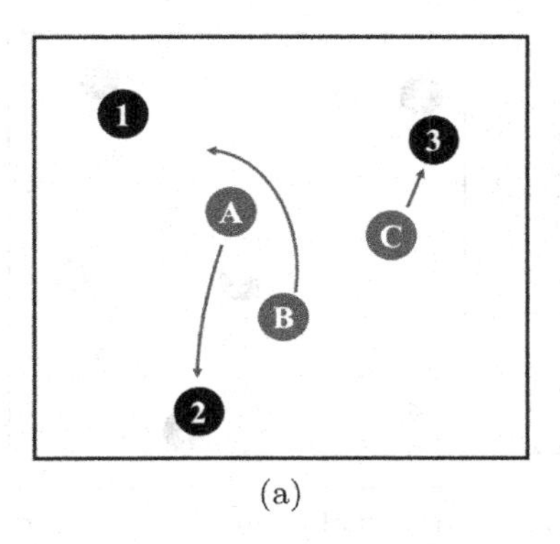

(a)

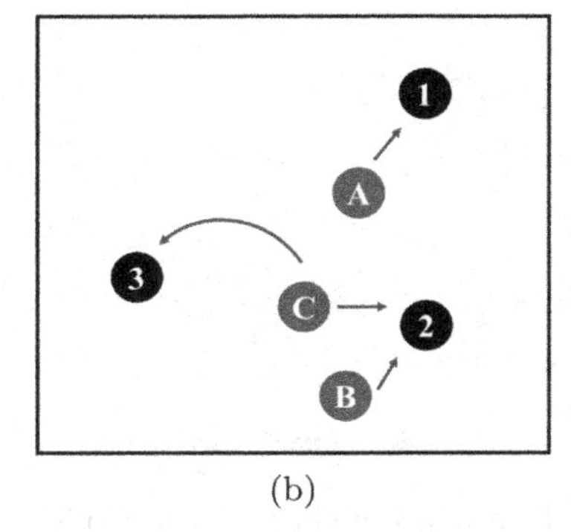

(b)

Fig. 4. Visualization trajectories of agents. (a) is the trajectory of MADDPG. (b) is the trajectory of COCC.

cooperation tasks well, where the average distance is 0.329, and collision rate is 2.64% when existing six agents and six landmarks in environment. Agents is able to occupy all the landmarks efficiently without missing them. From the discussion, one may conclude that communication between agents in the execution time is able to promote more reasonable decisions.

Comparing the average steps indicator, MADDPG with $N = 3$ requires more steps to search all landmarks than COCC. This is flawed when agents complete tasks that require real-time performance. The running process of the two methods is visualized in Fig. 4. It is shown that MADDPG weaker guides the agent to find the nearest landmark that is beneficial to team cooperation. On the contrary, they often make detours to look for landmarks that are relatively far away. Figure 4(a) shows the running process of MADDPG at a time step. At this time step, the fastest route to complete the cooperation task should be that agent A occupies landmark 1, agent B occupies landmark 2, and agent C occupies landmark 3. However, MADDPG prompts agents A and B to switch directions to find landmarks 2 and 1 respectively in fact. This not only increases the number of steps of completing the task but also increases the collision probability between agents. The collision rate of MADDPG is higher than that of COCC in each settings as shown in Table 1. However, COCC is able to help agents to find a suitable landmark quickly because this method communicates during execution. The running process of COCC is illustrated in the Fig. 4(b). When two agents B and C tend to find landmark 2 together at a time step, they will communicate and negotiate. After the negotiation, agent C will change the direction to find landmark 3 to avoid the collision of agents and gathering on the same landmark for agents. This phenomenon reflects that communication is a basic skill for multi-agent cooperation, helping agents to grasp broader information in the environment and the behavioral intentions of other agents. It is also seen from the side that the centralized critic can only help the agent to describe a stable environment to a certain extent. On this basis, proper communication enable the agent to complete the cooperation task and achieve a multiplier effect.

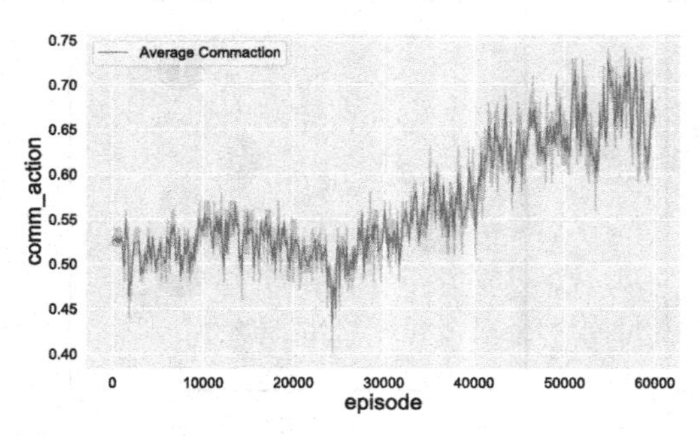

(a) Learning the Gating Communication action.

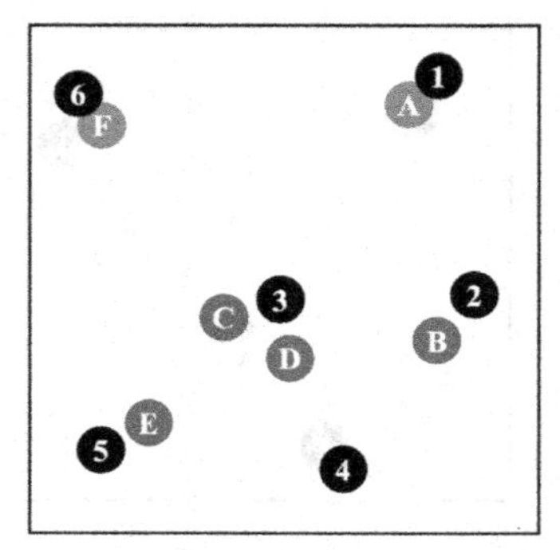

(b) Visualization of communication.

Fig. 5. Communication. (a) shows the varies of gating communication action over each episode during training. (b) visualize the communication of agents when $N, L = 6$

When to Communicate. COCC uses a gating mechanism with the centralized critic supporting to control communication to learn when to communicate. Figure 5(a) shows communicative action averaged over each episode for $N, L = 3$. Agents gradually learn to increase communication with each other during training phase in the cooperative task. Eventually, agents keeps the communication traffic within 60%–75% by learning when to communicate. The policy learned depending on partial communication is able to give better instruction for their behaviors.

Figure 5(b) visualizes the communication effect of the COCC after learning in a time step when the number of landmarks and agents is 6. Red circle represent the communicating agents, and purple circles represent the Non-communication agents. At this moment, agents A and F have reached landmark 1 and 6 respectively. They choose to keep silent because it is far away from other agents and landmarks. Agents C and D tend to find the same landmark 3, so they need to communicate and negotiate so that agent D makes a turn in time to occupy the landmark 4. Agents B, E are close to landmarks 2, 5 respectively, where communication is not very tight. But agents B, E is relatively close to agents C, D. It also needs to send information intermittently to inform other agents of their own situation and prevent them from conflicting. Therefore, agents B, C, D, and E keep talking by learning when to communicate.

5 Conclusions

In this paper, we propose the controlled communication model with the centralized critic, which uses gating mechanisms with hard attention to learn when to communicate. Learning communication under the centralized critic setting gets better performance and scalable compared with baseline. It is demonstrated that combing learning communication with the centralized critic is an effective method for multi-agent systems.

Acknowledgments. This work was supported in part by the National Key Research and Development Program of China (No. 2017YFB1001902), the National Natural Science Foundation of China (No. 61876151, 62032018) and the Fundamental Research Funds for the Central Universities (No. 3102019DX1005).

References

1. Chen, F., Ren, W., et al.: On the control of multi-agent systems: a survey. Found. Trends® Syst. Control **6**(4), 339–499 (2019)
2. d'Inverno, M., Luck, M.: An operational analysis of agent relationships. In: Understanding Agent Systems. Springer Series on Agent Technology. Springer, Heidelberg (2004). https://doi.org/10.1007/978-3-662-10702-7_4
3. Silver, D., et al.: Mastering the game of go with deep neural networks and tree search. nature **529**(7587), 484–489 (2016)
4. Fan, T., Long, P., Liu, W., Pan, J.: Distributed multi-robot collision avoidance via deep reinforcement learning for navigation in complex scenarios. Int. J. Robot. Res. **39**, 856–892 (2020)
5. Xiao, Y., Hoffman, J., Xia, T., Amato, C.: Learning multi-robot decentralized macro-action-based policies via a centralized Q-net. In: 2020 IEEE International Conference on Robotics and Automation (ICRA), pp. 10695–10701. IEEE (2020)
6. Kiran, B.R., Sobh, I., Talpaert, V., Mannion, P., Perez, P.: Deep reinforcement learning for autonomous driving: a survey. IEEE Trans. Intell. Transp. Syst. **23**, 4909–4926 (2022)
7. Lowe, R., Wu, Y., Tamar, A., Harb, J., Pieter Abbeel, O., Mordatch, I.: Multi-agent actor-critic for mixed cooperative-competitive environments. In: Advances in Neural Information Processing Systems, vol. 30, pp. 6379–6390 (2017)
8. Hernandez-Leal, P., Kartal, B., Taylor, M.E.: A survey and critique of multiagent deep reinforcement learning. Auton. Agent. Multi-Agent Syst. **33**(6), 750–797 (2019). https://doi.org/10.1007/s10458-019-09421-1
9. Sukhbaatar, S., Fergus, R., et al.: Learning multiagent communication with backpropagation. In: Advances in Neural Information Processing Systems, pp. 2244–2252 (2016)
10. Foerster, J., Assael, I.A., De Freitas, N., Whiteson, S.: Learning to communicate with deep multi-agent reinforcement learning. In: Advances in Neural Information Processing Systems, pp. 2137–2145 (2016)
11. Singh, A., Jain, T., Sukhbaatar, S.: Learning when to communicate at scale in multiagent cooperative and competitive tasks. In: International Conference on Learning Representations (2018)
12. Williams, R.J.: Simple statistical gradient-following algorithms for connectionist reinforcement learning. Mach. Learn. **8**(3), 229–256 (1992)
13. Das, A., et al.: TarMAC: targeted multi-agent communication. In: International Conference on Machine Learning, pp. 1538–1546 (2019)
14. Jiang, J., Lu, Z.: Learning attentional communication for multi-agent cooperation. In: Advances in Neural Information Processing Systems, pp. 7254–7264 (2018)
15. Lillicrap, T.P., et al.: Continuous control with deep reinforcement learning. arXiv preprint arXiv:1509.02971 (2015)

16. Iqbal, S., Sha, F.: Actor-attention-critic for multi-agent reinforcement learning. In: International Conference on Machine Learning, pp. 2961–2970. PMLR (2019)
17. Sutton, R.S., Barto, A.G.: Reinforcement Learning: An Introduction. MIT Press (2018)
18. Xu, K., et al.: Show, attend and tell: neural image caption generation with visual attention. In: International Conference on Machine Learning, pp. 2048–2057. PMLR (2015)
19. Hochreiter, S., Schmidhuber, J.: Long short-term memory. Neural Comput. **9**(8), 1735–1780 (1997)

Social Media and Online Communities

Academic Article Classification Algorithm Based on Pre-trained Model and Keyword Extraction

Zekai Zhou[1], Dongyang Zheng[1], Zihan Qiu[1], Ronghua Lin[1], Zhengyang Wu[1], and Chengzhe Yuan[2](✉)

[1] School of Computer Science, South China Normal University, Guangzhou, China
{2020022974,flash,hhhqzh,rhlin,wuzhengyang}@m.scnu.edu.cn

[2] School of Electronics and Information, Guangdong Polytechnic Normal University, Guangzhou, China
ycz@gpnu.edu.cn

Abstract. Text classification, which has extensive application in many fields, assigns a tag to a given piece of text. Academic articles are the most authoritative source of academic information and play an important role in the process of delivering latest academic information. On social media, these academic articles will generate considerable academic news, translated articles, tutorial articles, etc. How to classify these academic articles has become more and more important. In this paper, we employ pre-trained model for academic text classification. Further, to identify terminology in academic papers, we design a convolutional layer to capture local dependencies. We also introduce a max-pooling layer that can get the most important elements in the text. Considering that academic articles are usually long, we propose a fine-tuning technique based on keyword extraction for pre-trained model to obtain global information. We conduct experiments on the Fudan Text Classification Corpus and the SCHOLAT academic news dataset. The experimental results show that the proposed method outperforms the methods commonly used in recent years on both datasets.

Keywords: Text classification · Neural network · Pre-trained model · Keyword extraction

1 Introduction

Text classification, also known as text categorization, is the process of automatic classification or tagging according to a certain classification standard. The task of text classification is to assign a document to one or more tags. It has a wide range of applications including topic labeling [29], news categorization, humor recognition [2,3,25], suggestion mining [21], sentiment analysis [9,15], and so on.

This work was supported in part by the National Natural Science Foundation of China under Grant U1811263 and Grant 61772211.

Y. Sun et al. (Eds.): ChineseCSCW 2021, CCIS 1492, pp. 149–161, 2022.
https://doi.org/10.1007/978-981-19-4549-6_12

On academic website SCHOLAT, automatic text classification can help classify articles posted by users. People can easily find articles related to academic news, call for papers, and recruitment.

Categorizing academic articles will help scholars find the articles they want more quickly. However, automatic articles classification can be very challenging because of the gap between the human language and the binary representation. Academic articles are written by experts in a field. The people who are interested in a field can easily identify an article belong to that field or not. However, it is hard for them to classify articles to any field. Besides, academic articles will contain words and terms that are rarely seen elsewhere. Therefore, it is not an easy task for machine to classify the academic articles.

There is a lot of research focusing on automatic text classification. Recent studies use pre-trained model for text classification. BERT [5] is one of the most prominent pre-trained model and has improved the performance of text classification greatly. One of the drawbacks of BERT is the slow classification speed. Therefore, it is common to truncate the text to a certain length to improve the classification speed while maintaining the classification performance. However, truncating the text usually means that some global information is lost. To solve this problem, we propose a fine-tuning technique based on keyword extraction for pre-trained model to save global information.

In this paper, our main contributions are below.

First, a convolutional layer and a max-pooling layer are introduced to capture local dependencies, and thus important terminology in academic papers can be identified. For example, the word benzene ring. People can classify an article as a chemical document by seeing the word benzene ring in the literature, but for machines, benzene ring is a rare word, and it has not even learned the co-occurrence of benzene and ring.

Second, we propose a fine-tuning technique for pre-trained model. In our model, we extract keywords from the article to obtain global information. Keywords and truncated text will be input into the pre-trained model for fine-tuning. Keywords extracted from the article can reduce information loss when the article is truncated in order to use BERT.

We evaluate our method on the Fudan Text Classification Corpus and the SCHOLAT academic news dataset. The experimental results show that the proposed method can improve the F1 value on both datasets compared the RoBERTa [18], a robust version of BERT.

2 Background and Related Work

With the rapid growth of social media, academic articles and their related articles are able to be widely disseminated. Academic websites, such as www.cnki.net and www.scholat.com, provide many academic articles for people. For website operators, automatic text classification can reduce operational costs. For website users, automatic text categorization can help them find the articles they want more quickly, which improves user experience. How to categorize the large number of academic articles posted on social media has become more and more important. On social media, text classification can also be applied to a variety of other scenarios. For online shopping websites, classifying product description from user reviews can help users get a more objective description of the product [22]. For movie review websites, detecting spoiler from movie reviews can help users avoid spoiler before watching the movie [24]. However, there is less work related to academic text classification [16,26]. Therefore, we focus on studying academic text classification in this paper. We will explore the characteristics of academic text so we can classify them better.

Neural network methods have been rapidly developed in recent years. Many neural network models that can avoid explicitly extracting domain features are explored to automatically classify articles [20]. TextCNN [12], which uses the idea of convolutional neural networks, can learn local dependencies, but cannot capture very long contextual information. On the contrary, TextRNN [17] can learn long dependencies between words at the cost of greatly increased computation time. FastText [11], as the name suggests, allows for fast text classification. Its simple architecture can perform well in case of very large amount of data. On the basis of the models above, various tricks have been proposed. RCNN [13], an architecture applies a max-pooling layer over LSTM layer to obtain the maximum vector from LSTM layer, then a feedforward layer is applied on that vector to do the classification. DPCNN [10], just like other convolutional models in computer vision, stacks convolutional layers to capture long dependencies. Attention mechanism[1,23] will enhance weights of key elements. For long sequences, some researchers proposed hierarchical attention networks for classification [28]. It can also be applied directly to the raw input or to its higher level representation [6].

Before feeding the data to the classifiers, we need to transform the raw data to numeric data. Bag of words and their extensions are the most classical methods to transform a piece of text to vector. However, these methods always suffer sparseness because of the large size of vocabulary. In 2013, word2vec [19] became popular for most NLP task. As the name says, it converts words into fixed-dimensions vectors. However, the model architecture leads to the disadvantage of word2vec, that is word vectors lack context. In 2018, Google proposed a language model BERT that employs the encoder from transformer. There are two tasks for BERT, masked LM task can help learn word-level representation while next sentence prediction can capture contextual information. BERT has inspired many related works, such as RoBERTa [18], albert [14], XLM-RoBERTa [4] and XLNet [27]. BERT is the most advanced embedding model and can be fine-tuned to suit the task of text classification. The development of embedding models has greatly improved the performance of text classification.

3 Our Model

3.1 Problem Formulation

Before presenting our approach, we give the definition of the problem. The task of text classification is to assign a label to a piece of text and is a typical classification problem. Given a dataset $\{(x_1, y_1), (x_2, y_2), ..., (x_n, y_n)\}$ where x_i is composed of words. The goal is to find a classifier $H(x)$ that can minimize the cross entropy loss. The loss function is given by the Eq. (1) where M is the class number. y_{ic} will be 1 if the tag of x_i is c, otherwise 0. $H(x_i)_c$ represents the probability that the tag of x_i is c.

$$L = \frac{1}{N} \sum_{i=1}^{N} \sum_{c=1}^{M} y_{ic} log(H(x_i)_c) \tag{1}$$

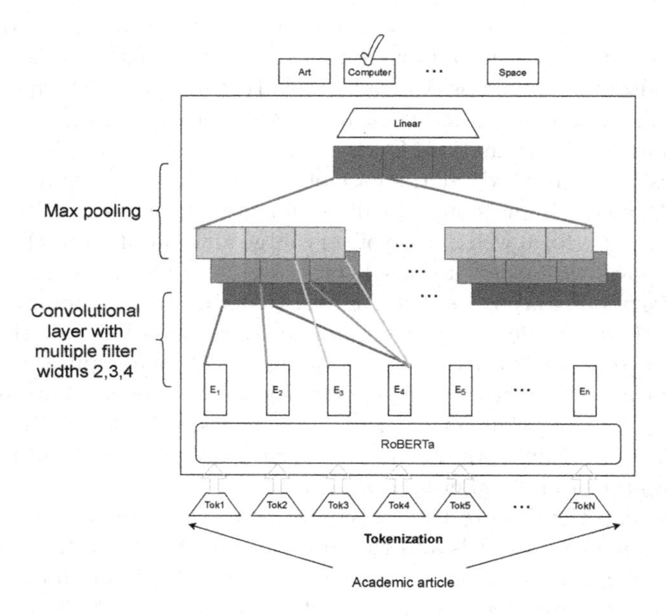

Fig. 1. The design of convolutional layer and max-pooling layer: a convolutional layer with different filter widths 2,3,4 is used to capture local dependencies and a max-pooling layer is adopted to gain key components. The pink lines for components indicate a filter width 4 and other cases are similar. (Color figure online)

3.2 The Design of Neural Network

In this paper, we employ the pre-train model RoBERTa for academic text classification. At the stage of tokenization, RoBERTa will separate the terminology into several words, so the co-occurrence dependencies between words may not be learned. In order to capture the co-occurrence dependencies, we design a convolutional layer and a max-pooling layer as depicted in the Fig. 1. For convolutional layer, a combination of different filter widths 2,3,4 is used in our model. We propose such a convolutional layer with the intention of capturing short-range dependencies. We also employ a max-pooling layer to gain key components from the output of convolutional layer. Different filter widths will gain different key components. These key components are concatenated to represent the whole text. Finally, a linear layer is used to perform the task of classification.

3.3 Fine-Tuning Technique Based on Keyword Extraction

To make full use of text and overcome the limitation of BERT, we propose a pre-trained model fine-tuning method based on keyword extraction as described in the Fig. 2. Compared with the Fig. 1, the most important difference is this model also needs to input keywords. We use a special character [SEP] in the BERT model to separate keywords and truncated text. Our proposed method can make full use of text by extracting keywords from the whole article. There are several methods can extract keywords from articles. We employ TF-IDF to extract keywords. TF-IDF is short for term frequency-inverse document frequency. TF-IDF can inflect how important a word is to a document in corpus. The equations below show the calculation of TF-IDF. In the Eq. (2), N_w denotes the number of a given word in a document while N denotes the total number of words in a document. In the Eq. (3), Y_w denotes the number of documents containing the word while Y denotes the total number of all documents. Finally, TF-IDF simply multiples TF and IDF in the Eq. (4).

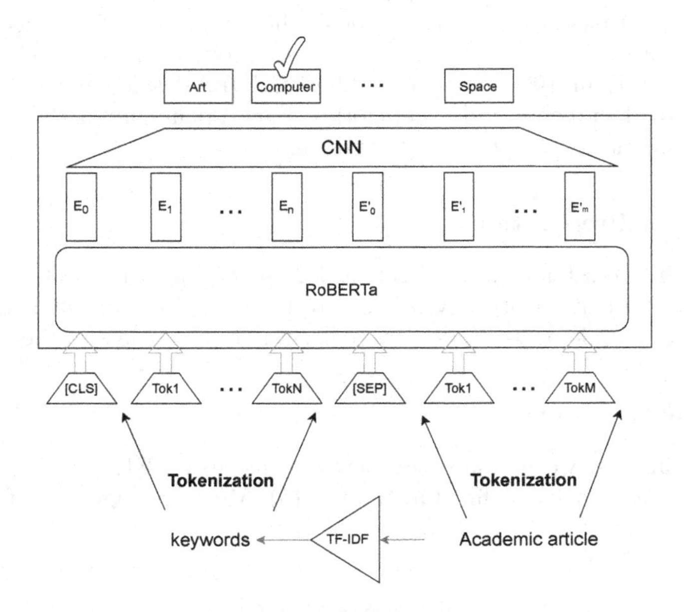

Fig. 2. A model based on pre-trained model and keyword extraction: The keywords will be extracted from academic article using TF-IDF. A special character [SEP] is used to separate keywords and article during the model fine-tuning phase.

$$TF = \frac{N_w}{N} \tag{2}$$

$$IDF = log(\frac{Y}{Y_w + 1}) \tag{3}$$

$$TF - IDF = TF \times IDF \tag{4}$$

Our proposed models are summarized as follows:

- **RoBERTaCNN**, a convolutional layer and a max-pooling layer are used to obtain higher level representation of the whole text on the basis of RoBERTa.
- **KE-RoBERTaCNN**, keywords are extracted to reduce the information loss on the basis of RoBERTaCNN. Keywords and truncated text are used to fine-tune the pre-trained model.

4 Experimental Setup

4.1 Dataset Description

Two dataset Fudan Text Classification Corpus and SCHOLAT academic news dataset are used to evaluate our methods. The Fudan Text Classification Corpus contains up to 19483 documents and 20 categories. There are 11 categories containing less than 100 documents. The SCHOLAT academic news dataset contains 9110 documents and 6 categories. Each document in the SCHOLAT academic news dataset contains title and content.

4.2 Dataset Preparation

we split each dataset into a train set, a dev set and a test set. Train set contains 60% documents. Both dev set and test set contain 20% documents. For SCHOLAT academic news dataset, we concatenate the title and the content.

4.3 Evaluation Metrics

In all experiments, we measure the performance using F1, specifically micro-average F1 is used. F1 is defined in the Eq. (5). Micro-average F1 is the sum of weighted F1s.

$$F1 = \frac{2 \times precision \times recall}{precision + recall} \tag{5}$$

4.4 Pre-experiments

In the early stage, we chose 8 simple models and 5 BERT-based models for our experiments.

The 8 simple models are CNN, RNN, RCNN, ARNN, ARCNN, DPCNN, FastText, Transformer(encoder only). We select the following representative models for further experiments based on their performance and popularity: CNN, RNN, RCNN, ARNN, FastText.

The 5 BERT-based models are ALBERT, BERT, RoBERTa, XLM-RoBERTa, XLNet. ALBERT, a lightweight BERT, reduces memory consumption by sharing weights, which helps model deployment, but does not perform as well as BERT. XLM-RoBERTa, optimized for cross-language scenarios, we observed that academic articles are usually mixed with multiple languages, so we also chose it for our experiments, but it did not show good performance in our application because the model structure is too complex and the data is not enough. XLNet, academic articles are usually very long, XLNet can handle more than 512 characters. However, we did not choose it because the training speed and testing speed were too slow. RoBERT, a more fully trained BERT, has shown the best performance in our preliminary experiments. Finally, we chose RoBERTa for further experiments.

4.5 Experiments

We carried out three experiments in total. The first experiment searches for the hyperparameter of sequence length and reasonably selects an appropriate length for model performance comparison. The second experiment is to show the effect of the convolutional layer. The third experiment verifies the effect of the pre-trained model fine-tuning techniques based on the keyword extraction algorithm.

Experiment 1: Selection of Sequence Length. The purpose of this experiment is to select an appropriate length for model comparison. Several models are selected to conduct the experiment. The models we selected are: CNN, RNN, RCNN, ARNN, FastText, RoBERTa. Experiment 1 is composed of two steps. First, we conduct the experiment using different sequence lengths. Then, we select an appropriate sequence length to compare the performance of different models.

Experiment 2: The Effect of Convolutional Layer. The aim of this experiment is to show the effect of convolutional layer. We conduct the experiment using RoBERTa and RoBERTaCNN.

Experiment 3: Keyword Extraction. The intention of this experiment is to explore the usefulness of the fine-tuning method based on keyword extraction. The number of keywords is a hyperparameter. So experiment 3 is divided into two steps. First, we conduct experiments on the public dataset Fudan Text

Classification Corpus to explore the effect of different number of keywords. In this experiment, the sequence length is set to the parameter recommended in experiment 1. Second, we use the best setting of the number of keywords to conduct experiments on two datasets.

5 Results and Discussion

5.1 Results

Experiment 1: Selection of Sequence Length. Experiment 1 can be divided into two steps. First, we conduct experiments on Fudan Text Classification Corpus to search the best parameter of sequence length. The result of the first step is shown is in the Fig. 3. Second, we use the best parameter to conduct experiment on two datasets to compare the performance of different models. The result of the second step is depicted in the Fig. 4.

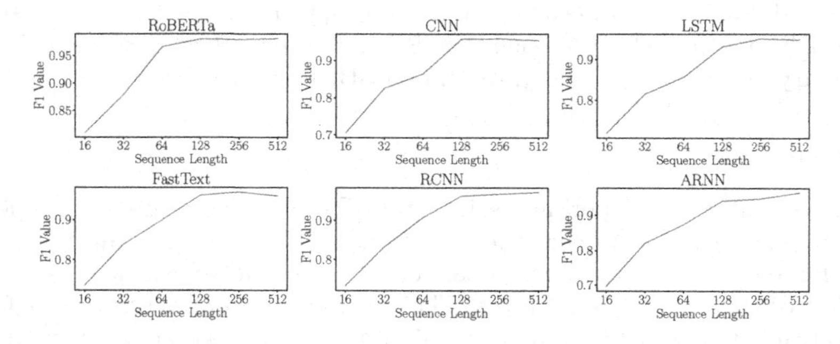

Fig. 3. The effect of sequence length on model performance: when the sequence length reaches a certain level, the improvement is no longer obvious

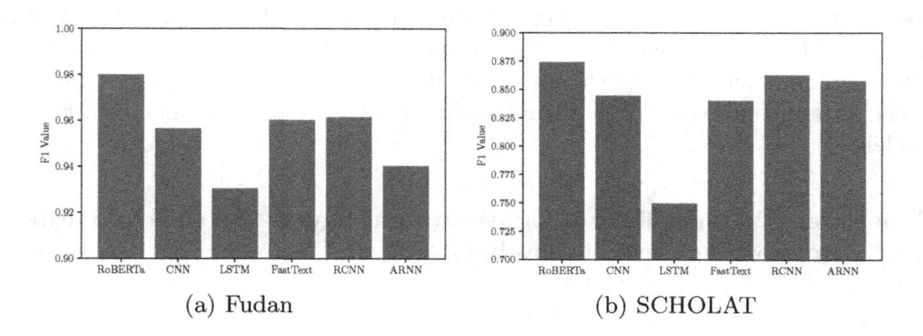

(a) Fudan (b) SCHOLAT

Fig. 4. Comparison of different models on two datasets: RoBERTa outperforms other commonly used models

Experiment 2: The Effect of Convolutional Layer. In experiment 2, both datasets are used. We compare the original RoBERTa and RoBERTaCNN. The result of the experiment is presented in the Table 1.

Table 1. The effect of convolutional layer: our proposed model RoBERTaCNN can outperform the original RoBERTa

F1 value		Sequence length					
Dataset and model		16	32	64	128	256	512
Fudan	RoBERTa	0.8089	0.8790	0.9660	0.9782	0.9793	0.9809
	RoBERTaCNN	**0.8106**	**0.8899**	**0.9718**	**0.9810**	**0.9838**	**0.9839**
SCHOLAT	RoBERTa	0.8500	0.8708	**0.8801**	0.8741	0.8685	**0.8735**
	RoBERTaCNN	**0.8709**	**0.8852**	0.8788	**0.8850**	**0.8916**	0.8704

Experiment 3: Keyword Extraction. The result of the first step is shown in the Fig. 5. The result of the second step is shown in Table 2. The Table 2 shows the comparison of RoBERTa, RoBERTaCNN and KE-RoBERTaCNN on two datasets.

Table 2. Comparison of three models on two datasets: our models can improve the performance compared to the original RoBERTa

F1 value	Dataset	
Model	Fudan	SCHOLAT
RoBERTa	0.9782	0.8741
RoBERTaCNN	0.9810	0.8850
KE-RoBERTaCNN	0.9832	0.8911

5.2 Discussion

Summary. Our proposed model KE-RoBERTaCNN performs best. The first experiment suggests to set sequence length to 128 and shows that RoBERTa performs best. The second experiment shows that the convolutional layer helps to improve the score. The third experiment shows that keyword extraction is indeed useful. Next, we will discuss the hyperparameters in the experiment, compare the performance of different models, analyze the role of the convolutional layer and the effect of keyword extraction.

Hyper-parameter: Sequence Length. We suggest to set sequence length to 128. The reason for this suggestion is below. In the Fig. 3, we can see that the F1 value improves a lot compared to 64 but improves slightly compared to 256. The training time and the test time will greatly increase because the cost time is proportional to the length of the sequence. For the reasons above, we recommend using 128.

Comparison of Different Models. We compare the performance of different models using sequence length 128. The result is shown in the Fig. 4. On Fudan Text Classification Corpus, the F1 value can improve nearly 2% compared to the best of other models. On SCHOLAT academic news dataset, we can see the similar result that the F1 value can improve 1% compared to the best of other models. The results of the experiment 1 suggest that the pre-trained model RoBERTa can indeed increase the performance of text classification.

The Effect of Convolutional Layer. The results of experiment is shown in the Table 1. On Fudan Text Classification Corpus, our proposed model can outperform the original RoBERTa. On SCHOLAT academic news dataset, our model is slightly lower in some cases, but can improve F1 value greatly in most cases. As a conclusion, the addition of convolutional layer can improve the performance of text classification.

Hyper-parameter: The Number of Keywords. The number of keywords is a hyper-parameter, we conduct experiment to search the best parameter. The results of experiment 3 is shown in the Fig. 5. The results show that proper number of keywords does help improve F1 value, but if too many keywords are used, F1 value decreases. From the results, we recommend using five keywords.

The Effect of Keyword Extraction. We compare KE-RoBERTaCNN with RoBERTaCNN. The KE-RoBERTaCNN uses five keywords in the experiment. The results of the experiment is shown in Table 2. We can see that extracting keywords can help improve F1 value on both datasets.

Comparison to the Original RoBERTa. Compared to the original RoBERTa, our proposed model KE-RoBERTaCNN can improve the F1 value on Fudan Text Classification Corpus and SCHOLAT academic news dataset.

5.3 Future Trend

Our experiment shows that the model can reach a prefect score on Fudan Text Classification Corpus. However, we observed that the model cannot perform as well on SCHOLAT academic news dataset. The same models with different data quality exhibit different performance. The problem is directly related to the model robustness. The study of model robustness can be divided into attacking and defending. On attacking models, adversarial samples are generated to attack BERT [?] [7]. On defending models, adversarial samples can be detected by replacing infrequent words. The model robustness should be evaluated. TextFlint [8], a robustness evaluation toolkit, provides comprehensive evaluation for model robustness. Recently, more and more studies focus on model robustness so it will become the future trend.

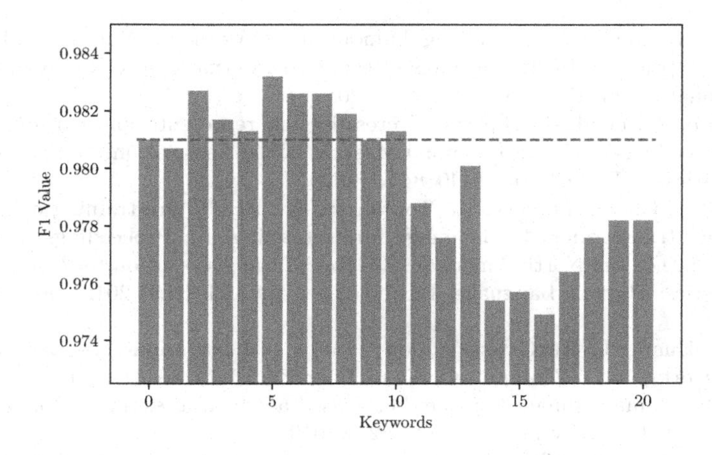

Fig. 5. The effect of the number of keywords: extracting five keywords can help improve F1 value most

6 Conclusion and Future Work

In this paper, we design a model based on pre-trained model and keyword extraction to improve the performance of text classification. The model we designed employs a convolutional layer to learn local dependencies and a max-pooling layer to capture key components. The introduction of the convolutional layer helps to identify the terminology in academic papers. Our proposed pre-trained model fine-tuning technique based on keyword extraction can improve scores. The final results show that our proposed model can improve the F1 value on Fudan Text Classification Corpus and SCHOLAT academic news dataset compared to the RoBERTa.

In the future, the model will be deployed online. First, we can try model pruning to reduce the testing time while keeping the performance. To model pruning, we need to evaluate the importance of network neurons and remove the least important neurons. The model needs to be fine-tuned on the dataset, repeat the whole process until the pruning stops. Besides, the model robustness should be comprehensively evaluated and improved. Our proposed fine-tuning method can improve the model robustness in a way, but further experiments will be required.

References

1. Bahdanau, D., Cho, K., Bengio, Y.: Neural machine translation by jointly learning to align and translate. In: 3rd International Conference on Learning Representations, ICLR 2015 (2015)
2. Blinov, V., Bolotova-Baranova, V., Braslavski, P.: Large dataset and language model fun-tuning for humor recognition. In: Proceedings of the 57th Conference of the Association for Computational Linguistics, ACL 2019, pp. 4027–4032 (2019)

3. Cattle, A., Ma, X.: Recognizing humour using word associations and humour anchor extraction. In: Proceedings of the 27th International Conference on Computational Linguistics, pp. 1849–1858 (2018)
4. Conneau, A., et al.: Unsupervised cross-lingual representation learning at scale. In: Proceedings of the 58th Annual Meeting of the Association for Computational Linguistics, ACL 2020, pp. 8440–8451 (2020)
5. Devlin, J., Chang, M., Lee, K., Toutanova, K.: BERT: pre-training of deep bidirectional transformers for language understanding. In: Proceedings of the 2019 Conference of the North American Chapter of the Association for Computational Linguistics: Human Language Technologies, NAACL-HLT 2019. pp. 4171–4186 (2019)
6. Du, C., Huang, L.: Text classification research with attention-based recurrent neural networks. Int. J. Comput. Commun. Control **13**(1), 50–61 (2018)
7. Garg, S., Ramakrishnan, G.: Bae: bert-based adversarial examples for text classification (2020). arXiv preprint arXiv:2004.01970
8. Gui, T., et al.: Textflint: Unified multilingual robustness evaluation toolkit for natural language processing (2021). arXiv preprint arXiv:2103.11441
9. Guy, I., Mejer, A., Nus, A., Raiber, F.: Extracting and ranking travel tips from user-generated reviews. In: Proceedings of the 26th international conference on world wide web, pp. 987–996 (2017)
10. Johnson, R., Zhang, T.: Deep pyramid convolutional neural networks for text categorization. In: Proceedings of the 55th Annual Meeting of the Association for Computational Linguistics, vol. 1(Long Papers), pp. 562–570 (2017)
11. Joulin, A., Grave, E., Bojanowski, P., Mikolov, T.: Bag of tricks for efficient text classification. In: Proceedings of the 15th Conference of the European Chapter of the Association for Computational Linguistics, EACL 2017, pp. 427–431 (2017)
12. Kim, Y.: Convolutional neural networks for sentence classification. In: Proceedings of the 2014 Conference on Empirical Methods in Natural Language Processing, EMNLP 2014, pp. 1746–1751 (2014)
13. Lai, S., Xu, L., Liu, K., Zhao, J.: Recurrent convolutional neural networks for text classification. In: Bonet, B., Koenig, S. (eds.) Proceedings of the Twenty-Ninth AAAI Conference on Artificial Intelligence, pp. 2267–2273 (2015)
14. Lan, Z., Chen, M., Goodman, S., Gimpel, K., Sharma, P., Soricut, R.: ALBERT: a lite BERT for self-supervised learning of language representations. In: 8th International Conference on Learning Representations, ICLR 2020 (2020)
15. Li, J., Li, Y., Wang, X., Tan, W.C.: Deep or simple models for semantic tagging? it depends on your data. Proc. VLDB Endowment **13**(12), 2549–2562 (2020)
16. Lin, R., Fu, C., Mao, C., Wei, J., Li, J.: Academic news text classification model based on attention mechanism and RCNN. In: Sun, Y., Lu, T., Xie, X., Gao, L., Fan, H. (eds.) ChineseCSCW 2018. CCIS, vol. 917, pp. 507–516. Springer, Singapore (2019). https://doi.org/10.1007/978-981-13-3044-5_38
17. Liu, P., Qiu, X., Huang, X.: Recurrent neural network for text classification with multi-task learning. In: Proceedings of the Twenty-Fifth International Joint Conference on Artificial Intelligence, IJCAI 2016, pp. 2873–2879 (2016)
18. Liu, Y., et al.: Roberta: a robustly optimized bert pretraining approach (2019). arXiv preprint arXiv:1907.11692
19. Mikolov, T., Chen, K., Corrado, G., Dean, J.: Efficient estimation of word representations in vector space. In: 1st International Conference on Learning Representations, ICLR 2013 (2013)

20. Minaee, S., Kalchbrenner, N., Cambria, E., Nikzad, N., Chenaghlu, M., Gao, J.: Deep learning based text classification: A comprehensive review (2020). arXiv preprint arXiv:2004.03705
21. Negi, S., Daudert, T., Buitelaar, P.: Semeval-2019 task 9: suggestion mining from online reviews and forums. In: Proceedings of the 13th International Workshop on Semantic Evaluation, SemEval@NAACL-HLT 2019, pp. 877–887 (2019)
22. Novgorodov, S., Guy, I., Elad, G., Radinsky, K.: Generating product descriptions from user reviews. In: The World Wide Web Conference, pp. 1354–1364 (2019)
23. Vaswani, A., et al.: Attention is all you need. In: Advances in Neural Information Processing Systems 30: Annual Conference on Neural Information Processing Systems 2017, pp. 5998–6008 (2017)
24. Wan, M., Misra, R., Nakashole, N., McAuley, J.J.: Fine-grained spoiler detection from large-scale review corpora. In: Proceedings of the 57th Conference of the Association for Computational Linguistics, ACL 2019, pp. 2605–2610 (2019)
25. Yang, D., Lavie, A., Dyer, C., Hovy, E.: Humor recognition and humor anchor extraction. In: Proceedings of the 2015 Conference on Empirical Methods in Natural Language Processing, pp. 2367–2376 (2015)
26. Yang, W., Zhang, H., Lin, J.: Simple applications of bert for ad hoc document retrieval (2019). arXiv preprint arXiv:1903.10972
27. Yang, Z., Dai, Z., Yang, Y., Carbonell, J.G., Salakhutdinov, R., Le, Q.V.: Xlnet: generalized autoregressive pretraining for language understanding. In: Advances in Neural Information Processing Systems 32: Annual Conference on Neural Information Processing Systems 2019, NeurIPS 2019, pp. 5754–5764 (2019)
28. Yang, Z., Yang, D., Dyer, C., He, X., Smola, A., Hovy, E.: Hierarchical attention networks for document classification. In: Proceedings of the 2016 conference of the North American chapter of the association for computational linguistics: human language technologies, pp. 1480–1489 (2016)
29. Yao, L., Mao, C., Luo, Y.: Graph convolutional networks for text classification. In: Proceedings of the AAAI Conference on Artificial Intelligence, vol. 33, pp. 7370–7377 (2019)

ResConvE: Deeper Convolution-Based Knowledge Graph Embeddings

Yongxu Long, Zihan Qiu, Dongyang Zheng, Zhengyang Wu, Jianguo Li, and Yong Tang(✉)

School of Computer Science, South China Normal University, Guangzhou 510631, Guangdong, China
{longyongxu,ytang}@m.scnu.edu.cn

Abstract. Link prediction on knowledge graphs (KGs) is an effective way to address their incompleteness. ConvE and InteractE have introduced CNN to this task and achieved excellent performance, but their model uses only a single 2D convolutional layer. Instead, we think that the network should go deeper. In this case, we propose the ResConvE model, which takes reference from the application of residual networks in computer vision, and deepens the neural network, and applies a skip connection to alleviate the gradient explosion and gradient disappearance caused by the deepening of the network layers. We also introduce the SKG-course dataset from Scholat for experiments. Through extensive experiments, we find that ResConvE performs well on some datasets, which proves that the idea of this method has better performance than baselines. Moreover, we also design controlled experiments setting different depths of ResConvE on FB15k and SKG-course to demonstrate that deepening the number of network layers within a certain range does help in performance improvement on different datasets.

Keywords: Knowledge graph embedding · Residual network · Knowledge graph · SCHOLAT · Link prediction

1 Introduction

Knowledge Graphs (KGs) are structured knowledge bases that are constructed by facts. One fact in KGs includes subject s, relation r, and object o, i.e. triplet (s, r, o), which means KGs are the collections of such triplets. Since Google announced its Knowledge Graph in 2012, many KGs such as WordNet [13], YAGO [21], Freebase [2], and SKG (Scholat Knowledge Graph) keep coming these years. They find various applications in quantities of area, for example, relation extraction, search, analytics, recommendation, and question answering [27,30].

However, the main problem faced by knowledge graphs when applied is incompleteness [5]. In particular, links in the KGs are missing, for example, 71% of users

This work was supported in part by the National Natural Science Foundation of China under Grant U1811263 and Grant 61772211.

Y. Sun et al. (Eds.): ChineseCSCW 2021, CCIS 1492, pp. 162–172, 2022.
https://doi.org/10.1007/978-981-19-4549-6_13

in Freebase are missing birthday information and 75% are missing nationality information [5]. To solve this problem, methods such as TransE [3] and TransH [28] are based on using existing subjects and relations in KGs, performing embedding operations to map them into a vector space, and making predictions, while the model learns parameters and constantly optimizes the score function.

In fact, however, real-world knowledge graphs are so large that link prediction [12] for such knowledge graphs requires not only the number of parameters to be considered, but also the computational cost. For shallow models such as TransE [3], TransH [28] and DisMult [29], the best way to make improvements is to increase the size of the embedding matrix, which may most likely result in an excessive number of parameters for large knowledge graphs [4]. ConvE [4] performs very well in several tasks, but the structure of the model is too shallow for CNNs and can not make sure that the features of the KG can be fully learned. Therefore, inspired by the shortcut idea in ResNets [8], we propose the ResConvE model to embed the KGs.

ResConvE adds several more convolutional layers to ConvE. However, in view of the problem of gradient disappearance and gradient explosion [1,6,7] caused by deeper networks, ResConvE uses a skip connection mechanisms, so that the model can achieve good results even if it is very deep.

Our contributions are as follows:

1. Inspired by ConvE, which treats entities and relationships as "images", we propose ResConvE, the first application of the idea of ResNets to link prediction models based on knowledge graph embedding, which deepens the neural network without losing the original ability to extract features, providing a new idea for link prediction model building.
2. ResConvE was evaluated on various link prediction datasets and proved to be more effective in most of the datasets. Meanwhile, we explored how effective deepening the neural network model was in improving performance.
3. The SKG-course dataset was introduced for link prediction tasks.

2 Related Work

Since the introduction of the TransE [3] model, a variety of link prediction models have emerged. Early models use the translation objective as the score function including the TransE [3] and TransH [28], and the DisMult [29] model, which is based on a bilinear diagonal, but although their models are effective, they do not deepen the neural network, which makes them less effective than our model.

The introduction of ConvE [4] provides a new way of thinking about link prediction models, and ConvE proposes to use 2D convolutional layers to build a neural network model for link prediction. InteractE [26] improves on ConvE: when entity vectors are spliced with relational vectors, the model first rearranges the concatenated vectors according to rules and then feeds them into a 2D convolutional for training.

ConvE has been very successful in introducing 2D convolutional layers. InteractE, based on ConvE, has improved on this by doing vectors processing before feeding in 2D convolutional layers, with good results. However, both of these end up using only one 2D convolutional layer, and there is no discussion of the effects of deepening the model in either of these articles. However, we believe that the structure is too simple and the network is not deep enough for CNNs, of which the advantage is feature extraction. Meanwhile, we know that making a simple stack of neural network layers to deepen it can lead to gradient explosion or gradient disappearance, hindering the emergence of loss function convergence and even making the accuracy degrade rapidly after reaching saturation. Inspired by the structure of the ResNets [8] network, we introduced a Shortcut mechanism in ResConvE to solve this problem. Shortcut refers to the back-propagation of the model by skipping one or more layers of connections and adding the data directly to the output of the mainstem, which is a stack of network layers, during forward propagation.

3 Background

Knowledge Graph: A knowledge graph $\mathcal{G}$ is a collection of triplets (s, r, o), consisting of relations r, subjects s and objects o. Figure 1 illustrates the structure of a triplet.

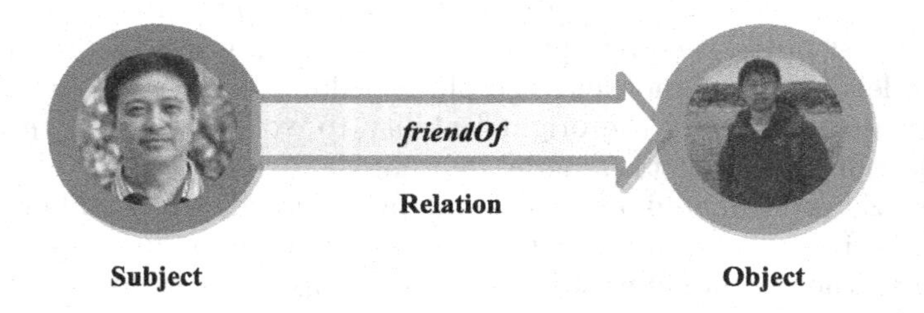

Fig. 1. A relation, a subject, and an object form a triplet.

Knowledge Graph Link Prediction: The main task of the link prediction in KGs is to make use of the existing facts in KGs to predict the new ones, which means we need the model to learn a score function ψ with an input triplet (s, r, o) whose score depends on the likelihood of the fact being true, to which is proportional.

Entities and relations will be encoded in most of the existing KGs embedding approaches. The validity of the output triplets will then be measured by a defined score function. Some score functions for existing models are presented in Table 1. Once the score function is defined, the model learns based on the inputs and outputs, thus continuously optimizing the model parameters.

ConvE: ConvE introduces 2D convolution into the model for KG link prediction. convE feeds entities and relationships into the 2D convolution layer after embedding them. The score function ψ for ConvE is given as follow:

$$\psi = f_a(vec(f_a(cat(\bar{emb_s}, \bar{emb_r}) \star \omega))W)emb_o \quad (1)$$

where $\bar{emb_s}$ and $\bar{emb_r}$ represent the tensors after 2D reshaping of the embedding matrices of subject emb_s and relation emb_r, while $\star$ denotes the convolution operation. e_o represents the embedding output, normally a matrix, of the object, and W represents the weight matrix to be learned. f_a refers to the activation function. 2D reshaping is considered to be useful for learning the representation of entities and relations.

Table 1. Some score functions for existing models. $\star$ represents convolution.

Model	Score function
TransE [3]	$\Vert emb_s + emb_r - emb_o \Vert_p$
DisMult [29]	$\langle emb_s, emb_r, emb_o \rangle$
ConvE [4]	$f_a(vec(f_a(cat(\bar{emb_s}, \bar{emb_r}) \star \omega))W)emb_o$
InteractE [26]	$f_a(vec(f_a(Perm(\mathcal{P}_k) \star \omega))W)emb_o$
ResConvE	$f_a(vec(f_a(cat(\bar{emb_s}, \bar{emb_r})) \star \omega + \sum f_a(cat(\bar{emb_s}, \bar{emb_r})) \star \omega')W)emb_o$

4 ResConvE

4.1 Overview

ConvE [4] indicates that using 2D convolution does boost the expressiveness of the model. The expressive ability of ConvE is further enhanced by reconstruction of the entity-relationship embedding before feeding into the convolutional layer for computation in InteractE [26]. From the experience of CNN in computer vision, we believe that a deeper network structure is conducive to capturing richer entity attributes and relationship features. So in order to extend the approach to capture entity-relationship features, ResConvE proposes the following two ideas:

1. **Deepening the neural network:** In contrast to ConvE and InteractE, which use only a single layer of 2D convolutional layers, ResConvE uses multiple convolutional layers to deepen the neural network, which is normally an important trick in the field of computer vision [18–20,22]. Inspired by ConvE's introduction of CNNs to the knowledge graph embedding task, we build on this approach to deepen the network, to extract features of entities and relationships better.
2. **Shortcut:** Many theories and practices have shown that if a neural network is only deepened, the gradients will eventually explode or vanish [7,8]. Inspired by ResNets [8], ResConvE introduces a shortcut mechanism, which allows the deepening of the neural network without compromising the model's capabilities.

4.2 Detail

The overall architecture of ResConvE is shown in Fig. 2. ResConvE learns a vector of dimension d to represent entities or relationships in KG. In forward propagation, the data input to ResConvE will be divided into two paths after the embedding operation, which are mainstem and shortcut, where the mainstem will have multiple convolutional layers, and the shortcut, with two 2D convolutional layers, makes sure that the model will not damage the original capability in the process of deepening.

Before Divided into Two Paths. In forward propagation, the model will initialize two embedding matrices for entities and relations respectively, and they will be embedded into embedding matrices. which are the low-dimensional representations of both. After that, the model will concatenate the embeddings $\bar{e_s}$ and $\bar{e_r}$ and feed them into the mainstem and shortcut at the same time.

Mainstem. After $\bar{e_s}$ and $\bar{e_r}$ enter Mainstem, they pass through a 2D convolutional layer with 16 1×1 filters, followed by several convolutional operations with 3×3 filters, normalisation [9] and activation using the *ReLu* [14,16,17] function. The amount of convolution filters of each layer is doubled compared to the previous one. At the end of these operations, the data is convolved using $n \times 1 \times 1$ filters (where n is the final number of channels) to fit the data from the shortcut and leave Mainstem.

Shortcut. The $\bar{e_s}$ and $\bar{e_r}$ from the other branch enter the shortcut, are normalized and dropout, and then fed into a 2D convolutional layer with $32 \times 3 \times 3$ filters for computation. The results of the computation are fed into the Mainstem for summation after being normalized and finally activated by *ReLu* [14].

Score Function. Formally, the score function for ResConvE can be defined as the following equation:

$$\psi = g_a(vec(f_a(cat(\bar{emb_s}, \bar{emb_r})) \star \omega + \sum f_a(cat(\bar{emb_s}, \bar{emb_r})) \star \omega')W)emb_o \quad (2)$$

where $\bar{emb_s}$ and $\bar{emb_r}$ represent the embedding tensors of the subjects and relations, emb_s and emb_r, after 2D reshapings, while $\star$ denotes the convolution operation. emb_o denotes the entity embedding matrices, W is the matrix of weights to be learned. f_a and g_a represent *ReLU* and *Logistic Sigmoid* respectively.

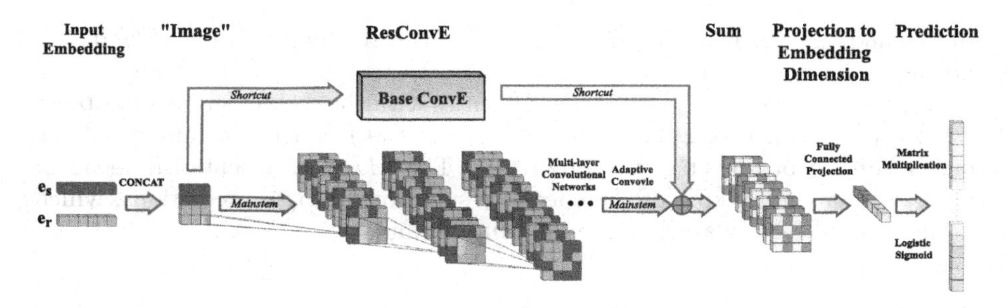

Fig. 2. After feeding the embedding of entities and relations into ResConvE, they will be reshaped and concatenated. Mainstem will then perform three convolutional calculations, while Shortcut will perform a ConvE-like convolutional operation and an adaptive convolutional calculation. The Tensor is then multiplied with the embedding space matrix and a logistic sigmoid is applied to generate a prediction.

5 Experiments

5.1 Knowledge Graph Datasets

FB15k: The FB15k [3] dataset consists of textual mentions of knowledge base relationship triples and Freebase entity pairs. There are 592,213 triples. The number of entities and relationships is 14,951 and 1,345 respectively. FB15K-237 [24] removed the inverse relations.

WN18: The WN18 [13] dataset, with 18 relationships and 40,943 entities, tends to obey a strict grading structure. WN18RR [4] is a new version of WN18 that has emerged from extensive research.

YAGO3-10: YAGO [21] is a KG composed of common knowledge facts extracted from Wikipedia to enhance WordNet. YAGO3-10 has 123,182 entities and 37 relations.

SKG-Course: The SKG-course dataset is derived from the knowledge graph of the *SCHOLAT course platform* (https://www.scholat.com/home.html?type=5), which has a total of 22,176 entities including users, courses, and classes, with the corresponding 4 relationships. Several Baseline models were replicated and trained and tested on this dataset.

5.2 Evaluation Protocol

We test performance through a widely used evaluation process [3,4,26]. We remove the subject or object from the complete triplets in the test set to create corrupted triplets of the form $(subject, relation, ?)$ or $(?, relation, object)$. The

Hits@k and the Mean Rank (MR), and the Mean Reciprocal Rank (MRR) are calculated for evaluation.

Hits@k represents the percentage of entities with the correct subject or object in the top k of all predictions. The MR represents the average ranking of the correct subject or object in the prediction. The MRR represents the mean of the inverse of the ranking of the correct results for multiple predictions, which can normally be calculated by the following formula:

$$MRR = \frac{1}{|pred|} \sum_{i=1}^{|pred|} \frac{1}{rank_i} \tag{3}$$

where $|pred|$ represents the total amount of predictions and $rank_i$ is the ranking of the correct object in the ith prediction.

5.3 Experimental Setup

To verify that ResConvE can extract features better when Mainstem is deepened, we set up three control groups. After the data had passed the first convolutional layer of Mainstem, we set up $\{1, 2, 3, 4\}$ convolutional layers for training respectively, where each convolutional layer had twice the number of filters as the previous one. The experiments were tested on the FB15k and SKG-course datasets respectively.

We tuned the hyperparameters by the performance of MRR. The dropout [9] of the embedding layers, 2D CNN layers and projection layer are set as $\{0.1, 0.2\}$, $\{0.2, 0.3\}$ and $\{0.2, 0.3\}$ respectively. The size of the embedding matrix and batch are set as $\{100 \times 100, 200 \times 200\}$ and $\{128, 256, 512\}$ respectively. We set the learning rate as $\{0.1, 0.001, 0.002, 0.003\}$. The label smoothing [23] coefficient is set as $\{0.1, 0.2\}$.

6 Result

6.1 Comparison of Performance

Comparison with Existing Methods. Besides the benchmarks dataset, we compared the performance of ResConvE with several existing methods on the SKG-course dataset to test generalization capabilities. We replicated several basic models to perform link prediction on the SKG-course to obtain training scores. Table 2, Table 3 and Table 4 summary the performance of ResConvE on the standard dataset and the SKG-course respectively. We find that ResConvE outperforms some metrics on FB15k, WIN18 dataset, and YAGO3-10, while all metrics are better on SKG-course. The results of ResConvE's link prediction on SKG-course are higher in MR metrics compared to ConvE and InteractE by 10.12%, 4.35%. On the validation set, ResConvE even outperformed ConvE and InteractE by 13.64% and 16.73% respectively.

Table 2. Performance on dataset FB15k and FB15k-237

	FB15k					FB15k-237				
	MR	MRR	Hits@10	Hits@3	Hits@1	MR	MRR	Hits@10	Hits@3	Hits@1
TransE [3]	125	–	0.471	–	–	–	0.290	0.470	–	0.290
TransD [10]	91	–	0.773	–	–	–	0.253	0.461	–	0.148
DistMult [29]	97	0.654	0.824	0.733	0.546	254	0.241	0.419	0.263	0.155
ComplEx [25]		0.692	0.840	0.759	0.599	339	0.247	0.428	0.275	0.158
ConvE [4]	51	0.657	0.831	0.723	0.558	244	0.325	0.501	0.356	0.237
InteractE [26]	–	–	–	–	–	172	0.354	0.535	–	0.263
ResConvE	60	**0.708**	**0.851**	**0.803**	**0.762**	272	0.312	0.486	0.341	0.225

Table 3. Performance on dataset WN18 and WN18RR

	WIN18					WIN18RR				
	MR	MRR	Hits@10	Hits@3	Hits@1	MR	MRR	Hits@10	Hits@3	Hits@1
DistMult [29]	902	0.022	0.936	0.914	0.728	–	0.43	–	–	0.39
ConvE [4]	374	0.943	0.956	0.946	0.935	4187	0.43	0.52	0.44	0.4
HHolE [11]	183	0.939	0.951	0.945	0.931	–	–	–	–	–
LogicENN [15]	357	0.923	0.948	–	–	–	–	–	–	–
InteractE [26]	–	–	–	–	–	5202	0.463	0.528	–	0.43
ResConvE	393	**0.943**	0.954	**0.949**	**0.936**	5006	0.424	0.491	0.435	0.393

Effect of Deepening the Mainstem. We analyzed whether deepening the mainstem would lead to better performance of ResConvE, i.e. by increasing the number of convolutional layers. We analyze this effect on the FB15k, SKG-course dataset respectively, which are shown in Table 5 and Table 6.

After deepening Mainstem, the model with 4 convolutional layers improved significantly for link prediction on dataset FB15k, with 23.08%, 30.27% and 17.54% on MR, MRR and $Hits$@10 respectively.

Meanwhile, We also found that deepening the convolutional layers on the SKG-course dataset resulted in significant improvements for each of the metrics in Table 6. The model with an increased number of convolutional layers of 3 is higher in MR and MRR by 18.739% and 3.101% respectively than the one with

Table 4. Performance on dataset SKG-course and YAGO3-10

	SKG-course					YAGO3-10				
	MR	MRR	Hits@10	Hits@3	Hits@1	MR	MRR	Hits@10	Hits@3	Hits@1
DisMult [29]	158	0.889	0.923	0.899	0.874	1107	0.500	0.660	0.550	0.410
ComplEx [25]	150	0.971	0.969	0.970	0.968	1127	0.490	0.660	0.540	0.400
ConvE [4]	112	0.931	0.961	0.942	0.915	1676	0.440	0.620	0.490	0.350
InteractE [26]	105	0.970	0.977	0.972	0.966	1671	0.541	0.620	–	0.462
ResConvE	**100**	**0.973**	**0.978**	**0.974**	**0.970**	2157	0.510	**0.664**	**0.558**	0.427

only 1 additional layer. But there is a limit to the optimization of this effect, and we learn that when the number is increased to 4, the increase is not as pronounced. What is clear, however, is that we believe our model is effective for deepening across a range of datasets, i.e. the generalization effect of our model does exist.

Table 5. Effect of deepening the mainstem on dataset FB15k

Number of convolution layers	MR	MRR	Hits@10	Hits@5	Hits@3	Hits@1
1	78	0.544	0.724	0.724	0.602	0.443
2	60	0.657	0.816	0.762	0.716	0.567
3	63	0.651	0.810	0.755	0.708	0.561
4	60	**0.708**	**0.851**	**0.803**	**0.762**	**0.626**

Table 6. Effect of deepening the mainstem on dataset SKG-course

Number of convolution layers	MR	MRR	Hits@10	Hits@5	Hits@3	Hits@1
1	123	0.944	0.969	0.963	0.957	0.929
2	113	0.947	0.969	0.964	0.958	0.937
3	**100**	**0.973**	0.978	**0.976**	0.973	**0.971**
4	101	0.972	**0.978**	0.975	**0.974**	0.968

Analysis of Experimental Results. Deepening the convolutional neural network to do the link prediction task seems effective. We have analyzed the reason for this: An embedding operation for low-dimensional representation usually means information compression [22]. ResConvE uses multiple CNN layers for modeling, which makes sure of fully learning and extracting the features of the entities and relationships while the Shortcut mechanism introduced by ResConvE ensures that the model is deepened without making the original performance worse.

7 Conclusion and Future Work

In this paper, a new method for KGs embedding, ResConvE, is proposed, which has a better capability of extracting the features of the KG by deepening the neural network, improving the model depth from the same type of ConvE and InteractE. At the same time, ResConvE borrowed the idea of skip connection on residual networks to alleviate the possible gradient disappearance and gradient explosion when the model is deepened and set up Mainstem and shortcut pathways on the model in learning data respectively. Through extensive experiments, we find that the idea of deepening neural networks has a role in optimizing performance. Moreover, we introduced the SKG-course dataset to demonstrate that this effect is not useful only on specific datasets, but has some generalization ability. We believe that ResConvE can be improved from more angles in the future. Although in this paper we have only made improvements in the depth of the neural network, we believe that improvements could perhaps be made in the width as well. If the model is constructed from CNNs, we believe that there are a large number of tricks in the field of computer vision that can be borrowed into the field of link prediction. We will look at this aspect.

References

1. Bengio, Y., Simard, P., Frasconi, P.: Learning long-term dependencies with gradient descent is difficult. IEEE Trans. Neural Netw. **5**(2), 157–166 (1994)
2. Bollacker, K., Evans, C., Paritosh, P.K., Sturge, T., Taylor, J.: Freebase: a collaboratively created graph database for structuring human knowledge. In: SIGMOD Conference (2008)
3. Bordes, A., Usunier, N., Garcia-Duran, A., Weston, J., Yakhnenko, O.: Translating embeddings for modeling multi-relational data. In: Neural Information Processing Systems (NIPS), pp. 1–9 (2013)
4. Dettmers, T., Minervini, P., Stenetorp, P., Riedel, S.: Convolutional 2d knowledge graph embeddings. In: Proceedings of the AAAI Conference on Artificial Intelligence, vol. 32 (2018)
5. Dong, X., et al.: Knowledge vault: a web-scale approach to probabilistic knowledge fusion. In: Proceedings of the 20th ACM SIGKDD International Conference on Knowledge Discovery and Data Mining, pp. 601–610 (2014)
6. Glorot, X., Bengio, Y.: Understanding the difficulty of training deep feedforward neural networks. In: Proceedings of the 13th International Conference on Artificial Intelligence and Statistics, pp. 249–256. JMLR Workshop and Conference Proceedings (2010)
7. He, K., Sun, J.: Convolutional neural networks at constrained time cost. In: Proceedings of the IEEE Conference on Computer Vision and Pattern Recognition, pp. 5353–5360 (2015)
8. He, K., Zhang, X., Ren, S., Sun, J.: Deep residual learning for image recognition. In: Proceedings of the IEEE Conference on Computer Vision and Pattern Recognition, pp. 770–778 (2016)
9. Ioffe, S., Szegedy, C.: Batch normalization: accelerating deep network training by reducing internal covariate shift. In: International Conference on Machine Learning, pp. 448–456. PMLR (2015)
10. Ji, G., He, S., Xu, L., Liu, K., Zhao, J.: Knowledge graph embedding via dynamic mapping matrix. In: Proceedings of the 53rd Annual Meeting of the Association for Computational Linguistics and the 7th International Joint Conference on Natural Language Processing (Volume 1: Long Papers), pp. 687–696 (2015)
11. Lalisse, M., Smolensky, P.: Augmenting compositional models for knowledge base completion using gradient representations. arXiv preprint arXiv:1811.01062 (2018)
12. Lin, Y., Liu, Z., Sun, M., Liu, Y., Zhu, X.: Learning entity and relation embeddings for knowledge graph completion. In: Proceedings of the AAAI Conference on Artificial Intelligence, vol. 29 (2015)

13. Miller, G.A.: WordNet: a lexical database for English. Commun. ACM **38**(11), 39–41 (1995)
14. Nair, V., Hinton, G.E.: Rectified linear units improve restricted Boltzmann machines. In: ICML (2010)
15. Nayyeri, M., Xu, C., Lehmann, J., Yazdi, H.S.: LogicENN: a neural based knowledge graphs embedding model with logical rules. arXiv preprint arXiv:1908.07141 (2019)
16. Nwankpa, C., Ijomah, W., Gachagan, A., Marshall, S.: Activation functions: comparison of trends in practice and research for deep learning. arXiv preprint arXiv:1811.03378 (2018)
17. Ramachandran, P., Zoph, B., Le, Q.V.: Searching for activation functions. arXiv preprint arXiv:1710.05941 (2017)
18. Simonyan, K., Zisserman, A.: Very deep convolutional networks for large-scale image recognition. arXiv preprint arXiv:1409.1556 (2014)
19. Srivastava, R.K., Greff, K., Schmidhuber, J.: Highway networks. arXiv preprint arXiv:1505.00387 (2015)
20. Srivastava, R.K., Greff, K., Schmidhuber, J.: Training very deep networks. arXiv preprint arXiv:1507.06228 (2015)
21. Suchanek, F.M., Kasneci, G., Weikum, G.: YAGO: a core of semantic knowledge. In: Proceedings of the 16th International Conference on World Wide Web, pp. 697–706 (2007)
22. Szegedy, C., et al.: Going deeper with convolutions. In: Proceedings of the IEEE Conference on Computer Vision and Pattern Recognition, pp. 1–9 (2015)
23. Szegedy, C., Vanhoucke, V., Ioffe, S., Shlens, J., Wojna, Z.: Rethinking the inception architecture for computer vision. In: Proceedings of the IEEE Conference on Computer Vision and Pattern Recognition, pp. 2818–2826 (2016)
24. Toutanova, K., Chen, D.: Observed versus latent features for knowledge base and text inference. In: Proceedings of the 3rd Workshop on Continuous Vector Space Models and Their Compositionality, pp. 57–66 (2015)
25. Trouillon, T., Dance, C.R., Welbl, J., Riedel, S., Gaussier, É., Bouchard, G.: Knowledge graph completion via complex tensor factorization. arXiv preprint arXiv:1702.06879 (2017)
26. Vashishth, S., Sanyal, S., Nitin, V., Agrawal, N., Talukdar, P.: InteractE: improving convolution-based knowledge graph embeddings by increasing feature interactions. In: Proceedings of the AAAI Conference on Artificial Intelligence, vol. 34, pp. 3009–3016 (2020)
27. Wang, Q., Mao, Z., Wang, B., Guo, L.: Knowledge graph embedding: a survey of approaches and applications. IEEE Trans. Knowl. Data Eng. **29**(12), 2724–2743 (2017)
28. Wang, Z., Zhang, J., Feng, J., Chen, Z.: Knowledge graph embedding by translating on hyperplanes. In: Proceedings of the AAAI Conference on Artificial Intelligence, vol. 28 (2014)
29. Yang, B., Yih, W., He, X., Gao, J., Deng, L.: Embedding entities and relations for learning and inference in knowledge bases. arXiv preprint arXiv:1412.6575 (2014)
30. Zou, X.: A survey on application of knowledge graph. J. Phys. Conf. Ser. **1487**, 012016 (2020)

Extractive-Abstractive: A Two-Stage Model for Long Text Summarization

Rui Liang, Jianguo Li(✉), Li Huang, Ronghua Lin, Yu Lai, and Dan Xiong

School of Computer Science of South China Normal University, Guangzhou 510631, Guangdong, China
jianguoli@m.scnu.edu.cn

Abstract. Currently, the mainstream text summarization techniques are divided into extractive and abstractive methods. Extractive method is suitable for long texts with a clear structure, while abstractive method is suitable for short texts. In this paper, we aim to address the problems of missing key words and incomplete overview that are usually caused by abstractive method in the face of long texts. To solve this problem, we propose a two-stage model that uses both extractive and abstractive methods for generating summaries. Firstly, we use multi-layer BiLSTM for long text summary extraction. Secondly, we use the classical UniLM as the base model while adding a novel copy mechanism to tackle out-of-vocabulary (OOV) problem and using the sparse softmax to avoid overfitting. Extensive experiments demonstrate that our models perform better than other baseline models, and our models can generate higher quality summaries.

Keywords: Text summarization · Extractive method · Abstractive method · BiLSTM · Copy mechanism · Sparse softmax

1 Introduction

With the development of Web 2.0, social networks have become indispensable in people's daily life. In this case, some vertical-domain social networks such as academic social networks have been emerged and developed rapidly. Taking SCHOLAT[1] (which is a kind of academic social networks) as an example, massive text data including personal blogs or news will be generated every day. This text data is in various formats, types and contents, which do not facilitate the recommendation of news and the detection of illegal news. Therefore, the generation of a summary of news is beneficial for us to identify high-quality news and anomalous news.

Text summarization can be divided into extractive summarization and abstractive summarization according to the implementation method. Most traditional methods belong to extractive summarization, which will extract the key sentences from the original text, and following rearrange and combine the sentences. However, extractive summarization has some common problems such as

[1] https://www.scholat.com.

Y. Sun et al. (Eds.): ChineseCSCW 2021, CCIS 1492, pp. 173–184, 2022.
https://doi.org/10.1007/978-981-19-4549-6_14

repeated generation and semantic incoherence. There are significant shortcomings in implicit features such as summary semantics and inter-sentence associations, and these shortcomings are especially evident in short texts.

Hence, the methods of abstractive become more and more popular. Generative summaries are closer to human abstract thinking and focus more on consistency and coherence. Abstractive summarization is mainly implemented by Seq2Seq [7] model and attention mechanism [27]. Seq2Seq mainly consists of an encoder and a decoder, which encodes the original text into a vector, subsequently extracts important information from the vector, processes the clips, and finally generates a text summary. However, abstractive summarization suffer from the problems that are duplication of words and OOV. Due to the "long distance dependency" problem, a significant part of the texts will be lost in the encoding stage. Moreover, abstractive summarization is proved to be less effective for long texts.

In this paper, a two-stage text summarization model is proposed that incorporates extractive and abstractive methods to make it more suitable for long texts. Specifically, we first use the sentence-level and document-level BiLSTM extraction models for summary extraction. We then take the extracted sentences as input to the abstractive model and finally generate the summaries. Our abstractive model adds a novel copy mechanism [28] to UniLM [6] to ensure the consistency of the summary and the text, which can also solve the OOV problem. Besides that, we use sparse softmax [18,22] to replace Softmax which can effectively avoid overfitting. The experimental results on NLPCC 2018 Shared Task Data [13] show that our model outperforms than other baseline models.

The main contributions of this work are summarized as follows.

1. We propose a two-stage model for long text summarization generation, which incorporates extractive and abstractive methods.
2. Our extractive model use multi-layer BiLSTM to obtain sentence and document representation, And use the simple formulation to facilitates interpretable of its decisions.
3. Based on UniLM, our abstractive model adds a novel copy mechanism and sparse softmax.

2 Related Work

In this section, we first briefly review the text summarization research in terms of both traditional methods and deep learning methods according to the different summarization methods. We then introduce the research progress about our proposed two-stage summarization method in this paper.

The earliest research on automatic text summarization was based on statistics, in detail, mainly on word frequency and sentence position. Thus, the words and sentences with high scores in the articles were composed as summaries. In 1958, Luhn proposed an automatic summarization method that tried to find the sentences which contained the most information [17]. The information content of a sentence can be measured by "keywords". If it contains more keywords, it means that the sentence is more important. In 1972, Jones et al. proposed the TF-IDF concept, which suggested that the importance of a word was related to the frequency of its occurrence in other documents [11]. In 2004, Mihalcea et al. applied PageRank for text summarization [20].

In recent decades, deep learning have been widely used in text summarization due to the capability of high-dimensional data computation and feature extraction. In 2006, Hinton et al. first used deep learning in text summarization by reducing the reliance on manual labor and allowing more efficient training [8]. In 2014, Sutskever et al. proposed a generic Sequence to Sequence Learning method that can assign different weight values to the output vector representation at each moment [26]. In 2017, Nallapati et al. used GRU-RNN to extract sentence features which can capture the hierarchical relationships between words, sentences and documents [21]. In 2019, the University of Edinburgh team first used BERT in summary extraction by leveraging the pre-training of BERT to obtain excellent results in summary extraction [15]. In 2019, Dong et al. proposed the UniLM model, which can conduct Seq2Seq tasks directly with a single BERT model, and set a new record in multiple text-generated datasets [6].

More recently, it has been found that it can yield better results by dividing the text summarization task into two phases, which are extraction and generation. In 2020, Zhong et al. tried to generate several semantic-represented summary candidates from the original text, and then converted the text summary into a matching problem in the semantic space [30]. In 2018, Cheng et al. learned sentence representation completion extraction through a joint convolutional neural network and a long and short-term memory network, and then implemented sentence rewriting through a pointer generation network [4]. In 2019, Bae et al. learned the representation of sentences by pre-training the language model to complete the extraction, and then used the same method as Cheng [4] to complete the sentence rewriting to generate the summary [1].

3 Our Model

Our model uses a two-stage generation, which divides the generated summary into two steps, where the first step is extraction and the second step is generation. Specifically, we use a classification approach for sentence extraction (represented as extractive model). Since the sentences can be divided into critical and non-critical sentences, we use sentence-level and document-leval based BiLSTM [3] to extract the critical sentences from the original document. The collection of critical sentences we call transitional document. The length of the transitional document is between the original document and the summarization, which retains most of the important information of the original document. Subsequently, we use the abstractive model to generate a summary based on the transitional documentation. Our abstractive model is based on UniLM [6] by extra introducing the copy mechanism [28] to ensure the consistency of the summary and the original text. Furthermore, we use sparse softmax [18,22] to avoid overfitting. The overall model is shown in Fig. 1.

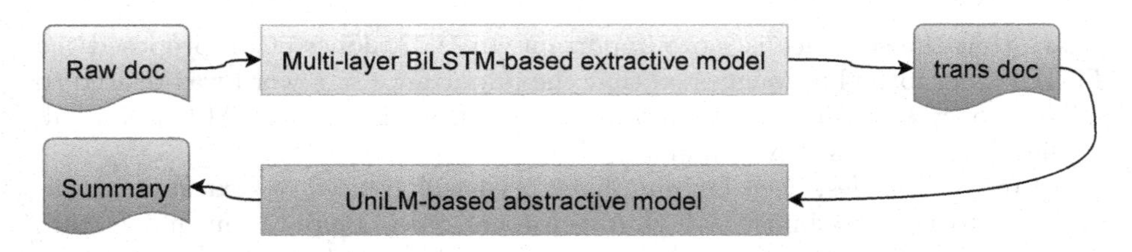

Fig. 1. The overall model.

3.1 Extractive Model

The purpose of extractive model is to convert the original text into an transitional document which can be suitable for processing by abstractive model. Our extractive model (shown in Fig. 2) is similar to SummaRuNNer [21] where the difference is that we use multi-layer BiLSTM for processing long texts into the transitional document.

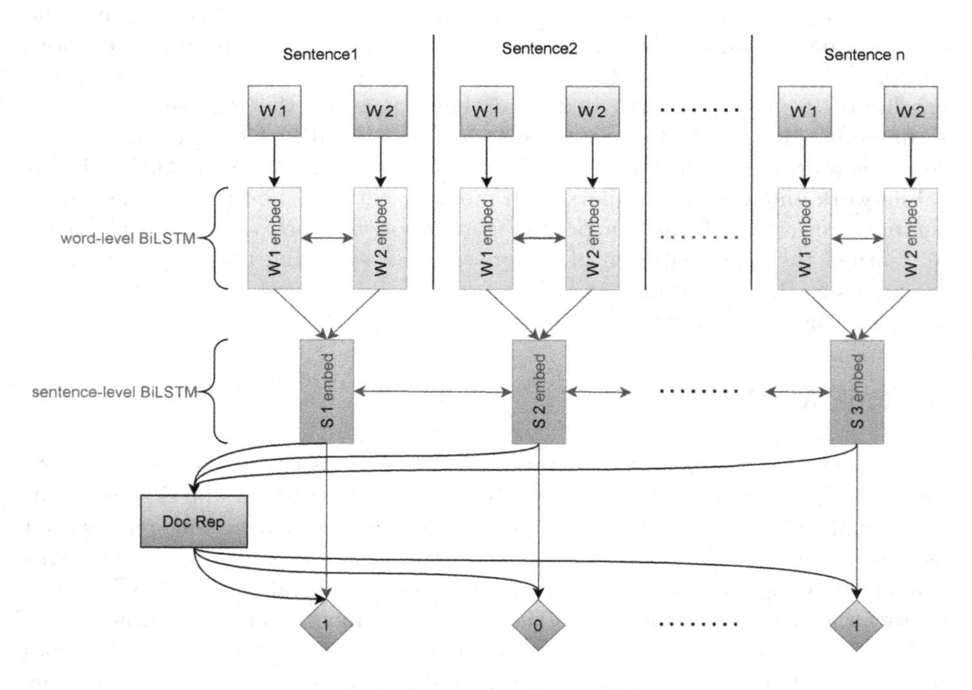

Fig. 2. The extractive model.

Our extractive model respectively uses word-level BiLSTM and sentence-level BiLSTM to obtain word representation and sentence representation. After that, the whole document representation is obtained by nonlinear transformation, as shown in Eq. (1).

$$\mathbf{doc} = \tanh\left(W_d \frac{1}{N_d} \sum_{j=1}^{N^d} \left[\mathbf{sent}_j^f, \mathbf{sent}_j^b\right]\right), \tag{1}$$

where **doc** is document representation, $\mathbf{N_d}$ is the number of sentences in the document, $\mathbf{sent}_j^f$ and $\mathbf{sent}_j^b$ are the hidden states corresponding to the $\mathbf{j}^{th}$ sentence of the forward and backward sentence-level BiLSTM respectively.

Finally, the sentence representation and document representation are feed into the classification layer to determine whether the sentence is a summary sentence or not. The classification layer will extract the sentences by considering the content, salience and novelty, as shown in Eq. (2).

$$\begin{aligned} P\left(y_j = 1 \mid \mathbf{sent}_j, \mathbf{sum}_j, \mathbf{doc}\right) = \sigma\,(W_c \mathbf{sent}_j \quad & \#(\text{ content }) \\ + \mathbf{sent}_j^T W_s \mathbf{doc} \quad & \#(\text{ salience }) \\ - \mathbf{sent}_j^T W_r \tanh\left(\mathbf{sum_j}\right)) \quad & \#(\text{ novelty }) \end{aligned} \tag{2}$$

where y_j is a binary variable which indicates whether the $\mathbf{j}^{th}$ sentence is part of summary, $\mathbf{sum}_j$ is the summary representation at $\mathbf{j}^{th}$ sentence, which is shown in Eq. (3).

$$\mathbf{sum}_j = \sum_{i=1}^{j-1} \mathbf{sent}_i P\left(y_i = 1 \mid \mathbf{sent}_i, \mathbf{sum}_i, \mathbf{doc}\right) \tag{3}$$

Besides that, our extractive model is intended to be the basis for the abstractive model. Since we need to extract the complete information, we design an extension algorithm to extend the original summary to allow the extraction model to better learn the features of the sentences. The algorithm pseudo code is shown in Algorithm 1. We stop the algorithm when the length of the selected sentences exceeds 300.

Algorithm 1. Original summary extension.

Input: selected, no-selected, text , sum
Output: select-id
1: pre-score←RougeScore(selected,text,sum), max-score←0, select-id←None
2: **for each** item **in** no-selected
3: selected←selected∪ item
4: new-score←RougeScore(selected,text,sum)
5: **if** new-score > pre-score
6: **return** *item*
7: **else if** new-score > max-score
8: max-score←new-score, id←item, selected←selected\item
9: **endif**
10: **return** select-id

In Algorithm 1, RougeScore is a function to calculate the rouge [14] score of the selected sentences and summaries. We mainly use Rouge-1 and Rouge-L to compute the rouge score, as shown in Eq. (4).

$$\text{RougeScore} = 0.4 \times \text{ Rouge-1 } + 0.6 \times \text{ Rouge-L} \tag{4}$$

3.2 Abstractive Model

Our abstractive model is mainly based on the UniLM model which yields a good performance in the field of text generation. We introduce a novel copy mechanism in the Encoder and Decoder stages of the abstractive model to ensure the correctness of the abstracts and to avoid professional errors in summarization. Furthermore, we use sparse softmax to replace the original softmax in the fine-tuning phase of the training process, which can effective avoid overfitting. The abstractive model is shown in Fig. 3.

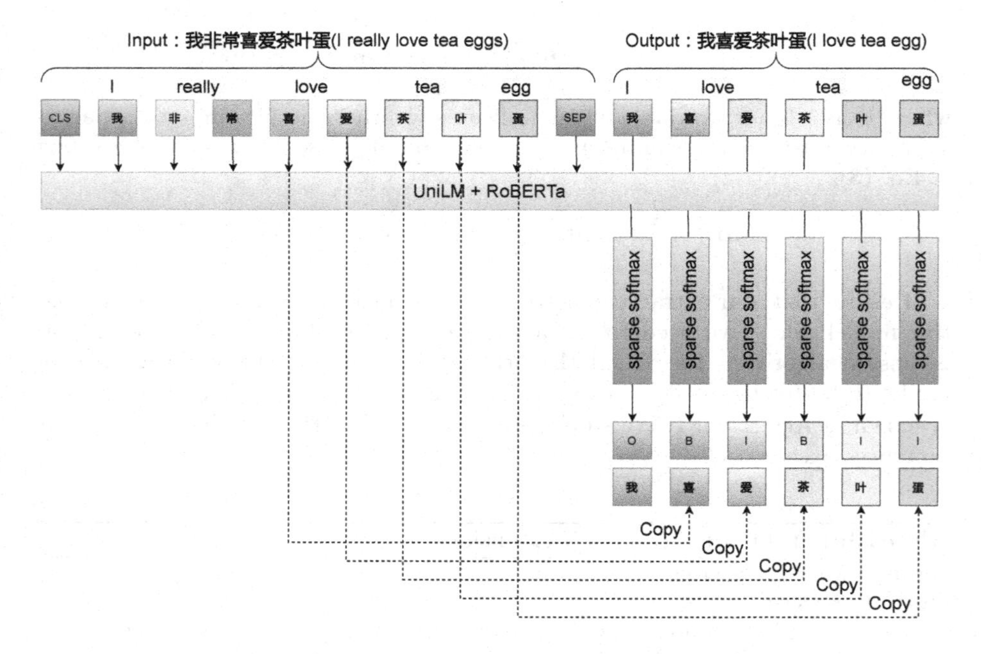

Fig. 3. The abstractive model.

One of the component of our abstractive model is UniLM, a Seq2Seq model proposed by Microsoft. UniLM is a multi-layer transformer network, which is based on the context of mask words to complete the prediction of mask words, that is, to complete the fill-in-the-blank task. We then use a sparse softmax to avoid overfitting. That is, we only save the top k probabilities in the calculation of the probability equation, and set the later ones to zero directly. In our model, we set k as 50. When calculating the cross-entropy, only the top k classes will be log-sum-up. The sparse softmax is shown in Eq. (5). The cross-entropy is calculated as shown in Eq. (6).

$$p_i = \begin{cases} \frac{e^{si}}{\sum_{j \in \Omega_k} e^{sj}}, i \in \Omega_k \\ 0, i \notin \Omega_k \end{cases} \tag{5}$$

$$\log \left(\sum_{i \in \Omega_k} e^{si} \right) - s_t \tag{6}$$

The original UniLM model use the fixed dictionary. Thus, if the text contains the keywords that are not in the dictionary, the summary generated by the model will be missing these keywords, which is called OOV problem. In order to solve the OOV problem, we use the copy mechanism proposed by Vinyals [28] and the BIO copy mechanism proposed by Su [25]. The original copy mechanism is shown in Eq. (7), and the BIO copy mechanism is shown in Eq. (8).

$$p\left(y_t \mid y_{<t}, x\right) \tag{7}$$

$$p\left(y_t, z_t \mid y_{<t}, x\right) = p\left(y_t \mid y_{<t}, x\right) p\left(z_t \mid y_{<t}, x\right) \tag{8}$$

In Eq. (8), $z_t \in \{\mathrm{B}, \mathrm{I}, \mathrm{O}\}$, where B means that the token is copied, I means that the token is copied and formed a continuous fragment with the previous token, O means that the token is not copied.

BIO copy mechanism can copy multiple tokens compared to the original copy mechanism. However, BIO copy mechanism copies those tokens that do not contain important information like "我 (I)". Hence, we use TF-IDF keyword extraction algorithm to remove the words that are not relevant to the summary. Then the solution of the label distribution of BIO copy mechanism becomes that if the continuous token is the same keyword, the token will be labeled as I, and the difference is shown in Table 1. For example, in Fig. 3, the TF-IDF algorithm is used to extract the keywords from the abstract, and the two keywords are "喜爱 (love)" and "茶叶蛋 (tea eggs)".

Table 1. The difference of BIO copy mechanism whether introducing TF-IDF.

Summary	I	love		tea		eggs
摘要	我	喜	爱	茶	叶	蛋
BIO copy mechanism	B	B	I	I	I	I
BIO copy mechanism + TF-IDF	O	B	I	B	I	I

4 Experimental Setup

In this section, we respectively describe the experimental dataset, evaluation metrics, and the details of model settings.

4.1 Experimental Dataset

Our experimental datasets are mainly used in the NLPCC 2017 and 2018 Shared Task3 Single Document Summarization [10,13]. After data cleaning, the specific statistical results are shown in Table 2.

Table 2. Statistical results for the dataset.

	docs	avg art len	max art len	min art len	avg sum len	max sum len	min sum len
Train	50000	964	22312	32	48	128	21
Test	2000	1324	17204	26	36	66	21

4.2 Evaluation Metrics

We use Rouge [14], a benchmark evaluation metric for text summarization domain to measure the quality of the abstracts. The basic idea is to compare the model-generated abstracts with the reference abstracts by calculating the basic units between them. We mainly use Rouge-N (Rouge-1 and Rouge-2) and Rouge-L as evaluation metrics to evaluate our model. The Rouge-N is shown in Eq. (9), and Rouge-L is shown in Eq. (10).

$$\mathrm{Rouge-N} = \frac{\sum_{S=\langle RS\}} \sum_{gram_n es} Count_{match}\left(gram_n\right)}{\sum_{s\in\{RS\rangle} \sum_{gram_n es} Count\left(gram_n\right)}, \tag{9}$$

where n is the $n-gram$ length, RS is the reference abstract, $Count_{\mathrm{match}}(gram_n)$ is the same number of $n-gram$ between the abstract and the reference abstract, $Count\left(gram_n\right)$ is the reference abstract in $n-gram$ number.

$$\mathrm{R}_{\mathrm{lcs}} = \frac{\mathrm{LCS}(X,Y)}{m}, \tag{10}$$

where $LCS(X,Y)$ is the length of the longest common subsequence between abstract to be tested and the reference abstract, m is the length of the reference abstract.

4.3 Model Settings

In extractive model, in order to speed up the training process and reduce memory requirements, we limited the vocabulary size to 80K and the max length of sentence to 60. We set embedding size to 300, batch size to 128, and epoch to 30. Word-level BiLSTM is set to 2 layers with hidden size of 100. Sentence-level BiLSTM is set to 2 layers with hidden size of 200. In addition, we use the adam optimizer and set the learning rate as 3e−4.

In abstractive model, we use chinese-roberta-wwm-ext [16] as our pre-trained model, and use the adam optimizer with learning rate of 1e−5. We set batch size as 8, epoch as 20, beam search as 3, and the max length of the generated summary to 150. The model is trained by v100 GPUs for about 21 h.

5 Results and Analysis

In this section, we evaluate the effectiveness of our model by comparing with baseline models and analyze the results. After that, we use the trained model to generate the text summarization by using a piece of news on SCHOLAT.

5.1 Results

We compare our model with the following baseline models.

1. LEAD: We select the first 70 words from the original text as the text summary [29].
2. Seq2Seq+Attention: Currently, the combination of Seq2Seq and attention is now the standard configuration for abstractive summarization [26,28].
3. Pointer-Generator: A model based on Seq2Seq+attention which uses pointer networks to solve OOV problems and coverage to discourages repetition [28].
4. BertSum: For the first time, the use of BERT in summary extraction makes full use of the prior knowledge generated by BERT pre-training [5].
5. UniLM: UniLM integrates Seq2Seq into the BERT framework by using the mask matrix to conduct summary generation tasks [6].

We re-run part of the baseline models and compare with our model. The experimental results are shown in Table 3.

Table 3. Performance of our model compared with baseline models.

	Model	Rouge-1	Rouge-2	Rouge-L
Extractive	LEAD	33.71	19.23	29.36
	BertSum	37.25	26.59	31.48
Abstractive	Seq2Seq+Attention	31.52	25.77	28.26
	Pointer-Generator	35.78	26.34	29.75
	UniLM	53.62	38.52	51.91
	Our model	**54.78**	**40.08**	**52.04**

5.2 Analysis

As can be seen from Table 2, the LEAD have a good performance. It is very surprising that such a super simple algorithm can achieve such good score. We can know that the key information of an article will appear roughly in the first few sentences. Compared with LEAD, we can find that using BERT for summary extraction can substantially improve the extraction accuracy. However, BERT is not suitable for long texts, so we turn to BiLSTM. Pointer-Generator is better than Seq2Seq+Attention, which indicates that the copy mechanism is necessary for summary generation. The excellent performance of UniLM shows that using

BERT and mask mechanism is a good solution for summary generation. By comparing our model with UniLM, our model has improved the rouge score with 1.16, 1.56, and 0.33% in Rouge-1, Rouge-2, and Rouge-L, respectively. We find that although most of the key information of the articles is concentrated in the first 512 words, some key information of the articles exists in the later part of the document. Since our model can retain the key information of the full text by the extraction method, the performance of our model will be more outstanding.

5.3 Application Case

We apply our trained model on the real news text on SCHOLAT, and the results are shown in Fig. 4. The length of the original text is 918, and the length of the transitional document obtained after the extractive model is 490. It shows that our abstractive model works well. Finally, our abstractive model generates a summary based on the transitional document. By comparing our model with UniLM, it is clear that our model generates a more concise and accurate summary. It shows that the extractive model can extract the key information from the text and avoid unimportant information from interfering with the results. In addition, it also shows that the copy mechanism can ensure the consistency of the text and avoid the OOV problem.

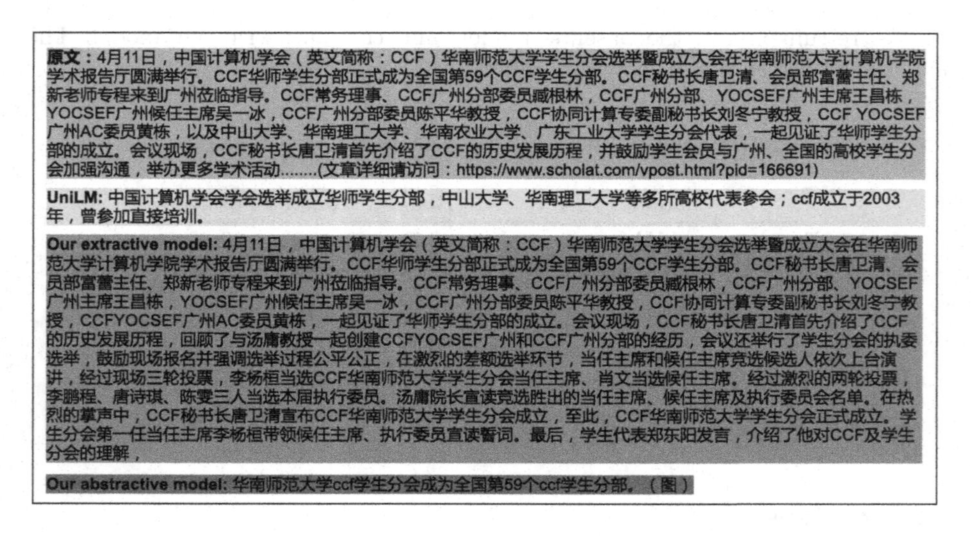

Fig. 4. An application case on a piece of news on SCHOLAT.

6 Conclusion and Future Work

In this paper, we propose a two-stage summarization model to solve the long document summarization problem by combining extractive and abstractive methods. The experimental results demonstrate the availability and effectiveness of our proposed model. For the extractive model, we obtain sentence representation and document representation by BiLSTM. We extract the sentences according to the content, salience and novelty. In addition, we design a summary extension model so that our transitional documents can be fully utilized in BERT. For the abstractive model, we introduce the novel BIO copy mechanism and sparse softmax to UniLM to improve the performance.

In future work, in order to further improve the practicability and accuracy of the model, we will focus on how to apply BERT to long text extraction and use multi-language BERT to process multi-language text.

Acknowledgement. This work was supported in part by the National Natural Science Foundation of China under Grant U1811263 and Grant 61772211.

References

1. Bae, S., Kim, T., Kim, J., Lee, S.G.: Summary level training of sentence rewriting for abstractive summarization. arXiv preprint arXiv:1909.08752 (2019)
2. Bao, G., Zhang, Y.: Contextualized rewriting for text summarization. arXiv preprint arXiv:2102.00385 (2021)
3. Chen, T., Xu, R., He, Y., Wang, X.: Improving sentiment analysis via sentence type classification using BiLSTM-CRF and CNN. Expert Syst. Appl. **72**, 221–230 (2017)
4. Cheng, J., Lapata, M.: Neural summarization by extracting sentences and words. arXiv preprint arXiv:1603.07252 (2016)
5. Cui, Y., et al.: Pre-training with whole word masking for Chinese BERT. arXiv preprint arXiv:1906.08101 (2019)
6. Dong, L., et al.: Unified language model pre-training for natural language understanding and generation. arXiv preprint arXiv:1905.03197 (2019)
7. Gehring, J., Auli, M., Grangier, D., Yarats, D., Dauphin, Y.N.: Convolutional sequence to sequence learning. In: International Conference on Machine Learning, pp. 1243–1252. PMLR (2017)
8. Hinton, G.E., Osindero, S., Teh, Y.W.: A fast learning algorithm for deep belief nets. Neural Comput. **18**(7), 1527–1554 (2006)
9. Hsu, W.T., Lin, C.K., Lee, M.Y., Min, K., Tang, J., Sun, M.: A unified model for extractive and abstractive summarization using inconsistency loss. arXiv preprint arXiv:1805.06266 (2018)
10. Hua, L., Wan, X., Li, L.: Overview of the NLPCC 2017 shared task: single document summarization. In: Huang, X., Jiang, J., Zhao, D., Feng, Y., Hong, Yu. (eds.) NLPCC 2017. LNCS (LNAI), vol. 10619, pp. 942–947. Springer, Cham (2018). https://doi.org/10.1007/978-3-319-73618-1_84
11. Jones, S.: KAREN: a statistical interpretation of term specificity and its application in retrieval. J. Documentation **28**(1), 11–21 (1972)
12. Kupiec, J., Pedersen, J., Chen, F.: A trainable document summarizer. In: Proceedings of the 18th Annual International ACM SIGIR Conference on Research and Development in Information Retrieval, pp. 68–73 (1995)
13. Li, L., Wan, X.: Overview of the NLPCC 2018 shared task: single document summarization. In: Zhang, M., Ng, V., Zhao, D., Li, S., Zan, H. (eds.) NLPCC 2018. LNCS (LNAI), vol. 11109, pp. 457–463. Springer, Cham (2018). https://doi.org/10.1007/978-3-319-99501-4_44

14. Lin, C.Y.: ROUGE: a package for automatic evaluation of summaries. In: Text Summarization Branches Out, pp. 74–81 (2004)
15. Liu, Y.: Fine-tune BERT for extractive summarization. arXiv preprint arXiv:1903.10318 (2019)
16. Liu, Y., et al.: Roberta: a robustly optimized BERT pretraining approach. arXiv preprint arXiv:1907.11692 (2019)
17. Luhn, H.P.: The automatic creation of literature abstracts. IBM J. Res. Dev. **2**(2), 159–165 (1958). https://doi.org/10.1147/rd.22.0159
18. Martins, A., Astudillo, R.: From SoftMax to SparseMax: a sparse model of attention and multi-label classification. In: International Conference on Machine Learning, pp. 1614–1623. PMLR (2016)
19. Mendes, A., Narayan, S., Miranda, S., Marinho, Z., Martins, A.F., Cohen, S.B.: Jointly extracting and compressing documents with summary state representations. arXiv preprint arXiv:1904.02020 (2019)
20. Mihalcea, R., Tarau, P.: TextRank: bringing order into text. In: Proceedings of the 2004 Conference on Empirical Methods in Natural Language Processing, pp. 404–411 (2004)
21. Nallapati, R., Zhai, F., Zhou, B.: SummaRuNNEr: a recurrent neural network based sequence model for extractive summarization of documents. In: Proceedings of the AAAI Conference on Artificial Intelligence, vol. 31 (2017)
22. Peters, B., Niculae, V., Martins, A.F.: Sparse sequence-to-sequence models. arXiv preprint arXiv:1905.05702 (2019)
23. Rush, A.M., Harvard, S., Chopra, S., Weston, J.: A neural attention model for sentence summarization. In: ACLWeb. Proceedings of the 2015 Conference on Empirical Methods in Natural Language Processing (2017)
24. Stiennon, N., et al.: Learning to summarize from human feedback. arXiv preprint arXiv:2009.01325 (2020)
25. Su, J.: Spaces: extractive-generative long text summaries (CAIL 2020) (2021). https://kexue.fm/archives/8046
26. Sutskever, I., Vinyals, O., Le, Q.V.: Sequence to sequence learning with neural networks. arXiv preprint arXiv:1409.3215 (2014)
27. Vaswani, A., et al.: Attention is all you need. arXiv preprint arXiv:1706.03762 (2017)
28. Vinyals, O., Fortunato, M., Jaitly, N.: Pointer networks. arXiv preprint arXiv:1506.03134 (2015)
29. Wasson, M.: Using leading text for news summaries: Evaluation results and implications for commercial summarization applications. In: 36th Annual Meeting of the Association for Computational Linguistics and 17th International Conference on Computational Linguistics, vol. 2, pp. 1364–1368 (1998)
30. Zhong, M., Liu, P., Chen, Y., Wang, D., Qiu, X., Huang, X.: Extractive summarization as text matching. arXiv preprint arXiv:2004.08795 (2020)

A Random-Walk-Based Heterogeneous Attention Network for Community Detection

Peng Zhang[1,2,3], Kun Guo[1,2,3], and Ling Wu[1,2,3(✉)]

[1] College of Computer and Data Science, Fuzhou University, Fuzhou 350116, China
{gukn,wuling1985}@fzu.edu.cn
[2] Fujian Provincial Key Laboratory of Network Computing and Intelligent Information Processing, Fuzhou, China
[3] Key Laboratory of Spatial Data Mining and Information Sharing, Ministry of Education, Fuzhou 350116, China

Abstract. Community detection in complex networks can find the community structure one of the most important properties of complex networks. Nodes in the same community have more dense connections than those in different communities, which can be utilized to analyze the function of complex networks. In addition, heterogeneous networks are ubiquitous in the real world. For example, academic networks have different types of nodes such as authors, papers, and conferences. Network representation learning is an important method to discover the complex nonlinear relationships between nodes in the network, which is of great help to community detection. Attention network is a typical network representation learning method, and it will pay attention to the important part in the network for the specific task. However, most existing heterogeneous NRL algorithms use metapaths to capture heterogeneous information, which requires prior knowledge to set metapaths in advance. This paper proposes a novel random-walk-based heterogeneous attention network (RHAN) for community detection on heterogeneous networks. Random walk is used to generate the neighbor nodes set of nodes, and heterogeneous information is considered by the intra-type attention and the inter-type attention, which is no need for metapaths. The experimental results on four widely used heterogeneous networks verify the effectiveness of RHAN.

Keywords: Community detection · Heterogeneous network · Random walk · Attention network · Network representation learning

1 Introduction

There are various networks in the real world, such as social networks [1], citation networks [2], traffic networks [3], and so on. Community structure is widely found in networks; nodes in one community have more connections than the ones in different communities [4]. Community detection can help us understand

Y. Sun et al. (Eds.): ChineseCSCW 2021, CCIS 1492, pp. 185–198, 2022.
https://doi.org/10.1007/978-981-19-4549-6_15

network structure and analyze network functions. For example, it can be used for personalized recommendations in e-commerce networks or predict protein functions in protein-protein networks [5].

Attention network is a typical method of network representation learning (NRL), which is a prevalent technology in mining network data recently. NRL generally aims to learn the low-dimension representation vector of nodes while preserving the original structure of the network. Usually, the dimension of the node representation vector is much smaller than the number of nodes in the network, which can reduce the cost of subsequent calculations. Meanwhile, the node representation vector can be easily analyzed by the traditional machine learning algorithms [6]. Many NRL algorithms have been proposed, which can be roughly divided into two types [7]: word2vec-based [8] (such as DeepWalk [9], LINE [10], and Node2Vec [11]) and graph-neural-network-based (such as GCN [2], GraphSAGE [1], GAT [12], and so on). However, these algorithms cannot deal with heterogeneous networks which have multi-types of nodes and edges. For example, the academic network shown in Fig. 1 has three node types: paper, author, and conference, and two edge types: author-paper and paper-conference.

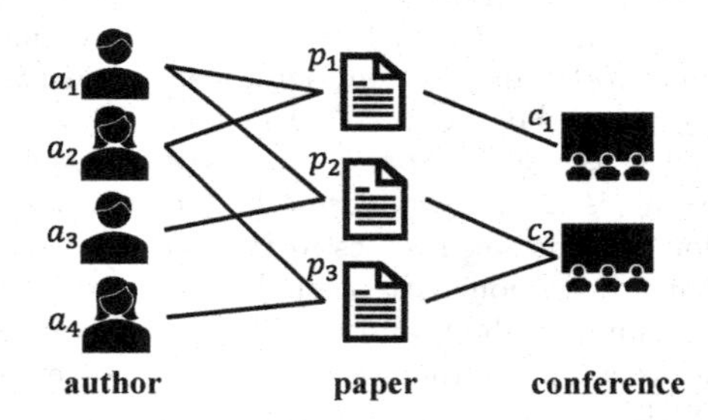

Fig. 1. An example of a heterogeneous network.

There have been studies on extending NRL algorithms to heterogeneous networks and most of them (such as metapath2vec [13], HIN2Vec [14], HAN [15], MAGNN [16], etc.) are designed based on the concept of metapaths [17]. A metapath is a sequence of node types describing a certain relationship among the nodes. This paper proposes a Random-walk-based Heterogeneous Attention Network (RHAN) for community detection, which does not require prior knowledge to set metapaths. We use the random walk to obtain a set of neighbor nodes closely related to the start node of the walk. Then, these nodes will be processed under classified by node type. In the neighbor nodes of the same kind, the attention layer (AL) [12] is applied to aggregate the representative vector of the type. For the symbolic vectors of each node type, another AL is used to obtain the final representation vector of the starting node of the walk. Attention within the same type can learn the importance of nodes in this category, and attention between different types can learn the importance of different kinds of neighbors to the starting node of the walk. Our main contributions can be summarized as follows.

(1) The strategy that processing nodes by the node types can consider the heterogeneous information, and it achieves higher precision in the subsequent community detection experiments than the homogeneous algorithms.
(2) The random-walk-based neighbor nodes set has the advantage of not requiring prior knowledge to set metapaths, and it has comparable results to those that require metapaths.
(3) We conduct comprehensive experiments on four widely used heterogeneous networks to evaluate the performance of the proposed algorithm. The experimental results show the effectiveness of RHAN by comparing it to various baseline algorithms.

2 Related Work

This section introduces homogeneous NRL algorithms and heterogeneous NRL algorithms.

2.1 NRL Algorithms on Homogeneous Networks

Perozzi et al. find the similarity between the degree distribution of network nodes and the frequency of text words [9]. Based on this, they propose DeepWalk, which uses the word2vec model [8] to generate the node representation vector. LINE and Node2Vec are also word2vec-based NRL algorithms, and both have made some improvements. Another main research direction of NRL algorithms is graph neural network (GNN). Kipf et al. propose a spectral-based GNN named Graph Convolution Network (GCN) [2]. However, spectral-based GNNs require the entire network as input. Then, spatial-based GNNs have been proposed, such as GraphSAGE, GAT, etc. In addition, GAE [18] uses GCN as an encoder to generate node representation vector and an inner product for unsupervised training. ARGA [19] improves GAE by adding adversarial constraints, which makes the node representation vector more robust.

2.2 NRL Algorithms on Heterogeneous Networks

Most of the existing heterogeneous network NRL algorithms use metapaths to consider heterogeneous information. For example, based on DeepWalk, metapath2vec uses a single metapath to guide the random walk and then inputs it into the Skip-Gram model [8] (an implementation of word2vec model) to generate the node representation vector. HIN2Vec learns the representations of nodes and metapaths with the node pairs connected by metapaths. HAN applies attention networks to aggregate node neighbor information and the information from different metapaths. MAGNN has designed a metapath instance encoder, which can consider the relationship between nodes within the metapath, not just the nodes at the beginning and the end of the metapath.

3 Preliminaries

Definition 1 (Heterogeneous Network). Given a network $G = (V, E, T)$, V is the set of nodes, E is the set of edges, and T is the set of node types. If $|T| > 1$, then G is a heterogeneous network.

Definition 2 (Metapath). A metapath is a sequence of node types. For example, $m = t_1 t_2 \cdots t_{l+1}$, where t_i is the type of node i. If a sequence of nodes $s_m = v_1 v_2 \cdots v_{l+1}$ meets the node type order in m, then s_m is an instance of m. For example, the APA (Author-Paper-Author) metapath in Fig. 1 means two co-authors of the paper. In comparison, the APCPA (Author-Paper-Conference-Paper-Author) metapath shows two authors who published a paper at the same conference.

Definition 3 (Community Detection). For a homogeneous network $G = (V, E)$, community detection is to find a disjoint nodes division that nodes in the same community will have more connections than those in different communities. As for a heterogeneous network, it is similar but requires same node type. For ease of evaluation, we take the node type with labels on each network as the target type for community detection.

4 RHAN

This section introduces our algorithm, which includes random walk, intra-type attention, and inter-type attention. The architecture of RHAN is shown in Fig. 2 and the pseudocode is in Algorithm 1. We also analyze the time complexity in the end.

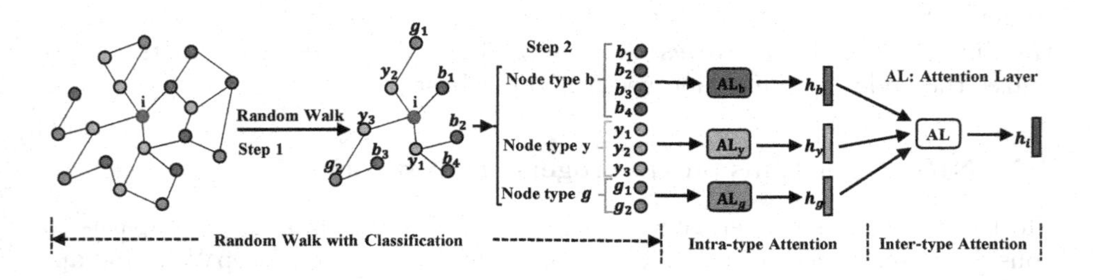

Fig. 2. The architecture of RHAN.

4.1 Random Walk with Classification

The core idea of the GNN-based NRL algorithm is to aggregate the information from nodes' direct neighbors, which can not distinguish the different types of nodes. To solve this problem, we use the random walk to capture the set of closely related neighbors, not only direct neighbors. Take node i in the network as an example. Our random walk includes the following two steps:

Step 1: We start from node i and do w_n random walks with a length of w_l. We record the nodes visited by the random walks and sort them in descending order according to the number of visits.

Step 2: From the set of nodes obtained from step 1, we select the nodes with high access frequency as N_i. The selected scale is controlled by the parameter β. For example, only the first 10% of the nodes will be picked when $\beta = 0.1$.

4.2 Intra-type Attention Computation

After we get the N_i, we will classify it by node type. In the same node type, the attention layer is used to aggregate information from each node (we call it intra-type attention) to obtain vectors representing this type of neighbor. Suppose the node type being processed is t, the attention coefficient a_{ij}^t of each pair of node i and its neighbor j will be learned as Eq. 1.

$$a_{ij}^t = \frac{\exp\left(\text{LeakReLU}\left(\vec{\alpha}^T\left[\mathbf{W}\vec{h_i}\|\mathbf{W}\vec{h_j}\right]\right)\right)}{\sum_{k\in N_i^t}\exp\left(\text{LeakyReLU}\left(\vec{\alpha}^T\left[\mathbf{W}\vec{h_i}\|\mathbf{W}\vec{h_k}\right]\right)\right)} \tag{1}$$

where $||$ means concatenation operation and $\mathbf{W}$ is the weight matrix used in linear transformation. LeakyRelu is a nonlinear activation function and vector $\vec{\alpha}$ will be learned by training. $\vec{h_i}$ is the hidden layer vector of node i. Then we can calculate the vector $\vec{z_i}^t$ that represents this type of neighbors as Eq. 2.

$$\vec{z_i}^t = \sigma\left(\sum_{j\in N_i^t} a_{ij}^t \mathbf{W}\vec{h_j}\right) \tag{2}$$

where σ means the activation function and $\mathbf{W}$ is another weight matrix. In addition, a multi-head attention mechanism is used for stable results and the final vector is the concatenation of each attention head.

4.3 Inter-type Attention Computation

Another attention layer is used to aggregate the representation vector of each node type, which we name inter-type attention. Finally, the attention coefficient of each node type is calculated by Eq. 3.

$$a_t = softmax\left(\frac{1}{|V|}\sum_{i\in V}\vec{\gamma}^T\cdot\sigma\left(\mathbf{W}\cdot\vec{z_i}^t+\vec{b}\right)\right) \tag{3}$$

where $\vec{\gamma}$ is one of the parameters that will be learned by training. $\mathbf{W}$ is the weight matrix and $\vec{b}$ is the bias vector. a_t will also be normalized via softmax function like Eq. 1. The final representation vector of node i is a weighted aggregation of $\vec{z_i}^t, t \in T$ as the same as Eq. 2.

4.4 Loss Function for Training

For unsupervised tasks like community detection, we define the following loss function for training.

$$\mathcal{L} = - \sum_{i \in V, t \in T, j \in N_i^t} p(j|i;\theta) \tag{4}$$

where N_i^t is the neighbors set of node i and the node type of the nodes in N_i^t is t. θ represents the parameters of model and $p(j|i;\theta)$ is defined as follows.

$$p(j|i;\theta) = \frac{exp(\vec{z_j} \cdot \vec{z_i})}{\sum_{k \in N_i^t} exp(\vec{z_k} \cdot \vec{z_i})} \tag{5}$$

However, it is impractical to accurately calculate Eq. 5 which is very time-consuming. We use negative sampling technique [10] to accelerate and the final loss function is shown in Eq. 6.

$$\mathcal{L}' = - \sum_{i \in V, t \in T, j \in N_i^t} (\log \sigma(\vec{z_j} \cdot \vec{z_i}) - \log(1 - \sigma(\vec{z_k} \cdot \vec{z_i})) \tag{6}$$

where the node k is the negative sample which not in N_i^t. The pseudocode of RHAN is shown in Algorithm 1.

Algorithm 1. RHAN

Input: The heterogeneous network $G = (V, E, T)$, the number of random walk w_n, the length of random walk w_l, the select scale of random-walk-based neighbors β.
Output: Node representation vectors $\{\vec{z_i}^t, \forall i \in V\}$.
1: **for** node i in V **do**
2: $\quad N_i = \text{RandomWalk}(w_n, w_l, \beta)$
3: **end for**
4: **for** node type t in T **do**
5: $\quad$ **for** node i in V **do**
6: $\quad\quad$ **for** node j in N_i^t **do**
7: $\quad\quad\quad$ Calculate a_{ij}^t according to Eq. 1.
8: $\quad\quad$ **end for**
9: $\quad\quad$ Calculate $\vec{z_i}^t$ according to Eq. 2.
10: $\quad$ **end for**
11: $\quad$ Calculate a_t according to Eq. 3.
12: **end for**
13: Aggregate $\vec{z_i}^t$ based on a_t to get $\vec{z_i}$.
14: Calculate loss according to Eq. 6.
15: Back propagation and update parameters using SGD.
16: **return** $\{\vec{z_i}^t, \forall i \in V\}$.

4.5 Time Complexity Analysis

The random walk from line 1 to line 3 in the pseudocode requires $O(w_n \times w_l \times |V|)$ time, which can be seen as linear to the number of nodes for usually $|V| >> w_n \times w_l$. For lines 4 to 12 in the pseudocode, the time for running a single graph attention layer is $O(|V| \times N_F \times N_F^{'} + |E| \times N_F^{'})$, where N_F and $N_F^{'}$ are the numbers of input and output dimension, respectively. The number of node types is relatively small and can be ignored here. In total, the time complexity of RHAN is linear to the number of nodes and edges.

5 Experiments

This section first describes the experimental settings, including the used datasets, evaluation metrics, baseline algorithms, and parameter settings. Then there are experimental results and analysis, including parameter experiments, ablation experiments, and accuracy experiments.

5.1 Datasets

We use four widely used heterogeneous networks to evaluate the performance of RHAN. The details of each network are as follows.

DBLP. DBLP [20] is an academic network. We extract a subnetwork of DBLP, which contains 500 authors, 2467 papers, and 20 conferences. Author nodes are labeled to represent the four research fields of database, data mining, machine learning, and information retrieval. We use Kmeans [21] to cluster the author node representation vectors generated by the algorithms to complete the task of community detection.

AMiner. AMiner [13] is also an academic network. We extract a subset of AMiner, which contains 1266 papers, 100 conferences, and 3303 authors. Papers are divided into ten kinds, such as software engineering, computer networks, etc. So we take papers as target nodes for community detection.

Amazon. Amazon [22] is a shopping network. The subnetwork of Amazon we used contains 989 items, 6131 users, and 162 bands. Items have labels, and we use them for community detection.

Yelp. Yelp [23] is a online review network. The network we used in the experiments contains 2614 businesses, 1286 users, and nine stars. We use labeled business nodes as the target of the community detection task.

5.2 Evaluation Metrics

We use two clustering evaluation metrics, normalized mutual information (NMI) [24] and adjusted rand index (ARI) [25], to evaluate the quality of community detection results. Their specific calculation methods are as follows.

NMI. Normalized mutual information (NMI) is defined as follows.

$$\mathrm{NMI}(U,V) = \frac{\mathrm{MI}(U,V)}{\sqrt{\mathrm{H}(U)\mathrm{H}(V)}} \tag{7}$$

$$\mathrm{MI}(U,V) = \sum_{i=1}^{|U|}\sum_{j=1}^{|V|} \frac{|U_i \bigcap V_j|}{N} log\left(\frac{N|U_i \bigcap V_j|)}{|U_j||V_j|}\right) \tag{8}$$

$$\mathrm{H}(U) = -\sum_{i=1}^{|U|} \frac{|U_i|}{N} log(\frac{|U_i|}{N}) \tag{9}$$

where U and V represent the results of two kinds of community division, N is the number of nodes. The closer the NMI value is to 1, the better the community division result.

ARI. Adjusted rand index (ARI) is defined as follows.

$$\mathrm{ARI} = \frac{\mathrm{RI} - \mathbf{E}[RI]}{max(RI) - \mathbf{E}[RI]} \tag{10}$$

$$\mathrm{RI} = \frac{a+b}{C_2^{n_{samples}}} \tag{11}$$

where C is the true community assignment and $C_2^{n_{samples}}$ is the total number of possible pairs in the network. Represent the predicted community assignment as K, then a is defined as the number of pairs of elements that are in the set in C and in the same set in K, b is defined as the number of pairs of elements that are in different sets in C and in different sets in K. $\mathbf{E}[RI]$ is the expected value of RI. The value of ARI is also closer to 1, the better.

5.3 Baseline Algorithms

We compare RHAN with different NRL algorithms, including word2vec-based homogeneous NRL algorithms, GNN-based homogeneous NRL algorithms, and heterogeneous NRL algorithms. The list of baseline algorithms is shown as follows.

DeepWalk. It uses random walks to generate sequences of nodes and then input them into the Skip-Gram model to train the node representation vectors.

LINE. It designs the first-order and second-order similarity between nodes and uses these as the goal to learn the representation vectors of nodes.

Node2Vec. It improves the random walk strategy in DeepWalk to generate higher quality node sequences, and then it also uses the Skip-Gram model to generate the node representation vectors.

ARGA and ARVGA. Adversarial restrictions are added based on GAE and VGAE, respectively, which makes the generated node representation vectors more robust.

Metapath2vec. It uses a metapath to guide random walks on heterogeneous networks and input the generated node sequences into the Skip-Gram model to learn node representation vectors.

HIN2Vec. It designs the objective function based on the node pairs connected by the metapaths to learn node representation vectors.

HAN. Based on the metapaths, node-level and semantic-level attention are designed, and the information of the neighbors reachable by the metapaths is aggregated to generate the node representation vectors.

MAGNN. It uses the metapaths instance encoder based on the relational rotation in the complex space, which makes the information embedded in the sequence structure of the metapaths can be considered into the node representation vectors.

5.4 Parameter Settings

The dimension of the node representation vectors generated by all algorithms is set to 64. Then we apply Kmeans to obtain communities from the node representation vectors, and the number of clusters K is set to each network's actual number of communities. Finally, we run ten times Kmeans and take the average as the final result. As for the specific parameter settings of each algorithm, they are as follows.

LINE: The total number of samples t is set to $w_n \times w_l \times |V|$ and we use the objective function including the first-order and second-order similarity.

Node2Vec: We pick the optimal combination of the parameter p and q from 0.5, 1, 2.

metapath2vec: The best performing metapath is taken on each network. For the algorithms including random walk (DeepWalk, Node2Vec, metapath2vec,

HIN2Vec, RHAN), the number of walks w_n, the walk of length w_l and the window size w_w are set to 10, 40, and 5, respectively. For the other metapath-based algorithms (HIN2Vec, HAN, MAGNN), the metapaths of length three are used. The other parameters of the algorithms are set to the values suggested by the authors.

5.5 Parameter Experiment

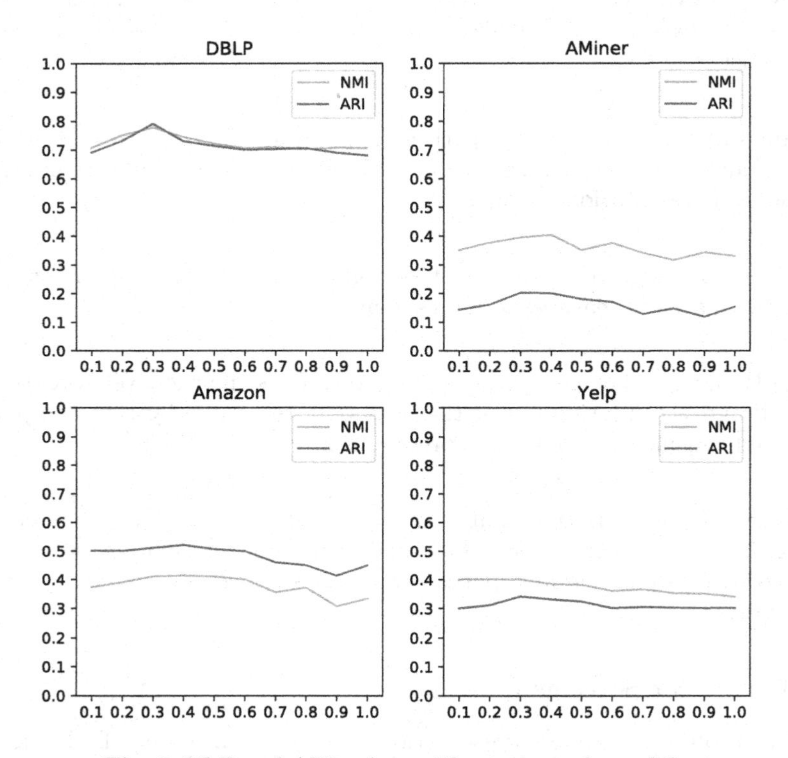

Fig. 3. NMI and ARI values with varying values of β.

The parameter β controls the ratio of selecting the node's neighbor set. For example, only the top 10% of the nodes with the highest number of visits will be selected when $\beta = 0.1$. Figure 3 shows the NMI and ARI values with varying values of β on the four widely used heterogeneous networks. We can find that the values of NMI and ARI will increase with the increase of β when β is less than 0.3. Because the set of neighbor nodes will increase with β and the node vector can learn more information. However, when the $\beta > 0.3$, the values of NMI and ARI will decrease as β increases. This is because the set of neighbor nodes will contain noisy nodes, making the node vector fused with wrong information. Therefore, we set the value of β to 0.3 in subsequent experiments.

5.6 Ablation Experiment

Table 1. NMI and ARI values of RHAN and RAN.

		DBLP	AMiner	Amazon	Yelp
NMI	RAN	0.6821	0.3476	0.3769	0.3178
	RHAN	**0.7790**	**0.3956**	**0.4121**	**0.4011**
ARI	RAN	0.6816	0.1647	0.4625	0.3013
	RHAN	**0.7920**	**0.2033**	**0.5016**	**0.3412**

After using random walk and β to get the neighbor set of nodes, we use a classification processing strategy to consider heterogeneous information. In order to verify the effectiveness of this strategy, we designed an ablation experiment. RAN is RHAN without the classification strategy, which treats different types of nodes as the same type. The experimental results are in Table 1, and better performance is shown in bold. It can be seen that the results of RHAN are better than those of RAN, indicating that our strategy of fusing heterogeneous information is helpful.

5.7 Accuracy Experiment

Table 2. NMI and ARI values of the algorithms.

	NMI				ARI			
	DBLP	AMiner	Amazon	Yelp	DBLP	AMiner	Amazon	Yelp
DeepWalk	0.6874	0.3546	0.3755	0.3072	0.6959	0.1626	0.4233	0.3095
LINE	0.6914	0.3755	0.3820	0.3087	0.7082	0.1404	0.4608	0.3395
Node2vec	0.6957	0.3573	0.3960	0.3102	0.7173	0.1387	0.4734	0.3109
ARGA	0.6768	0.3397	0.3698	0.2646	0.6946	0.1709	0.4897	0.2853
ARVGA	0.6709	0.3121	0.3554	0.2699	0.6997	0.1639	0.4847	0.2915
metapath2vec	0.7104	0.3926	0.4229	0.3847	0.7270	0.1822	0.5046	0.3144
HIN2Vec	0.7493	0.3973	0.4003	0.4001	0.7554	0.1941	0.5006	0.3001
HAN	0.7504	0.4063	**0.4345**	0.3942	0.7644	0.2007	**0.5175**	0.3452
MAGNN	0.7757	0.4056	0.4319	**0.4156**	0.7902	0.2000	0.5105	**0.3512**
RHAN	**0.7790**	**0.4151**	0.4112	0.4011	**0.7920**	**0.2033**	0.5116	0.3412

We compare RHAN with various NRL algorithms, and the experimental results are in Table 2 where bold represents the best performance on each network. DeepWalk, LINE, Node2Vec, ARGA, and ARVGA are homogeneous NRL algorithms, while the remaining baseline algorithms are heterogeneous NRL algorithms. The homogeneous NRL algorithms will ignore the node type and treat the heterogeneous network as a homogeneous network to learn the representation vector of nodes. So we can see that the performance of most homogeneous algorithms is not as good as that of heterogeneous algorithms. Most heterogeneous algorithms use metapaths to capture and utilize heterogeneous information. metapath2vec uses a single metapath to guide random walk and feeds to the Skip-Gram model. HIN2Vec designs objective functions based on the node pairs connected by metapaths. HAN considers the different importance of different metapaths. MAGNN can use the information of the nodes inside the metapaths. HAN and MAGNN have more comprehensive mining of the metapaths, which explains their better performance. However, these algorithms all require prior knowledge to pre-specify the metapaths, and the choice of metapaths will significantly impact the result. We propose RHAN that uses random walk to consider heterogeneous information. It does not need prior knowledge to set metapaths and can achieve comparable results as shown in Table 2.

6 Conclusions

We propose a new heterogeneous representation learning algorithm named RHAN for community detection, and it does not require pre-specify metapaths. Instead, the random walk is employed to capture the neighbor nodes set of nodes. Furthermore, intra-type attention and inter-type attention are designed to consider heterogeneous information. The experiments demonstrate the effectiveness of RHAN. However, RHAN's loss function is not closely related to the downstream community detection task. In the future, we intend to design an end-to-end unified model which can directly give the community assignment. Compared with the two-stage approach of first learning the node representation vector and then clustering, it is more applicable to different networks.

Acknowledgements. This work was supported by the National Natural Science Foundation of China under Grant No. 61672159, No. 61672158, No. 61300104 and No. 62002063, the RGC Theme-based Research Scheme [T41-603/20-R], the Fujian Collaborative Innovation Center for Big Data Applications in Governments, the Fujian Industry-Academy Cooperation Project under Grant No. 2017H6008 and No. 2018H6010, the Natural Science Foundation of Fujian Province under Grant No. 2018J07005, No. 2019J01835, No. 2020J05112 and No. 2020J01494, the Fujian Provincial Department of Education under Grant No. JAT190026, the Fuzhou University under Grant 510872/GXRC-20016 and Haixi Government Big Data Application Cooperative Innovation Center.

References

1. Hamilton, W.L., Ying, R., Leskovec, J.: Inductive representation learning on large graphs. arXiv preprint arXiv:1706.02216 (2017)
2. Kipf, T.N., Welling, M.: Semi-supervised classification with graph convolutional networks. arXiv preprint arXiv:1609.02907 (2016)

3. Zhang, J., Shi, X., Xie, J., Ma, H., King, I., Yeung, D.-Y.: GaAN: gated attention networks for learning on large and spatiotemporal graphs. arXiv preprint arXiv:1803.07294 (2018)
4. Newman, M.E.J.: Detecting community structure in networks. Eur. Phys. J. B **38**(2), 321–330 (2004). https://doi.org/10.1140/epjb/e2004-00124-y
5. Jin, D., et al.: Incorporating network embedding into Markov random field for better community detection, vol. 33, no. 01, pp. 160–167 (2019)
6. Zhang, Z., Cui, P., Wang, X., Pei, J., Yao, X., Zhu, W.: Arbitrary-order proximity preserved network embedding. In: Proceedings of the 24th ACM SIGKDD International Conference on Knowledge Discovery & Data Mining, pp. 2778–2786 (2018)
7. Yang, Z., Ding, M., Zhou, C., Yang, H., Zhou, J., Tang, J.: Understanding negative sampling in graph representation learning. In: Proceedings of the 26th ACM SIGKDD International Conference on Knowledge Discovery & Data Mining, pp. 1666–1676 (2020)
8. Mikolov, T., Sutskever, I., Chen, K., Corrado, G., Dean, J.: Distributed representations of words and phrases and their compositionality. arXiv preprint arXiv:1310.4546 (2013)
9. Perozzi, B., Al-Rfou, R., Skiena, S.: DeepWalk: online learning of social representations. In: Proceedings of the 20th ACM SIGKDD International Conference on Knowledge Discovery and Data Mining, pp. 701–710 (2014)
10. Tang, J., Qu, M., Wang, M., Zhang, M., Yan, J., Mei, Q.: LINE: large-scale information network embedding. In: Proceedings of the 24th International Conference on World Wide Web, pp. 1067–1077 (2015)
11. Grover, A., Leskovec, J.: node2vec: scalable feature learning for networks. In: Proceedings of the 22nd ACM SIGKDD International Conference on Knowledge Discovery and Data Mining, pp. 855–864 (2016)
12. Veličković, P., Cucurull, G., Casanova, A., Romero, A., Lio, P., Bengio, Y.: Graph attention networks. arXiv preprint arXiv:1710.10903 (2017)
13. Dong, Y., Chawla, N.V., Swami, A.: metapath2vec: scalable representation learning for heterogeneous networks. In: Proceedings of the 23rd ACM SIGKDD International Conference on Knowledge Discovery and Data Mining, pp. 135–144 (2017)
14. Fu, T.-Y., Lee, W.-C., Lei, Z.: HIN2Vec: explore meta-paths in heterogeneous information networks for representation learning. In: Proceedings of the 2017 ACM on Conference on Information and Knowledge Management, pp. 1797–1806 (2017)
15. Wang, X., et al.: Heterogeneous graph attention network. In: The World Wide Web Conference, pp. 2022–2032 (2019)
16. Fu, X., Zhang, J., Meng, Z., King, I.: MAGNN: metapath aggregated graph neural network for heterogeneous graph embedding. Proc. Web Conf. **2020**, 2331–2341 (2020)
17. Sun, Y., Han, J., Yan, X., Yu, P.S., Wu, T.: PathSim: meta path-based top-k similarity search in heterogeneous information networks. Proc. VLDB Endow. **4**(11), 992–1003 (2011)
18. Kipf, T.N., Welling, M.: Variational graph auto-encoders. arXiv preprint arXiv:1611.07308 (2016)
19. Pan, S., Hu, R., Long, G., Jiang, J., Yao, L., Zhang, C.: Adversarially regularized graph autoencoder for graph embedding. arXiv preprint arXiv:1802.04407 (2018)
20. Gao, J., Liang, F., Fan, W., Sun, Y., Han, J.: Graph-based consensus maximization among multiple supervised and unsupervised models. In: Advances in Neural Information Processing Systems 22, pp. 585–593 (2009)

21. Lloyd, S.: Least squares quantization in PCM. IEEE Trans. inf. Theor. **28**(2), 129–137 (1982)
22. Zhang, C., Song, D., Huang, C., Swami, A., Chawla, N.V.: Heterogeneous graph neural network. In: Proceedings of the 25th ACM SIGKDD International Conference on Knowledge Discovery & Data Mining, pp. 793–803 (2019)
23. Shi, C., Hu, B., Zhao, W.X., Philip, S.Y.: Heterogeneous information network embedding for recommendation. IEEE Trans. Knowl. Data Eng. **31**(2), 357–370 (2018)
24. Tian, F., Gao, B., Cui, Q., Chen, E., Liu, T.-Y.: Learning deep representations for graph clustering. In: Proceedings of the AAAI Conference on Artificial Intelligence, vol. 28, no. 1 (2014)
25. Hubert, L., Arabie, P.: Comparing partitions. J. Classif. **2**(1), 193–218 (1985)

Attributed Network Embedding Based on Attributed-Subgraph-Based Random Walk for Community Detection

Qinze Wang[1,2,3], Kun Guo[1,2,3](✉), and Ling Wu[1,2,3]

[1] Fujian Provincial Key Laboratory of Network Computing and Intelligent Information Processing, Fuzhou University, Fuzhou 350108, China
gukn@fzu.edu.cn, wuling1985@fzu.edu.cn

[2] College of Mathematics and Computer Science, Fuzhou University, Fuzhou 350108, China

[3] Key Laboratory of Spatial Data Mining and Information Sharing, Ministry of Education, Fuzhou 350108, China

Abstract. The random-walk-based attribute network embedding methods aim to learn a low-dimensional embedding vector for each node considering the network structure and node attributes, facilitating various downstream inference tasks. However, most existing attribute network embedding methods base on random walk usually sample many redundant samples and suffer from inconsistency between node structure and attributes. In this paper, we propose a novel attributed network embedding method for community detection, which can generate node sequences based on attributed-subgraph-based random walk and filter redundant samples before model training. In addition, an improved network embedding enhancement strategy is applied to integrate high-order attributed and structure information of nodes into embedding vectors. Experimental results of community detection on synthetic network and real-world network show that our algorithm is effective and efficient compared with other algorithms.

Keywords: Attribute network embedding · Attributed-subgraph-based random walk · Community detection

1 Introduction

Attributed networks are ubiquitous in real-world systems, such as protein, social and citation networks. These networks usually have two features. First, they are typically consist of dense clusters of nodes called communities [1]. Second, it is difficult to ensure the quality of network embedding by relying only on network structure. The attributes of node can provide complementary information [2] to enhance the precision of network embedding jointly. Community detection on the attributed networks, as a sophisticated and challenging topic, has been widely applied in many fields, such as social circle detection, protein structure detection, traffic flow control, etc.

Y. Sun et al. (Eds.): ChineseCSCW 2021, CCIS 1492, pp. 199–213, 2022.
https://doi.org/10.1007/978-981-19-4549-6_16

Attributed network embedding (ANE) aims to learn a low-dimensional vector space while preserving network structure and attributes of origin networks. The existing ANE algorithms can be divided into three categories: the algorithms base on matrix factorization [3,4], base on the random walk [5,6], and base on the neural network [7,8]. The algorithms base on random walk capture the relationship between nodes by walking through the network under the guidance of node attributes and structure, and Skip-Gram model [9] is used to obtain the embedding vectors. Although the existing works have achieved considerable success, they still face some challenges. First, traditional random walk redundantly samples the low-information nodes that are easy to find out community affiliation. It leads the Skip-Gram model to repeatedly learn the relationships between the node and the context nodes in a node sequence. Second, existing ANE methods usually calculate transition probabilities combining nodes attributes and structure. A random walk process may sample both nodes with similar structure and nodes with similar attributes simultaneously, which may cause the nodes in the node sequence to dissimilar with its context because there may exist inconsistency in node structure and attributes. For example, nodes v_i and v_j have similar attributes but are far apart in network structure. Suppose a random walk process jumps from v_i to v_j. In that case, the subsequent sample will most likely be limited in the neighborhood of node v_j, which will result in a lower similarity between the subsequent nodes and v_i and make the learned embedding vector of node v_i inaccurate. Third, although most existing ANE works have considered the role of high-order structure proximity, high-order proximity on attributes is also important for attributed network embedding.

We propose an ANE algorithm based on Attributed-Subgraph-based Random Walk (ANE_ASRW) in the paper to solve the problems above. ANE_ASRW does not need node labels, which can be well applied to community detection tasks. First, we design different strategies to capture the network topology and node attributes. Specifically, for the extraction of network structure, we develop a node-information-based sequence filtering strategy to select high-information sequences. In addition, we construct an attributed subgraph for each node and use a random walk strategy to extract attribute information. Second, inspired by NEU [10], an improved fast network embedding enhancement technology is adopted to embed the high-order structure and attribute information after obtaining the embedding vectors. Finally, we feed them to K-Means [11] to detect community structure. The main contributions of this paper are as follows:

1. The information-based sequence filtering strategy reduces sampling of low-information sequences, making Skip-Gram model focus on the learning of complex node relationships.
2. The random walk strategy based on attribute subgraphs can accurately sample nodes with similar attributes independently of the network structure, which solves the problem of different contexts in node sequences.
3. The improved network embedding enhancement strategy can accurately incorporate high-order attribute similarity and structure proximity, improving network embedding quality.

4. We evaluate ANE_ASRW on several synthesized and real-world networks. The results show that ANE_ASRW algorithm is superior to the comparison algorithms in community detection.

The rest of this paper is organized as follows: The related work is represented in Sect. 2. Section 3 describes the ANE_ASRW algorithm in detail. The experimental results and conclusions are given in Sect. 4 and 5 respectively.

2 Related Work

The typical network embedding algorithms include DeepWalk [5], LINE [12], SDNE [13], Node2Vec [6], etc. These methods only utilize network structure to learn the embedding vectors, which can not cope with the attribute network. Unlike these, Attributed Network Embedding (ANE) methods can utilize both network structure and node attribute to learn embedding vectors jointly. The ANE algorithms can be divided into three categories.

(1) Matrix-factorization-based algorithms: TADW [14] proves that DeepWalk is equivalent to matrix factorization and then improves the matrix to incorporate node attributes information. AANE [15] learns embedding vectors based on the decomposition of attribute similarity matrix with regularization constraints from network edges. BANE [16] constructs a Weisfeiler-Lehman matrix to combine structure and attribute information and then obtains embedding vectors by decomposing the matrix. The main challenge of this category is scalability because the matrix factorization operation is time-consuming.
(2) Graph-neural-network-based algorithms: GCN [17] applies average aggregation strategy to capture the relationships in the local neighborhood. GAE and VGAE [18] both use the GCN as an encoder to learn network embedding. Based on the GAE and VGAE, ARGA and ARVGA [19] both adopt adversarial training principles to enforce embedding vectors to match a prior Gaussian distribution. The main challenges of this category are the large scale of network parameters and the existence of the over-fitting problem.
(3) Random-walk-based algorithms: TriDNR [20] not only performs the random walk to preserve topological proximity, but also the preservation of node attributes also rely on the random walk in the origin network. There may exist inconsistency between attributes similarity and structural proximity because many nodes with similar attributes may be far apart in network structure. MIRand [21] adopts a random walk based on the information of nodes in multi-layer graph to capture network structure and node attribute information. RoSANE [2] integrates network structure and attributes to reconstruct a dense network. And the random walk is executed on the reconstruction graph to learn the embedding vectors. Similar to TriDNR, MIRand and RoSANE algorithms may sample nodes with similar attributes but dissimilar structure to the same sequence, resulting in dissimilar node pairs in the context window.

3 ANE_ASRW Algorithm

The framework of ANE_ASRW is shown in Fig. 1. It is mainly composed of four parts. First, different strategies are designed for network structure and attribute information extraction. Specifically, on the one hand, we first perform a random walk to generate sequences with structure information. Then an information-based sequence filtering strategy is used to delete redundant samples with low information. On the other hand, we construct an attribute subgraph for each node and perform the same random walk to extract the node sequences relying on the node attribute. Second, the Skip-Gram model generates embedding vectors, which contain network structure and attribute similarity information. Third, a network embedding enhancement strategy is improved to embed high-order structures and attribute information to the embedding vectors. Finally, we use K-Means to obtain community divisions.

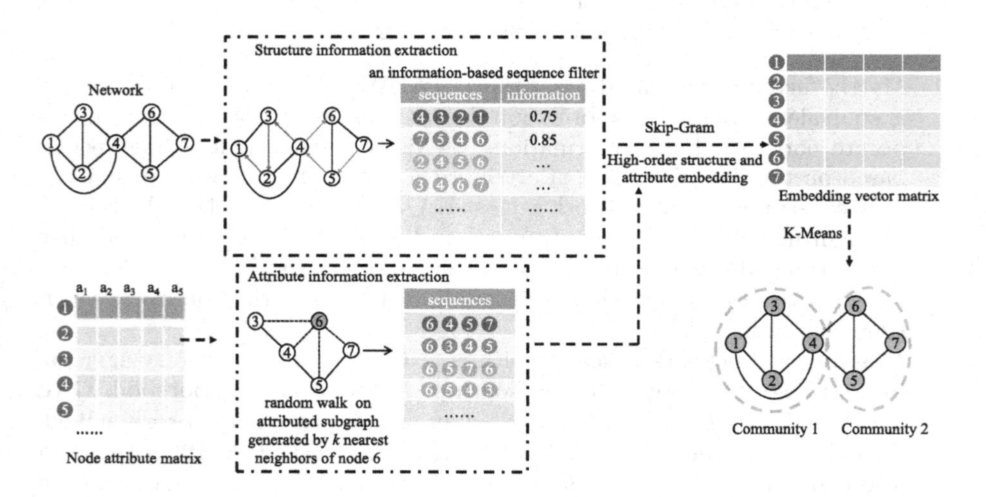

Fig. 1. Framework of ANE_ASRW

3.1 Structure Information Extraction

The traditional random walk of attribute networks combines attribute and structure similarity to calculate transition probability and sample nodes with similar attributes or similar structure to the same sequence, resulting in many dissimilar pairs of nodes in a context window. To avoid the problem, we separate attribute information extraction from the extraction of network structure to ensure all nodes in the same sequence are either similar on attributes or structure. In experiments, we set a total number of samples and use a hyperparameter λ to represent the weight of network structure and node attributes.

The advantage of an unbiased random walk is the diversity of sampled node sequences. Therefore, we adopt an unbiased random walk to capture the network structure. However, this approach may sample many redundant nodes which contain little information and are useless for accuracy improvement. In this paper, we assume that the nodes that are easy to determine the community affiliation are low-information because the model may only need a few iterations to determine their community affiliation accurately. The node information is calculated according to Eq. (1), where $d(v_j)$ represents node degree of v_j and $N(v_i)$ represents the neighborhood of v_i. Generally, a high-degree node is high-information because of the difficulty of determining its community affiliation.

$$I(v_i) = -\sum_{v_j \in N(v_i)} \frac{1}{d(v_j)} \log \frac{1}{d(v_j)}, \tag{1}$$

We design a node-information-based sequence filtering strategy to delete sequences composed of many low-information nodes, making the Skip-Gram model focus on the learning of complex relationships between a high-information node and others. In order to evaluate the information of node sequence, we first define the concept of window information because a sequence is composed of windows according to Skip-Gram, and sequence information can be deduced from the window information. Specifically, the window information is defined in Eq. (2),

$$I(w_{v_i}) = \sum_{v_j \in w(v_i)} (I(v_i) + I(v_j))\,(1 - sim(v_i, v_j)), \tag{2}$$

where w_{v_i} represents a window centered on node v_i and $sim(v_i, v_j)$ represents the structural similarity between v_i and v_j. It includes both first-order and second-order similarities, is calculated according to Eq. (3), where $N(v_i)$ is the neighbors of node v_j.

$$sim(v_i, v_j) = \begin{cases} 1.0, & v_j \in N(v_i) \\ Jaccard(v_i, v_j), & v_j \in \{N(v_k) | v_k \in N(v_i)\}. \end{cases} \tag{3}$$

We set the similarity between the node and its first-order neighbor to 1.0 because low-order proximity is the cornerstone of the high-order [10]. In addition, the Jaccard coefficient calculated in Eq. (4) is used as the similarity between a node and its second-order neighbor.

$$Jaccard(v_i, v_j) = \frac{N(v_i) \cap N(v_j)}{N(v_i) \cup N(v_j)}. \tag{4}$$

The window information includes two parts. First, it is positively correlated with the number of high-information nodes in the window. Second, the window information is composed of the node pair information. The node pairs with low structure similarity have a more significant contribution, where the node pair information is defined as the sum of node information for efficiency. Finally, the sequence information is defined as the sum of windows information, as shown in Eq. (5).

$$I(S) = \sum_{w \in S} I(w). \tag{5}$$

Based on this definition, it is difficult to determine the relationship between nodes in a high-information window using network structure alone. The core idea of the filtering strategy is to preserve the sequences composed of high-information windows and filter out the low-information sequences. It will facilitate the Skip-Gram model to focus on the learning of complex relationships. In this paper, we set a ratio threshold to filter out low-information sequences.

3.2 Attribute Information Extraction

The extraction strategy for attribute information includes two steps. First, we use the approximate nearest neighbor method [22] to efficiently search k nearest neighbors of a node according to node attribute, where the attribute similarity is measured by the dot product of node attribute vectors. The attributed subgraph of node v_i consists of node v_i and its k nearest neighbors. For example, there is an attributed subgraph of node v_6 as shown in Fig. 2. The neighbors indicated by the dashed line are the k nearest attributed neighbors of node v_6, and the solid line is the edge in the original network. Second, a random walk process is performed on the attribute subgraph to generate node sequences. Assuming that the attributes have been normalized to numerical values, we first evaluate the node attribute richness as follows:

$$R(x_i) = \frac{||x_i||}{\frac{1}{N}\sum_{j=1}^{N}||x_j||}, \tag{6}$$

where x_i is the attribute of node v_i and N is the size of the network. Intuitively, the size of the attributed graph of node v_i is positively correlated with the attribute richness of node v_i, and it is necessary to extract a large number of samples from a larger attributed subgraph to better capture attribute information. Therefore, we adaptively calculate the corresponding attribute subgraph size k_i and random walk length l_i for each node according to node attribute richness according to Eq. (7) and Eq. (8).

$$k_i = k * R(x_i), \tag{7}$$

$$l_i = l_{max} * R(x_i), \tag{8}$$

where k denotes the maximum number of nearest neighbors and l_{max} is the maximum random walk length.

3.3 Embedding Vector Generation

We combine the Skip-Gram model with negative sampling [23] to generate network embedding vectors. Skip-Gram learns the vector of nodes by maximizing the co-occurrence probability of words in the same window, as follows:

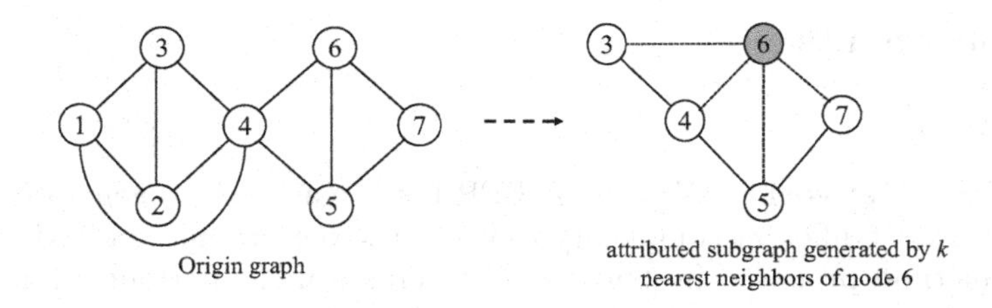

Fig. 2. An toy example of attributed subgraph of node v_6

$$p(\{v_{i-w}, ..., v_{i+w}\} \setminus v_i \mid \phi(v_i)) = \prod_{j=i-w, j \neq i}^{i+w} p(v_j \mid \phi(v_i)), \tag{9}$$

where w represents the size of windows, and $\phi(v_i)$ represents the node v_i's center vector.

3.4 Embedding Vector Enhancement

NEU [10] proves that high-order structure information can enhance the accuracy of network embedding algorithms. Inspired by this idea, we try to combine high-order attribute similarity information into the node embedding vectors similarly. In order not to cost extra time to construct the structural similarity matrix M and the attribute similarity matrix F, we use the by-products of Sects. 3.1 and 3.2, such as the results of $sim(v_i, v_j)$ and dot similarity of attributes. It can be noticed that matrix M and F are sparse in real-world networks. The embedding vector enhancement is an iterative process as shown in Eq. (10):

$$R = R + \lambda MR + (1 - \lambda)FR, \tag{10}$$

where R is the embedding vector matrix, λ is a hyperparameter that balances structure and attribute.

3.5 Time Complexity Analysis

Assuming the number of walk γ, the length of walk t, the size of window w, the dimension of node vector d and the number of node n, the time complexity of two random walk strategies and the generation of node vector is dominated by the Skip-Gram, which is $O(\gamma twn(d + d \log n))$. The complexity of k approximate nearest neighbor method is $O(nk \log n)$ approximately. The improved network embedding enhancement strategy is $O(nd)$, which is consistent with the NEU method [10]. Thus, the overall time complexity of ANE_ASRW is $O(\gamma twn(d + d \log n))$, which is linear to the network size n.

4 Experiments

4.1 Datasets

Synthesized Networks. We use the LFR benchmark [24] to generate synthesized networks of different complexity to verify the performance of ANE_ASRW. And the method proposed in [25] is used to create node attributes for all networks. The parameters of LFR are given in Table 1 and the parameters of the synthetic network are described in Table 2.

Table 1. Parameters of artificial networks

Parameter	Description
N	Number of nodes
k	Average degree of nodes
k_{max}	Maximum degree of nodes
c_{min}	Minimum size of communities
c_{max}	Maximum size of communities
mu	Mixing parameter

Table 2. Synthesis networks

Network	Parameter configuration
D1	$mu = 0.1...0.7, N = 1000, k = 20, k_{max} = 50, c_{min} = 10, c_{max} = 100$

Real-World Networks. The networks[1] include two paper citation networks Cora and Citeseer, the Wikipedia web pages network Wiki, the blog network BlogCataLog, and four university networks Cornell, Texas, Washington, and Wisconsin, which are shown in Table 3.

4.2 Baseline Algorithms

1. RoSANE [2]: The algorithm integrates network topology and node attributes to reconstruct a dense network, and a designed random walk is performed on the reconstruction network to learn the embedding vectors.
2. TADW [14]: The algorithm first proves that DeepWalk is equivalent to matrix decomposition, and then includes node attribute information in the process of matrix decomposition.

[1] https://linqs.soe.ucsc.edu/data.

Table 3. Real-world networks

Network	Number of nodes	Number of edges	Average degree	Communities
Cora	2708	5278	3.89	7
Citeseer	3264	4536	2.78	6
Wiki	2363	11596	9.81	17
BlogCataLog	5196	171743	66.11	6
Cornell	195	286	2.93	5
Texas	187	298	3.19	5
Washington	230	417	3.63	5
Wisconsin	265	479	3.62	5

3. ARGA and ARVGA [19]: ARGA adds adversarial training principle to GAE to make the embedding vectors match a prior Gaussian distribution. ARVGA is the variational version of ARGA.
4. AANE [15]: The algorithm learns embedding vectors based on the decomposition of attribute similarity matrix with regularization constraints from network edges.

4.3 Parameter Settings

For ANE_ASRW and RoSANE algorithms, the number of walks, the length of walk and the size of window are set to 10, 80, 10 respectively. For TADW, ARGA, ARVGA and AANE algorithms, we use the default parameter settings used in their paper. For all algorithms, the node vector dimension is fixed at 128.

4.4 Evaluation Metric

Normalized Mutual Information (NMI) [26] is an important metric for community detection. The larger the NMI, the closer the division X and Y. For the division X and Y, the specific calculation of NMI is defined as follows:

$$NMI = \frac{\frac{1}{2}[H(X) - H(X|Y) + H(Y) - H(Y|X)]}{H(X) + H(Y)}, \qquad (11)$$

where $H(X)$ is the information entropy of X and $H(X|Y)$ denotes the difference between the entropy of X conditioned on Y.

4.5 Parameter Experiment

The ANE_ASRW algorithm includes three parameters: r, λ, and k. In this section, comprehensive experiments in synthesis networks and real-world networks are conducted to investigate the influence of parameters on accuracy. In all experiments, we use K-Means to detect communities and calculate NMI to measure the performance of community detection. Considering the randomness of algorithms, we run the algorithms ten times in each network and report the average value of NMI.

Parameter r. Parameter r plays as a threshold to control the ratio of filtered sequence. Specifically, some sequences with rich information will be filtered when r is too large.

The experimental result on the D1 networks and four real-world networks is shown in Fig. 3 and Fig. 4, respectively. The accuracy of ANE_ASRW increases with the value of r when $r \leq 0.4$ as can be seen from the Fig. 3. When $r > 0.4$, many high-information sequences will be deleted, resulting in decreased algorithm accuracy. Figure 4 shows a more significant increase trend when $r \leq 0.4$ because there exist a large amount of low-information sequences in the complex real-world networks. According to the experimental results, we found the stable interval of r is $[0.3, 0.5]$. Therefore, we search for the optimal value of r in $[0.3, 0.5]$.

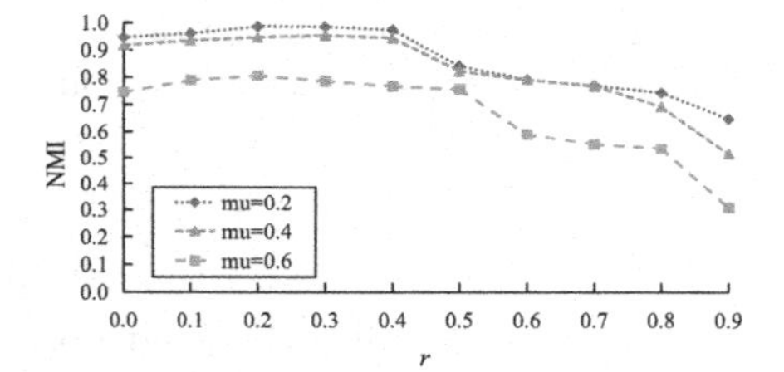

Fig. 3. Parameter sensitivity of parameter r on the D1 networks

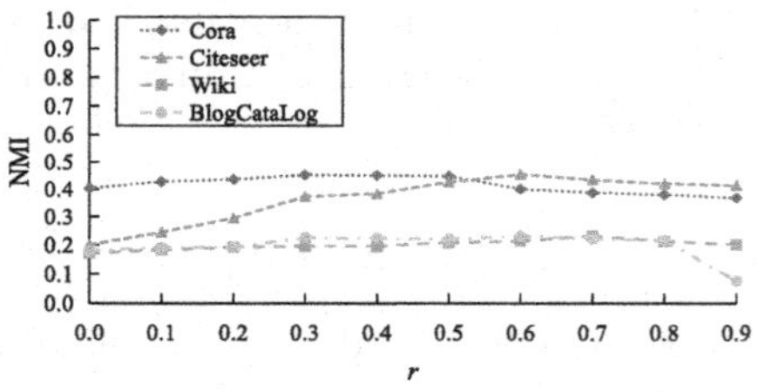

Fig. 4. Parameter sensitivity of parameter r on real-world networks

Parameter λ. Parameter λ is used to control the balance of the node attribute and structure in the network. The sampling strategy will generate more sequences containing the network structure information when λ is too large, ignoring the information extraction of node attributes.

The performance of ANE_ASRW with varying λ on the D1 and four real networks is shown in Fig. 5 and Fig. 6, respectively. Figure 5 shows an upward trend with the increase of the value of λ when $\lambda \leq 0.7$. However, it can be seen from Fig. 6, the increase of algorithm accuracy is almost stable when λ is close to 0.4. It is because real networks have a more ambiguous community structure than synthetic networks and often require more attribute information to detect communities more accurately. Therefore, we set the value of λ in $[0.3, 0.7]$ in the remaining experiments.

Parameter k. Parameter k is used to determine the maximum number of node neighbors in the attributed subgraph. On the one hand, many nodes with attributes similar to the current node will be ignored if the value of k is small. On the other hand, if the value of k is large, it will cause a great time overhead.

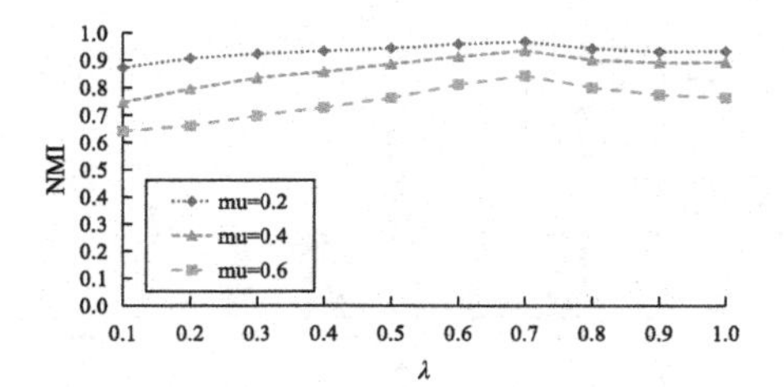

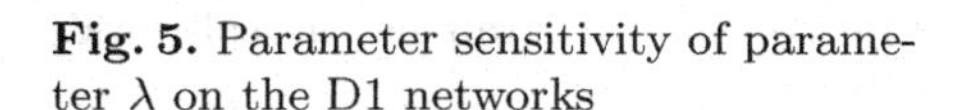
Fig. 5. Parameter sensitivity of parameter λ on the D1 networks

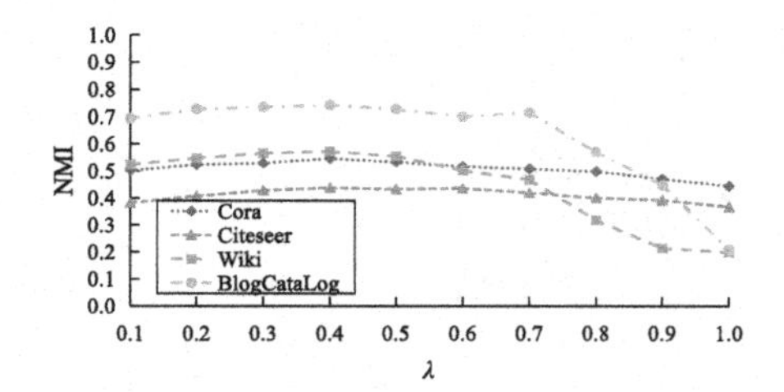

Fig. 6. Parameter sensitivity of parameter λ on real-world networks

The experimental result on the D1 and four real networks are shown in Fig. 7 and Fig. 8, respectively. From Fig. 7 and Fig. 8, the accuracy of ANE_ASRW increases with the log of k and remain stable when the log of k is close to 6. Therefore, the value of k is fixed at 64.

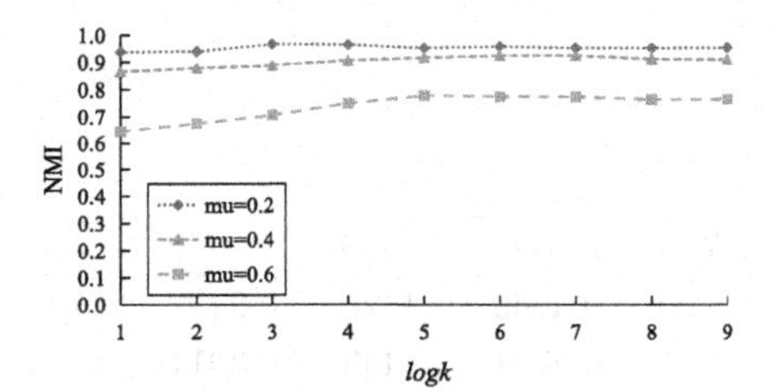

Fig. 7. Parameter sensitivity of parameter k on the D1 networks

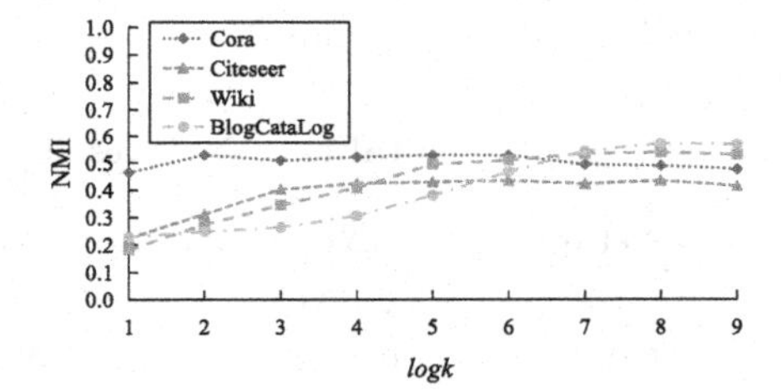

Fig. 8. Parameter sensitivity of parameter k on real-world networks

4.6 Accuracy Experiment

We compared the accuracy of ANE_ASRW with baseline algorithms on the synthetic and real-world networks.

Results on Synthesis Networks The NMI results of varying mu on the D1 networks is shown in Fig. 9. From the Fig. 9, as the value of mu increases, the accuracy of each algorithm decreases gradually. It is because as the value of mu increases, the community boundaries become more indistinguishable. ANE_ASRW performs better in all networks for three reasons. First, the filtering strategy deletes many low-information sequences, which makes the Skip-Gram model focus on the learning of complex relationships. Second, the random walk on attributed subgraph strategy captures the attributed similarity of nodes accurately. Third, the network embedding enhancement strategy embeds high-order structure and attribute similarity to the embedding vectors. They all improve the accuracy of the node embedding.

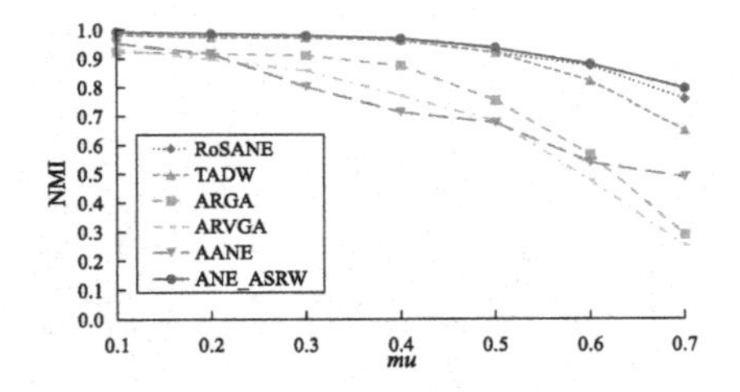

Fig. 9. NMI results with varying *mu* on the D1 networks

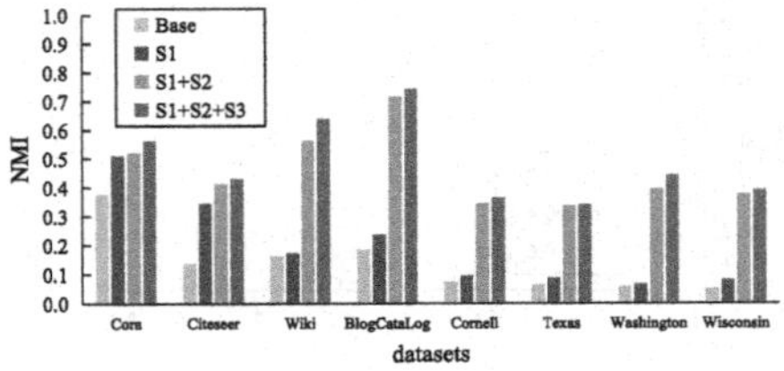

Fig. 10. Effect of strategies on real-world networks

Results on Real-World Networks. Table 4 shows the NMI results of the algorithms on real-world networks. From Table 4, the precision of ANE_ASRW is higher than that of the baseline algorithms except for Citeseer and Wisconsin networks, which reflects the effectiveness of the information-based sequence filtering, attributed-subgraph-based random walk and the embedding enhancement strategy. Although the accuracy of the ANE_ASRW in the two networks is inferior to RoSANE and AANE algorithms, it can also get a better result than other baseline algorithms.

Table 4. NMI results on real-world networks

Network	ANE_ASRW	RoSANE	TADW	ARGA	ARVGA	AANE
Cora	**0.5613**	0.5521	0.5348	0.4949	0.4924	0.2477
Citeseer	0.4317	**0.4438**	0.4104	0.2701	0.3495	0.3011
Wiki	**0.6385**	0.4213	0.3833	0.4179	0.3607	0.4705
BlogCataLog	**0.7431**	0.7105	0.5521	0.1719	0.2071	0.3591
Cornell	**0.3638**	0.0895	0.0904	0.0982	0.1124	0.3224
Texas	**0.3415**	0.0774	0.1041	0.0775	0.0912	0.3293
Washington	**0.4425**	0.1389	0.1544	0.1147	0.1045	0.3691
Wisconsin	0.3916	0.1051	0.0931	0.0611	0.0903	**0.4151**

4.7 Ablation Experiment

We evaluate the effect of each strategy on real-world networks. The information-based sequence filtering strategy, the attribute-subgraph-based random walk, and the network embedding enhancement strategy are denoted by S1, S2, S3, respectively. In Fig. 10, S1 denotes the structure version of the proposed algorithm that only uses strategy 1, S1+S2 represents the version that combines attributes with the structure version, S1+S2+S3 denotes the final version includes the proposed three strategies. It is worth noting that base indicates the baseline without the three strategies above. As shown in Fig. 10, the final version performs best, followed by S1+S2, and the worst performance is the baseline without any strategies. Therefore, the results prove the effectiveness of the combination of S1 and S2 and S3 strategies in attributed network embedding.

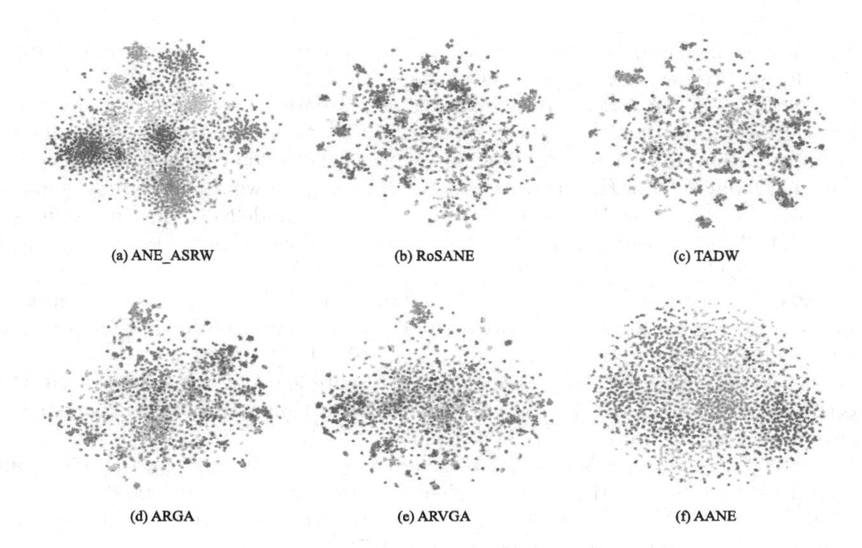

Fig. 11. Embedding vectors visualization on the Wiki network

4.8 Embedding Vectors Visualization

T-SNE [27] is used to convert the high-dimensional embedding vectors to a two-dimensional space to investigate the quality of community detection. The visualization results on the Wiki are shown in Fig. 11. Nodes in different communities are represented in different colors. From the Fig. 11, the embedding vectors generated by ANE_ASRW exhibits more obvious aggregation effects than other baseline algorithms. There is a more obvious boundary between the communities, which helps the clustering methods to distinguish better.

5 Conclusions

This study proposes an ANE algorithm based on attributed-subgraph-based random walk for community detection. ANE_ASRW can make the Skip-Gram focus on the learning of complex relationships by the proposed sequences filtering strategy. At the same time, it also can construct attributed subgraph accurately, which helps the random walk capture the attribute information more precisely. Furthermore, it integrates high-order attributes and structural proximity information into the embedding vectors. Experimental results on different networks show that the ANE_ASRW algorithm is more efficient and effective than the baseline algorithms. In the future, we will extend ANE_ASRW to heterogeneous attribute networks to make it suitable for real world scenarios.

References

1. Girvan, M., Newman, M.E.: Community structure in social and biological networks. Proc. Natl. Acad. Sci. **99**(12), 7821–7826 (2002)

2. Hou, C., He, S., Tang, K.: RoSANE: robust and scalable attributed network embedding for sparse networks. Neurocomputing **409**, 231–243 (2020)
3. Cao, S., Lu, W., Xu, Q.: GraRep: learning graph representations with global structural information. In: Proceedings of the 24th ACM International on Conference on Information and Knowledge Management, pp. 891–900 (2015)
4. Qiu, J., Dong, Y., Ma, H., Li, J., Wang, K., Tang, J.: Network embedding as matrix factorization: unifying DeepWalk, LINE, PTE, and node2vec. In: Proceedings of the 11th ACM International Conference on Web Search and Data Mining, pp. 459–467 (2018)
5. Perozzi, B., Al-Rfou, R., Skiena, S.: DeepWalk: online learning of social representations. In: Proceedings of the 20th ACM SIGKDD International Conference on Knowledge Discovery and Data Mining, pp. 701–710 (2014)
6. Grover, A., Leskovec, J.: node2vec: scalable feature learning for networks. In: Proceedings of the 22nd ACM SIGKDD International Conference on Knowledge Discovery and Data Mining, pp. 855–864 (2016)
7. Scarselli, F., Gori, M., Tsoi, A.C., Hagenbuchner, M., Monfardini, G.: The graph neural network model. IEEE Trans. Neural Netw. **20**(1), 61–80 (2008)
8. Hamilton, W.L., Ying, R., Leskovec, J.: Inductive representation learning on large graphs. arXiv preprint arXiv:1706.02216 (2017)
9. Mikolov, T., Chen, K., Corrado, G., Dean, J.: Efficient estimation of word representations in vector space. arXiv preprint arXiv:1301.3781 (2013)
10. Yang, C., Sun, M., Liu, Z., Tu, C.: Fast network embedding enhancement via high order proximity approximation. In: IJCAI, pp. 3894–3900 (2017)
11. Lloyd, S.: Least squares quantization in PCM. IEEE Trans. Inf. Theor. **28**(2), 129–137 (1982)
12. Tang, J., Qu, M., Wang, M., Zhang, M., Yan, J., Mei, Q.: LINE: large-scale information network embedding. In: Proceedings of the 24th International Conference on World Wide Web, pp. 1067–1077 (2015)
13. Wang, D., Cui, P., Zhu, W.: Structural deep network embedding. In: Proceedings of the 22nd ACM SIGKDD International Conference on Knowledge Discovery and Data Mining, pp. 1225–1234 (2016)
14. Yang, C., Liu, Z., Zhao, D., Sun, M., Chang, M.: Network representation learning with rich text information. In: 24th International Joint Conference on Artificial Intelligence (2015)
15. Huang, X. Li, J., Hu, X.: Accelerated attributed network embedding. In: Proceedings of the 2017 SIAM International Conference on Data Mining, pp. 633–641. SIAM (2017)
16. Yang, H., Pan, S., Zhang, P., Chen, L., Lian, D., Zhang, C.: Binarized attributed network embedding. In: 2018 IEEE International Conference on Data Mining (ICDM), pp. 1476–1481. IEEE (2018)
17. Kipf, T.N., Welling, M.: Semi-supervised classification with graph convolutional networks. arXiv preprint arXiv:1609.02907 (2016)
18. Kipf, T.N., Welling, M.: Variational graph auto-encoders. arXiv preprint arXiv:1611.07308 (2016)
19. Pan, S., Hu, R., Long, G., Jiang, J., Yao, L., Zhang, C.: Adversarially regularized graph autoencoder for graph embedding. arXiv preprint arXiv:1802.04407 (2018)
20. Pan, S., Wu, J., Zhu, X., Zhang, C., Wang, Y.: Tri-party deep network representation. Network **11**(9), 12 (2016)
21. Bandyopadhyay, S., Biswas, A., Kara, H., Murty, M.: A multilayered informative random walk for attributed social network embedding. Front. Artif. Intell. Appl. **325**, 1738–1745 (2020)

22. Arya, S., Mount, D.M., Netanyahu, N.S., Silverman, R., Wu, A.Y.: An optimal algorithm for approximate nearest neighbor searching fixed dimensions. J. ACM (JACM) **45**(6), 891–923 (1998)
23. Mikolov, T., Sutskever, I., Chen, K., Corrado, G., Dean, J.: Distributed representations of words and phrases and their compositionality. arXiv preprint arXiv:1310.4546 (2013)
24. Lancichinetti, A., Fortunato, S., Radicchi, F.: Benchmark graphs for testing community detection algorithms. Phys. Rev. E **78**(4), 046110 (2008)
25. Huang, B., Wang, C., Wang, B.: NMLPA: uncovering overlapping communities in attributed networks via a multi-label propagation approach. Sensors **19**(2), 260 (2019)
26. Danon, L., Diaz-Guilera, A., Duch, J., Arenas, A.: Comparing community structure identification. J. Stat. Mech. Theor. Exp. **2005**(09), P09008 (2005)
27. Van der Maaten, L., Hinton, G.: Visualizing data using t-SNE. J. Mach. Learn. Rese. **9**(11), 2579–2605 (2008)

Adaptive Seed Expansion Based on Composite Similarity for Community Detection in Attributed Networks

Wenju Chen[1,2,3], Kun Guo[1,2,3], and Yuzhong Chen[1,2,3](✉)

[1] College of Computer and Data Sciences, Fuzhou University, Fuzhou 350116, China
{gukn,yzchen}@fzu.edu.cn
[2] Fujian Provincial Key Laboratory of Network Computing and Intelligent Information Processing, Fuzhou, China
[3] Key Laboratory of Spatial Data Mining and Information Sharing, Ministry of Education, Fuzhou 350116, China

Abstract. Community detection is a fundamental tool to uncover organizational principles in complex networks. With the proliferation of rich information available for real-world networks, it is useful to detect communities in attributed networks. Recently, many algorithms consider combinating node attributes and network topology, and the effect of these methods is better than using only one information source. However, the existing algorithms still have some shortcomings. First, only a few algorithms can process both categorical type and numerical type at the same time. Second, the contribution between attributes and topology cannot be adjusted adaptively. Third, most algorithms do not consider combining the high-order structure with the node attributes. Therefore, we propose an adaptive seed expansion algorithm based on composite similarity to solve these problems(ASECS). We generate a new weighted KNN graph according to the composite similarity. The composite similarity combines high-order structure similarity, low-order structure similarity and attributed similarity by weighting them. Our method can adaptively adjust the contribution between topology and node attributes. Moreover, the designed attributed similarity function can process both categorical and numerical attributes. Finally, we find the seed nodes on the weighted KNN graph and expand the seed nodes to communities. The superiority of our algorithm is demonstrated on many networks.

Keywords: Community detection · Attributed network · Motif · Seed expansion · Composite similarity

1 Introduction

The real-world networks is quite complex, so it is too difficulty to analyze the whole network directly and get valuable information. Community detection can divide networks into different communities , and entities in same community often share same characteristics [1]. The exist studies [2] have shown that the

Y. Sun et al. (Eds.): ChineseCSCW 2021, CCIS 1492, pp. 214–227, 2022.
https://doi.org/10.1007/978-981-19-4549-6_17

recognition of community can help uncover hidden information of the network, as well as support some high practical value applications such as protein analysis [3] and recommender system [4].

Traditional community detection methods usually fully use network topology characteristics [1]. Besides topology, node content is another common information source in the real-world networks. Previous study [6] find that network attributes can help to get some different knowledge beyond topology. At present, many methods have considered using topology and attributes at the same time [2,7,8]. However, these methods cannot process both categorical and numerical attributes. Moreover, topology and node attributes are two heterogeneous information sources, which may have inconsistent features or noise. How to counterpoise the contribution of topology and node attributes adaptively is a challenge.

In addition, the above method only uses the low-order structure without considering high-order structure, such as motifs, they are tiny subgraphs that are ubiquitous and regarded as composition modules for networks. Recently, there are methods to combine node attributes with high-order structures [9], but the method relies too much on high-order structures and will appear fragmentation issue [10]. In [10,11], the fragmentation issue is solved, but they are not considering the attribute information in the network.

In this paper, to fully use the abundant information in the attributed network, we propose an adaptive seed expansion algorithm based on composite similarity. First, we calculate the attributed similarity, low-order structure similarity, and high-order structure similarity of nodes. Second, according to the designed weighted function, the three similarities are combined, and the composite similarity between nodes is obtained, which contains various information. Based on the composite similarity of nodes, we generate a weighted KNN graph that retains Top-K similarity neighbors for each node. Finally, we find the seed nodes on the weighted KNN graph and expand the seed nodes to communities. There are three main contributions in this paper:

(1) The proposed composite similarity contains attributed and high and low order structure information, which can enhance the quality of seed nodes and greatly improves the performance of community detection.
(2) The designed weighted function can adjust the contribution ratio between attributes and structure adaptively by network clustering coefficient.
(3) The proposed attributed similarity calculation method can process both categorical and numerical attributes, which help to get the complex relationship between nodes.

2 Related Work

2.1 Community Detection Based on Seed Expansion

Seed expansion algorithms mainly focus on seed selection and community extension strategies. Lancichinetti et al. [5] proposed the well-known algorithm LFM, which randomly selects nodes from the network as seeds and proposes community fitness as a quality function to optimize. Lee et al. [12] proposed GCE, which selects the k-cliques as seeds, and uses the same extension strategy as the LFM algorithm. Guo et al. [13] proposed InfoNode algorithm, which extends fitness funtion by the internal force between nodes to fully explore local information. Zhang et al. [14] proposed the CFCD algorithm, which uses core similarity to select seed node and takes core fitness as the objective function.

2.2 Community Detection in Attributed Networks

At present, many researchers have demonstrated that combining topology structure and node attributes can enhance the quality of community detection. Zhou et al. [7] converted the original graph G into the attributed augmented graph G_A with attribute nodes, which added fictitious attribute nodes to represent different attribute values and then further used the random walk to measure the tightness between nodes. Yang et al. propose CESNA [8] model, which uses the community membership to model the interaction between topology and attributes, and updates all model parameters using the block coordinate rising method. Wang et al. [2] proposed SCI model with two parameters and proposed an effective updating rule to evaluate the parameters with convergence guarantee. He et al. proposes NEMBP [6] model; NEMBP combines nested EM algorithm and confidence propagation to train the community and attributes. Then, it extracts and explores the potential correlation between them to divide the community. Besides, there are also methods focusing on network embedding [15,16].

2.3 Community Detection Based on High-Order Structure

In recent years, some studies have tried to use network high-order features for community detection. In [9], according to attributes of nodes included in the motif, the so-called homogeneity value is provided for the structure motif identified by the network. Then the homogeneity value is stored by a particular adjacency matrix. Finally, put the matrix into the existing algorithm based on similarity. In [10], defined the hypergraph fragmentation problem, generated a new graph by edge enhancement strategy to overcome the hypergraph fragmentation problem, and then divided the new graph to obtain the high-order community structure. In [11], a modularity-based method of micro-unit is designed, which can find overlapping community structures by using low-order connectivity patterns and high-order connectivity patterns. In [17], designed a reweighted network with unified low and high order structure, and then detected communities on the reweighted network.

3 Preliminaries

We use $G = (V, E, A)$ to represent an attributed network where V represents the set of nodes and E represents the set of edges. As for A, A is the attributed matrix of nodes. There are some basic definitions in our algorithm:

Definition 1 (Motif). Motif is a kind of interconnection mode between nodes, and its number in real-world networks is obviously high than in the random network [18]. The motif with p nodes and q edges is defined as $\mathcal{M}(p,q)$ for clarity. We utilize the FANMOD [19] tool to identify motif. It can identify all variants of the size-k motif, then the corresponding z-score will be reported.

Definition 2 (High-order Structure Similarity). The high-order structure similarity is defined as:

$$M_{\mathcal{M}}(u,v) = I_{u,v}^{\mathcal{M}} \tag{1}$$

where $I_{u,v}^{\mathcal{M}}$ represents the number of the motif instances $\mathcal{M}$ that contain node u and node v. The high-order structural similarity between nodes is stored in the high-order structure similarity matrix. In Fig. 1, we take the size-4 motif to illustrate how to construct a high-order structure similarity matrix. As shown in Fig. 1, node 1 and node 2 appear in two motif instances (0–1–2–3 and 1–2–3–4), so their high-order similarity is 2.

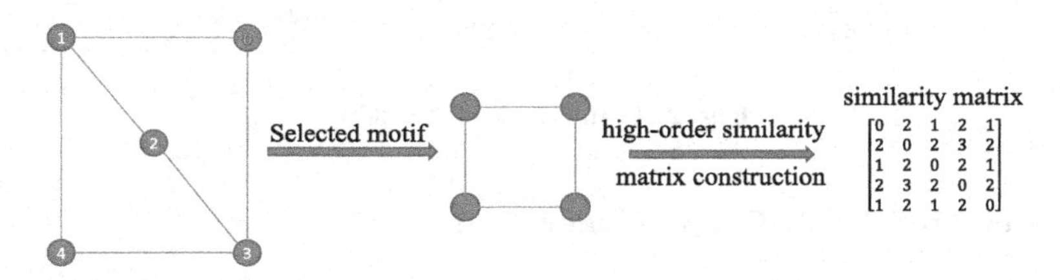

Fig. 1. The construction of the high-order similarity matrix

Definition 3 (Low-order Structure Similarity). We define the Jaccard coefficient [20] between two nodes as the low-order structure similarity:

$$J(\mathrm{u},\mathrm{v}) = \frac{|N(u) \cap N(v)|}{|N(u) \cup N(v)|} \tag{2}$$

where $N(u)$ represents the neighbor set of node u.

Definition 4 (Clustering Coefficient). The clustering coefficient [21] for each node in complex networks is defined as:

$$C_v = \frac{2e_v}{k_v\left(k_v - 1\right)} \tag{3}$$

where k_v represents the number of neighbors of node v, e_v represents the number of edges between neighbors of node v. Similarly, for the whole network, the average clustering coefficient [22] is defined as follow:

$$C = \frac{1}{|V|} \sum_{i \in V} C_i \tag{4}$$

4 ASECS

Figure 2 shows the framework of ASECS. ASECS algorithm is mainly divided into two stages: weighted KNN graph generation and community detection by seed expansion. The details of two stages are given as follows.

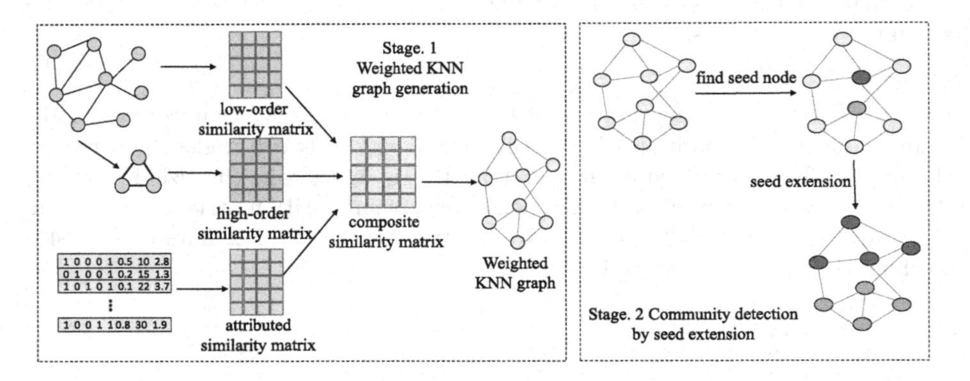

Fig. 2. Framework of ASECS

4.1 Weighted KNN Graph Generation

In this stage, we propose an attributed similarity calculation method that can process both categorical and numerical attributes. And we define the attributed similarity between two nodes as follow:

$$\mathrm{S(u, v)} = S_c(u, v) + S_n(u, v) \tag{5}$$

where S_c represents the categorical attributed similarity, S_n represents the numerical attributed similarity, which are defined as follows:

$$S_c(u, v) = \frac{\sum_{j=1}^{S_1} \eta\left(T_{uj}, T_{vj}\right)}{s_1} \tag{6}$$

$$\eta\left(T_{uj}, T_{vj}\right) = \begin{cases} 1, T_{uj} = T_{vj} = 1 \\ 0, \quad \text{others} \end{cases} \tag{7}$$

$$S_n(u, v) = \exp\left(-\frac{\sum_{j=1}^{s_2} \left(X_{uj} - X_{vj}\right)^2}{2\delta^2}\right) \tag{8}$$

where T_{uj} represents that node u has categorical j, X_{uj} represents that node u has numerical attribute j. For categorical attributed similarity, we calculate the number of common attributes of the two nodes and divide by the number of attributes to normalize. We calculate the Euclidean distance of numerical attributes for numerical attributed similarity and map it to [0, 1] by RBF kernel function.

To fully utilize the information in the network, ASECS combine attributed similarity, low-order structure similarity, and high-order structure similarity of nodes by the designed adaptive weighting function as the composite similarity between nodes. Before, we normalize the high-order structure similarity, low-order structure similarity, and attribute similarity of nodes:

$$M_{\mathcal{M}}(u,v) = \frac{M_{\mathcal{M}}(u,v)}{M_{\mathcal{M}}(u,v)_{\max}} \tag{9}$$

$$J(\mathrm{u},\mathrm{v}) = \frac{\mathrm{J}(\mathrm{u},\mathrm{v})}{\mathrm{J}(\mathrm{u},\mathrm{v})_{\max}} \tag{10}$$

$$\mathrm{S}(\mathrm{u},\mathrm{v}) = \frac{\mathrm{s}(\mathrm{u},\mathrm{v})}{\mathrm{s}(\mathrm{u},\mathrm{v})_{\max}} \tag{11}$$

The composite similarity is defined as:

$$\mathrm{W}(\mathrm{u},\mathrm{v}) = C^{\rho} M_{\mathcal{M}}(u,v) + C^{\rho}\mathrm{J}(\mathrm{u},\mathrm{v}) + (1-C)^{\rho}\mathrm{S}(\mathrm{u},\mathrm{v}) \tag{12}$$

where C is average clustering coefficient which defined in Eq. (4). For attributed networks, sometimes network topology is dominant, sometimes node attribute information is dominant. Clustering coefficient can measure the degree of aggregation of nodes in networks. We use it to adapt to the contribution of balanced node attributes and topology structure. When the value of C is larger, there are more edges in the network, and the connections between nodes are relatively close. In this case, nodes of the network have more obvious aggregation phenomenon in topological perspective, and we should make the weight of the topology account for a larger proportion. The network is sparse when C is small. It is usually more necessary to rely on attributed information to guide community division, making the weight of attribute information account for a large proportion. ρ is a parameter, the larger the value can make dominant information have a larger proportion.

Finally, for each node, we retain top-k composite similarity neighbors, and the weighted KNN graph is generated. Details of weighted KNN graph generation are shown in Algorithm 1.

4.2 Community Detection by Seed Expansion

In this stage, we use seed-expansion-based algorithms to do community detection. For the algorithm based on seed expansion, the accuracy of the final community detection depend on whether the algorithm finds high-quality seed nodes. In general, seed nodes should have considerable influence and should be far away from each other. Therefore, we designed a function for node influence considering the above two requirements:

$$\mathrm{I}\ (\mathrm{u}) = \mathrm{S}_{\mathrm{dict}}(\mathrm{u}) * \mathrm{D}(\mathrm{u}) \tag{13}$$

$$\mathrm{D}(\mathrm{u}) = \sum_{v \in N(u)} W_{u,v} \tag{14}$$

Algorithm 1. Generating weighted KNN graph

Input: $G(V, E)$, attributed matrix$A$, motif instance set $I^{\mathcal{M}}, \rho, \sigma, K$.
Output: Weighted KNN graph $G^{'}$.
1: $E^{'} = \oslash$;
2: Calculate C via (4);
3: Construct high-order structure similarity matrix M via (9);
4: Construct low-order structure similarity matrix J via (10);
5: Construct attributed similarity matrix S via (11);
6: Construct composite similarity matrix W via (12);
7: **for** $i = 1$ in $|V|$ **do**
8: **for** each $v_j \in TopK(v_i, K, W)$ **do**
9: $E^{'} = E^{'} \cup (v_i, v_j)$
10: **end for**
11: **end for**
12: return $G^{'}$;

$$S_{dict}(\mathrm{u}) = \sum_{v \in \mathrm{seeds}} 1 - J(u, v) \tag{15}$$

where $D(u)$ represents the weighted degree of node u, $S_{dict}(u)$ represents the distance sum between node u and selected seed nodes.

We select the node which has the largest value of influence among nodes that are not extended as the seed node. After that, the seed node will be the initial community. When to community expansion, ASECS calculates the fitness with composite similarity of current community to each neighboring node. The fitness with composite similarity is defined as:

$$F_c = \frac{W_{in}^c}{(W_{in}^c + W_{out}^c)^\alpha} \tag{16}$$

where W_{in}^c and W_{out}^c represent the sum of the internal and external composite similarity of community c respectively.

ASECS select the node that can maximize fitness, and add it to the current community. Then, the set of neighbors of the community will be updated. The step of community expansion stops when the maximum fitness value of the neighboring nodes is negative. The details of community detection by seed expansion are given in Algorithm 2.

4.3 Complexity Analysis

Suppose the graph has q communities, r represents the average size of communities, k represents the average degree of the weighted KNN graph, and the motif size is p. In the first stage, we need $O(n^p)$ time to find each p-tuple of nodes in the graph. In the second stage, for each local community detection, the average time cost of seed selection is $O(k \times logn)$; For community extension, its time cost is $O(r^2 \times k)$, and the total time cost of Algorithm 2 is $O(t \times k(logn + r^2))$.

Algorithm 2. Community detection by seed expansion

Input: Weighted KNN graph $G^{'}$,α.
Output: Community set C.

1: $G_{free} = G^{'}, Seeds = \oslash, C = \oslash$;
2: **while** $V_{free} \neq \oslash$ **do**
3:   **for** each v in V_{free} **do**
4:     calculate $Influence(v)$ via (13)
5:   **end for**
6:   $seed = \{max(Influence(v)|v \in V_{free}\}$;
7:   $Seeds = Seeds \cup \{seed\}$;
8:   $c = \{seed\}$;
9:   **while** $N(c) \neq \oslash$ **do**
10:     **for** each v in $N(c)$ **do**
11:       calculate $F_c^v = F_{c\cup\{v\}} - F_c$ according to Eq. (12);
12:     **end for**
13:     $f_{max} = max(F_c^v|v \in N(c))$;
14:     **if** $f_{max} > 0$ **then**
15:       $c = c \cup \{v_{max}\}$;
16:     **else**
17:       break;
18:     **end if**
19:     $N(c) = N(c) \cup N(v_{max})$;
20:   **end while**
21:   $V_{free} = V_{free} - V_c$;
22:   $C = C \cup c$;
23: **end while**
24: return C;

So, the time cost of ASECS is $O(n^p)$. Usually, we use the 3-motif, and the time cost is $O(n^3)$.

The space cost of ASECS is $O(n^2)$ because the similarity between two nodes needs to be stored.

5 Experiments

5.1 Datasets Description

Real-World Networks. We select 11 real-world networks, 8 networks come from subsets of the three social networks available at SNAP. They are Facebook-348, Facebook-414, Facebook-698, Facebook-1912, Google-429402, Google-387583, Twitter-356963, Twitter-745823. And 2 networks come from WebKB, which are Washington and Wisconsin. The last network Cora is a citation network, which includes paper reference relationships and content. These networks have different characteristics, their statistics are given in Table 1.

Table 1. Description of real-world networks

Network	Number of nodes	Number of edges	Number of attributes	Average clustering coefficient	Average degree	Communities
Facebook-348	227	3192	161	0.54	28.1	14
Facebook-414	159	1693	105	0.67	21.2	7
Facebook-698	66	270	48	0.73	8.2	13
Facebook-1912	755	30025	480	0.64	79.6	46
Google-429402	273	6618	21	0.64	48.4	2
Google-387583	457	15148	264	0.65	66.2	10
Twitter-356963	134	495	247	0.37	7.4	11
Twitter-745823	249	4372	1174	0.48	35.1	13
Cora	2708	5278	1433	0.24	3.9	7
Washington	230	417	1703	0.2	3.6	5
Wisconsin	265	479	1703	0.21	3.6	5

Artificial Networks. We generate two groups of artificial networks by the tool which named LFR-benchmark [23] to verify the robustness of the algorithm in different networks. The method proposed in [24] is used to generate node attributes for all networks. We generate 300 categorical attributes and 20 numerical attributes for each artificial network. Table 2 shows the specific parameter settings of each network.

Table 2. Parameter settings of artificial networks

Network	Parameter settings
D1	$N = 1000$, $k = 20$, $k_{max} = 50$, $C_{min} = 10$, $C_{max} = 100$, $\mu = 0.1 - 0.7$, $on = 0.1$, $om = 2$
D2	$N = 1000$, $k = 20$, $k_{max} = 50$, $C_{min} = 10$, $C_{max} = 100$, $\mu = 0.3$, $on = 0.1 - 0.6$, $om = 2$

5.2 Experimental Settings

Baseline Methods. To verify the superiority of ASECS, we select three algorithms based on seed expansion and two algorithms based on auto-encoder as baseline algorithms. Namely LFM [5], DEMON [25], InfoNode [13], ARGA [16], ARVGA [16]. For all algorithms, we set the parameters that suggested by authors. And we use the triangle motif to construct the high-order similarity.

Evaluation Measure. In our experiments, the overlapping version of the Normalized Mutual Information (ONMI) [26] is selected as the evaluation measure. The values closer to 1 indicate good performance.

The ONMI is defined as:

$$\mathrm{ONMI}(A,B)=1-\frac{1}{2}\left(\sum_{i}^{|A|}\frac{\mathrm{H}\left(A_i,B\right)}{|A|\mathrm{H}\left(A_i\right)}+\sum_{j}^{|B|}\frac{\mathrm{H}\left(B_j\mid A\right)}{|B|\mathrm{H}\left(B_j\right)}\right) \tag{17}$$

where $\mathrm{H}\left(A_i\right)$ represents the entropy of the i-th community A_i, $\mathrm{H}\left(A_i|B\right)$ represents the entropy of A_i with respect to B.

5.3 Parameter Experiments

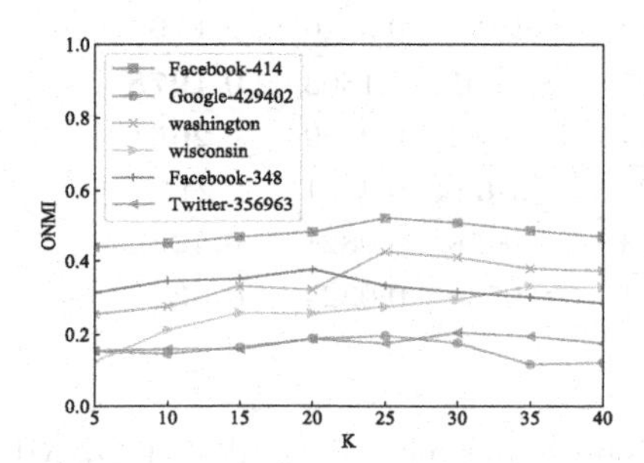

Fig. 3. Experiments on parameter K

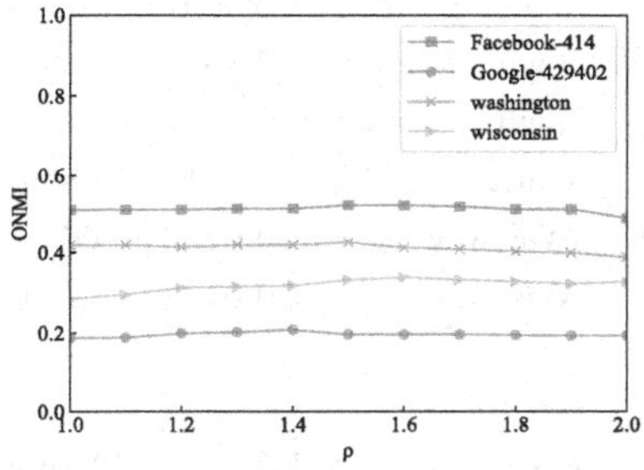

Fig. 4. Experiments on parameter ρ

Parameter K. ASECS use the parameter K to select the top K nodes with the large composite similarity as the neighbors for each node in generating weighted KNN network stage. The larger the values of the K, the denser the weighted KNN graph. Figure 3 shows the experimental results on some real-world networks. With the increase of K, ONMI values first increases and then decreases, and the best values of K mainly in [20, 25, 30]. The regular is in line with common sense. In the weighted KNN graph, too few edges may lead to the underutilization of information, while too many edges may involve noise. We suggest to set the value of K between 20 and 30.

Parameter ρ. ASECS use the parameter ρ to adjust the contribution ratio between attributes and topology. The larger the value of the ρ, the larger the contribution ratio of the dominant information. The value of ρ is between 1 and 2. As seen in Fig. 4, most networks perform well when ρ=1.5, so the default value of ρ is set to 1.5.

5.4 Accuracy Experiments

Table 3. ONMI values on real-world networks

Network	LFM	InfoNode	DEMON	ARGA	ARVGA	ASECS
Facebook-348	0.3113	0.3279	0.2498	0.1918	0.1942	**0.3765**
Facebook-414	0.4797	0.5127	0.4483	0.4561	0.4415	**0.5217**
Facebook-698	0.4081	0.4185	0.4255	0.3876	0.3959	**0.4955**
Facebook-1912	0.3016	0.2908	0.3142	0.1713	0.1228	**0.3414**
Google-429402	0.0187	0.0401	0.0335	0.0171	0.0336	**0.1965**
Google-387583	0.0651	0.0757	0.0662	0.0837	0.0726	**0.1196**
Twitter-356963	0.1671	0.1369	0.0885	0.1201	0.1303	**0.1978**
Twitter-745823	0.1162	0.1235	0.0764	0.0386	0.0429	**0.2051**
Cora	0.1218	0.0858	0.0133	**0.3644**	0.3521	0.2447
Washington	0.0222	0.0699	0.0173	0.0658	0.0821	**0.4255**
Wisconsin	0.0124	0.0244	0.0091	0.0530	0.0324	**0.3316**

Accuracy on Real-World Networks. The experimental results of ONMI measure on real-world networks are shown in Table 3. As we can see, ASECS gets the best results on all networks except Cora. This is because ASECS integrates high-order and low-order topology information and attribute information and adaptively adjusts the contribution of information according to the clustering coefficient of the network. The social networks usually have many triangle structures, so using the triangle motif can obviously make the result of community detection better. In networks of Washington and Wisconsin, attributed information is dominant, so the seed-expansion-based algorithms do not perform well in these networks. And the auto-encoder-based methods cannot adjust the contribution adaptively, so they also perform poorly in these networks. ARVGA and ARGA have got nice results on Cora, because Cora has many connected branches. The method based on auto-encoder is more suitable to deal with this situation.

Accuracy on Artificial Networks. The experimental results of ONMI measure on the D1 networks are shown in Fig. 5. With the parameter μ increasing, the accuracy of all algorithms will decrease. Because when the community structure becomes fuzzy, it is difficult for algorithms to identify boundaries between communities. As shown in Fig. 5, ASECS performs well in all values of μ. The accuracy of ASECS is significantly high than other algorithms, and the decline is slight. This is because ASECS effectively combines high-order and low-order topology information and attribute information. When the topology structure is clear, ASECS can fully use of high-order structure and low-order structure to

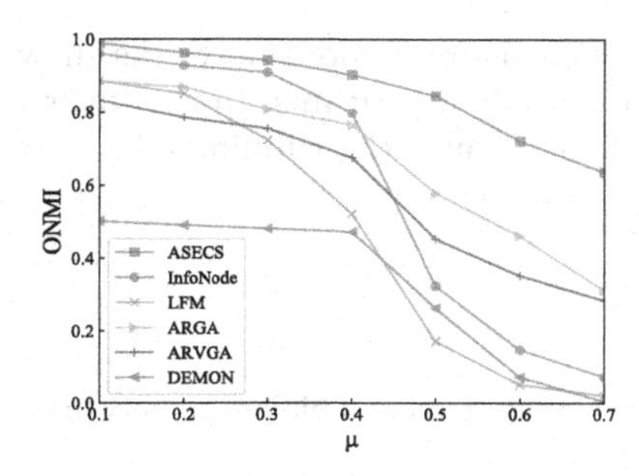

Fig. 5. ONMI results on D1 networks

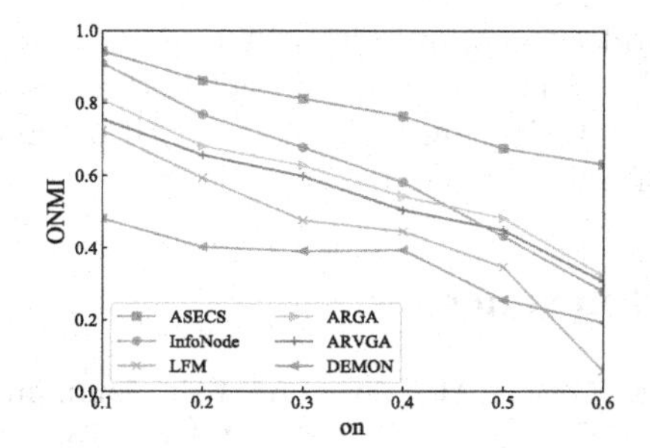

Fig. 6. ONMI results on D2 networks

detect community. When the topology information is chaotic, ASECS can adaptively reduce the contribution of topology information and improve the contribution of node attribute information. Besides, we design a new attributed similarity calculation method that considered numerical attributes. Although ARGA and ARVGA can combine topology and attributes, they do not use high-order structure information and cannot automatically adjust contribution degree. The other three methods based on seed expansion cannot use node attribute information and perform poorly when the value of μ is big.

Figure 6 shows the experimental results of ONMI measure on D2 networks. With the increase of the proportion of overlapping nodes on networks, the accuracy of algorithms decreases to a certain extent. ARGA and ARVGA are not specially designed for overlapping communities, so the accuracy decreases obviously when the overlapping ratio is very big. LFM uses a random seed selection strategy, and it is difficult for low-quality seeds to correctly divide nodes into multiple communities, so the accuracy of LFM decreases significantly. ASECS can combine multi-information and uses the algorithm based on seed expansion, which can also have a good performance when the proportion of overlapping nodes is large.

6 Conclusions

In this paper, we propose an adaptive seed expansion algorithm based on composite similarity. First, we calculate the low-order structure similarity, high-order structure similarity, and attributed similarity of nodes, respectively. The calculation of attribute similarity considers both numerical type and categorical type. Second, we design a weighted function to combine multi-similarity as the composite similarity of nodes to generate a weighted KNN graph. The designed weighted function can adaptively adjust the contribution ratio between topology and node attributes according to the clustering coefficient of the network. Finally, we find the seed nodes on the weighted KNN graph and expand the seed nodes to communities. A large number of experiments on various networks show that our algorithm can get better result than baseline algorithms. In the future, parallelization technologies will be considered to enhance the running efficiency of our algorithm.

References

1. Girvan, M., Newman, M.E.: Community structure in social and biological networks. Proc. Natl. Acad. Sci. **99**(12), 7821–7826 (2002)
2. Wang, X., Jin, D., Cao, X., Yang, L., Zhang, W.: Semantic community identification in large attribute networks. In: Proceedings of the AAAI Conference on Artificial Intelligence, vol. 30, no. 1 (2016)
3. Ikeda, K., Hattori, G., Ono, C., Asoh, H., Higashino, T.: Twitter user profiling based on text and community mining for market analysis. Knowl. Based Syst. **51**, 35–47 (2013)
4. Yan, C., Huang, Y., Wan, Y., Liu, G.: Community-based matrix factorization model for recommendation. In: International Conference on Cloud Computing and Security, pp. 464–475. Springer (2018)
5. Lancichinetti, A., Fortunato, S., Kertész, J.: Detecting the overlapping and hierarchical community structure in complex networks. New J. Phys. **11**(3), 033015 (2009)
6. He, D., Feng, Z., Jin, D., Wang, X., Zhang, W.: Joint identification of network communities and semantics via integrative modeling of network topologies and node contents. In: Proceedings of the AAAI Conference on Artificial Intelligence, vol. 31, no. 1 (2017)
7. Zhou, Y., Cheng, H., Yu, J.X.: Graph clustering based on structural/attribute similarities. Proc. VLDB Endowm. **2**(1), 718–729 (2009)
8. Yang, J., McAuley, J., Leskovec, J.: Community detection in networks with node attributes. In: IEEE 13th International Conference on Data Mining, vol. 2013, pp. 1151–1156. IEEE (2013)
9. Li, P.-Z., Huang, L., Wang, C.-D., Huang, D., Lai, J.-H.: Community detection using attribute homogenous motif. IEEE Access **6**, 47707–47716 (2018)
10. Li, P.-Z., Huang, L., Wang, C.-D., Lai, J.-H.: Edmot: an edge enhancement approach for motif-aware community detection. In: Proceedings of the 25th ACM SIGKDD International Conference on Knowledge Discovery & Data Mining, pp. 479–487 (2019)
11. Huang, L., Chao, H.-Y., Xie, Q.: Mumod: a micro-unit connection approach for hybrid-order community detection. Proc. AAAI Conf. Artif. Intell. **34**(01), 107–114 (2020)
12. Lee, C., Reid, F., McDaid, A., Hurley, N.: Detecting highly overlapping community structure by greedy clique expansion. arXiv preprint arXiv:1002.1827 (2010)
13. Guo, K., He, L., Chen, Y., Guo, W., Zheng, J.: A local community detection algorithm based on internal force between nodes. Appl. Intell. **50**(2), 328–340 (2020)
14. Zhang, J., Ding, X., Yang, J.: Revealing the role of node similarity and community merging in community detection. Knowl. Based Syst. **165**, 407–419 (2019)
15. Wang, C., Pan, S., Long, G., Zhu, X., Jiang, J.: Mgae: marginalized graph autoencoder for graph clustering. In: Proceedings of the 2017 ACM on Conference on Information and Knowledge Management, pp. 889–898 (2017)

16. Pan, S., Hu, R., Long, G., Jiang, J., Yao, L., Zhang, C.: Adversarially regularized graph autoencoder for graph embedding. arXiv preprint arXiv:1802.04407 (2018)
17. Li, P.-Z., Huang, L., Wang, C.-D., Lai, J.-H., Huang, D.: Community detection by motif-aware label propagation. ACM Trans. Knowl. Discov. Data **14**(2), 1–19 (2020)
18. Milo, R., Shen-Orr, S., Itzkovitz, S., Kashtan, N., Chklovskii, D., Alon, U.: Network motifs: simple building blocks of complex networks. Science **298**(5594), 824–827 (2002)
19. Wernicke, S., Rasche, F.: Fanmod: a tool for fast network motif detection. Bioinformatics **22**(9), 1152–1153 (2006)
20. Jaccard, P.: Étude comparative de la distribution florale dans une portion des alpes et des jura. Bull. Soc. Vaudoise Sci. Nat. **37**, 547–579 (1901)
21. Watts, D.J., Strogatz, S.H.: Collective dynamics of 'small-world' networks. Nature **393**(6684), 440–442 (1998)
22. Kemper, A.: Valuation of Network Effects in Software Markets: A Complex Networks Approach. Springer, Heidelberg (2009). https://doi.org/10.1007/978-3-7908-2367-7
23. Lancichinetti, A., Fortunato, S., Radicchi, F.: Benchmark graphs for testing community detection algorithms. Phys. Rev. E **78**(4), 046110 (2008)
24. Huang, B., Wang, C., Wang, B.: Nmlpa: uncovering overlapping communities in attributed networks via a multi-label propagation approach. Sensors **19**(2), 260 (2019)
25. Coscia, M., Rossetti, G., Giannotti, F., Pedreschi, D.: Demon: a local-first discovery method for overlapping communities. In: Proceedings of the 18th ACM SIGKDD International Conference on Knowledge Discovery and Data Mining, pp. 615–623 (2012)
26. McDaid, A.F., Greene, D., Hurley, N.: Normalized mutual information to evaluate overlapping community finding algorithms. arXiv preprint arXiv:1110.2515 (2011)

MDN: Meta-transfer Learning Method for Fake News Detection

Haocheng Shen, Bin Guo(✉), Yasan Ding, and Zhiwen Yu

Northwestern Polytechnical University, Xi'an, China
hcshen@mail.nwpu.edu.cn, {guob,zhiwenyu}@nwpu.edu.cn

Abstract. The rapid development of social media has brought convenience to people's lives, but at the same time, it has also led to the widespread and rapid dissemination of false information among the population, which has had a bad impact on society. Therefore, effective detection of fake news is of great significance. Traditional fake news detection methods require a large amount of labeled data for model training. For emerging events (such as COVID-19), it is often hard to collect high-quality labeled data required for training models in a short period of time. To solve the above problems, this paper proposes a fake news detection method MDN (**M**eta **D**etection **N**etwork) based on meta-transfer learning. This method can extract the text and image features of tweets to improve accuracy. On this basis, a meta-training method is proposed based on the model-agnostic meta-learning algorithm, so that the model can use the knowledge of different kinds of events, and can realize rapid detection on new events. Finally, it was trained on a multi-modal real data set. The experimental results show that the detection accuracy has reached 76.7%, the accuracy rate has reached 77.8%, and the recall rate has reached 85.3%, which is at a better level among the baseline methods.

Keywords: Fake news detection · Meta-learning · Multimodal feature extraction

1 Introduction

In recent years, the development of Internet social media platforms at home and abroad has also led to the rapid spread of a large number of fake news on the platform. A large number of fake news confuses the public and interferes with people's judgment of the facts. Fake news detection [1] is essentially a classification problem. The purpose is to judge whether a news is true or false based on the potential characteristics of different news. How to automatically and effectively detect false news has gradually attracted attention from academia and industry.

The current methods of fake news detection mainly combine the content characteristics of the message (such as text and comment information) [2, 3], communication characteristics (such as forwarding structure) [4] and interactive characteristics (such as the number of likes), and apply convolutional neural networks (CNN) [5] perform feature extraction and classification and recognition. However, traditional methods require large amounts of annotation data to train the model. For emerging events (such as the

Y. Sun et al. (Eds.): ChineseCSCW 2021, CCIS 1492, pp. 228–237, 2022.
https://doi.org/10.1007/978-981-19-4549-6_18

COVID-19, terrorist attacks, etc.), as shown in Fig. 1, it is often difficult to collect high-quality relevant data in a short time. Therefore, the model's recognition of new events that have not appeared in the training set is usually bad [7, 8].

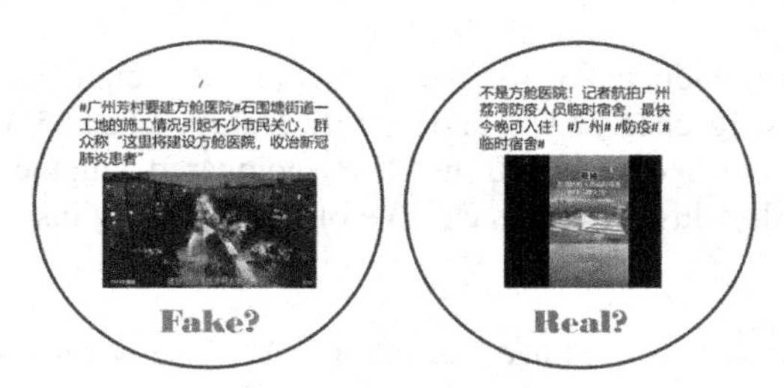

Fig. 1. A The detection of new events.

Since fake news spreads quickly on social networks and has a wide range of influence, it often has a serious impact on society in a short period of time. Therefore, It's essential to detect fake news in real time. In existing methods, whether feature-based detection methods or social context-based detection methods [4–7] are assumed that the detection model has data for the entire life cycle of a news event, and can train a model based on aggregated features to detect fake news. Therefore, the key issues studied in this paper face the following challenges:

(1) In the early stage of propagation, the amount of data is small and the labels are sparse. Existing methods still require a large amount of data to obtain an effective detection model. How to effectively detect fake news in the early stage is a big challenge.
(2) Because of the different feature distributions of events, existing methods are less effective in detecting new events. How to use the tweet data of existing historical events for knowledge transfer, so as to guide specific events, make the model generalized, and respond to new events is a challenge.
(3) In addition to the text content in tweets, more users are used to using images or videos to post tweets. Traditional detection methods only use text to detect. How to combine multimodal data for detection is also a challenge.

Meta-learning is a few-shot transfer learning method [9–11]. The key idea is using the network to acquire meta-knowledge through training on a large number of similar tasks, and quickly learn on new tasks [12], Compared with traditional transfer learning, it has the advantage of fewer training samples, so we use meta-learning to address the above challenges.

In this paper, we propose a fake news detection method based on meta-learning, namely MDN. It combines the text and image information of the tweet, the meta-learning method is used to construct a fake news detection model on new events with a small number of samples. The model can realize knowledge transfer such as model parameters by constructing subtask sequences, and finally can realize early detection of fake news. The main contributions are as follows:

(1) We propose a fake news detection method MDN based on meta-learning, which can learn the knowledge of historical event tweets and perform early and effective detection on new events.
(2) We build a multi-modal feature extractor, which can extract high-dimensional features of tweet text and images for detection.
(3) Experiments are conducted on the Weibo dataset, the accuracy rate reached 76.7%, the precision rate reached 77.8%, and the recall rate reached 85.3%. In addition, the training time is reduced by 57.8% and 48.7% compared with the two baseline models, indicating that this method is capable of fast and early inspection of emerging events.

The rest of this paper is organized as follows: We briefly review related works in Sect. 2, and we introduced details of MDN in Sect. 3. In Sect. 4 we show the experiment results. Finally, we conclude our work in Sect. 5.

2 Related Work

2.1 Fake New Detection

In recent years, researchers have proposed a variety of fake news detection methods. Content-based methods extract vocabulary features [2, 3], syntactic features [13] and topic features [14] for classification; Social context-based methods judges whether the news is false by analyzing user characteristics [15] and the propagation characteristics [6, 16]. For the early detection of news, Ma et al. [17] proposed the DSTS model. Bian et al. [18] proposed a bidirectional graph convolutional neural network Bi-GCN to capture the propagation characteristics of news.

To solve the problem of lacking labels in the early stage of fake news propagation, Sampson et al. [19] evaluated the credibility of the original news by combining the text content of the news with the third-party web content information. Wang et al. [8] proposed EANN, which reduces the specific features of the event and retaining the detection of the general feature of the events.

2.2 Few-Shot Learning

Because of large amounts parameters, neural networks often need lots of data for training. Therefore, whether a small amount of labeled data can be used to achieve better training results has become a very important topic, which has attracted great attention.

Model-based methods quickly updates the parameters on few samples. Santoro et al. [20] proposed the use of memory enhancement to solve few-shot learning tasks. Koch et al. [21] proposed Siamese Network, which reuses the features extracted by the network for few-shot learning. Match Network [22] build different encoder for training set and test set, and output the weighted sum of the predicted values between support set and query set.

Meta-learning is a few-shot learning method to update parameters on a small number of samples. Finn et al. [5] proposed the MAML, which has generalization performance with a small number of iterative steps. Compared with other methods, meta-learning performs better in a wider range of tasks and lower computational cost. Therefore, we use the idea of meta-learning, and learns the data of previous knowledge to detect new events.

3 Methodology

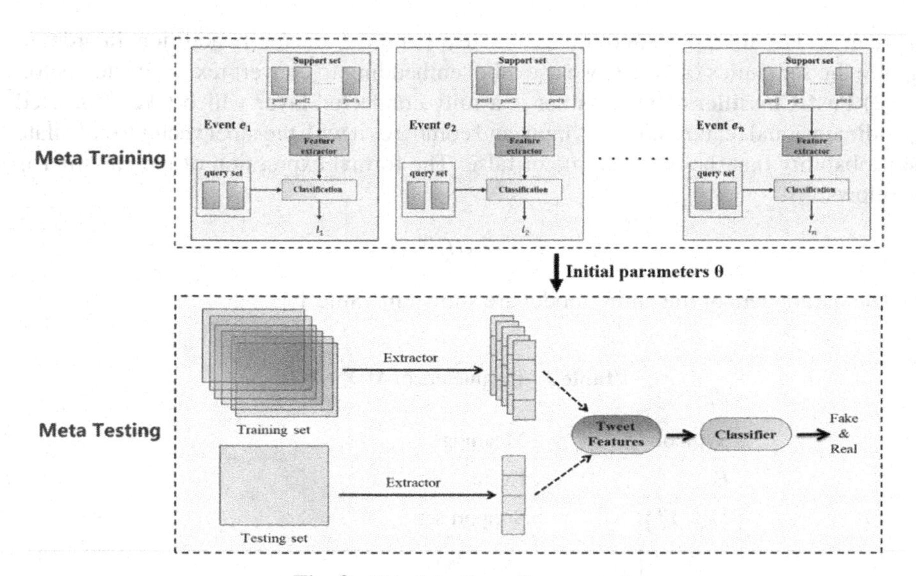

Fig. 2. The detection of new events.

3.1 Problem Statement

Fake news in this paper refers to text or images that are spread via social media and proved to be false. For each event e, given support dataset $\{X_e^s, Y_e^s\}$ and query dataset $\{X_e^q, Y_e^q\}$. This paper use the knowledge from past events to get a better initialization parameter in the meta-training stage, as follows:

$$\theta_e = \theta - \sum\nolimits_{i=0}^{n} Loss(\{X_e^q, Y_e^q\}) \quad (1)$$

where n is epoch of meta-training, $Loss(\cdot)$ is the loss function, on meta-testing stage, we search for a predicted function $F_{\theta_e}(s) \to \{0, 1\}$ for query set $\{X_e^q, Y_e^q\}$:

$$F_{\theta_e}(s) = \begin{cases} 1, & \text{if } s \text{ is a fake post} \\ 0, & \text{otherwise} \end{cases} \quad (2)$$

3.2 The MDN Framework

The framework of MDN mainly consists of two parts, as shown in Fig. 2.

(1) **Feature Extractor**: The feature extractor has two parts: The first is text feature extractor, which uses the Text-CNN [25] network to extract the unstructured text content feature [24]; the other part is the image feature extractor uses the VGG-16 [26] to extract the feature of the input image, as follows:

$$F_{ei} = q(h_1(X_{ei}) \oplus h_2(X_{ei})) \tag{3}$$

where h_1 and h_2 are represent of image and text feature extractor. In addition, in order to capture the semantics of labels, we use label embedding to convert text labels to vector.

(2) **Post Classifier**: The classifier is a fully connected layer which takes extracted high-dimensional feature vector as input and compares it with the label vector to calculate the probability that the tweet is true or false. The formal expression of the classifier is as follows:

$$\hat{y}^s_{ei} = softmax(f(F_{ei}), y_{ei}) \tag{4}$$

The parameters of the entire model are shown in Table 1:

Table 1. Parameters of MDN

Symbol	Meaning
E	Event
$\{X^s_e, Y^s_e\}$	Support set
$\{X^q_e, Y^q_e\}$	Query set
θ	Parameter set
$h(\cdot)$	Feature extractor
F_{ei}	Features
α	Inner learning rate
β	Outer learning rate
$f(\cdot)$	Classification function
θ_e	Initial parameter set

3.3 Training of MDN

The training of MDN consists two stages: The first stage is meta-training, we input the context data into the model and the prediction result is output. The loss can be calculated by comparing the predicted label and the actual label:

$$\nabla_\theta L_{ei}\left(\hat{y}^s_{ei}\right) \tag{5}$$

where θ is the parameters of the model, L_{ei} is the loss function of inner loop. After calculating the loss, for each event e update the θ:

$$\theta_i^{'} = \theta - \alpha \nabla_\theta L_{ei}\left(\hat{y}_{ei}^s\right) \tag{6}$$

The next stage is meta-training, after meta-training we get an event-specific parameter set θ_e, which is used as the initialization parameter. Taking the query set X_e^q and the entire support set $\left\{X_e^s, Y_e^s\right\}$ as input, and get the label set $\hat{Y}_e^q$, and then compared with the actual label set of the query set to calculate the loss:

$$\theta_e \leftarrow \theta - \beta \nabla_\theta \sum\nolimits_E L\left(\theta_i^{'}\right) \tag{7}$$

where L is the loss function of outer loop.

Through meta-learning we get a better initialization parameter set, and can quickly detect the new events. The algorithm is shown in Table 2:

Table 2. Algorithm of the MDN

Algorithm	MDN
Input	$p(E)$: distribution over events α: step size of inner loop β: step size of outer loop
Output	initial parameter set θ_e

1: randomly initialize θ
2: **while not done do**
3: Sample batch from events $E_i \sim p(E)$
4: **for all** E_i **do**
5: Extract features from image and tweets $F_{ei} = q(X_{ei})$
6: Compute predicted label $\hat{y}_{ei}^s = softmax(f(F_{ei}), y_{ei})$
7: Evaluate $\nabla_\theta L_{ei}(\hat{y}_{ei}^s)$
8: Compute gradient descent of inner loop:
$$\theta_i' = \theta - \alpha \nabla_\theta L_{ei}(\hat{y}_{ei}^s)$$
9: **end for**
10: Update $\theta_e \leftarrow \theta - \beta \nabla_\theta \Sigma_E L(\theta_i')$
11: **end while**

4 Experiments

4.1 Datasets

This paper uses the Weibo Dataset [24], the details is shown in Table 3:

Table 3. Statistics of dataset

Data	Number
Fake tweets	4050
Real tweets	3558
Images	7606

The dataset consists of 7608 tweets from February 2012 to December 2015, the fake tweets are verified on Weibo platform. In addition to text, Weibo Dataset also contains additional images and social context information.

4.2 Model Training

In this paper we use pytorch1.6 to train the model, the learning rate is 0.01, and the kernel size of text-CNN is 5, the embedding dimension of post classifier is 32. In the training of the model, we set the maximum epoch to 500, and the k in the meta-training, which is the number of samples per step is set to 5. The training loss curve and the valid curve are as follows:

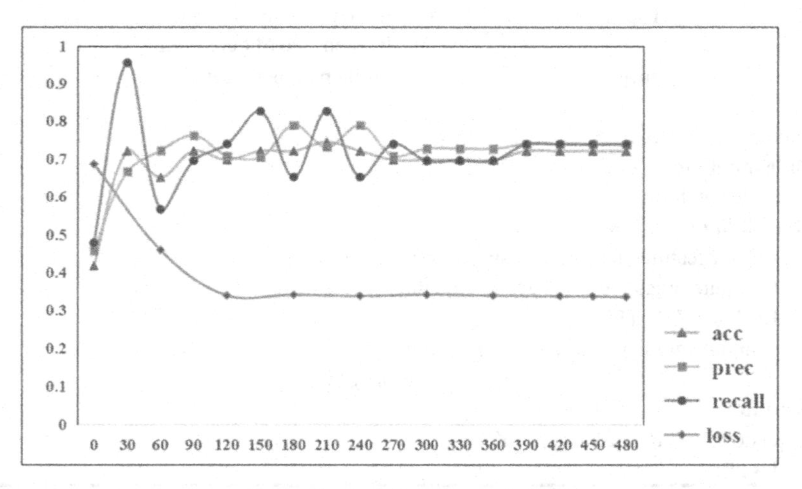

Fig. 3. The training curve of MDN

It can be seen from Fig. 3 that the loss drops from 0.686 to 0.336, and the loss stabilizes when the epoch reaches about 100, which proves that the training performs well. It is verified that all indicators are stabilized after 300 rounds, which reflects the performance of the model.

4.3 Detection Results

To evaluate the performance, we select four methods for comparison: DNN [23], Text-CNN [25], att-RNN [25] and EANN.

- DNN: It uses nonlinear fitting capabilities of multiple fully connected layers to learn the text features of news.
- Text-CNN: It uses CNN to model news content, and extracts potential language features of different granularities through different filters.
- att-RNN: It use the RNN and attention mechanism to fuse multi-modal features for effective fake news detection.
- EANN: It uses an event discriminator to reduce the difference in the feature of different events, aims to get the transferable features that can detect fake news of the target event.

This paper mainly uses the idea of meta-learning, and its purpose is to optimize the parameters of the model through a series of sub-tasks in the meta-training process, so as to have a good performance in the test.

Table 4. Accuracy on Weibo Datasets of different model

	Accuracy	Precision	Recall	F1-score
DNN [23]	0.6476	0.7674	0.5440	0.6372
Text-CNN [25]	0.7102	0.7772	0.7051	0.7244
att-RNN [25]	0.7407	0.7550	0.6942	0.7436
EANN [8]	0.7443	0.7688	0.6838	0.7451
MDN	**0.7674**	**0.7778**	**0.8533**	**0.8484**

According to the test result shown in Table 4, it is found that the accuracy of MDN has reached 76.7%, which is better than the baselines, which proves that the model can adjust the parameters according to the auxiliary data of different events, and strengthen the detection effect on small samples of specific events.

In addition, in order to verify the ability to detect in time, we use the same data set to experiment on different models. The training time is defined as the time from the start of the model training to the time when the training is completed. The results are shown in Table 5:

Table 5. Training time of different model

	att-RNN	EANN	MDN
Accuracy	0.745	0.754	0.767
Training time/s	1797	1478	758

Because of the reduction of the training data set, the prediction accuracy has been reduced to a certain extent. Compared with att-RNN model, the training time of this model has been reduced by 57.8%, and compared with the EANN by 48.7%, maintaining a high prediction. The accuracy rate greatly reduces the running time and training cost, indicating that the model is capable of early detection of new events.

5 Conclusion

This paper proposes a fake news detection method MDN based on meta-transfer learning, which extracts text and image features in tweets. In the meta-training, the tweet data of different events is used as auxiliary data for event adaptive training to learn a better initialization parameter set, and gradient descent is performed on the tweets of new events to get higher accuracy rate. At the same time, we compare the training time of different methods. The result shows that our model has early detection abilities and can quickly detect new events.

References

1. Allport, G.W., Postman, L.: The Psychology of Rumor. Russell & Russell Pub, New York (1965)
2. Castillo, C., Mendoza, M., Poblete, B.: Predicting information credibility in time-sensitive social media. Internet Res. **23**(5), 560–588 (2013)
3. Kwon, S., Cha, M., Jung, K., Chen, W., Wang, Y.: Prominent features of rumor propagation in online social media. In: 2013 IEEE 13th International Conference on Data Mining, pp. 1103–1108. IEEE (2013)
4. Ma, J., Gao W., Wong, K.-F.: Rumor detection on twitter with tree-structured recursive neural networks. Association for Computational Linguistics (2018)
5. Finn, C., Abbeel, P., Levine, S.: Model-agnostic meta-learning for fast adaptation of deep networks. In: Proceedings of the 34th International Conference on Machine Learning, vol. 70 (2017). JMLR.org
6. Ma, J., et al.: Detecting rumors from microblogs with recurrent neural networks, p. 3818 (2016)
7. Guo, B., et al.: The future of false information detection on social media: new perspectives and trends. ACM Comput. Surv. (CSUR) **53.4**, 1–36 (2020)
8. Wang, Y., et al.: EANN: event adversarial neural networks for multi-modal fake news detection. In: Proceedings of the 24th ACM SIGKDD International Conference on Knowledge Discovery & Data Mining (2018)
9. Pan, S.J., Yang, Q.: A survey on transfer learning. IEEE Trans. Knowl. Data Eng. (TKDE) **22**(10), 1345–1359 (2009)
10. Tzeng, E., Hoffman, J., Darrell, T., et al.: Simultaneous deep transfer across domains and tasks. In: Proceedings of the IEEE International Conference on Computer Vision (CVPR), pp. 4068–4076 (2015)
11. Li, Z., Wei, Y., Zhang, Y., et al.: Hierarchical attention transfer network for cross-domain sentiment classification. Proc. AAAI Conf. Artif. Intell. (AAAI). **32**(1), 5852–5859 (2018)
12. Vanschoren, J.: Meta-learning: a survey. arXiv preprint arXiv:1810.03548 (2018)
13. Acerbi, A.: Cognitive attraction and online misinformation. Palgrave Commun. **5**(1), 1–7 (2019)

14. Ito, J., Song, J., Toda, H., et al.: Assessment of tweet credibility with LDA features. In: Proceedings of the 24th International Conference on World Wide Web (WWW), pp. 953–958 (2015)
15. Shu, K., Wang, S., Liu, H.: Understanding user profiles on social media for fake news detection. In: Proceedings of IEEE Conference on Multimedia Information Processing and Retrieval (MIPR), pp. 430–435 (2018)
16. Wu, K., Yang, S., Zhu, K.Q.: False rumors detection on Sina Weibo by propagation structures. In: Proceedings of IEEE 31st International Conference on Data Engineering (ICDE), pp. 651–662 (2015)
17. Ma, J., Gao, W., Wei, Z., Lu, Y., Wong, K.-F.: Detect rumors using time series of social context information on microblogging websites. In: Proceedings of the 24th ACM International on Conference on Information and Knowledge Management, pp. 1751–1754. ACM (2015)
18. Bian, T., Xiao, X., Xu, T., et al.: Rumor detection on social media with bi-directional graph convolutional networks. Proc. AAAI Conf. Artif. Intell. (AAAI) **34**(01), 549–556 (2020)
19. Ma, J., Gao, W., Mitra, P., et al.: Detecting rumors from microblogs with recurrent neural networks (2016)
20. Santoro, A., Bartunov, S., Botvinick, M., et al.: One-shot learning with memory-augmented neural networks. arXiv preprint arXiv:1605.06065 (2016)
21. Koch, G., Zemel, R., Salakhutdinov, R.: Siamese neural networks for one-shot image recognition. In: ICML Deep Learning Workshop, vol. 2 (2015)
22. Vinyals, O., Blundell, C., Lillicrap, T., Wierstra, D., et al.: Matching networks for one shot learning. In: Advances in Neural Information Processing Systems, pp. 3630–3638 (2016)
23. Ma, J., et al.: Detecting rumors from microblogs with recurrent neural networks. In: Proceedings of the Twenty-Fifth International Joint Conference on Artificial Intelligence (New York, New York, USA) (IJCAI 2016), pp. 3818–3824. AAAI Press (2016)
24. Gupta, A., Kumaraguru, P., Castillo, C., Meier, P.: TweetCred: real-time credibility assessment of content on Twitter. In: Aiello, L.M., McFarland, D. (eds.) SocInfo 2014. LNCS, vol. 8851, pp. 228–243. Springer, Cham (2014). https://doi.org/10.1007/978-3-319-13734-6_16
25. Simonyan, K., Zisserman, A.: Very deep convolutional networks for large-scale image recognition. arXiv preprint arXiv:1409.1556 (2014)
26. Bojanowski, P., Grave, E., Joulin, A., Mikolov, T.: Enriching word vectors with subword information. Trans. Assoc. Comput. Linguist. **5**, 135–146 (2017). ISSN 2307-387X

Local Community Detection Algorithm Based on Core Area Expansion

Pengyun Ji[1,2], Kun Guo[1,2,3](✉), and Zhiyong Yu[1,2,3]

[1] College of Computer and Data Science, Fuzhou University, Fuzhou 350108, China
{gukn,yuzhiyong}@fzu.edu.cn

[2] Fujian Provincial Key Laboratory of Network Computing and Intelligence Information Processing, Fuzhou University, Fuzhou 350108, China

[3] Key Laboratory of Spatial Data Mining and Information Sharing, Ministry of Education, Fuzhou 350108, China

Abstract. Local community detection is an innovative method to mine cluster structure of extensive networks, that can mine the community of seed node without the need for global structure information about the entire network, as distinct from the global community detection algorithm, it is efficient and costs less. However, a key problem with this field is that the location of seed nodes affects the performance of the algorithm to a great extent, and it is easy to add abnormal nodes to the community, the robustness of the algorithm is low. In this study, we proposed a novel algorithm named CAELCD. First, find the high-quality seed node of the community starting from the initial seed node, so as to avoid the seed-dependent problem. Second, generate the community's core area and expand to get the local community, which solves the problem that the expansion from a single seed prefers to add the wrong nodes. Experiments on the parameter, accuracy and visualization of the CAELCD are designed on networks with different characteristics. Experimental results demonstrate that CAELCD has superior performance and high robustness.

Keywords: Complex network · Local community detection · Core node · Community expansion

1 Introduction

Local community [1] is a special collection of nodes that contains the specified node. The popularity of big data and other technologies have brought about the rapid expansion of network scale, the comprehensive node connection structure of some complex networks is difficult or almost impossible to obtain. In many cases, we only care about the community of a specific node(in the following, we call it seed), such as friend recommendation. Benefit from low demand for network topology information, the local methods are faster and more effective than the global methods. Therefore, local community detection came into being.

Y. Sun et al. (Eds.): ChineseCSCW 2021, CCIS 1492, pp. 238–251, 2022.
https://doi.org/10.1007/978-981-19-4549-6_19

Existing related algorithms are classified into methods based on local expansion [1–3]; cluster structure [4,5] and personalized PageRank [6,7], etc. The common local methods have the following defects. First, the position of the seeds has a profound impact on the performance of the algorithms. The local community's quality expanded from the seed of core position is higher, but the quality expanded from boundary seed is bad, that is, the seed-dependent problem. Second, in the initial stages of the methods based on local expansion, if the community expansion is from a single seed, abnormal nodes were inclined to be selected to add to the node set, so then, the excavated community are inconsistent with reality.

This study proposed CAELCD, which first combines high-order structure, edge clustering coefficient, and PageRank to find the high-quality seed, then generates the community's core area and expands it to get the local community. The main contributions are as follows:

(1) The strategy of finding high-quality seeds makes our algorithm immune to effect of initial seed location, solved the seed-dependent problem, and improved the robustness of algorithm.
(2) Expanding community from core area solves the problem that the methods based on local expansion tend to choose low-degree nodes to add to the local community, which leads to the error of community expansion. The accuracy of the algorithm is significantly improved.
(3) Compared with the existing algorithms on the networks with different characteristics, the results confirm that the methods proposed in this paper can effectively solve the above problems and outperforms other algorithms.

2 Related Work

Clauset et al. [1] defined the local modularity R according to the community's structure characteristics, they adopt a greedy optimization of local modularity R to mine local communitiy. Luo et al. [2] defined analogous local modularity M from the perspective of the degree of nodes inside and around the subgraph. Lancichinetti et al. [3] detined the fitness function that used to judge the clarity of node clusters. Chen et al. [8] gave a concept of local degree central nodes and proposed LMD algorithm. The clusters expands from the local degree center nodes associated with the initial seed in this method. Luo et al. [9] proposed two novel methods: DMF-R and DMF-M, which are sliced into different stages by dynamic membership functions. Meng et al. [10] proposed a FuzLhocd algorithm which expands the community by looking for higher-order graph structures in the network. But this method is powerless for sparse networks and time-consuming. Guo et al. [11] proposed LCDMD and designed a new modularity density to mine the core part of a node set, which solved the seed-dependent problem to a certain extent.

Cui et al. [4] proposed a new method by exploiting the high-density connection property of k-core. Due to the high density of k-core, the community obtained by the algorithm is usually incomplete. Huang et al. [5] proposed a similar algorithm. Extending from the seed, any two edges in the current community either participate in forming a triangle, or start from one edge and reach the other edge through a group of triangles, but the method is time-consuming.

Kloumann et al. [6] used a personalized PageRank model to determine the community division of a group of nodes, but this method requires prior knowledge of the benchmark community. Hollocou et al. [7] proposed WalkSCAN based on PageRank. This algorithm first embeds the graph and then uses the improved PageRank to separate the nodes belonging to different communities in the embedded space.

3 Methodology

3.1 Concepts and Definitions

Definition 1 (Jaccard Similarity). Jaccard similarity [12] is a measurement index of the similarity between adjacent nodes:

$$J(i,j) = \frac{|N(i) \cap N(j)|}{|N(i) \cup N(j)|} \tag{1}$$

where the neighbors of a given node s are represented by $N(s)$.

Definition 2 (Motif Degree). The motif degree [10] represents the quantity of times that preset node participates in constructing M, M in this paper is the triangular motif. Motif degree represents the tightness of the connection between a node and its neighbors. Its definition is as follows:

$$MD(i) = |\{M \in G\}|i \in M \;| \tag{2}$$

Definition 3 (Edge Clustering Coefficient). The clustering coefficient [13] characterizes the degree of closeness between the two endpoints of a given edge, denoted as $C(ij)$:

$$C(ij) = \frac{z_{ij}}{s_{ij}} = \frac{|N(i) \cap N(j)|}{\min\left[(k_i - 1), (k_j - 1)\right]} \tag{3}$$

where z_{ij} represents the quantity of triangles that edge (i,j) actually participates in constructing, s_{ij} represents the maxcount of triangles that edge (i,j) may participate in, k_i represents the degrees of node i. To avoid the denominator being 0 when the node degree is 1, we add 1 to its value, and the equation becomes:

$$C(ij) = \frac{|N(i) \cap N(j)|}{\min(k_i, k_j)} \tag{4}$$

Definition 4 (Aggregation Degree). Aggregation degree is used to measure the degree of aggregation between a node and its neighbors, denoted as $AD(i)$:

$$AD(i) = MD(i)\frac{\sum_{j,k \in N(i)} C(jk)}{|e| + 1} \tag{5}$$

where $|e|$ represents the totality of edges among the adjacent nodes of node i. The aggregation degree measures the closeness of the connection among the node itself and its neighbors, as well as the closeness among the neighbors.

Definition 5 (Core Degree). The core degree of the node is used to measure the potential of a node to be regarded as a core node of the community. The higher the potential, the more likely it is to be a core node. Here we improved the PageRank algorithm [14] to calculate the core degree:

$$CD(i) = \beta \sum_{j \neq i} \frac{AD(i)\omega_{ij}}{\sum_{q=1}^{n} AD(q) \sum_{k \in N(i)} \omega_{ik}} + \frac{1-\beta}{n} \tag{6}$$

where ω_{ij} represents the weight of the edge (i, j). In this paper, the edge weight is 1, n represents the node numbers in the network, β is the damping coefficient. The original PageRank algorithm thinks that all nodes have the same initial PR value. However, some important nodes should be set with a larger PR value at the beginning of the iteration for social networks. Therefore, we normalize the PR value of the graph nodes according to their aggregation degree.

Definition 6 (Fitness Function). Fitness function [3] is a metric of the connection tightness among node clusters:

$$F(C) = \frac{k_{in}^{C}}{\left(k_{in}^{C} + k_{out}^{C}\right)^{\alpha}} \tag{7}$$

where k_{in}^{C} and k_{out}^{C} represent the sum of internal and external degree, respectively. The scope of node cluster is regulated by resolution parameter α. The fitness of node v to community C is defined as:

$$F_v(C) = F(C \cup \{v\}) - F(C) \tag{8}$$

if $F_v(C)>0$, meaning that the structure of C will become more compact after v add to it.

Definition 7 (Membership of Node to Community). The probability that node i is part of community C is defined as the membership of i to C:

$$M_i(C) = \frac{|N(i) \cap N(C)|}{\min(|N(i)|, |N(C)|)} \tag{9}$$

3.2 The Proposed Algorithm

CAELCD mainly includes two stages: Core nodes selection and Core area expansion. The framework of CAELCD is shown in Algorithm 1. In the following part of this section, the processing of the algorithm will be described in detail.

Algorithm 1. CAELCD

Input: Network G, seed s, α, d. // α and d are parameters
Output: Local community C of seed s.

1: $H = \emptyset$, $C_s = \emptyset$;
2: $H =$ CoreNodesSelection(G, s, d);
3: **for** node v in H **do**
4:     $C_i =$ CoreAreaExpansion(G, v, α);
5:     **if** s in C_v **then**
6:         Put C_v in C_s;
7:     **end if**
8: **end for**
9: **if** $|C_s| = 0$ **then**
10:     $C =$ CoreAreaExpansion(G, s, α);
11:     return C;
12: **end if**
13: **if** $|C_s| = 1$ **then**
14:     $C = C_s\,[0]$;
15:     retuen C;
16: **else**
17:     $C =$ Max(C_s); // Local community expanded from max CD node
18:     return C;
19: **end if**

Core Nodes Selection. Seed's position profoundly influences the accuracy of the extended community. Generally speaking, starting from the core node that is in the community center, it is more likely to get a high-quality community. Intuitively, the core node is generally located in the community center, so the core node should be closely connected with its neighbors. In addition, the neighbor nodes of the core node generally belong to the same community. Therefore, neighbor nodes are closely related to each other. To sum up, core nodes have two characteristics: Closely connected to their neighbor nodes; their neighbor nodes are closely linked to each other.

For this reason, we proposed AD to measure the closeness of the connection between nodes and their neighbors from the above two aspects. However, simply using AD to select the community's core node will ignore the interaction between the initial seed and the potential core node. It is likely to take the core nodes of other communities as the new seeds. Therefore, we modified the PageRank algorithm and proposed the concept of CD. For social networks, nodes in the core position should be set higher initial PR value. So, we use the AD of nodes to normalize the PR value. If a node has a larger AD, it will be given a larger initial PR value. Then CAELCD constantly adjusts the PR value in accordance with the connection structure. Finally, the two nodes with the supreme CD will be put into the candidate core node-set. In the core area expansion stage, the core node will be used as the new seed to start expanding the community. The pseudo-code of the Core nodes selection stage is given in Function 1.

Function 1. CoreNodesSelection

Input: Network G, seed s, d.
Output: Core nodes set H.

```
1: i = 0, H = ∅;
2: Dict_ad = ∅, Dict_cd = ∅; // Initialize dictionary
3: subG = BFS(G, d); // d is the depth of BFS
4: for v in subG do
5:     Dict_ad(v) = AD(v); // Calculate AD of v using Eq. (5)
6: end for
7: for v in subG do
8:     Dict_cd(v) = CD(v); // Calculate CD of v using Eq. (6)
9: end for
10: while i < 2 do
11:     N_maxCD = Max(k | k ∈ Dict_cd); // Select the node k with the largest CD
12:     H = H ∪ {N_maxCD};
13:     Remove N_maxCD from Dict_cd;
14:     i+ = 1;
15: end while
16: return H;
```

Core Area Expansion. After getting the new seeds, the algorithm expands the local community starting from them. Based on greedy optimization of community quality metrics, the classical algorithms began to expand cluster from a single node. This method usually adds abnormal nodes to the community. For example, in optimizing the local modularity R or M, the node with the lowest degree will be given priority to join the community. In this situation, if the seed is located on the boundary of two communities, it will be possible to add nodes from other communities to the current local community.

To solve this problem, CAELCD first calculates the Jaccard similarity between the seed and all its neighbors. Then two highest similarity nodes are selected to join the community to form the core area of the community. After that, the expansion of the community starts from the core area, calculate the fitness increment of the community's neighbors, select the node corresponding to the maximum increment value to join the community, and repeat this process until the fitness function reaches the maximum value. Finally, the community structure has been formed. When the fitness function cannot continue to increase, we use membership of node to community to optimize the local community. The pseudo-code of the core area expansion phase is given in Function 2.

3.3 Complexity

Presume that a network has n nodes and m edges, the average number of node degrees is d, a community contains on average s nodes. Assuming that BFS has detected v nodes, $v \ll n$, core nodes selection stage's time complexity is $O((t+1) \times v^2)$, where t is the iteration times; In the stage of core area expansion, the time complexity is $O(s \times d)$. Consequently, the total time complexity is

Function 2. CoreAreaExpansion

Input: Network G, Core node v, α.
Output: C_v of node v.

1: $C_v = \{v\}$, $i = 0$;
2: $Dict = \emptyset$; // Initialize dictionary
3: Get the neighbor nodes $N(v)$ of node v;
4: **for** node j in $N(v)$ **do**
5: $\quad Dict(j) = J(v, j)$ // Calculate Jaccard Similarity with v using Eq. (1)
6: **end for**
7: **while** $i < 2$ **do**
8: $\quad N_{maxSim} = Max(k \mid k \in Dict)$; // Select the node with the most similarity
9: $\quad C_v = C_v \cup \{N_{maxSim}\}$;
10: $\quad$ Remove N_{maxSim} from $Dict$;
11: $\quad i+ = 1$;
12: **end while**
13: Get the neithbor nodes $N(C_v)$ of C_v;
14: **while** $N(C_v) \neq \emptyset$ **do**
15: $\quad$ **for** node m in $N(C_v)$ **do**
16: $\quad\quad$ Calculate $F_m(C_v)$ using Eq. (8);
17: $\quad$ **end for**
18: $\quad f_{max} = Max(F_m(C_v) \mid m \in N(C_v))$;
19: $\quad$ **if** f_{max}>0 **then**
20: $\quad\quad C_v = C_v \cup \{v_{maxF}\}$; // v_{maxF} is the node corresponding to f_{max}
21: $\quad\quad N(C_v) = N(C_v) \cup N(v_{maxF})$;
22: $\quad$ **else**
23: $\quad\quad$ break;
24: $\quad$ **end if**
25: **end while**
26: **while** $N(C_v) \neq \emptyset$ **do**
27: $\quad$ **for** node n in $N(C_v)$ **do**
28: $\quad\quad$ Calculate $M_n(C_v)$ using Eq. (9);
29: $\quad$ **end for**
30: $\quad m_{max} = Max(M_n(C_v) \mid n \in N(C_v))$;
31: $\quad$ **if** m_{max}>0.5 **then**
32: $\quad\quad C_v = C_v \cup \{v_{maxM}\}$; // v_{maxM} is the node corresponding to m_{max}
33: $\quad\quad N(C_v) = N(C_v) \cup N(v_{maxM})$;
34: $\quad$ **else**
35: $\quad\quad$ break;
36: $\quad$ **end if**
37: **end while**
38: return C_v;

$O((t+1) \times v^2 + s \times d)$. In addition, considering the worst case, the algorithm needs to store the complete adjacency information of the current community, so the space complexity is $O(n \times d)$.

4 Experiments

4.1 Datesets

We selected a series of networks with different structural characteristics for experimentation. Tables 1 and 2 list the specific parameters.

Table 1. Real-world networks

Network	Nodes	Edges	Average degree	Communities
Karate [15]	34	78	4.59	2
Dolphins [16]	62	159	5.13	2
PolBooks [17]	105	441	8.40	3
Football [18]	115	616	10.66	13
Amazon [19]	334863	925872	5.53	75149
Ca-GrQc [20]	5242	14496	5.53	/
Ca-HepTh [20]	9987	51971	10.41	/
RoadNet-CA [21]	1088092	3083796	5.67	/
RoadNet-PA [21]	1965206	5533214	5.63	/
RoadNet-TX [21]	13979917	3843320	5.57	/

Table 2. Artificial networks

Parameter	Values
N	1000
k	20
k_{max}	50
C_{min}	10
C_{max}	100
μ	0.1–0.6

4.2 Comparison Algorithms and Evaluation Metrics

Comparison Algorithms. We selected six algorithms for comparison: Clauset [1], LWP [2], MLC [22], MULTICOM [23], LMD [8], and LCDMD [11].

Evaluation Metrics. Two customary metrics are used to test algorithm performance: F-Score [24] and Conductance [25].

The specific definition of F-score is as follows:

$$Precision = \frac{|C_{Found} \cap C_{True}|}{|C_{Found}|}$$
$$Recall = \frac{|C_{Found} \cap C_{True}|}{|C_{True}|} \quad (10)$$
$$F - Score = \frac{2 * Precision * Recall}{Precision + Recall}$$

where C_{Found} represents the community mined by the algorithm, and C_{True} represents the community that actually contains the seed. F-Score reflects the overall performance of the algorithm, its value range is [0, 1].

Conductance [25] is used to verify the algorithm when the actual community partition is unknown:

$$Conductance = \frac{\sum_{u \in C} \sum_{u \notin C} \varphi(u, v)}{\min(\mathrm{vol}(C), \mathrm{vol}(\bar{C}))} \quad (11)$$

$\varphi(u, v)$ indicates whether two nodes are directly connected. If there has an edge, its value is 1; otherwise, the value is 0. vol(C) is the volume of set C, i.e. the total number of edges with at least one end belonging to set C. The value range of Conductance is [0, 1], and the smaller its value, the closer the cluster is to the reality.

4.3 Parameter Experiments

The fitness function used in CAELCD is the same as that in the LFM [3] algorithm, so the value range of the α is consistent with it, both being 0.8–1.2, for each network, CAELCD selects the best value for α from this interval. The parameter d is used to control the depth of the BFS. The larger the d, the greater the time cost. The value range of d in this experiment is 1–6. In this experiment, the first five real-world networks with actual community partition in Table 1 and the S1 group networks in Table 2 are used.

Real-world Networks' Results. As shown in Fig. 1a. For Karate and PolBooks, the algorithm's accuracy increases gradually with the increase of d. The accuracy reaches the maximum when $d = 3$, then remains stable. For Amazon, the accuracy is the highest at $d = 1$, when the value of d continues to increase, the accuracy decreases gradually, then increases slightly. For Dolphins and Football, the algorithm's accuracy is less affected by the parameter d.

Artificial Networks' Results. As shown in Fig. 1b. When the value of μ is 0.1 to 0.3, F-Score can reach 1.0 for all d values. When $\mu \geqslant 0.4$, the algorithm's accuracy first increases with the increase of d, when $d = 2$, the accuracy is the optimum, then the accuracy gradually decreases with the increase of d, finally stabilizes. Therefore, for artificial networks, CAELCD has the highest accuracy when $d = 2$.

In summary, d has a slight influence on the accuracy of the real-world network, and $d = 2$ is the best for the artificial network. The value of the parameter d can be set to 2.

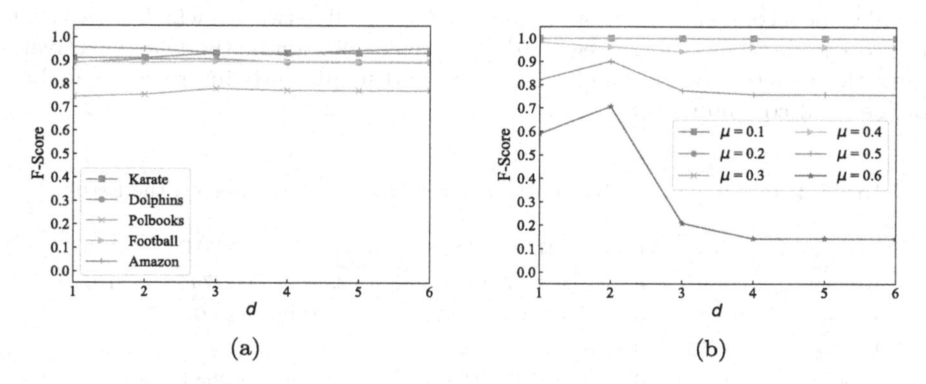

Fig. 1. Parameter experimental results. (a) Real-world networks. (b) Artificial networks.

4.4 Accuracy Experiments

In the accuracy experiment, take the average of all nodes in the network as the experimental result.

Real-World Networks' Results. First, F-Score is used to measure the accuracy of the top 5 networks with actual community partition in Table 1. Table 3 gives the experimental results.

Table 3. F-score values on real-world networks

Networks	Clauset	LWP	MLC	MULTICOM	LMD	LCDMD	CAELCD
Karate	0.6835	0.7320	0.8478	0.3938	0.8201	0.8345	**0.9318**
Dolphins	0.4265	0.5078	0.6105	0.6919	0.7775	0.7235	**0.9054**
PolBooks	0.5719	0.5863	0.6531	0.1770	0.7497	0.7301	**0.7789**
Football	0.6876	0.8568	0.4506	0.5352	0.7267	0.8134	**0.8948**
Amazon	0.8629	0.8575	0.8397	0.1699	0.8614	0.8172	**0.9069**

As can be seen, the F-Score of CAELCD has achieved the highest values on all real-world networks. This is because the core nodes selection strategy of the algorithm can select high-quality seeds to expand the community. On the other hand, the new community outreach strategy avoids the introduction of abnormal nodes. LMD and LCDMD also get better accuracy than other baseline algorithms, but LMD does not consider the relationship between nodes; LCDMD is still expanding community from the initial seed.

Second, Conductance is used to measure the accuracy of all real-world networks in Table 1. Table 4 gives the experimental results. CAELCD achieves the smallest or second smallest Conductance value for all networks, which shows that the core nodes selection strategy and core area expansion method can effectively mine the structure of community boundary and significantly improve the quality of the local community.

Table 4. Conductance values on real-world networks (the smaller, the better)

Networks	Clauset	LWP	MLC	MULTICOM	LMD	LCDMD	CAELCD
Karate	0.2543	0.2539	0.3977	0.5463	0.1850	0.2774	**0.1705**
Dolphins	0.3519	0.3825	0.3760	0.2552	**0.1235**	0.2221	0.1431
PolBooks	0.2750	0.2255	0.3232	0.1272	0.0866	0.1269	**0.0745**
Football	0.3341	0.3070	0.4515	0.2974	0.3286	**0.2884**	0.2921
Amazon	0.1037	0.1366	0.2168	0.3972	**0.0845**	0.2081	0.0942
CA-GrQc	0.2928	0.5571	0.4110	0.2795	0.2478	0.2053	**0.1664**
CA-HepTh	0.3515	0.5746	0.5038	0.4967	0.3084	0.2769	**0.2382**
RoadNet-CA	0.2980	0.3885	0.4302	0.1698	0.2066	0.1884	**0.1503**
RoadNet-PA	0.3027	0.4281	0.4638	0.1720	0.2212	0.2025	**0.1585**
RoadNet-TX	0.2623	0.4434	0.4094	0.1762	0.1940	0.1835	**0.1488**

Artificial Networks' Results. Figure 2 illustrates the accuracy results of each algorithm on artificial datasets. From the change trend in the figures, it can be observed that with the increase of parameters μ, the F-score of each algorithm decrease, and the Conductance increase, respectively. This is due to the fact that it becomes more difficult to accurately mine local communities when the community boundaries are blurred. Even so, the accuracy of CAELCD is still significantly better than other algorithms. Because CAELCD can find higher quality seeds from the given seed, it can still identify the local community accurately when the community boundary becomes fuzzy.

4.5 Visualization Experiments

In order to intuitively demonstrate the performance that CAELCD mine local communities starting from different quality seeds, we conduct a visual analysis on Karate and Dolphins network. For each network, we select the core node, the boundary node, and the hub node at the junction of the communities as the original seeds for local community detection. Figures 3 and 4 reveal the results, where the red and the orange node represents the seed and the local community mined by CAELCD, respectively.

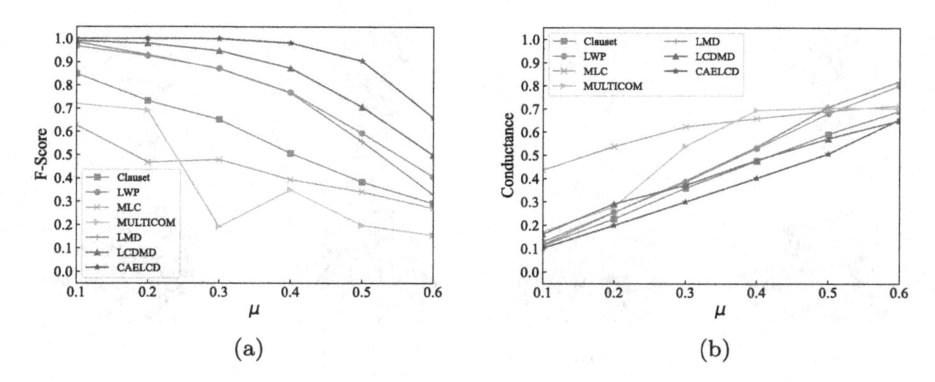

Fig. 2. Accuracy results on artificial networks. (a) F-score values. (b) Conductance values.

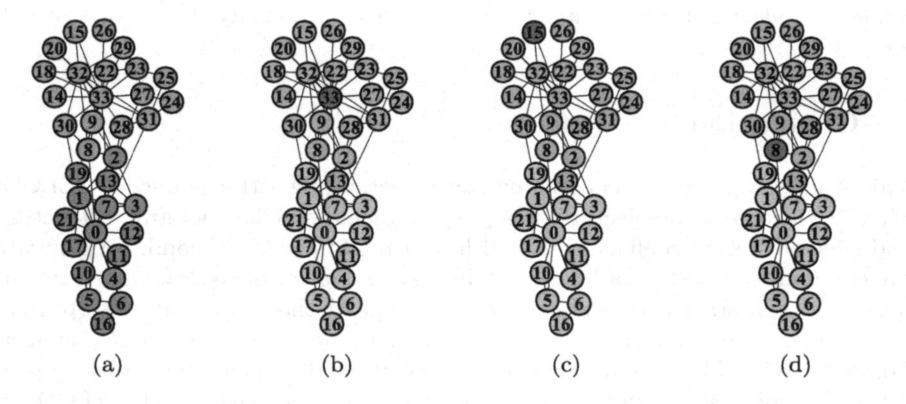

Fig. 3. Experimental result of local community visualization on Karate network. (a) Ground truth of Karate. (b) The community identified from core node. (c) The community identified from boundary node. (d) The community identified from hub node.

As can be seen, for nodes in multiple locations in the community, CAELCD can detect communities that are almost the same as the actual community. In the Karate network, for $8th$ node located at the junction of two communities, CAELCD can correctly find the community to which it belongs. In the Dolphins network, even for the $60th$ boundary node whose degree is 1, the local community identified by our method is still the same as the $57th$ node at the central position of the community. This proves that CAELCD is able to significantly solve the seed-dependent problem.

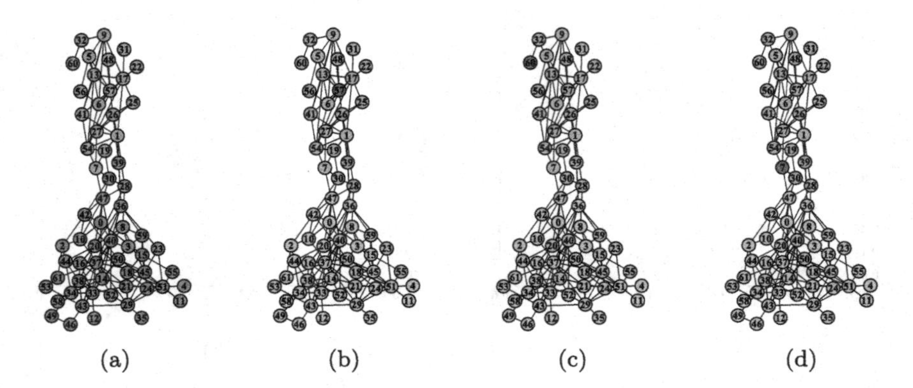

Fig. 4. Experimental result of local community visualization on Dolphins network. (a) Ground truth of Dolphins. (b) The community identified from core node. (c) The community identified from boundary node. (d) The community identified from hub node.

5 Conclusion

This study proposed a local community detection algorithm named CAELCD. First, in the core node selection stage, we combine the high-order graph structure and edge clustering coefficient to find higher quality seeds; Second, we generate the community's core area by calculating the similarity of seed and their neighbors and then start from the core area to expand the community by optimizing the fitness function. Experiments on the parameter, accuracy and visualization of the CAELCD are designed on real-world and artificial networks, respectively. Experimental results demonstrate the advanced nature of CAELCD. In real-world scenarios, the structure of networks changes dynamically over time. Therefore, it is also necessary to extend the current research to be applicable to scenarios with dynamic changes in network structure. In addition, using big data frameworks to parallelize and accelerate algorithms is also a problem worth exploring.

References

1. Clauset, A.: Finding local community structure in networks. Phys. Rev. E **72**(2), 026132 (2005)
2. Luo, F., Wang, J.Z., Promislow, E.: Exploring local community structures in large networks. Web Intell. Agent Syst. Int. J. **6**(4), 387–400 (2008)
3. Lancichinetti, A., Fortunato, S., Kertész, J.: Detecting the overlapping and hierarchical community structure in complex networks. New J. Phys. **11**(3), 033015 (2009)
4. Cui, W., Xiao, Y., Wang, H., Wang, W.: Local search of communities in large graphs. In: Proceedings of the 2014 ACM SIGMOD International Conference on Management of Data, pp. 991–1002 (2014)

5. Huang, X., Cheng, H., Qin, L., Tian, W., Yu, J.X.: Querying k-truss community in large and dynamic graphs. In: Proceedings of the 2014 ACM SIGMOD International Conference on Management of Data, pp. 1311–1322 (2014)
6. Kloumann, I.M., Kleinberg, J.M.: Community membership identification from small seed sets. In: Proceedings of the 20th ACM SIGKDD International Conference on Knowledge Discovery and Data Mining, pp. 1366–1375 (2014)
7. Hollocou, A., Bonald, T., Lelarge, M.: Improving pagerank for local community detection. arXiv preprint arXiv:1610.08722 (2016)
8. Chen, Q., Wu, T.-T., Fang, M.: Detecting local community structures in complex networks based on local degree central nodes. Phys. A Statist. Mech. Appl. **392**(3), 529–537 (2013)
9. Luo, W., Zhang, D., Jiang, H., Ni, L., Hu, Y.: Local community detection with the dynamic membership function. IEEE Trans. Fuzzy Syst. **26**(5), 3136–3150 (2018)
10. Meng, T., Cai, L., He, T., Chen, L., Deng, Z.: Local higher-order community detection based on fuzzy membership functions. IEEE Access **7**, 128510–128525 (2019)
11. Guo, K., Huang, X., Wu, L., Chen, Y.: Local community detection algorithm based on local modularity density. Appl. Intell. 1–16 (2021)
12. Jaccard, P.: Étude comparative de la distribution florale dans une portion des alpes et des jura. Bull. Soc. Vaudoise Sci. Nat. **37**, 547–579 (1901)
13. Radicchi, F., Castellano, C., Cecconi, F., Loreto, V., Parisi, D.: Defining and identifying communities in networks. Proc. Natl. Acad. Sci. **101**(9), 2658–2663 (2004)
14. Page, L., Brin, S., Motwani, R., Winograd, T.: The pagerank citation ranking: bringing order to the web. Tech. Rep. Stanford InfoLab (1999)
15. Zachary, W.W.: An information flow model for conflict and fission in small groups. J. Anthropol. Res. **33**(4), 452–473 (1977)
16. Lusseau, D., Schneider, K., Boisseau, O.J., Haase, P., Slooten, E., Dawson, S.M.: The bottlenose dolphin community of doubtful sound features a large proportion of long-lasting associations. Behav. Ecol. Sociobiol. **54**(4), 396–405 (2003)
17. Newman, M.E., Girvan, M.: Finding and evaluating community structure in networks. Phys. Rev. E **69**(2), 026113 (2004)
18. Girvan, M., Newman, M.E.: Community structure in social and biological networks. Proc. Natl. Acad. Sci. **99**(12), 7821–7826 (2002)
19. Yang, J., Leskovec, J.: Defining and evaluating network communities based on ground-truth. Knowl. Inf. Syst. **42**(1), 181–213 (2015)
20. Leskovec, J., Kleinberg, J., Faloutsos, C.: Graph evolution: densification and shrinking diameters. ACM Trans. Knowl. Discov. Data. **1**(1), 2-es (2007)
21. Leskovec, J., Lang, K.J., Dasgupta, A., Mahoney, M.W.: Community structure in large networks: natural cluster sizes and the absence of large well-defined clusters. Internet Math. **6**(1), 29–123 (2009)
22. Kamuhanda, D., He, K.: A nonnegative matrix factorization approach for multiple local community detection. In: 2018 IEEE/ACM International Conference on Advances in Social Networks Analysis and Mining (ASONAM), pp. 642–649. IEEE (2018)
23. Hollocou, A., Bonald, T., Lelarge, M.: Multiple local community detection. ACM SIGMETRICS Perform. Evaluat. Rev. **45**(3), 76–83 (2018)
24. Jianhua, L., Xiaofeng, W., Peng, W.: Review on community detection methods based on local optimization. Bullet. Chin. Acad. Sci. **2**, 238–247 (2015)
25. Andersen, R., Chung, F., Lang, K.: Local graph partitioning using pagerank vectors. In: 47th Annual IEEE Symposium on Foundations of Computer Science (FOCS 2006), pp. 475–486. IEEE (2006)

Federated Clique Percolation for Overlapping Community Detection on Attributed Networks

Mingyang Wei[1,2,3], Kun Guo[1,2,3](✉), and Ximeng Liu[1]

[1] College of Computer and Data Science, Fuzhou University, Fuzhou 350108, China
gukn@fzu.edu.cn

[2] Fujian Provincial Key Laboratory of Network Computing and Intelligence Information Processing, Fuzhou University, Fuzhou 350108, China

[3] Key Laboratory of Spatial Data Mining and Information Sharing, Ministry of Education, Fuzhou 350108, China

Abstract. Community detection is a popular research topic in complex network analysis, which can be applied in many real-world scenarios such as disease prediction. With the increase of people's awareness of privacy protection, more and more laws enforce the protection of sensitive information while transferring data. The anonymization-based community detection methods have to sacrifice accuracy for privacy protection. In this paper, we first propose a standalone clique percolation algorithm to detect overlapping communities on attributed networks. A clique similarity metric is designed to percolate cliques accurately. Second, we develop a federated clique percolation algorithm to detect overlapping communities on distributed attributed networks. Perturbation strategy and homomorphic encryption are used to protect network privacy. The experiments on real-world and artificial datasets demonstrate that the federated clique percolation algorithm achieves identical results to the standalone ones and realizes higher accuracy than the simple distributed ones without federating learning.

Keywords: Community detection · Federated learning · Clique percolation · Vertex perturbation · Homomorphic encryption

1 Introduction

Communities exist universally in many real-world networks, such as scientific collaboration and social networks. A community in a network refers to a group of entities closely formed by vertices and their links. Community is overlapping when vertices belong to multiple different communities. Community detection is hot research in complex network analysis because it plays a vital role in many practical scenarios, such as analyzing criminal activities, predicting disease, or building user portraits [7]. A type of algorithm that relies on discovering and percolating cliques (complete subgraphs) has been proposed to detect overlapping

Y. Sun et al. (Eds.): ChineseCSCW 2021, CCIS 1492, pp. 252–266, 2022.
https://doi.org/10.1007/978-981-19-4549-6_20

communities [15,25]. Moreover, Literatures [6,18,23] extend the clique percolation method to weighted and attributed networks.

Privacy disclosure events raise user privacy fears in recent years. In response, governments have designated a series of strict laws to protect user privacy. For instance, the European Union enacted the General Data Protection Regulation (GDPR) on data protection for all individuals [8]. With the increasing awareness of privacy protection for governments and individuals, graph data should also be protected when conducting network data mining tasks. The existing privacy-preserving community detection algorithms mainly rely on differential privacy and anonymization technology. Differential privacy [12,24] is a method based on strict mathematical proofs that usually contain network and algorithm perturbation. Anonymous approaches [5,14] let the identity of vertex or link can not be separated from its counterparts by generalizing sensitive data. In summary, these methods sacrifice accuracy for community detection because the structure of the input network or the result of the algorithms' output is modified. Federated learning [27] is a distributed privacy-preserving machine learning paradigm. It allows multiple participants to train a global model jointly based on all their local data without leaking anyone's privacy. Only the encrypted intermediate results are shared among the participants. Due to the advantages of supporting distributed privacy-preserving machine learning, federated learning is the basis of the distributed privacy-preserving community detection problem studied in this paper.

In the paper, we first propose a standalone clique percolation algorithm to detect overlapping communities on attributed networks. The designed clique similarity metric helps to detect communities more accurately. Second, a federated clique percolation algorithm was proposed to detect overlapping communities on distributed attributed networks. Perturbation strategy and homomorphic encryption are used to protect network privacy simultaneously. The experiments demonstrate that the federated algorithm is the correct extension of its standalone counterpart and achieves higher accuracy than the simple distributed clique percolation algorithm. The main contributions of this paper are summarized as follows:

(1) The vertex and clique perturbation strategies can protect the degree of the vertices without affecting the correctness of community detection.
(2) The homomorphic encryption technique protects the vertex attributes of each participant from being leaked to other participants and the coordinator. At the same time, the federated algorithm ensures the correctness of community detection.
(3) The designed clique similarity metric integrates links and attributes between vertices, which improved the accuracy of community detection than traditional clique percolation algorithms.
(4) Experimental results on real-world and artificial datasets demonstrate the effectiveness and correctness of our algorithms.

2 Related Work

2.1 Clique Percolation Algorithms

The idea of clique percolation is that a community is composed of a group of cliques that can be reached from each other through a series of adjacent cliques. Palla et al. [25] proposed CPM based on the maximal clique in graph theory, which joins the interconnected maximal cliques to build communities. Farkas et al. [6] proposed CPMw, which extended CPM to weighted networks by introducing a subgraph intensity threshold. Kumpula et al. [15] proposed a sequential algorithm named SCP for fast clique percolation by replacing maximal clique with k-clique. Liu et al. [18] proposed CMC that can discover complexes from the weighted PPI network. CMC integrates link weight and vertex relationships while conducting maximal cliques percolation. Mougel et al. [23] proposed the algorithm to find communities on isomorphic vertex sets of the attributed network. Only when the number of common attributes is greater than a threshold, the vertex set can be regarded as isomorphic.

2.2 Privacy-Preserving Community Detection

The existing privacy-preserving community detection algorithms mainly rely on anonymizing network structures and attributes. Dev et al. [5] constructed a probability graph that contains likelihood information to detect high precision communities. Kumar et al. [14] combined the fuzzy sets approach with naive anonymization for privacy-preserving community detection. Nguyen et al. [24] proposed input perturbation LouvainDP and algorithm perturbation ModDivisive to analyze complex networks secretly. The two methods suitable for different scenarios to detect community. Ji et al. [12] proposed a DPCD scheme to protect the privacy of both network topology and node attributes while detecting social communities. DPCD can achieve ϵ-differential privacy theoretically.

2.3 Federated Learning on Graphs

Present works for federated learning on graphs mainly focus on supervised learning tasks and have accuracy loss. Mei et al. [22] proposed an SGNN model to conduct vertex classifying. The center server aggregates encrypted network data to train the SGNN model. Lalitha et al. [16] consider training a machine learning model over a network of vertices. All vertices collectively learn an optimal global model. At the same time, metadata need not be shared between vertices. Zheng et al. [28] proposed an ASFGNN model to decouple the training of GNN. Loss computing was learned by clients federally.

3 Preliminaries

3.1 Problem Definition

Suppose that multiple participants collaborate to detect communities from their local networks. The local network is denoted as $G(V, E, \mathbf{A})$, where $V =$

$\{v_1, v_2, \ldots, v_n\}$ denotes vertex set with $|V| = n$, E represents link set with $e_{ij} = (v_i, v_j) \in E$ if a link exists between v_i and v_j, and $\mathbf{A} = [\boldsymbol{a_1}, \boldsymbol{a_2}, \ldots, \boldsymbol{a_n}]^{\mathrm{T}} \in \mathbb{R}^{n \times w}$ is an attribute matrix, where $\boldsymbol{a_i} \in \mathbb{R}^w$ denotes an attribute vector of v_i.

Given n_p participant networks $G_p = \{G_1, G_2, \ldots, G_{n_p}\}$, the goal of distributed privacy-preserving community detection on G_p is to partition the vertices in G_p into n_c clusters $\{C_1, C_2, \ldots, C_{n_c}\}$ without disclosing the network privacy which defined in Sect. 3.2, where $|C_i \cap C_j| \geq 0, i, j \in \{1, 2, \ldots, n_c\}$.

3.2 Network Privacy

We define network privacy under the semi-honest model [1] that each participant strictly follows the procedure of an algorithm and never conspires with others to pry into any participant's privacy. We summarize network privacy that should be protected in distributed privacy-preserving community detection. (1) *The structure of a network* including vertex degrees and the existence of links. The vertex degree distribution of a complex network is highly skewed. Therefore, an adversary can find out a vertex's position quickly if he knows its degree. The existence of a link also gives an adversary a hint on the location of an unknown vertex connecting to a compromised one. (2) *The attributes of vertices and links.* Vertex attributes are usually different in a complex network. Therefore, an adversary can recognize a vertex easily if he knows all the attributes of the vertex. A link can be located similarly.

4 Proposed Method

4.1 SCPAN

We develop a Standalone Clique Percolation algorithm on Attributed Networks (SCPAN) to detect communities, which is used as a template to build the federated clique percolation algorithm. SCPAN contains three stages, as shown in Algorithm 1.

Stage 1: k-clique computation. Detect all k-cliques of a network.

Stage 2: k-clique percolation. Build a clique network G_c with all k-cliques as its vertex set. For any pair of k-cliques sharing at least one vertex, a link between the two k-cliques will be added to G_c while the similarity of the two k-cliques is greater than a threshold α. We design a metric to measure the similarity between two k-cliques according to the links and attributes of vertices between two k-clique, which is obtained from Eq. 1.

Stage 3: Community generation. Recognize all connected components of G_C and output each connected component as a community.

$$s_a(C_p, C_q) = \sqrt{\frac{\sum\limits_{v_i \in (C_p - C_q)} \sum\limits_{v_j \in C_q} s(\boldsymbol{a_i}, \boldsymbol{a_j})}{|C_p - C_q| \cdot |C_q|} \cdot \frac{\sum\limits_{v_i \in (C_q - C_p)} \sum\limits_{v_j \in C_p} s(\boldsymbol{a_i}, \boldsymbol{a_j})}{|C_q - C_p| \cdot |C_p|}} \tag{1}$$

Algorithm 1: SCPAN

Input: Network $G(V, E, \mathbf{A})$, size of clique k, threshold α
Output: Community set C

```
   // STAGE 1:k-CLIQUE COMPUTATION
1  S ← set of k-cliques from G
   // STAGE 2:k-CLIQUE PERCOLATION
2  G_C(V_C, E_C) ← ∅
3  V_C ← V_C ∪ S
4  for C_p, C_q ∈ S do
5      Calculate s(C_p, C_q) according to Eq. (1)
6      if s(C_p, C_q) ≥ α then
7          E_C ← E_C ∪ {(C_p, C_q)}
   // STAGE 3:COMMUNITY GENERATION
8  C ← Find all connected components of G_C
9  return C
```

$$s(\boldsymbol{a}_i, \boldsymbol{a}_j) = \mathrm{Ind}(v_i, v_j) + \frac{|\boldsymbol{a}_i| - ||\boldsymbol{a}_i \oplus \boldsymbol{a}_j||_1}{2 \cdot |\boldsymbol{a}_i|} \tag{2}$$

where C_p and C_q represent two different k-cliques, v_i and v_j denote two vertices in different k-cliques. $s(\boldsymbol{a}_i, \boldsymbol{a}_j)$ is the attribute similarity between v_i and v_j. $\mathrm{Ind}(v_i, v_j)$ is the identity function that returns 1 if $(v_i, v_j) \in E$ and 0 othewise. $\oplus$ is the XOR operator, $|\cdot|$ represents the size of a set , $||\boldsymbol{a}||_1$ denotes $\ell 1$ norm.

4.2 FCPAN

We design a Federated Clique Percolation algorithm to detect overlapping communities on distributed Attributed Networks (FCPAN) , which is based on SCPAN. The framework of FCPAN is shown in Fig. 1. FCPAN is composed of 4 stages, as shown in Algorithm 2.

Stage 1: Vertex ID matching. The Private Set Intersection (PSI) protocol proposed in [4] is run among the participants to align vertex IDs that represent overlapping vertices. PSI is used to match the identities (IDs) of the overlapping elements of different collections, guaranteeing that the elements that do not exist in other collections will not be leaked simultaneously.

Stage 2: k-clique computation. The participants cooperate in computing k-cliques with the coordinator to obtain k-cliques across participants. Moreover, the coordinator needs to get the encrypted attributes of the vertices for clique percolation. Therefore, A participant is randomly selected to generate homomorphic encryption key pair (pk,sk) and sends the key pair to all other participants through Secure Shell (SSH) before the stage. An improved Paillier homomorphic encryption system [20] is used in this paper, which can determine whether two elements are equal under ciphertexts. The stage is composed of 5 steps:

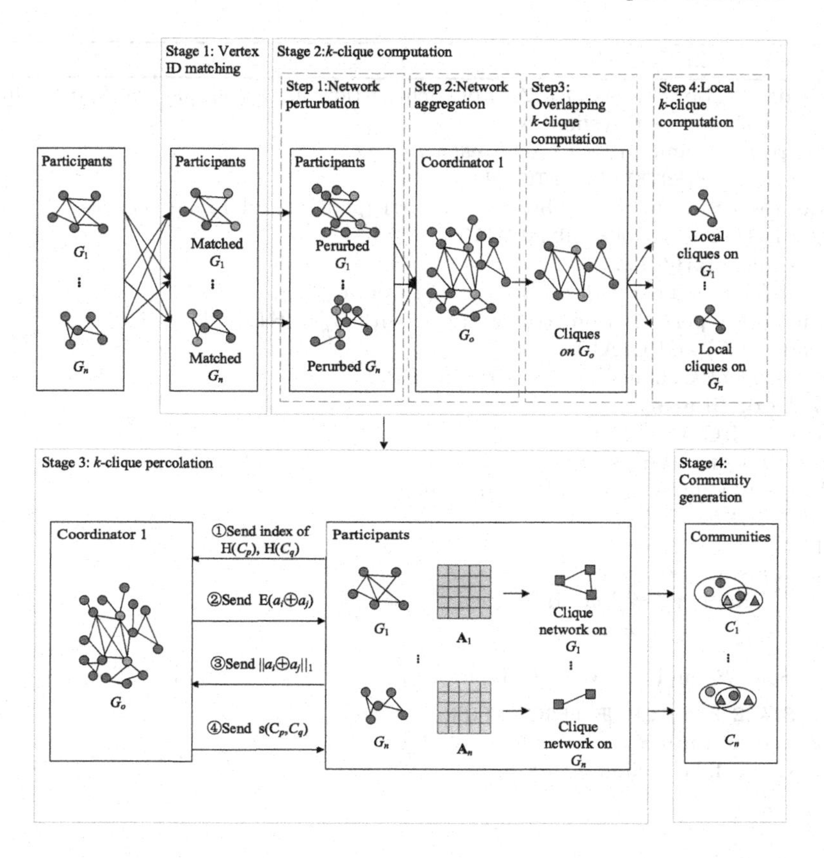

Fig. 1. Framework of FCPAN

*S*TEP 1: Each participant P_h first extracts a subnetwork $G_h^e = (V_h^e, E_h^e)$ containing only the overlapping vertices and their neighbors from his local network. Second, each participant confuses vertex's degrees in G_h^e by adding fake vertices. We name the strategy vertex perturbation that satisfies k-degree-anonymity criterion [2]. Figure 2 illustrates the perturbation of three overlapping vertices (in grey color) and their neighbors (in green color). The red vertices are the added fake vertices. Third, the vertex ID of G_h^e is obfuscated by a hash function such as SHA256 [13] to get $\mathrm{H}(G_h^e)$, where H() denotes the hash function. Finally, the perturbed $\mathrm{H}(G_h^e)$ and $\mathrm{E}(\mathbf{A_h^e})$ are sent to the coordinator, where the $\mathbf{A_h^e}$ represent the attribute matrix of G_h^e and E() is the encryption function.

k-degree-anonymity is a widely used privacy protection criterion that requires that at least $k-1$ vertices have the same degree as a sensitive vertex. We take Fig. 2 as an example to show how our vertex perturbation strategy satisfies the k-degree-anonymity criterion, k takes 3 in the example. First, we get the original degree sequence [3,2,1,1,1] of the original network by descending order. Second, the method proposed in [19] is used to calculate anonymous degree

Algorithm 2: FCPAN

Input: Network $G_h(V_h, E_h, \mathbf{A}_h)$ of each participant P_h, clique size k, attribute similarity threshold α
Output: Community set C_h for each P_h

```
// STAGE 1:VERTEX ID MATCHING
1  Run the PSI protocol to obtain each participant's overlapping vertices
   // STAGE 2:k-CLIQUE COMPUTATION
   // Participant P_h:
2  Identify the subnetwork G_h^e = (V_h^e, E_h^e) from G_h
3  Run vertex perturbation strategy to perturb G_h^e //STEP 1
4  Send H(G_h^e) and E(A_h^e) to coordinator
5  S'_h ← set of k-cliques in G_h ∪ S_h //STEP 4
   // Coordinator:
6  G_o ← ∪H(G_h^e) //STEP 2
7  S_o ← set of k-cliques in G_o //STEP 3
8  Assign unique index for all k-cliques in S_o
9  for h = 0 to n_p do
10     S_h ← ∅
11     for s ∈ S_o do
12         if s ∩ V_h^e ≠ ∅ then
13             S_h ← S_h ∪ s
14     Send S_h and indices of k-cliques to P_h
   // STAGE 3:k-CLIQUE PERCOLATION
   // Participant P_h:
15 Filter out fake k-cliques fron S_h
16 G_h^C(V_h^C, E_h^C) ← ∅
17 V_h^C ← V_h^C ∪ S_h
18 for h = 0 to n_p do
19     for C_p, C_q ∈ S_h do
20         Calculate s(C_p, C_q) according to Function 1
21         if s(C_p, C_q) ≥ α then
22             E_h^C ← E_h^C ∪ {(C_p, C_q)}
   // STAGE 4: COMMUNITY GENERATION
23 For each P_h, C_h ← connected components of G_h^C
24 return C_h
```

sequence [3,3,3,3,3] that satisfies 3-degree-anonymity. Note that the degree of each vertex in the anonymous degree sequence is greater than or equal to its original degree. Third, fake neighbors (in red color) are added to each vertex until their degree equals the degree in the anonymous degree sequence. All true vertices satisfied 3-degree-anonymity in the intermediate network so far. Fourth, All fake vertices have 1-degree in the intermediate network. The coordinator can restore the original network by removing these unique vertices. Therefore, We add links between fake vertices to generate 1-degree and 2-degree fake vertices.

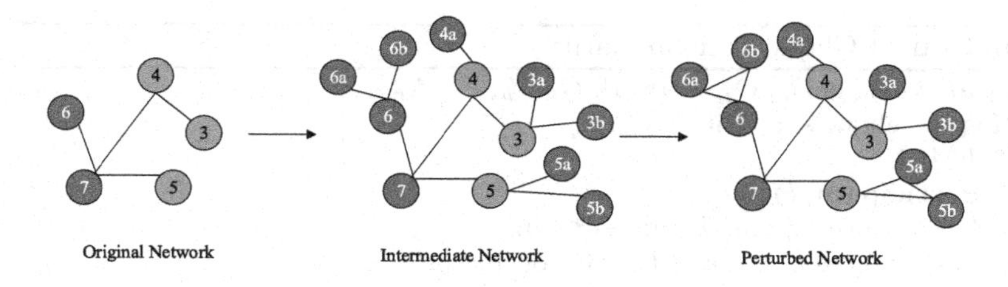

Fig. 2. Vertex perturbation (Color figure online)

At the same time, the operation generates fake k-cliques that can protect true k-cliques that can not be distinguished. We can see that the degree sequence of the perturbed network is [3,3,3,3,3,2,2,2,2,1,1,1], which satisfies 3-degree anonymity. The benefits are twofold: (1) the degrees of vertices in the perturbed network are protected; (2) the true k-cliques can be protected by confusion.

STEP 2: Coordinator aggregates all received subnetworks into a network G_o and detect *overlapping k-cliques* on it in the same way as stage 1 of SCPAN.

STEP 3: Coordinator assigns unique index for all *overlapping k-cliques* and then sends the k-cliques and indices back to each participant. Only the k-cliques related to the subnetwork sent by the participant are sent to him, which prevents a participant from knowing other participants' privacy.

STEP 4: After receiving the *overlapping k-cliques* and their index, each participant performs the same k-clique computation on his local network like that in stage 1 of SCPAN to obtain the *local k-cliques.*

Stage 3: k-clique percolation. Each participant conducts k-clique percolation with the coordinator. Before the stage, the participant filters fake k-cliques that contain fake vertices, which are generated in stage 2. For any two pair of k-cliques C_p and C_q, the process of calculating $s(C_p, C_q)$ is composed of 4 steps:

STEP 1: If both C_p and C_q are in *local k-cliques*, clique similarity $s(C_p, C_q)$ can be calculated locally. Otherwise, *local k-cliques* and index of *overlapping k-cliques* will be sent to the coordinator to calculate $s(S_p, S_q)$. Vertices in *local k-cliques* and their attribute vector will be hashed and encrypted before they are sent to the coordinator.

The coordinator can get the lower bound of the degree of vertices in *local k-cliques.* Therefore, the participant sends fake k-cliques composed of vertices in true *local k-cliques* and fake vertices to the coordinator to perturb degree of vertices in *local k-cliques.* Fake k-cliques is not involved in the stage of k-clique percolation. Therefore, it will not affect the correctness of community detection. We name the strategy clique perturbation.

STEP 2: Coordinator calculates $\boldsymbol{a}_i \oplus \boldsymbol{a}_j$ under ciphertexts and sends it to the participant.

STEP 3: The participant decrypts the vector $\mathrm{E}\,(\boldsymbol{a}_i \oplus \boldsymbol{a}_j)$ and sends $||\boldsymbol{a}_i \oplus \boldsymbol{a}_j||_1$ to coordinator.

Function 1: Clique Pair Similarity

Input: k-clique C_p, C_q, Network $G_h(V_h, E_h, \mathbf{A_h})$
Output: clique similarity $s(C_p, C_q)$

```
1  //STEP 1
   // Participant P_h:
2  if C_p, C_q ∈ overlapping k-cliques then
3      Send indices of C_P and C_q to coordinator
4  else if C_p ∈ overlapping k-cliques then
5      Run clique perturbation strategy to perturb C_q
6      Send index of C_p, H(C_q) and E(A_(C_q)) to coordinator
7  else if C_q ∈ overlapping k-cliques then
8      Run clique perturbation strategy to perturb C_p
9      Send H(C_p), E(A_(C_p)) and index of C_q to coordinator
10 else
11     Calculate s(C_p, C_q) according to Eq. (1)
12 Decrypt E(a_i ⊕ a_j) and send ||a_i ⊕ a_j||_1 to coordinator //STEP 3
   // Coordinator:
13 Calculate a_i ⊕ a_j under ciphertexts and send E(a_i ⊕ a_j) to
14  participant //STEP 2
15 Calculate s(C_p, C_q) according to ||a_i ⊕ a_j||_1 and Eq. (1), then send it to
   participant //STEP 4
16 return s(C_p, C_q)
```

*S*TEP 4: Coordinator calculates $s(C_p, C_q)$ by Eq. (1) with $||\boldsymbol{a}_i \oplus \boldsymbol{a}_j||_1$ and sends it to the participant.

Stage 4: Community generation. Each participant generates communities in the same way as stage 3 of SCPAN.

4.3 Complexity Analysis

We first derive the time complexity of FCPAN. The PSI protocol matching overlapping vertices for any two pair of participants in stage 1 requires $O(n \times n_p^2)$. Stage 2 contains 4 steps. The step 1 for netowrk perturbation is $O(n \times k)$ because calculating anonymous degree sequence requires $O(n \times k)$ [19]. In step 2, detecting k-cliques on the aggregated network requires $O((m \times n_p)^{k/2})$ time because the worst time complexity to find k-cliques in a network with m links is $O(m^{k/2})$ [3]. The time for sending k-cliques to participant in step 3 require $O(n \times (m \times n_p)^{k/2})$. In step 4, each participant finds local k-cliques without overlapping vertices that occupies $O(n)$ time. Therefore, the total time complexity of stage 2 is $O(n \times (m \times n_p)^{k/2})$. The time for clique percolation in stage 3 requires $O(m \times n_p)$. In stage 4, the time for community generation is $O((m \times n_p)^k)$. In summary, the time complexity of FCPAN is $O(n \times {n_p}^2 + n \times (m \times n_p)^{k/2} + m \times n_p + (m \times n_p)^k)$, which can be reduced to $O((m \times n_p)^k)$ because $m > n$.

Second, we derive the space complexity of FCPAN. In stage 1, the space for storing the overlapping vertices require $O(n)$. Stage 2 for storing k-cliques and attribute matrix requires $O((m \times n_p)^{k/2})$ and $O(n \times n_p \times w)$ space, respectively. The clique networks occupy $O((m \times n_p)^k)$ space in stage 3 and the space complexity of stage 4 is the same as stage 3. In summary, the space complexity of FCPAN is $O(n + (m \times n_p)^{k/2} + (n \times n_p \times w) + (m \times n_p)^k + (m \times n_p)^k)$, which can be reduced to $O((m \times n_p)^k)$ because $m > n$.

5 Experiments

5.1 Datasets

We use 6 real-world datasets and 2 artificial datasets to evaluate our algorithms' performance. Table 1 shows the details of the datasets, where n, m, $avgd$, and w denote the number of vertices, the number of links, the average vertex degree of a network, and the size of the attribute vector of vertex, respectively.

Table 1. Description of datasets

Datasets	Parameters
fb1	$n = 52,\ m = 146,\ avgd = 5.6,\ w = 42$
fb2	$n = 61,\ m = 270,\ avgd = 8.9,\ w = 48$
tw1	$n = 126,\ m = 496,\ avgd = 7.9,\ w = 247$
parklot	$n = 274,\ m = 9189,\ avgd = 67.1,\ w = 11$
cora	$n = 2708,\ m = 5278,\ avgd = 3.9,\ w = 1433$
citeseer	$n = 3264,\ m = 4536,\ avgd = 2.8,\ w = 3730$
D1	$n = 5000,\ avgd = 15,\ \mu = 0.1 \sim 0.6,\ w = \lvert\text{true communities}\rvert$
D2	$n = 1000 \sim 5000,\ avgd = 15,\ \mu = 0.4,\ w = \lvert\text{true communities}\rvert$

Real-world datasets: The real-world datasets include 6 attributed networks. The facebook-3980, facebook-698, and twitter-356963 datasets are obtained from SNAP[1], the cora and citeseer datasets come from LINQS[2], and the parking lot dataset is collected from Hong Kong Transport Department[3]. Each real community for parking lot represents a collection of areas with close distance. The three datasets from SNAP and the parking lot dataset are abbreviated as fb1, fb2, tw1, and parklot.

Artificial datasets: We use LFR benchmark [17] to generate two groups of artificial datasets, D1 and D2. The values of μ of the networks in D1 reflect the degree of mixture of the communities. Communities are more and more challenging to be detected as μ increases. Furthermore, we generate vertex attributes for D1 and D2 networks by the method proposed in [10].

[1] http://snap.stanford.edu/data.
[2] https://linqs.soe.ucsc.edu/data.
[3] https://data.gov.hk/sc.

5.2 Experimental Setup

The vertex-cut-based method proposed in [9] is used to split a network into 2, 4, 6, 8, and 10 subnetworks to simulate the participants' networks. Each vertex of participants' networks has complete attributes.

Overlapping Normalized Mutual Information (ONMI) [21] and Extended Modularity (EQ) [26] are used to measure the algorithms' accuracy. We merge the communities in different participants to calculate the ONMI and EQ values because each participant outputs communities locally. For FCPAN, the communities detected on each participant's network were collected. Those communities between participants will be merged if the similarity of the two k-cliques in different communities large than α. For the simple distributed clique percolation scene without federated learning, we first collect the communities detected on each participant's network and merge the communities in different participants whose similarity measured by the Jaccard index [11] is greater than or equal to 0.5.

In all the experiments, the method proposed in [3] is used to find k-cliques and the value of parameter k in k-clique percolation is varied from 3 to 5 to obtain comparable results which is suggested by [25]. In the parameter experiment, the value of α is varied from 0.1 to 1.5.

5.3 Parameter Experiment

We conducted the parameter experiment to investigate the impact of the parameter α on the accuracy of FCPAN. Figures 3 and 4 shows the results of the parameter experiment. On the real-world datasets, FCPAN obtains the maximum values of ONMI and EQ when $\alpha \in [0.7, 1.1]$. Moreover, the common interval of α for FCPAN to reach the highest ONMI and EQ values is $[0.7, 1.0]$ on the artificial dataset. A great value of α will lead to those k-cliques with a high similarity that cannot be merged. On the contrary, a too-small value of α will lead to the excessive merging of k-cliques. We selected the value of α in the range of $[0.7, 0.1]$ and reported the best results in the remaining experiments.

5.4 Consistency Experiment

We verified the correctness of the FCPAN by comparing the accuracy and the detected communities with SCPAN on the same real-world and artificial datasets. As shown in Fig. 5, the ONMI and EQ values of FCPAN are identical to those of SCPAN on all datasets with varying numbers of participants. Figure 6 also shows that the communities detected by FCPAN on the parklot dataset and D1 network with $\mu = 0.5$ are all identical to those detected by SCPAN. Vertices in the same communities are drawn in the same colors. Consequently, the experimental results indicate that FCPAN is a precise extension of the SCPAN. Vertex perturbation strategy without generating redundant k-cliques does not affect the correctness of the clique percolation process. Furthermore, the property of homomorphic encryption makes the result of ciphertext calculation consistent

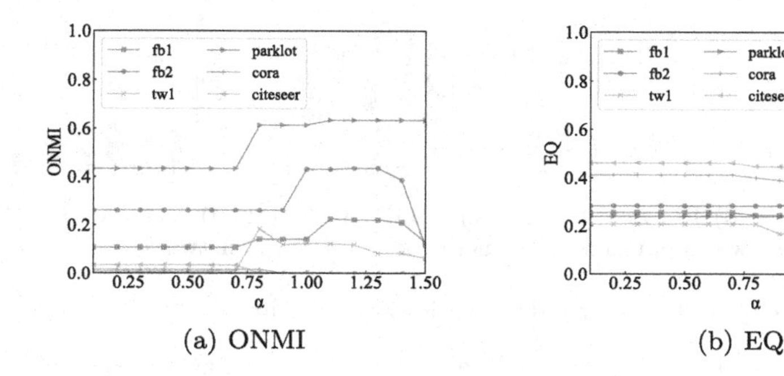

(a) ONMI (b) EQ

Fig. 3. Results of the parameter experiment on the real-world datasets

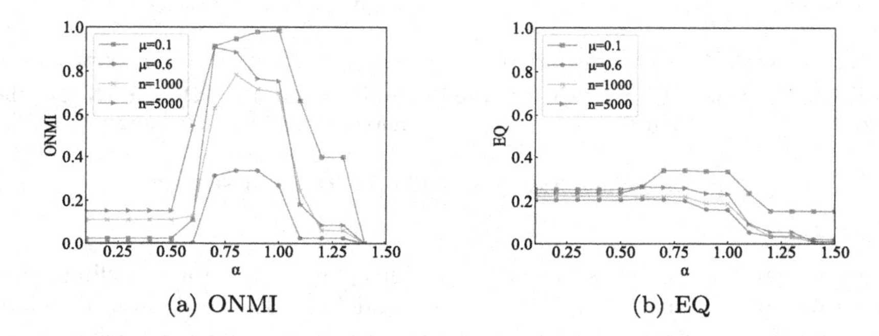

(a) ONMI (b) EQ

Fig. 4. Results of the parameter experiment on the artificial datasets

with that of plaintext, which ensures that there is no accuracy loss when calculating attribute similarity. Based on the experimental results, we can conclude that the FCPAN does not cause accuracy loss in privacy-preserving community detection.

5.5 Ablation Experiment

We compared the accuracy of FCPAN with the simple Distributed Clique Percolation algorithm on Attributed Networks (DCPAN) without federated learning to verify the effectiveness of our algorithm. In DCPAN, each participant runs the SCPAN algorithm to detect communities on his local network instead of cooperating. As showing in Fig. 7, on *tw*1 and D1 network with $\mu = 0.5$, the ONMI and EQ values of FCPAN both are higher than those of DCPAN as the number of participants changes. With the mechanism of federated learning, each participant can complete k-cliques information corresponding to his overlapping vertices and obtain attribute similarity of vertices in different participants. However, DCPAN achieves unsatisfactory results on detected communities due to lacking k-clique information where each participant's overlapping vertices contain. The above experimental results reveal the advantage of federated learning.

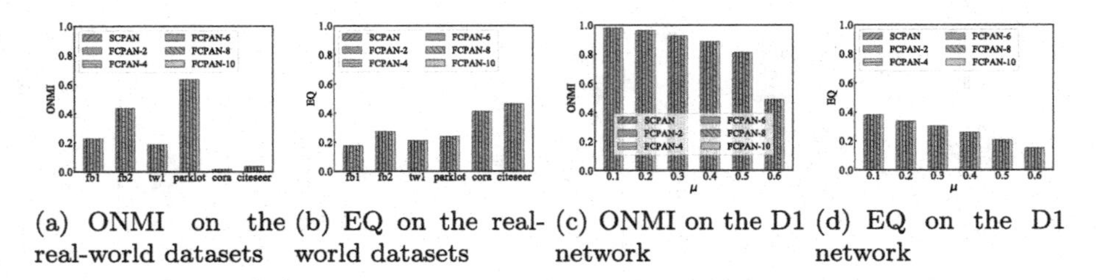

(a) ONMI on the real-world datasets (b) EQ on the real-world datasets (c) ONMI on the D1 network (d) EQ on the D1 network

Fig. 5. Accuracy of the consistency experiment

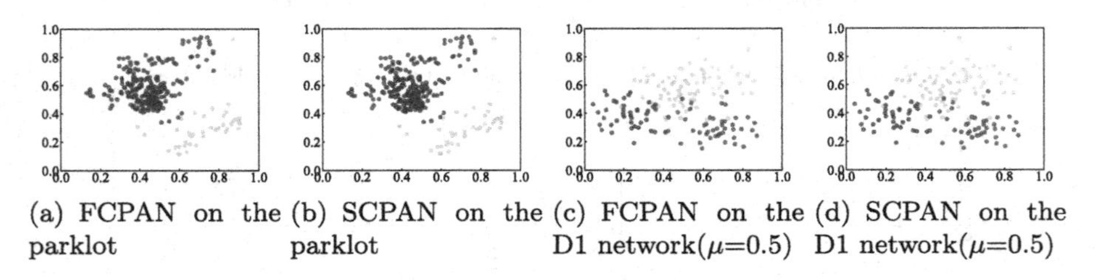

(a) FCPAN on the parklot (b) SCPAN on the parklot (c) FCPAN on the D1 network(μ=0.5) (d) SCPAN on the D1 network(μ=0.5)

Fig. 6. Communities detected by FCPAN and SCPAN

Federated learning combines the data of various participants for modeling, which can obtain better results than independent modeling with local datasets. At the same time, it can protect the data owned by each participant from being leaked.

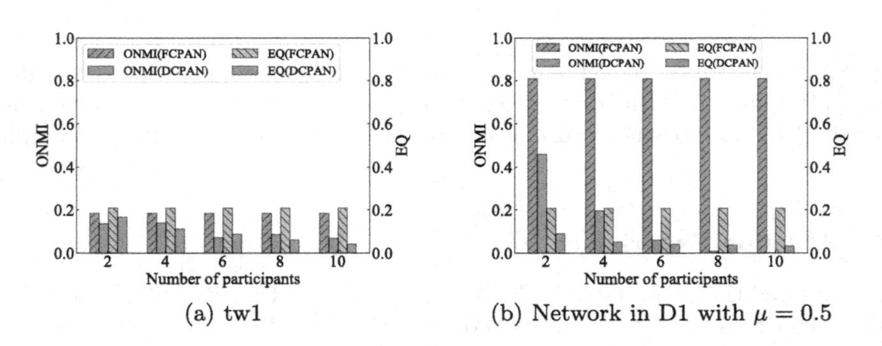

(a) tw1 (b) Network in D1 with $\mu = 0.5$

Fig. 7. Accuracy of FCPAN and DCPAN

6 Conclusions

In this paper, we first propose a clique percolation algorithm to detect communities on attributed networks. The designed clique similarity metric improves the quality of clique percolation. Second, the federated clique percolation algorithm is developed for privacy-preserving overlapping community detection on distributed attributed networks. Perturbation strategy and homomorphic encryption protect network privacy strictly under the semi-honest model without accuracy loss. Comprehensive experiments on real-world and artificial datasets prove the correctness and efficiency of our algorithms. In the future, We will study federated transfer learning and develop federated network data mining algorithms on it.

Acknowledgements. This work was supported by the National Natural Science Foundation of China under Grant No. 61672159, No. 61672158, No. 61300104 and No. 62002063, the Fujian Collaborative Innovation Center for Big Data Applications in Governments, the Fujian Industry-Academy Cooperation Project under Grant No. 2017H6008 and No. 2018H6010, the Natural Science Foundation of Fujian Province under Grant No. 2018J07005, No. 2019J01835 and No. 2020J05112, the Fujian Provincial Department of Education under Grant No. JAT190026, the Fuzhou University under Grant 510872/GXRC-20016 and Haixi Government Big Data Application Cooperative Innovation Center.

References

1. Brickell, J., Shmatikov, V.: Privacy-preserving graph algorithms in the semi-honest model. In: Roy, B. (ed.) ASIACRYPT 2005. LNCS, vol. 3788, pp. 236–252. Springer, Heidelberg (2005). https://doi.org/10.1007/11593447_13
2. Casas-Roma, J., Herrera-Joancomartí, J., Torra, V.: k-Degree anonymity and edge selection: improving data utility in large networks. Knowl. Inf. Syst. **50**(2), 447–474 (2017)
3. Chiba, N., Nishizeki, T.: Arboricity and subgraph listing algorithms. SIAM J. Comput. **14**(1), 210–223 (1985)
4. De Cristofaro, E., Tsudik, G.: Practical private set intersection protocols with linear complexity. In: Sion, R. (ed.) FC 2010. LNCS, vol. 6052, pp. 143–159. Springer, Heidelberg (2010). https://doi.org/10.1007/978-3-642-14577-3_13
5. Dev, H.: Privacy preserving social graphs for high precision community detection. In: Proceedings of the 2014 ACM SIGMOD International Conference on Management of Data (SIGMOD 2014), pp. 1615–1616. ACM (2014)
6. Farkas, I., Ábel, D., Palla, G., Vicsek, T.: Weighted network modules. New J. Phys. **9**(6), 180 (2007)
7. Fortunato, S., Hric, D.: Community detection in networks: a user guide. Phys. Rep. **659**, 1–44 (2016)
8. Goddard, M.: The EU General Data Protection Regulation (GDPR): European regulation that has a global impact. Int. J. Mark. Res. **59**(6), 703–705 (2017)
9. Gonzalez, J.E., Low, Y., Gu, H., Bickson, D., Guestrin, C.: Powergraph: distributed graph-computation on natural graphs. In: Proceedings of the 10th USENIX Symposium on Operating Systems Design and Implementation (OSDI 2012), pp. 17–30. USENIX Association (2012)
10. Huang, B., Wang, C., Wang, B.: NMLPA: uncovering overlapping communities in attributed networks via a multi-label propagation approach. Sensors **19**(2), 260 (2019)

11. Hwang, C.M., Yang, M.S., Hung, W.L.: New similarity measures of intuitionistic fuzzy sets based on the Jaccard index with its application to clustering. Int. J. Intell. Syst. **33**(8), 1672–1688 (2018)
12. Ji, T., Luo, C., Guo, Y., Wang, Q., Yu, L., Li, P.: Community detection in online social networks: a differentially private and parsimonious approach. IEEE Trans. Comput. Soc. Syst. **7**(1), 151–163 (2020)
13. Kasgar, A.K., Agrawal, J., Shahu, S.: New modified 256-bit MD5 algorithm with SHA compression function. Int. J. Comput. Appl. **67**(7), 1804–6 (2012)
14. Kumar, S., Kumar, P., Bhasker, B.: Privacy preserving graph publishing using fuzzy set. In: 2016 12th International Conference on Natural Computation, Fuzzy Systems and Knowledge Discovery (ICNC-FSKD), pp. 1233–1238. IEEE (2016)
15. Kumpula, J.M., Kivelä, M., Kaski, K., Saramäki, J.: Sequential algorithm for fast clique percolation. Phys. Rev. E **78**(2), 026109 (2008)
16. Lalitha, A., Kilinc, O.C., Javidi, T., Koushanfar, F.: Peer-to-peer federated learning on graphs. arXiv preprint arXiv:1901.11173 (2019)
17. Lancichinetti, A., Fortunato, S., Radicchi, F.: Benchmark graphs for testing community detection algorithms. Phys. Rev. E **78**(4), 046110 (2008)
18. Liu, G., Wong, L., Chua, H.N.: Complex discovery from weighted PPI networks. Bioinformatics **25**(15), 1891–1897 (2009)
19. Liu, K., Terzi, E.: Towards identity anonymization on graphs. In: Proceedings of the 2008 ACM SIGMOD International Conference on Management of Data (SIGMOD 2008), pp. 93–106. ACM (2008)
20. Liu, X., Lu, R., Ma, J., Chen, L., Qin, B.: Privacy-preserving patient-centric clinical decision support system on Naive Bayesian classification. IEEE J. Biomed. Health Inf. **20**(2), 655–668 (2015)
21. McDaid, A.F., Greene, D., Hurley, N.: Normalized mutual information to evaluate overlapping community finding algorithms. arXiv preprint arXiv:1110.2515 (2011)
22. Mei, G., Guo, Z., Liu, S., Pan, L.: SGNN: a graph neural network based federated learning approach by hiding structure. In: Proceedings of the 2019 IEEE International Conference on Big Data (Big Data 2019), pp. 2560–2568. IEEE (2019)
23. Mougel, P.-N., Rigotti, C., Gandrillon, O.: Finding collections of k-clique percolated components in attributed graphs. In: Tan, P.-N., Chawla, S., Ho, C.K., Bailey, J. (eds.) PAKDD 2012. LNCS (LNAI), vol. 7302, pp. 181–192. Springer, Heidelberg (2012). https://doi.org/10.1007/978-3-642-30220-6_16
24. Nguyen, H.H., Imine, A., Rusinowitch, M.: Detecting communities under differential privacy. In: Proceedings of the 2016 ACM on Workshop on Privacy in the Electronic Society (WPES 2016), pp. 83–93. ACM (2016)
25. Palla, G., Derényi, I., Farkas, I., Vicsek, T.: Uncovering the overlapping community structure of complex networks in nature and society. Nature **435**(7043), 814–818 (2005)
26. Shen, H., Cheng, X., Cai, K., Hu, M.B.: Detect overlapping and hierarchical community structure in networks. Phys. A: Statist. Mech. Appl. **388**(8), 1706–1712 (2009)
27. Yang, Q., Liu, Y., Chen, T., Tong, Y.: Federated machine learning: concept and applications. ACM Trans. Intell. Syst. Technol. **10**(2), 1–19 (2019)
28. Zheng, L., Zhou, J., Chen, C., Wu, B., Wang, L., Zhang, B.: Asfgnn: automated separated-federated graph neural network. Peer-to-Peer Netw. Appl. **14**(3), 1692–1704 (2021)

A New Academic Conference Information Management System Based on Social Network

Wen Xiao, Liantao Lan(✉), Jiongsheng Guo, Ronghua Lin, Yu Lai, and Yong Tang

South China Normal University, Guangzhou, China
lanlt@m.scnu.edu.cn

Abstract. SCHOLAT is a free, massive and comprehensive academic social network platform, which aims to realize reliable data sharing. In this paper, a new conference information management system based on SCHOLAT is designed and developed. In addition to containing typical functions of conference information management system, the system also provides individual services, including convenient online conference space, personalized domain, association with SCHOLAT ecosystem and customized access control. A hybrid classification algorithm for SCHOLAT user's academic emails is proposed to enhance the high quality services of the system. Benefit from convenient and helpful services, the conference system has grown rapidly in popularity since it was deployed and has been applied to lots of academic conferences up to now.

Keywords: Social network · Conference system · SVM · Text processing · SCHOLAT · Classification algorithm

1 Introduction

SCHOLAT [21] is a free, massive and comprehensive academic social network platform, which aims to realize reliable data sharing. At present SCHOLAT has more than 100,000 active users and produces more than 100 million academic information every year. The platform provides a series of high quality service for scholars, including academic ecosystem, cooperative teaching, academic portal, knowledge graph, "SCHOLAT+", scientific research, cooperation and so on. Here scholars represent not only teachers, but also students and all the people who love leaning.

Surfing on SCHOLAT with a registered account, one can build personalized academic homepage of your own, in which you can manage your courses, teams, academic archives, friend circles, academic moments and messages. For example, Fig. 1 shows the personal homepage of Professor Tang, who is the founder of SCHOLAT. It is worth noting that all the registered scholars have been strictly certified in several ways and important information are required to be verified.

In this paper, a novel academic conference information management system is designed and developed. The system is an already online project under "SCHOLAT+" and aims to provide good, convenient and efficient services for conference organizers and

Y. Sun et al. (Eds.): ChineseCSCW 2021, CCIS 1492, pp. 267–280, 2022.
https://doi.org/10.1007/978-981-19-4549-6_21

participants. The model of the system consists of three modules: academic email classification module, user association with SCHOLAT module, and conference navigation module.

Besides the typical functions of conference information management system, the system also provides some individual services, mainly including convenient online conference space, personalized domain, association with SCHOLAT ecosystem and customized access control. In addition, a hybrid classification algorithm for academic emails is proposed to enhance the high quality services of the system. Since the system deployed and launched three months ago, it has been successfully applied to thousands of academic conferences domestic and abroad.

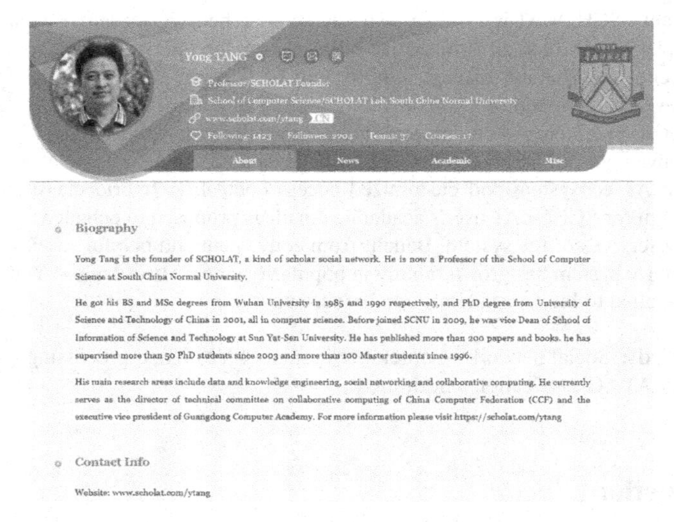

Fig. 1. Professor Tang's personal homepage

The remainder of this paper is organized as follows. Section 2 reviews the related work about conference information systems. In Sect. 3, the model design of our conference information management systems is described in detail. Then the logic of the model and three modules is presented. A hybrid classification algorithm based on SVM and TF-IDF algorithms is also designed to improve the system's performance. Section 4 gives the implement of the system. A brief conclusion is drawn in Sect. 5.

2 Related Work

In recent years, with the development of the internet and economy, the number of academic conferences has grown rapidly and the era of information explosion has come. There are thousands of academic conferences held in China every year. Therefore, a convenient and efficient academic conference information management system (**ACIMS**) is of great significant.

There are currently many ACIMSs on the internet, for example, EasyChair [17], Aconf [18], Allconferences [19] and so on. EasyChair is a famous ACIMS, which provides convenient and efficient conference management services. Aconf supports academic conference information sharing via mainstream foreign social media. Allconferences supports for sharing academic information by email and conference reminder function. In addition to, Twitter [9] introduced social elements into conference information collection. However, the known conference information management systems can not provide comprehensive navigation function, reliable data and personalized services for users.

Despite the traditional academic conference what mentioned above, the function of processing academic conference information has also been extended and enhanced. For example, Hamed [2] applied probabilistic topic modeling based on Gibbs sampling algorithms for semantic mining from eight conference publications in computer science from the DBLP dataset. Then Brusilovsky [7] linked information and people in a social system for academic conferences. Hornick [6] introduced recommendation engine to address the recommendation conference sessions problem from a new disjoint set. Bert [10] decided to perform an analysis of Twitter's use during the 7th European Public Health (EPH) Conference. Wang [3] proposed the power equipment fault information acquisition system based on internet of things. Jie [15] proposed conference navigation system construction based on standardized documentations. However, the functions of the above systems are relatively single, which cannot really apply to the systems.

Email is an efficient communication channel for a larger digital campaign. The significance of email can be seen with the rapid growth in volume of emails [1]. According to Radicati Group's report, in the first quarter of 2017 there were average 269 billion emails sent per day [20]. Utilization of email lies at the heart of the information society. Hu [13] built up a Chinese spam filtering approach with semantics-based text classification technology. Machine learning algorithms have been widely applied in text classification and usually perform well. Riato [16] used the support vector machine algorithm to improve the accuracy of text classification. In the early days, **SVM** [4, 12], **TF-IDF** [5], **TextRank** [5, 11] and **Structure-tags** [15] was deployed in text classification as a method to enhance email classification performance. Later, Venkatraman [1] used semantic similarity approach, Borg [8] used machine learning and word embeddings and Rianto [16] used stemming method to strength the effects. Academic email data contains many important academic information which can be used to facilitate the conference management. In the paper, academic email classification is employed to extract the important information.

3 Model Design

The model of our system consists of three modules: academic email classification module, user association with SCHOLAT module, and conference navigation module. The model framework of the system is shown in Fig. 2.

The academic email classification module classifies all the academic emails by a hybrid classification algorithm based on **TF-IDF** and **SVM**. With the classification, academic conferences, courses, teams and potential friends are recommended through SCHOLAT platform.

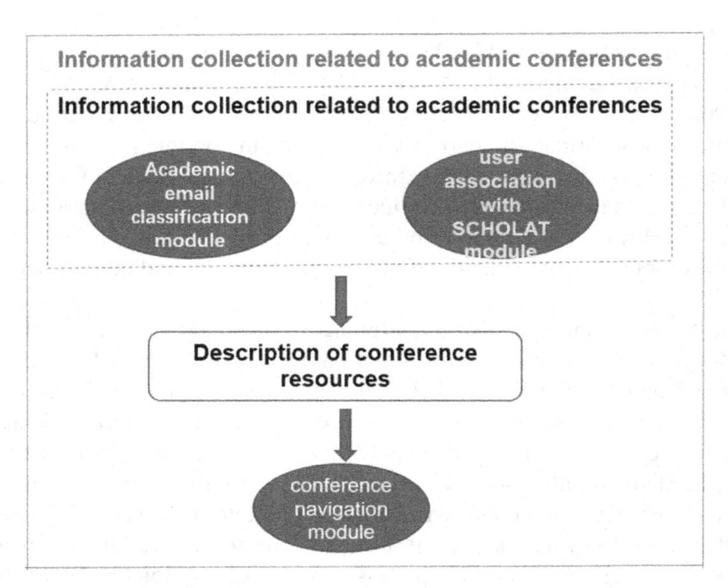

Fig. 2. The model framework

The user association with SCHOLAT module aims to collect and exchange academic conference information with SCHOLAT. Conference organizers and participants can export themselves' data on SCHOLAT that conference required, and SCHOLAT recommends conferences to its users.

The conference navigation module provides conference searching and exhibition services. Users can apply for a conference with personalized domain and customized access control. Once the application is approved, the conference will be assigned an individual conference space and association.

3.1 Academic Email Classification Module

Academic emails from users in social networks are an alternative source of information about academic conferences. Two problems need to be solved. One is how to identify whether the emails are related to academic conferences, another is how to extract the key information of academic conferences from the emails. This section describes the classification module of academic email collection from scholar network. The diagram of academic email classification module is shown in Fig. 3.

(1) Email Type Classification

We first introduce the academic email screening method based on keyword rules. The method is implemented by a form of probabilistic inference. Each rule is given a different score that trained by the assignment algorithm on the email dataset. The results from which the sum of the triggered rule scores compare with the set threshold value determine the email category. The following will introduce the keyword rule-based academic email screening in detail from three aspects. The first is how to select the candidate keywords.

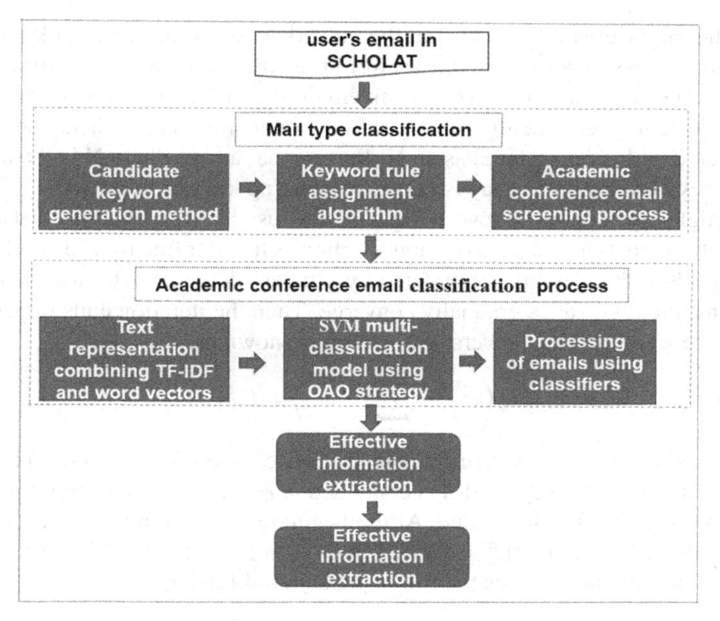

Fig. 3. The email classification module

The second is how to use the assignment algorithm to assign values to keyword rules. The third is how to use the keyword rules to filter the target emails.

Candidate Keyword Generation Method. Identified keywords from the email dataset can represent the characteristics of the academic conference text. Firstly, it requires data preprocessing of the dataset. The data preprocessing includes word separation, deactivation and stemming extraction. After completed preprocessing the keywords are selected from the dataset. The criteria for word separation includes two aspects, the one that candidate word is denoted a word with high frequency contained the text of academic conferences, and the other needs to consider eliminating the influence of common words. The specific three steps are as follows.

1) All academic conference email documents are combined into one document, and the N words with the highest word frequency in the document are selected as candidate words.

2) All the email documents are classified into two categories: academic and non-academic. Let AC_w be the number of academic email documents containing word w and NAC_w the number of non-academic email documents containing word w.

3) A word w is chosen as a candidate keyword if

$$AC_w/(AC_w + NAC_w) > T\% \tag{1}$$

where T is a predetermined threshold value.

Keyword Rule Assignment Algorithm is Explained. After a candidate keyword is identified, it is determined as a rule. The criterion for whether the rule is triggered is whether the target email text contains the candidate keyword. The work we need to accomplish is to assign a score to the keyword rule and calculate a score for each email text. The final score determines whether an email is an academic email or not.

A simple neural network algorithm generates keywords weights. The rule triggering case of each email is represented as an N-dimensional array, where N is the number of candidate keywords and the array is the corresponding keyword rule set.

The single-layer neural network algorithm is used in this module that applies a transformation function and an activation function as transfer functions. In each iteration of the algorithm, the weights are updated by gradient descent. After a certain number of iterations, the loss values gradually converge. Then the iteration ends to get the final keyword rule weights. The conversion function is shown as:

$$f_w\left(x^i\right) = \sum\nolimits_{j=1}^{n} w_j x_j^i + bias, \tag{2}$$

where suppose r_i is the keyword rule and m_i is the user email. Let w_j represents the value of keyword rule r_j. If the keyword r_j is contained in m_i, it is $x_j^i = 1$, otherwise is $x_j^i = 0$. The *bias* is the uniform offset value. At the beginning, w_j is randomly generated. The bias is initialized between -0.5 and 0.5.

The activation function uses the following sigmoid function:

$$h_w\left(x^i\right) = \frac{1}{1 + e^{-f_w(x^i)}} \tag{3}$$

The degree of error is calculated using cross entropy as a loss function, and the calculation is shown as follow.

$$loss(w) = -\frac{1}{m}\sum\nolimits_{i=1}^{m} \log_{10}\left(h_w\left(x^i\right)\right) + \left(1 - y^i\right)\log_{10}\left(1 - h_w\left(x^i\right)\right) \tag{4}$$

The weights are updated using the gradient descent method for each iteration. The weight w_j and the *bias* update is shown as follow.

$$w_j = w_j + \frac{1}{m}\sum_{i=1}^{m}\left(f_w\left(x^i\right) - y^i\right) * rate \tag{5}$$

$$bias = bias + \frac{1}{m}\sum\nolimits_{i=1}^{m}\left(f_w\left(x^i\right) - y^i\right) * rate \tag{6}$$

where the *rate* indicates the learning rate of the weight update. The appropriate number of iterations and learning rate are determined experimentally.

After the algorithm completes the iteration, the score of the keyword rule is calculated.

$$eScore_j = threshold * \frac{2w_j}{1 - 2bias} \tag{7}$$

where the *threshold* indicates the score threshold. When the final score exceeds this threshold, the email is judged as an academic conference email. We delete the keyword rules with a score of 0 to get the final keyword rule set, the final format of the single rule is {keyword: score}.

Academic Conference Email Screening Process. For unknown emails, the total score is initialized to 0 in the data preprocessing stage. The words are taken out after the word separation and see if they meet the keyword rules. The total score is calculated, and then it is judged whether the threshold is exceeded. If it is exceeded, it is identified as academic conference email, otherwise as non-academic conference email.

(2) Academic Conference Field Classification Process

The classification algorithm is based on statistical. The core method is to extract the features of the classified data and choose the best matching method according to the features. The following will introduce the professional domain recognition scheme for academic conference emails from three aspects:

1. Text representation combining TF-IDF and word vectors,
2. SVM multi-classification model using OAO strategy,
3. Processing of academic conference emails using classifiers.

Text Representation Combining TF-IDF and Pre-trained Word Vectors. It can preserve all semantic features of word vectors and importance features of words of email documents. Firstly, the email documents are preprocessed. Then, the TF-IDF value is calculated for each word in the email documents word list. Let *tf(t,d)* be the frequency of word t occurring in email documents d. N is the total number of training texts and n_t is the number of documents containing word t. The denominator is called the normalization factor. Meanwhile, for each word in the list of document words, the corresponding representation is queried from a pre-trained word vector dictionary. If it does not exist, it is filled with the mean vector of the dictionary. The word vector dimension is predefined. The TF-IDF value $w(t, d)$ of word t in the email documents d is defined to be:

$$w(t,d) = \frac{tf(t,d)*\log(N/n_t + 0.01)}{\sqrt{\sum_{t\in d}\left[tf(t,d)*\log(N/n_t + 0.01)\right]^2}} \tag{8}$$

Finally, $w(t, d)$ represents the level of importance that the word t contained in email documents d and wv_t represents word vector of word t. the document vectors v are calculated:

$$v = \sum\nolimits_{t\in d} w(t,d) * wv_t \tag{9}$$

SVM (Support Vector Machines) Multi-classification Model Using OAO Strategy. We use the classifier to solve the multiple classification problem. SVM is constructed between any two classes of samples using a one-to-one (OVO) approach. To avoid skewing of the data, one sample at a time is selected as a positive or negative sample. The category number determines the classifier, if categories is k, then SVM model is constructed in $k*(k-1)/2$. When classifying an unknown sample, all the sub-classifiers need to be tested. The highest score is considered the category of the test data after accumulating scores for each category.

Processing of Academic Conference Emails Using Classifiers. The flowchart is shown in Fig. 4. After preprocessing data, the word text is converted into vectors by TF-IDF and word vectors. Then the trained SVM model is used to predict the professional domain classification.

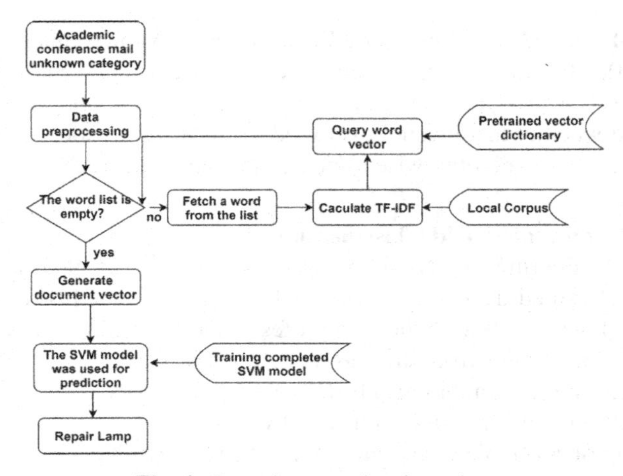

Fig. 4. Domain extraction flow chart

(3) Effective Information Extraction

Extracting effective information as email label can help users get rid of noise and get the needed information directly. The core is to identify the data that users are interested in. The data resource is the unstructured or semi-structured text, the goal is that transform it into a more structured and semantically clear format. We use regular expression to extract effective information and achieve semi-automatic extraction. Regular expression is a powerful tool for extracting the string data in computer operation and verification. The extracted information includes conference abbreviation, date information, conference location and other information.

The dates conclude a start date, essay deadlines date, hire general date, the final deadline date and the date of the conference held. The date format in English words or Chinese words are needed be noticed, the final date result is expressed as year/month/day. To extract the place where the conference was held and get the formatted address information in the email, you need to enter a dictionary of countries, locations, etc. in advance. Other valuable data include the website of the conference organizer, contact email, other contact details, and the level of the conference which is classified according to "SCI", "EI" and "ISTP".

3.2 Information Collection Based on SCHOLAT

In the real process of data collection, it exists many problems, such as the diversity of data sources, the guarantee of data reliability and repeated data. SCHOLAT is an important information source of conference organizational structure data (The parameters of conference interaction refer to the positioning and operation of academics' roles in the organizational structure of the conference). The detailed parameters of specific scholars are shown as follow Table 1. The detailed attributes of scholars mainly include conference interaction parameters and the characteristics of scholars' social parameters. The information can help the participants to acknowledge each other and establish relationships for communication.

Scholar Attribute
The specific attribute is shown in Table 1.

Table 1. The information of organization

Interaction feature	Personal attributes
User	Username
Role	Contact
Permission	Affiliated colleges

a) Conference interaction parameter data: The data records the conference management role of each user, and the role information is a specific administrator identity under the multi-layer administrators. The relationship between scholars and conferences is reflected in the role of the organization to which the scholar belongs and the dynamic power distribution. The relationship among conference interaction attributes is dynamic.

b) Scholars themselves have a variety of attributes, such as the name, university the scholar belongs to, the contact information and so on. The scholar information are collected from SCHOLAT. The process is shown in Fig. 5.

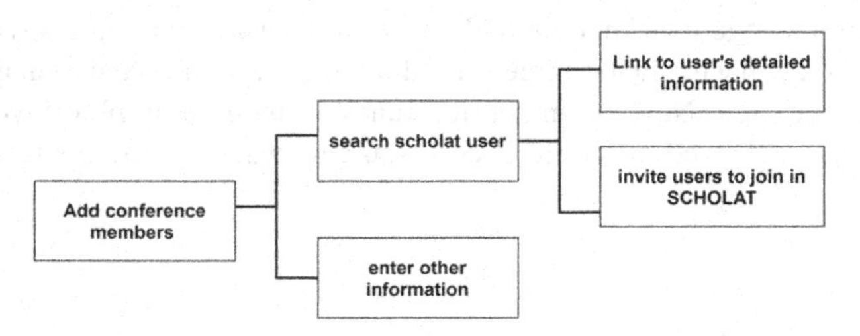

Fig. 5. User's data linked with SCHOLAT

Information Acquisition Manner
Multi-parameter uncertainty is a key problem in complex system design. Whether and how to obtain more information to reduce uncertainty are important for system-level decision-making. The key is that the cost of information and its influence on the overall design effectiveness are different. To improve the overall utility and the sensitivity of each conference parameter, at the same time reduce the cost of information acquisition, we adopt the information acquisition method based on social network.

a) An information acquisition way based on social network: it obtains the specific parameter data of related scholar data through UserID which connect with SCHOLAT. It is accessed to get low cost and high accuracy.

b) Heterogeneous data acquisition: lots of unstructured data are collected, such as academic developments in text and conference agenda documents in PDF format. These data contain a large amount of fresh and authoritative information. After screening and integration, the data is associated with the conference organizational structure information.

3.3 Conference Navigation Module

Academic conference resources include conference names, organization names and other contents. Due to the complexity of academic conference resource, it is necessary to standardize academic information documents to ensure that the data normalization is consistent and normative. Finally, it will get a powerful academic conference navigation system with high data quality, which can uniformly represent and cluster related conference contents.

Navigation Services. In this section, it describes attribute in uniform and standardized way, which includes conference name, organization name, the name of the organizers, home, the conference time and conference place. Meanwhile it expresses the conference association and clustering, such as the continuous relationship (continuous conference), parallel relationship (such as the joint conference). For academic conference resources, the metadata description scheme is used to express the attributes of the various specifications. For name, conference name specification document is used to a unified expression and the various conference relationship. It achieves the multi-dimensional expression and correlation clustering by mining of academic conference resources deeply.

As shown in the Fig. 6, the module framework of the navigation services is divided into navigation management module and navigation service module. Navigation management aims to form a navigation tree according to certain rules and manage the layout of the navigation page. Navigation service function include graphic layout and content layout, that organized all conference resources wisely to the navigation page. It

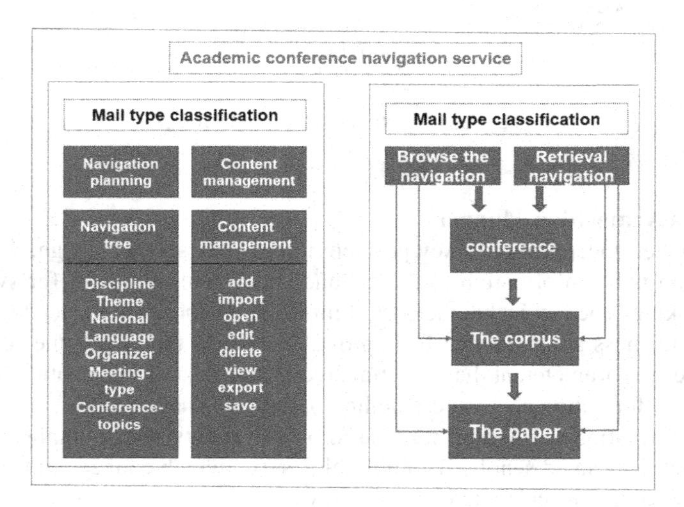

Fig. 6. Navigation service formation mechanism

organizes and clusters the resource information of all academic conferences or some of the retrieved academic conferences, and then generates navigation trees, finally realizing multi-dimensional and multi-level navigation. Navigation management is used to define, name, set and adjust the navigation tree. A navigation tree is a data object defined by the navigation system. The navigation tree is generated by the navigation model based on the relevant attributes entered. The content management function is mainly used to pre-customize navigation page templates, describe navigation pages. We set the location and layout of content such as headings, category trees, tables in a scripting language that can be recognized and understood.

Conference Resource Description. Academic conference resources mainly involve six entity element sets. It includes conference, conference proceedings, conference papers, conference organization structure and normative document element sets. The description framework is shown in the Fig. 7, which objectively describes the mutual relations among each entity element set. The implementation of multi-level navigation model of academic conference navigation is based on these relations.

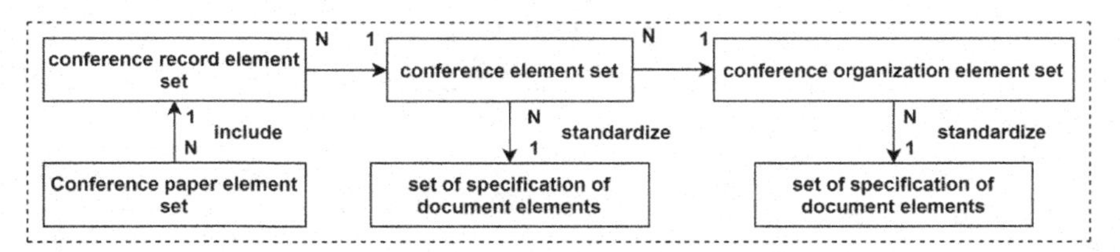

Fig. 7. Conference resource metadata description

After defining the elements of the entity element set, the specific constructing process is divided into three steps. Firstly, data are extracted, including information of conference organization, papers and related conference. Then, data cleaning is performed to remove stop words from all the papers by using a stop list. Finally, the entity sets are concatenated according to the relationships, then generate a conference criteria library. The library includes a conference name criteria file, a name criteria file for the main organizers of the conference, a name criteria file for the organizers of the conference, a standard name/original name comparison file and a conference relationship file, etc.

4 System Implementation

The academic conference system proposed in this paper is designed and implemented based on SCHOLAT. Now the model has been practiced and verified. The system has been successfully deployed and launched three months ago. It applied B/S system architecture, using Java as the main development language and database management system called MYSQL. The operating environment requirements were not harsh, the system could be used in the Intranet stand-alone or the Internet browser. The specified implement of system is shown in Figs. 8 and 9.

Fig. 8. Conference navigation service module

Fig. 9. HCC 2021 conference details page display

Through the above research, this system provided a kind of accurate and practical conference information management tool for most scholars. Through an online questionnaire survey on whether users have facilitated academic communication through the system and whether users are satisfied with the system. The final survey results showed that users are satisfied with the collection and service functions. After logging into the home page, users could see the formatted navigation of conferences according to various dimensions such as conference theme and year, which made it is easy to search the conference they need quickly. After entering the conference details page, users could view information about the conference theme, key dates (such as submission time, deadline, acceptance time), contact email and information about the conference organization. The conference organization information is a complex data structured, which was shared with the SCHOLAT network to guarantee the authenticity, authority and freshness of the scholars' data. At the same time, a security permission mechanism was set up in the background. It set permissions flexibly according to the user's role setting in multiple layers, blocked illegal operations and ensured the safe operation of its system. It could be seen that the system has a strong practicality. This system could deeply mine and organize the information of academic conferences and realize the standardized navigation of conference information. It has good application prospects.

5 Conclusion

This paper proposed a conference information collection and service model based on social network. Firstly, SCHOLAT-based information collection module improved the efficiency of setting up conferences data. Then we had established a navigation service for conferences search, which not only realized the associated clustering of conference information, but also accomplished multi-dimensional information retrieval. In addition, the academic email sorting feature improved the system performance. The feedback from users showed that the system had good accuracy and practicability, which could better help scholars manage and retrieve the information of scientific conference. In the future work, the association and clustering relationship between conferences can be applied for further study, including the connection relationship of academic conferences, constructing the knowledge map of academic conferences and making conference portraits, so as providing better services for scholars.

Acknowledgement. This work was supported by the National Natural Science Foundation of China under Grant U1811263, Grant 61772211(Y. Tang), NSFC under Grant 11901210 and China Postdoctoral Science Foundation under Grant 2019M652924 (L. Lan).

References

1. Venkatraman, S., Surendiran, B., Kumar, P.: Spam e-mail classification for the Internet of Things environment using semantic similarity approach. J. Supercomput. **76**(2), 756–776 (2020)
2. Jelodar, H., Wang, Y., Xiao, G., et al.: Recommendation system based on semantic scholar mining and topic modeling on conference publications. Soft. Comput. **25**(1), 3675–3696 (2021)
3. Wang, R., Li, M.: Power equipment fault information acquisition system based on Internet of Things. EURASIP J. Wirel. Commun. Netw. **2021**(1), 1–22 (2021). https://doi.org/10.1186/s13638-021-01942-2
4. Vapnik, V.: SVM method of estimating density, conditional probability, and conditional density. In: IEEE International Symposium on Circuits and Systems, pp. 749–752 (2000)
5. Lu, Y., Zhang, P., Zhang C.: Research on news keyword extraction technology based on TF-IDF and TextRank. In: IEEE/ACIS 18th International Conference on Computer and Information Science (ICIS), pp. 452–455 (2019)
6. Hornick, M.F., Tamayo, P.: Extending recommender systems for disjoint user/item sets: the conference recommendation problem. IEEE Trans. Knowl. Data Eng. **24**(8), 1478–1490 (2012)
7. Brusilovsky, P., On, S.J., López, C., Parra, D., Jeng, W.: Linking information and people in a social system for academic conferences. New Rev. Hypermedia Multimedia **23**(2), 81–111 (2017)

8. Borg, A., Boldt, M., Rosander, O., et al.: E-mail classification with machine learning and word embeddings for improved customer support. Neural Comput. Appl. **33**(6), 1881–1902 (2021)
9. Lee, M.K., Yoon, H.Y., Smith, M., et al.: Mapping a Twitter scholarly communication network: a case of the association of internet researchers' conference. Scientometrics **112**(2), 767–797 (2017)
10. Bert, F., Paget, D.Z., Scaioli, G.: A social way to experience a scientific event: Twitter use at the 7th European public health conference. Scand. J. Public Health **44**(2), 130–136 (2015)
11. Wen, Y., Hui, Y., Zhang, P.: Research on keyword extraction based on Word2Vec weighted TextRank. In: 2016 2nd International Conference on Computer and Information Science (ICIS), pp. 452–455 (2016)
12. Ola, A., Nizar, B.: A study of spam filtering using support vector machines. Artif. Intell. Rev. **34**(1), 73–108 (2010)
13. Hu, W., Du, J., Xing, Y.: Spam filtering by semantics-based text classification. In: Proceedings of the 8th International Conference on Advanced Computational Intelligence, pp. 89–94 (2016)
14. Wenniger, G., Dongen, T, V., Aedmaa, E., et al.: Structure-tags improve text classification for scholarly document quality prediction. In: SDP@EMNLP, pp. 158–167 (2020)
15. Jie, Z., Yang, D.Q., Xing, W.: Research of conference navigation system construction based on standardized documentations. Digit. Libr. Forum **7**, 20–24 (2015)
16. Rianto, Mutiara, A.B., Wibowo, E.P., et al.: Improving the accuracy of text classification using stemming method, a case of non-formal Indonesian conversation. J. Big Data **8**(1), 16 (2021)
17. The EasyChair conference system. http://easychair.org. Accessed 14 June 2021
18. The Allconferences conference management system. https://www.allconferences.com/. Accessed 14 June 2021
19. The Aconf conference management system. https://www.aconf.org/. Accessed 14 June 2021
20. The History of Spam, Switzerland (2014). https://www.internetsociety.org/sites/default/files/HistoryofSpam.pdf
21. Tang, F., Jia, Z., He, C., et al.: SCHOLAT: an innovative academic information service platform. In: Cheema, M., Zhang, W., Chang, L. (eds.) Databases Theory and Applications, ADC 2016. LNCS, vol. 9877, pp. 453–456. Springer, Cham (2016). https://doi.org/10.1007/978-3-319-46922-5_38

A Full Information Enhanced Question Answering System Based on Hierarchical Heterogeneous Crowd Intelligence Knowledge Graph

Lei Wu, Bin Guo(✉), Hao Wang, Jiaqi Liu, and Zhiwen Yu

School of Computer Science, Northwestern Polytechnical University, Xi'an 710072, China
{leiwu,wanghao456}@mail.nwpu.edu.cn, {guob,jqliu, zhiwenyu}@nwpu.edu.cn

Abstract. With the development of deep learning technology, generative question answering models based on neural networks have gradually become a mainstream research direction in academia and industry. The current question answering models fail to make full use of the multi-level knowledge embedded in the learned corpus, and the interpretability and robustness of the models in the face of attack samples have certain shortcomings. From the perspective of information theory, this paper constructs the semantic, pragmatic and syntactic knowledge contained in the large amount of crowd intelligence corpora obtained from the Internet platform into a hierarchical and heterogeneous natural language knowledge graph. The graph-based full information enhanced question answering model (GFIQA) is proposed, and the hierarchical heterogeneous knowledge graph is incorporated in the model. Through the crowd intelligence knowledge interpretation module, knowledge-enhanced generation module and single-layer anisotropic decoder, the relevant knowledge in the crowd intelligence natural language knowledge graph is appropriately selected based on the attention mechanism, and the ability of question understanding and answer generation is improved. The experimental results show that the GFIQA model has a large improvement in PPL, BLEU, and ENC (PPL: −11.76, BLEU: +0.126, ENC: + 0.232) compared with the baseline model, and can generate fluent and smooth answers with reasonable grammatical modifications and rich semantics.

Keywords: Question answering system · Knowledge graphs · Information theory · Attention mechanism

1 Introduction

The study of question answering system has both theoretical significance and application value. It is expected to change the shape of human machine interaction in the future. As early as the birth of the concept of artificial intelligence, Turing [1] pointed out that the ability to correctly answer arbitrary human questions is an important basis for judging

Y. Sun et al. (Eds.): ChineseCSCW 2021, CCIS 1492, pp. 281–294, 2022.
https://doi.org/10.1007/978-981-19-4549-6_22

whether a machine has intelligence or not. The emergence of artificial intelligence assistants such as Microsoft Ice and Apple Siri marks the development of question answering system towards the application stage.

A purely data-driven question answering model can extract statistical features of the data but cannot really understand the information in the learned corpus, and the interpretability and robustness of the model in the face of attack samples have certain shortcomings. It has gradually become a consensus among researchers to use deep learning techniques to learn from the data and enhance the performance of question answering models by introducing suitable external knowledge [2–6]. However, most of the existing question answering models based on knowledge bases or knowledge graphs can only focus on how to incorporate knowledge at the semantic level, which has limitations in the use of knowledge.

To address the above problems, this paper proposes a graph-based full-information enhanced question answering model, which enhances the performance of model using the rich knowledge contained in the crowd intelligence data. There is a huge amount of structured data, semi-structured data and unstructured text contributed by large-scale managers, crowdsourced users and ordinary users on the Internet, and these crowd intelligence data contain rich crowd intelligence information, which is an important data and knowledge source to support the implementation of question answering system. Firstly, we process the original group intelligence data to eliminate the influence of the uncertainty of some crowd behaviors and explore the rich group intelligence knowledge in it; Secondly, we select suitable structures (knowledge graphs and syntax graphs) to organize the crowd intelligence knowledge according to its hierarchical and heterogeneous characteristics in order to retain the comprehensive crowd intelligence information in an orderly way; Finally, we design several modules for the question answering model based on the attention mechanism, so that it has the ability to make full use of the heterogeneous crowd intelligence knowledge.

In summary, the main work and contributions of this paper contain the following aspects.

1. Defining a hierarchical heterogeneous crowd intelligence natural language knowledge graph (HHCIKG) from an information theoretic perspective, which organizes the multidimensional heterogeneous knowledge implied in crowd intelligence data in a graph structure.
2. Proposed a graph-based full information enhanced question answering model (GFIQA) to enhance the semantic understanding of questions and the quality of generated answers by incorporating the knowledge in the hierarchical heterogeneous knowledge graph through the crowd intelligence knowledge interpretation module, knowledge-enhanced generation module and single-layer anisotropic decoder.
3. Taking Xi'an tourism domain as an example, a crowd intelligence dataset (including group wise question and answer dataset and group wise natural language knowledge graph) is constructed for experiments. The experimental results verify the effectiveness of the method in this paper and show that each knowledge utilization module has an important contribution to the model performance.

2 Our Approach

We define a hierarchical and heterogeneous crowd intelligence knowledge graph from the perspective of information theory, based on which a full information enhanced question answering model is designed.

2.1 Crowd Intelligence Natural Language Knowledge Graph

In order to effectively capture and utilize the multidimensional heterogeneous crowd intelligence knowledge contained in the massive crowd intelligence data on the Internet platform, this paper explicitly constructs the crowd intelligence knowledge as a hierarchical heterogeneous crowd intelligence natural language knowledge graph.

From an information theory perspective, due to the differences in thinking, uncertainty of behavior and diversity of expression, the information utilization rate of the practice of mining the semantic-level crowd intelligence knowledge from crowd intelligence data only and storing it as a particular structure is not high. Therefore, as shown in Fig. 1, our approach achieves full information capture and portrayal of crowd intelligence knowledge from three levels: semantic, pragmatic and syntactic, and constructs a hierarchical and heterogeneous crowd intelligence natural language knowledge graph. The semantic and pragmatic information is organized and constructed in the form of a knowledge graph, capturing objective facts and consensus on crowd topics and emotions, including general objective knowledge of specific domains and subjective description of things by crowd. The syntax information section is organized in the form of syntax graph which focus on the differences in expressions of crowd in different contexts, including the contextual associations and local grammatical modification relations of the content related to the questions and answers.

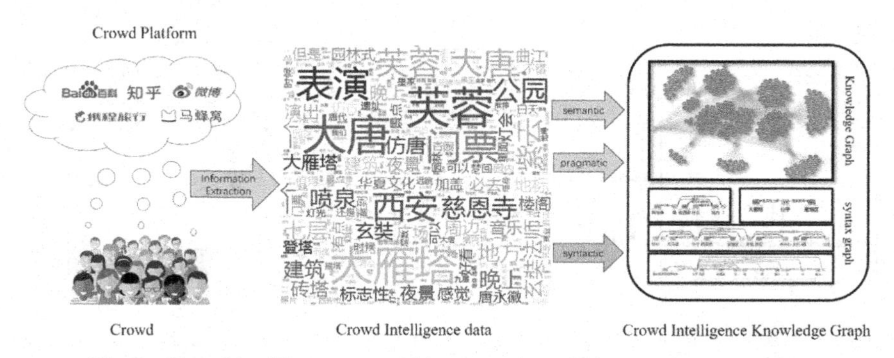

Fig. 1. Crowd intelligence natural language knowledge graph construction

2.2 Overall Architecture

The task addressed by the GFIQA model proposed in this paper is to achieve full-information-enhanced high-quality responses using external knowledge of the graph structure: given an input question X and a crowd intelligence natural language knowledge graph G (including a knowledge graph $G2$ and a syntactic graph $G2$), the model needs to select appropriate knowledge from the graph to aid in the understanding of the question in order to generate high-quality responses Y with appropriate syntactic logic and accurate and rich semantic information. i.e., estimate the probability distribution (1):

$$P(Y|X, G) = \prod_{t=1}^{m} P(y_t|y < t, X, G) \tag{1}$$

The overall architecture of GFIQA is shown in Fig. 2. The knowledge subgraph extraction module $F(X, G)$ takes the words in the question text as keywords and retrieves the knowledge subgraph from the whole crowd intelligence natural language knowledge graph. The crowd intelligence knowledge interpretation module further obtains the information in the knowledge subgraph: the subgraph is represented as a knowledge graph vector through the graph attention mechanism, the knowledge interpretation module connects the word vector, grammar vector, and knowledge graph vector as the input vector of the encoder. The encoder represents the input sequence as a hidden vector containing bidirectional sequence information through BiGRU. And the decoder first generates the appropriate root node in the answer based on the attention mechanism combined with the syntax graph information, then generates the vocabulary from the forward and backward directions of the root node based on the attention mechanism through two GRUs, respectively. During the decoding process the knowledge-enhanced generation module generates ordinary vocabulary or vocabulary in the knowledge graph through the graph attention mechanism.

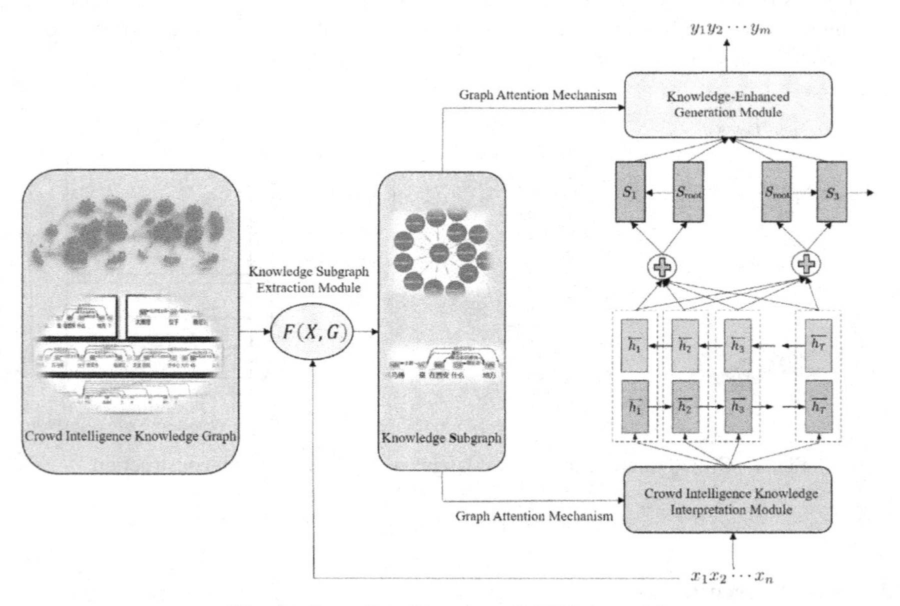

Fig. 2. Overall architecture of GFIQA model

2.3 Encoder

Crowd Intelligence Knowledge Interpretation Module

The crowd intelligence knowledge interpretation module enhances the semantic and pragmatic information of a vocabulary by introducing its corresponding knowledge graph vector and explicitly enhances its grammatical information by introducing its grammatical vector.

The crowd intelligence knowledge interpretation module uses the attention mechanism to obtain knowledge graph vectors containing semantic or pragmatic information and grammar vectors containing syntactic information from the extracted subgraphs. In the example "Where is the Terracotta Warriors in Xi'an?" shown in Fig. 3, the retrieved knowledge graph subgraph is transformed by the attention mechanism into a graph vector (triangle) with a local grammatical feature vector (square) corresponding to the vocabulary in the retrieved grammar graph, stitched after the word vector of the corresponding vocabulary.

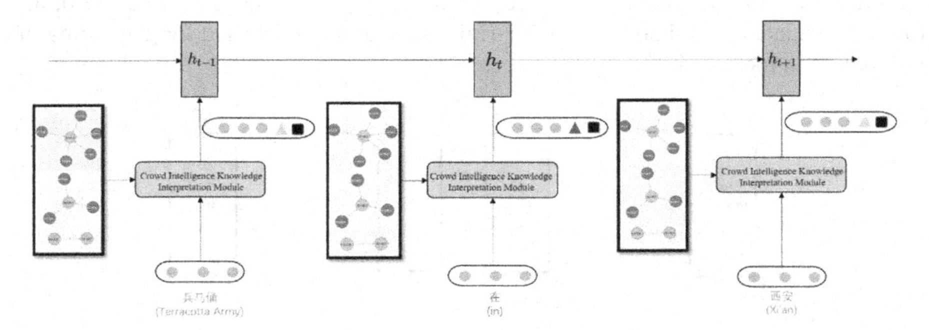

Fig. 3. The crowd intelligence knowledge interpretation module

The knowledge interpretation module computes the graph vector g_{1i} of the knowledge graph subgraph using the graph noticing mechanism. After connecting the word vector $w(x_t)$ with the knowledge graph vector g_{1i} and the grammar vector g_{2i}, the full-information enhanced word embedding representation $e(x_t)$ is obtained and then input to the BiGRU unit of the encoder (Eq. 2) to obtain the hidden vector.

$$h_t = BiGRU(h_{t-1}, e(x_t)) \tag{2}$$

Graph Attention Mechanism

We use the graph attention mechanism [7] to generate the corresponding vector representation for the crowd intelligence knowledge subgraph, the semantic and pragmatic information is encoded as vectors by considering all structures in the graph. Take the knowledge triple vector $K(g_i) = \{K_1, K_2 ... K_N\}$ in the knowledge subgraph g_i as input, the graph vector $\boldsymbol{g}_i$ is generated as shown in Eq. (3).

$$\boldsymbol{g}_i = \sum\nolimits_{n=1}^{N_i} \alpha_n^s [\boldsymbol{h}_n; \boldsymbol{t}_n]$$

$$\alpha_n^s = \frac{exp\left(\beta_n^s\right)}{\sum_{j=1}^{N_j} exp\left(\beta_j^s\right)} \tag{3}$$

$$\beta_n^s = (\boldsymbol{W_r r_n})^T \tanh(\boldsymbol{W_h h_n} + \boldsymbol{W_t t_n})$$

where $\boldsymbol{h_n}, \boldsymbol{r_n}, \boldsymbol{t_n}$ are the elements in the knowledge triple and $\boldsymbol{W}$ is the weight matrix.

2.4 Decoder

Knowledge-Enhanced Generation Module

The knowledge-enhanced generation module enhances the generated answers by making full use of the retrieved crowd intelligence natural language knowledge subgraphs. As shown in Fig. 4, taking the generated response "Terracotta Warriors is located in Lintong District" as an example, the knowledge-enhanced generation module can use the graph attention mechanism to select the retrieved knowledge graph (shaded), and further calculate the probability that the triple in the graph is selected for generating the vocabulary in the answer.

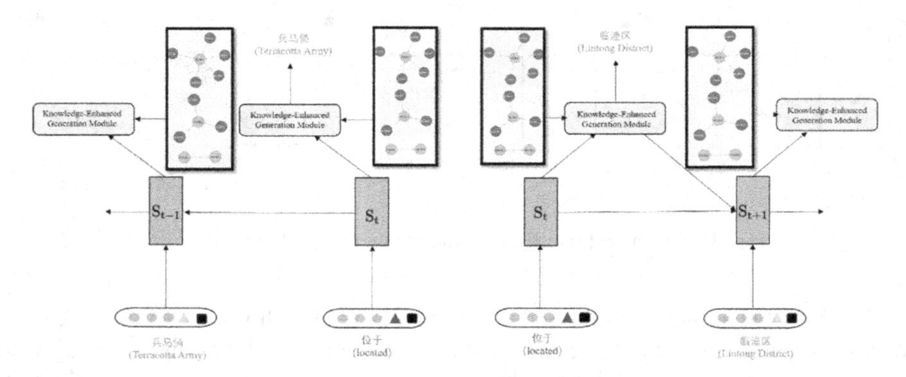

Fig. 4. The knowledge-enhanced generation module

The decoder firstly generates a vocabulary as the root node of the reply based on the attention mechanism combined with the syntax graph information in priority (as "located" in Fig. 4), and then generates the first half and the second half of the answer from the root node forward and backward respectively, and the state update rule of the right decoder is shown in Eq. (4).

$$s_{t+1} = GRU\left(s_t, \left[\boldsymbol{c_t}; \boldsymbol{c_t^{g_1}}; \boldsymbol{c_t^{k}}; \boldsymbol{c_t^{g_2}}; \boldsymbol{e}\left(\boldsymbol{y_t}\right)\right]\right) \tag{4}$$

Similarly, the state update rule for the left decoder is shown in Eq. (5)

$$s_{t-1} = GRU\left(s_t, \left[\boldsymbol{c_t}; \boldsymbol{c_t^{g_1}}; \boldsymbol{c_t^{k}}; \boldsymbol{c_t^{g_2}}; \boldsymbol{e}\left(\boldsymbol{y_t}\right)\right]\right) \tag{5}$$

where c_t is the context vector used in updating the decoder state, $e(y_t) = [w(y_t); k_j; g_{2j}]$ is the word vector $w(y_t)$ of the output vocabulary and the previously focused triple vector k_j and the connection of the grammar vector g_{2j}. $c_t^{g_2}$ is the contextual grammar vector of interest, $c_t^{g_1}$ and c_t^k are the context vectors of the knowledge graph vector and the knowledge triad vector of attention, respectively. These context vectors are obtained using the graph attention mechanism, where $c_t^{g_1}$ and c_t^k are computed with reference to CCM [8].

For the grammar part of the crowd intelligence natural language knowledge graph, the attention mechanism is applied to focus on the most important syntax graph among all graphs retrieved using the key vocabulary. And the probability of using all the grammar vector information in each relevant syntax graph is calculated.

$$c_t^{g_2} = \sum_{i=1}^{N_2} \alpha_{ti}^{g2} g_{2i}$$

$$\alpha_{ti}^{g2} = \frac{exp\left(\beta_{ti}^{g2}\right)}{\sum_{j=1}^{N_2} exp\left(\beta_{ti}^{g2}\right)} \quad (6)$$

$$\beta_{ti}^{g2} = V_{2b}^T \tanh\left(W_{2b} s_t + U_{2b} g_{2i}\right)$$

where α_{ti}^{g2} is the probability of choosing to use the syntax graph vector g_{2i} at time step t, and the grammar context vector $c_t^{g_2}$ is the weighted sum of the syntax graph vectors, and its summation weight measures the association between the decoder state s_t and the syntax graph vector g_{2i}.

Finally, the decoder generates the current response based on the following distribution (7):

$$P(y_t) = \begin{bmatrix} (1 - \gamma_t)P_o(y_t) \\ \gamma_t P_k(y_t)] \end{bmatrix} \quad (7)$$

where $P_o(y_t = w_o) = softmax(W_o a_t)$ is the probability of generating ordinary vocabulary and $P_k(y_t = w_k) = \alpha_{ti}^{g1} \alpha_{tj}$ is the probability of selecting vocabulary in the knowledge graph. $\gamma_t = sigmoid\left(V_0^T a_t\right)$ seeking a balance between them.

Loss Function

The loss function, shown in Eq. (8), is the cross entropy between the distribution of the predicted output and the actual distribution of the sample. The supervisory term uses the choice of vocabulary type $q_t \in \{0, 1\}$ for the actual responses to guide the vocabulary selection (ordinary or knowledge vocabulary) of the knowledge enhancement generator.

$$L(\theta) = -\sum_{t=1}^{m} p_t \log(o_t) - \sum_{t=1}^{m} (q_t \log(\gamma_t) + (1 - q_t)\log(1 - \gamma_t)) \quad (8)$$

3 Experimental

3.1 Experimental Setup

Experimental Dataset

In this paper, we construct a crowd intelligence dataset, including the training dataset of the model and the necessary auxiliary dataset, taking the Xi'an tourism domain as an example.

Q&A Dataset

The question and answer data in the Q&A dataset are mostly in the form of one question and many answers as shown in Table 1, and some questions are in the form of one question and one answer. The training set contains 32,820 question-answer pairs, the validation set contains 4,000 question-answer pairs, and the test set contains 8,000 question-answer pairs. The test set was divided into four categories by using the crowd intelligence knowledge graph, namely, high-frequency vocabulary test set (HFV), low-frequency vocabulary test set (LFV), complex grammatical sentence test set (CGS), and simple grammatical sentence test set (SGS) to test how our model performs in the face of the presence of common or rare knowledge in the question and the different levels of grammatical complexity.

Table 1. Q&A dataset display

Name	Text	Id	Type
Question	兵马俑是在西安什么地方?	3.1	兵马俑
Answer1	兵马俑位于西安市临潼区，距离西安市中心大约45公里。	3.2	西安
Answer2	从西安火车站有公交直接到马俑博物馆下车就是景点。	3.3	地址

Crowd Intelligence Knowledge Graph

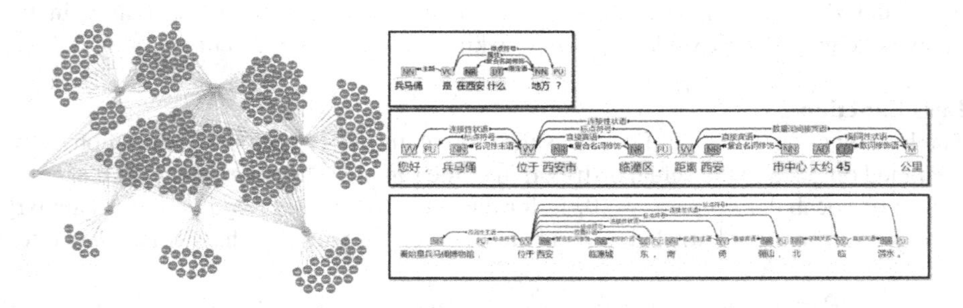

Fig. 5. Crowd intelligence knowledge graph display

The semantic and pragmatic parts of the crowd intelligence natural language knowledge graph are jointly constructed in the form of a knowledge graph (some nodes and relationships are shown in Fig. 5-left), which contains semantic information of attractions as well as pragmatic information of attractions, can provide both semantic and pragmatic knowledge support for question answering.

The grammar part of the crowd intelligence natural language knowledge graph is constructed as a syntax graph based on the Q&A dataset data, containing 89,640 subgraphs, each of which represents the grammatical knowledge of the question or answer text in the form of a list of five tuples shown in Table 2, and its visualization is shown in Fig. 5-right.

Table 2. Syntax graph data format

Id	Syntax structure data
3.1	[(是,VC,兵马俑,NN,top),(是,VC,地方,NN,attr),(地方,NN,在西安,NR,nn),(地方,NN,什么,DT,det),(是,VC,?,PU,punct)]
3.2	[(您好,VV,位于VV,conj),(位于,VV,兵马俑,NN,nsubj),(位于,VV,临潼区,NR),(临潼区,NR,西安市,NR,dobj),(位于,VV,距离,VV,conj)…]
3.4	[(位于,VV,秦始皇兵马俑博物馆,NN,nsubj),(位于,VV,东,LC,loc),(东,LC,临潼城,NR,lobj),(临潼城NR,西安,NR,nn),(位于VV,倚,VV,conj)…]

Implementations

The hardware environment is based on a 64-bit Linux operating system and uses an NVIDIA GeForce GTX 1080 Ti graphics card with 11G of video memory to accelerate the computation. The software is based on the GPU version of the deep learning framework Tensorflow_gpu developed using Python 3.6.

The word vector dimension is 300 (Wikipedia Chinese corpus pre-trained), the knowledge graph vector dimension based on TransE representation is 100, and the syntax graph vector dimension based on GAT representation is 100. The encoder is a single-layer BiGRU, the decoder is two single-layer GRUs with different directions, each layer contains 128 hidden units, and the hidden unit dimension is 500. Batch size is set to 100, and the learning rate is 0.00001.

Evaluation Metrics

In this paper, three automatic evaluation metrics, PPL (Perplexity), BLEU (Bilingual evaluation understudy) and ENC (Entity number score), are used to evaluate the model comprehensively.

- Perplexity evaluates the model at the content level of the generated responses, i.e., whether the content of the responses is relevant to the topic and whether the grammar is correct. A lower value of PPL indicates that the generated responses have fewer grammatical errors and are more fluent and natural in expression.
- BLEU is an index calculated based on the n-gram accuracy rate, which can well reflect the fluency of the generated replies and has a good consistency with the results of manual evaluation.

- ENC is the number of entities contained in each response, which measures the ability of the model to extract semantic and pragmatic information from the crowd intelligence knowledge graph to enhance the quality of generated responses.

3.2 Comparative Experiment

Three suitable models were selected as baseline models for comparison experiments with GFIQA model.

- Seq2Seq model, the classical codec framework generation model, which is widely used in question and answer systems.
- CopyNet model [9], which can generate responses by copying words from triples of knowledge graphs or selecting words from vocabularies, and has the ability to utilize information in knowledge graphs.
- The CCM model [8], which can fuse the knowledge in the knowledge graph using the graph attention mechanism to improve the semantic richness of the generated responses.

Table 3. Results of the comparison experiment

Test set	Metric	Model			
		S2S	CopyNet	CCM	GFIQA
PPL		53.05	46.71	45.18	**33.42**
BLEU		0.452	0.536	0.560	**0.686**
ENC		0.618	0.859	0.935	**1.167**
HFV	PPL	52.41	44.97	43.69	**31.89**
	BLEU	0.461	0.562	0.587	**0.694**
	ENC	0.615	0.845	0.924	**1.143**
LFV	PPL	55.03	48.15	47.08	**35.37**
	BLEU	0.440	0.509	0.535	**0.687**
	ENC	0.621	0.873	0.946	**1.191**
SGS	PPL	48.54	42.73	40.58	**32.57**
	BLEU	0.472	0.579	0.593	**0.688**
CGS	PPL	56.22	50.97	49.35	**33.85**
	BLEU	0.437	0.495	0.524	**0.675**

As shown in Table 3, GFIQA achieves the best results in both PPL and BLEU, indicating that GFIQA can better understand the question text and utilize the crowd intelligence knowledge of the graph structure to generate higher quality answer. In addition, the higher ENC indicates that GFIQA selects more entities from the crowd intelligence knowledge graph during the generation process, which can verify to a certain extent that higher quality answer is generated by more external knowledge as well as entity information facilitation, validating the correctness of the full information enhancement theory.

The difference in performance on the four test sets reveals to some extent the mechanism by which GFIQA takes effect. GFIQA performed better on the HFV test set than LFV (PPL: 31.89 vs. 35.37; BLEU: 0.694 vs. 0.687) because high-frequency common words could be more adequately trained in the training set. GFIQA extracts knowledge in the crowd intelligence knowledge graph more frequently in the LFV test set than in the HFV test set (ENC. 1.191 vs. 1.143), indicating that GFIQA needs and will automatically acquire more external knowledge to obtain semantic and pragmatic information to enhance question understanding and answer generation when the question text itself is difficult to provide sufficient semantic information. GFIQA performs better than CGS on the SGS test set (PPL: 32.57 vs. 33.85; BLEU: 0.688 vs. 0.675) due to the greater overall frequency of syntactically structured corpus in the training set and the fact that less training data is required for the model to achieve better performance for simple statements than for complex structured statements. In particular, GFIQA performs significantly better than the baseline model on the CGS test set, further validating that the approach of introducing external grammatical information through grammar vectors enables the model to maintain better performance in the face of question and answer data with complex grammatical structures.

3.3 Ablation Experiment

We analyze the impact of each module on the model performance through ablation experiment. On the basis of GFIQA, we remove the graph attention mechanism in the crowd intelligence knowledge interpretation module and the knowledge-enhanced generation module, remove the decoder based on the attention mechanism to generate root nodes preferentially based on the syntax graph (i.e., the decoder is ablated to a normal GRU), and eliminate the grammar vector of the syntax graph in the crowd intelligence knowledge interpretation module to obtain GFIQA-G1, GFIQA- ROOT, and GFIQA-G2 models, which represent the elimination of semantic and pragmatic information utilization mechanism, elimination of partial (answer text generation part) syntactic information utilization mechanism, and elimination of all syntactic information utilization mechanism on the basis of GFIQA model, respectively.

The results of the ablation experiment are shown in Table 4. The changes in the PPL and BLEU of GFIQA-G1 reflect a significant degradation in the performance of the model with the removal of the knowledge graph attention module, and the significant decrease in ENC intuitively reveals that the decrease in performance is strongly correlated with the number of entities, i.e., the external semantic and pragmatic information they represent. The GFIQA-G1 model that eliminates the knowledge graph attention module loses the ability to utilize external knowledge in the crowd intelligence knowledge graph, making its ability in semantic understanding and knowledge enhancement significantly degraded.

The performance of GFIQA-ROOT also degrades considerably (PPL: +6.42, BLEU: −0.07), indicating that the strategy of using the attention mechanism to preferentially

Table 4. Results of ablation experiment

Metric	Model			
	GFIQA	GFIQA-G1	GFIQA-ROOT	GFIQA-G2
PPL	33.42	42.71	39.84	45.18
BLEU	0.686	0.573	0.616	0.560
ENC	1.167	0.742	0.971	0.935

generate the root node before using it as a starting point to generate the whole sentence in both directions is effective. While the performance loss of GFIQA-G2 that completely eliminates the grammatical information exploitation mechanism is noted (PPL: +11.76, BLEU: −0.126), more than half of the positive effect on the quality of answer after the introduction of grammatical information is provided by the ROOT mechanism, indicating that the introduction of grammatical graph vector without changes in the model structure plays a limited role, which points the way to subsequent work. It is worth noting that removing the grammatical information utilization mechanism also decreases the ENC results, indicating that the mechanism of using grammatical information in GFIQA will at the same time positively contribute to the performance of the semantic information understanding module. Relatively, the absence of information understanding at one level of the corpus will weaken the effectiveness of using other levels of information, illustrating the need for full information enhancement.

4 Related Work

How to incorporate external knowledge in the form of knowledge graphs in text generation tasks [2–6, 8–12] such as question answering has been extensively studied, however, these works can only utilize information at the semantic level, resulting in low information utilization and low robustness of the model. In contrast, GFIQA can also utilize pragmatic knowledge in the form of knowledge graph as well as grammatical knowledge in the form of syntax graph to achieve full information enhancement.

Syntax graphs are directed acyclic graph that represent syntactic relationships between words. By making rational use of syntax graphs, models in tasks such as relation extraction [13–15], sentiment classification [16, 17], and machine translation [18, 19] improve their understanding of grammatical knowledge and obtain better performance. However, to the best of our knowledge, no work has yet been done to integrate knowledge of grammatical structures well into question-and-answer models.

There are three main different ways to encode syntax graphs for incorporation into various types of models: (1) linearized representation of grammar graphs and encoding using sequential models (e.g., RNN) [19]; (2) encoding based on path (inter-word, word-root node, etc.) distances in grammar graphs [13, 20]; and (3) aggregating information from dependencies using graph neural networks [15, 18, 21]. In this paper, we refer to [22] to encode local grammatical information of words in syntax graphs as vectors and then select appropriate grammatical knowledge based on the graph attention mechanism.

5 Conclusion

In this paper, we define a hierarchical heterogeneous crowd intelligence natural language knowledge graph from an information theoretic perspective, organize the multidimensional heterogeneous information in the crowd intelligence data in a graph structure, and design a full information enhanced question answering model that can simultaneously utilize syntactic, pragmatic, and semantic information in the crowd intelligence natural language knowledge graph. The experimental results on Q&A dataset show that compared with the baseline models, GFIQA can generate high quality responses with rich semantic information and reasonable syntactic logic, especially in the case of more complex syntactic structures.

In the future, we will further improve the model's ability to exploit full information (e.g., designing encoder module that can explicitly use syntactic structure information). The full-information enhancement idea proposed in this paper is universal for many kinds of natural language processing tasks, and we can try to apply it to other types of text generation tasks.

Acknowledgements. This project was supported by the National Outstanding Young Scientists Foundation of China (62025205), the National Key Research and Development Program of China (2019QY0600), and the National Natural Science Foundation of China (61960206008, 61725205).

References

1. Turing, A.M.: Computing machinery and intelligence. Creative Comput. **6**(1), 44–53 (1980)
2. Lukovnikov, D., Fischer, A., Lehmann, J., et al.: Neural network-based question answering over knowledge graphs on word and character level. In: Proceedings of the 26th International Conference on World Wide Web, pp. 1211–1220 (2017)
3. Saha, A., Pahuja, V., Khapra, M.M., et al.: Complex sequential question answering: towards learning to converse over linked question answer pairs with a knowledge graph. In: Thirty-Second AAAI Conference on Artificial Intelligence (2018)
4. Huang, X., Zhang, J., Li, D., et al.: Knowledge graph embedding based question answering. In: Proceedings of the Twelfth ACM International Conference on Web Search and Data Mining, pp. 105–113 (2019)
5. Sawant, U., Garg, S., Chakrabarti, S., et al.: Neural architecture for question answering using a knowledge graph and web corpus. Inf. Retriev. J. **22**(3), 324–349 (2019)
6. Sheng, M., et al.: DSQA: a domain specific QA system for smart health based on knowledge graph. In: Wang, G., Lin, X., Hendler, J., Song, W., Xu, Z., Liu, G. (eds.) WISA 2020. LNCS, vol. 12432, pp. 215–222. Springer, Cham (2020). https://doi.org/10.1007/978-3-030-60029-7_20
7. Veličković, P., Cucurull, G., Casanova, A., et al.: Graph attention networks. arXiv preprint arXiv:1710.10903 (2017)
8. Zhou, H., Young, T., Huang, M., et al.: Commonsense knowledge aware conversation generation with graph attention. In: IJCAI, pp. 4623–4629 (2018)
9. Zhu, W., Mo, K., Zhang, Y., et al.: Flexible end-to-end dialogue system for knowledge grounded conversation. arXiv preprint arXiv:1709.04264 (2017)
10. He, H., Balakrishnan, A., Eric, M., et al.: Learning symmetric collaborative dialogue agents with dynamic knowledge graph embeddings. arXiv preprint arXiv:1704.07130 (2017)

11. Wu, S., Li, Y., Zhang, D., et al.: Diverse and informative dialogue generation with context-specific commonsense knowledge awareness. In: Proceedings of the 58th Annual Meeting of the Association for Computational Linguistics, pp. 5811–5820 (2020)
12. Koncel-Kedziorski, R., Bekal, D., Luan, Y., et al.: Text generation from knowledge graphs with graph transformers. arXiv preprint arXiv:1904.02342 (2019)
13. Xu, Y., Mou, L., Li, G., et al.: Classifying relations via long short term memory networks along shortest dependency paths. In: Proceedings of the 2015 Conference on Empirical Methods in Natural Language Processing, pp. 1785–1794 (2015)
14. Guo, Z., Zhang, Y., Lu, W.: Attention guided graph convolutional networks for relation extraction. arXiv preprint arXiv:1906.07510 (2019)
15. Sun, K., Zhang, R., Mao, Y., et al.: Relation extraction with convolutional network over learnable syntax-transport graph. Proc. AAAI Conf. Artif. Intell. **34**(05), 8928–8935 (2020)
16. Huang, B., Carley, K.M.: Syntax-aware aspect level sentiment classification with graph attention networks. arXiv preprint arXiv:1909.02606 (2019)
17. Zheng, Y., Zhang, R., Mensah, S., et al.: Replicate, walk, and stop on syntax: an effective neural network model for aspect-level sentiment classification. Proc. AAAI Conf. Artif. Intell. **34**(05), 9685–9692 (2020)
18. Bastings, J., Titov, I., Aziz, W., et al.: Graph convolutional encoders for syntax-aware neural machine translation. arXiv preprint arXiv:1704.04675 (2017)
19. Aharoni, R., Goldberg, Y.: Towards string-to-tree neural machine translation. arXiv preprint arXiv:1704.04743 (2017)
20. Chen, K., Wang, R., Utiyama, M., et al.: Syntax-directed attention for neural machine translation. Proc. AAAI Conf. Artif. Intell. **32**(1), 4792–4799 (2018)
21. Chen, Y., Wu, L., Zaki, M.J.: Reinforcement learning based graph-to-sequence model for natural question generation. arXiv preprint arXiv:1908.04942 (2019)
22. Xia, Q., Li, Z., Zhang, M., et al.: Syntax-aware neural semantic role labeling. Proc. AAAI Conf. Artif. Intell. **33**(01), 7305–7313 (2019)

Exploring the Content Sharing Practice Across Social Network Sites

Baoxi Liu, Peng Zhang(✉), Tun Lu, and Ning Gu

School of Computer Science, Fudan University, Shanghai, China
{bxliu18,zhangpeng_,lutun,ninggu}@fudan.edu.cn

Abstract. Almost all popular social network sites (SNSs) provide the content sharing functionality to support users to distribute text messages, music, videos and some other online resources. As most SNS users are involved in multiple sites to meet their diverse needs, we wonder how adopting a new SNS to share content affects sharing behavior in an existing one. For this problem, we conducted an empirical study by utilizing two Chinese popular SNSs - Weibo and Douban as research sites, and investigated how content sharing in a new SNS affected that in an existing one, and what kinds of users' content sharing behaviors in the existing site could be more or less likely to be affected. Our results indicate that adopting a new site to share content is beneficial for content quantity but detrimental to content quality (number of likes) in the existing SNS in a short time period. We also find that longer-time users in the existing SNS correspond to less improvement on content quantity but less decline on content quality after initializing content sharing in a new site, and the existence of a broader audience, more interactions through a more restricted span of interaction relationships and more groups is helpful to stimulate improvement of content quantity and alleviate setback of content quality, which provides valuable insights for SNS practitioners to maintain and promote user content contribution.

Keywords: Social network sites · Content sharing · Cross-SNS analysis · Weibo · Douban

1 Introduction

Nowadays, social network sites (SNSs) like Facebook, Twitter and Weibo have been popular and important platforms in our daily life. Almost all popular SNSs provide the content sharing functionality to support users to distribute and exchange online resources. For each SNS, users' sustained content sharing is important for its survival and success as SNS content is entirely contributed by individuals [1]. Content disclosing is also critical for SNS users to be involved into a social world as it can enhance their self-expression as well as social relationship and interaction by intensifying feelings of closeness, enjoyment and intimacy among them [2].

Most SNS users tend to utilize multiple sites simultaneously in order to meet diverse needs for SNS use like information seeking, relationship maintenance

Y. Sun et al. (Eds.): ChineseCSCW 2021, CCIS 1492, pp. 295–311, 2022.
https://doi.org/10.1007/978-981-19-4549-6_23

and communicating. A survey [21] conducted by Pew Research Center in 2018 has found that over 73% users use more than one social medias among Twitter, Instagram, Pinterest, LinkedIn and Facebook. As these sites all support users to share content, the users are involved in a composite sharing context constituted by multiple social medias. Thus many previous studies have combined two or more SNSs to analyze user content sharing behavior like the platform preference of content sharing, similarity and topic relevance of content, sharing channel choice, etc. The existing conclusions concentrate on two folds. On one hand, there exist some significant differences of user content among different SNSs in a general view, which is in accordance with the faceted identity theory stating that people have multi-faceted identities, and enable different aspects of their personalities depending on the social context. For example, [15] found that the topics of tweets in Twitter concentrated on news, opinions or other general user interests, while posts in Instagram were mostly related to joyful and happy moments of personal lives. On another hand, a user's content sharing behavior is not isolated among different platforms, and there exist some similarities or relevances. For example, a person usually posts the same content to different SNSs by taking advantage of cross-posting or cross-network sharing [14]; [20] suggest that people sometimes disclose content with similar topics while without the same language to different SNSs to ensure the content can be perceived appropriate by each platform and audience, etc. It is driven by their opposing desires to guard platform boundaries in order to maintain separate spaces but also feel the need to relax platform boundaries to allow audience and content to permeate others.

Although the above research has provided lots of insights for user content sharing behavior understanding across different social network sites, it mostly concentrates on basic characteristics such as content difference and content similarity, while the complex features have not been investigated, one of which is the interplay of user content sharing among different SNSs. For SNS use, users' time and efforts are generally limited [24]. How can people's content sharing in one SNS affect that in other sites? For example, when a person starts content sharing practice in a new SNS, if it would be detrimental to her/his content sharing frequency and quality in an existing one. To study such questions has significant meanings for SNSs. As the number of SNSs increases, no platform can monopolize people's time and develop in isolation. It is critical for each SNS to comprehend how users' participating into the other sites affects their activity and retention in that. By studying the interplay of user content sharing between different SNSs, each SNS can gain insights into how users' content sharing behavior in that can be influenced by their content distributing in other sites, how to identify the users that are more likely to be affected proactively and what prevention and remedial measures can be conducted for their sustained content contribution, which is significant for the SNS's survival and success. However, exploring the interplay of user content sharing between different SNS is non-trivial. It is essentially a causal inference problem which cannot be solved by nowadays prevalent machine learning methods like Bayesian Network as they are designed for association mining but not causality inferring. Moreover, as SNSs are generally isolated to each other, obtaining a longitudinal data set containing user generated content in two or more SNSs wherein the users overlap is very difficult. These aspects above result that combining two or more SNSs to study the interplay of user content sharing is very challenging.

In our research, we conducted an empirical study to investigate how content sharing in a new SNS affected that in an existing one, and what kinds of users' content sharing in the existing site was more or less likely to be affected. For these two research questions, we utilized two Chinese popular SNSs - Weibo[1] and Douban[2] as research sites and made causality analysis by utilizing methods like Regression Discontinuity Design (RDD), Panel Data analysis, etc. Our analysis focused on two most important dimensions of SNS content: content quantity and content quality (number of likes). Content quality is difficult to measure directly. In our research, we utilize number of likes as an indicator of content quality as previous research has suggested the number of likes to be a valid proxy for content quality in online communities [6]. After analyzing the 6,229 overlapping users (users that have accounts in both Weibo and Douban) with 20,899,266 Weibo posts and 469,295 Douban posts, we find adopting a new SNS to share content can be beneficial for content quantity but detrimental to content quality in an existing site. Our results also suggest that longer-time users in the existing SNS tend to correspond to less improvement on content quantity but less decline on content quality after initializing content sharing in a new site, and the existence of a broader audience, more interactions through a more restricted span of interaction relationships and more groups is helpful to stimulate improvement of content quantity and alleviate setback of content quality. These findings provide SNS practitioners valuable insights into how to maintain and promote user content contribution confronting with nowadays situation wherein most SNS users are involved into multiple platforms. On one hand, as users' content quantity in an existing SNS can be improved in a short period after beginning sharing content in a new site, it provides the SNS an opportunity to enhance these users' content quantity and achieve development with other emerging platforms. While on another hand, practitioners of the existing SNS should pay attention to the content quality in the meanwhile as the content sharing practice in a new SNS can be detrimental to content quality in that. First, SNS practitioners should lay more emphasis on longer-time users as they tend to contribute high-quality but less contents as well as users with less followers, less groups and less interactions especially through strong ties because their content quantity and content quality are more sensitive to sharing content in a new site. Second, making it easy for users to communicate through strong relationships and recommending more groups to them are suggested as such mechanisms can stimulate improvement of content quantity and alleviate setback of content quality, which is crucial for the users' sustained content contribution in the SNS.

---

[1] https://weibo.com/.
[2] https://www.douban.com/.

The following sections are organized as below. In Sect. 2, we review related research about content sharing in SNSs and cross-SNS analysis and give our research questions. The research sites and data set are exhibited in Sect. 3, and research methods including model and variables are given in Sect. 4. In Sects. 5 and 6, we exhibit the results, highlight their implications for SNSs. Finally, The conclusions are given in Sect. 7.

2 Related Work

As content sharing is critical for both SNSs and users, lots of research has focused on it to comprehend users' motivations and investigate incentive mechanisms for sharing. According to previous research, both intrinsic and extrinsic factors motivate users to contribute content in SNSs. Intrinsic motivations mean users' inherent satisfactions like enjoyment, reciprocity, commitment to the platform [1], while extrinsic factors refer to expected benefits including social relationship maintenance [13], self-development [1], social capital enhancement [13], etc. Many research focuses on exploring incentive mechanisms to stimulate content contribution in SNSs. [22] found that the others' frequent and pleasant feedbacks like commenting to posts, tagging and re-posting could promote a SNS user's further content sharing. [19] showed that users would like to share more contents if SNSs could provide functionalities to help them exactly target desired audiences.

Moreover, there has been some research studying correlations between content sharing and some other user attributes or behaviors. [2] found that female users were more likely to post messages than males. [1,16] suggests that users characterized with longer tenure in a platform tend to have less content contribution. Besides, [11] studied correlations between users' content sharing and social network dynamics, which indicates post size in Twitter increases with both follower size and friend size but eventually saturates as a function of number of followers.

For cross-SNS user content analysis, many studies have been conducted from perspectives of content divergence, content sharing similarity and channel choice for sharing. [17] investigated users' content difference by utilizing Pinterest and Twitter. It indicates users tend to have a wider range of content categories in Twitter than in Pinterest. [15] studied the topic difference of user content between Twitter and Instagram, and found that topics in Twitter were mainly related to news, opinions or other general user interests, while that in Instagram were mainly about joyful and happy moments of their personal lives. As mentioned above, some prior research has also found there exist some similarities and relevances of user content among different SNSs such as similarity of topic, location, time, etc. [20]. These studies show that users generally consider audiences and platform norms like privacy, security and institutional policy, and choose the most appropriate sites for sharing. If sharing in one SNS cannot meet a user's needs, she/he would like to combine channels to usably share content with their desired audience.

To conclude, a great deal of research has focused on content sharing behavior analysis and cross-SNS user behavior analysis. However, to the best of our knowledge, the effects of adopting a new SNS to share content to the content sharing practice in an existing site have been unknown. It motivates us to focus on the following research questions in this paper.

RQ 1: How does adopting a new SNS to share content affect content sharing in an existing site?

- RQ 1a: How does adopting a new SNS to share content affect content quantity in an existing site?
- RQ 1b: How does adopting a new SNS to share content affect content quality (number of likes) in an existing site?

RQ 2: What kinds of users' content sharing in the existing SNS is more or less likely to be affected?

- RQ 2a: What kinds of users' content quantity in the existing SNS is more or less likely to be affected?
- RQ 2b: What kinds of users' content quality (number of likes) in the existing SNS is more or less likely to be affected?

3 Research Sites and Data Collection

Weibo and Douban are two of the most popular social network sites in China. According to a report released in 2018, Weibo has 431 million users, which is slightly more than Twitter. Douban is also very popular especially among young adults. It has more than 160 million registered users up to 2018, and nearly 300 million users are active in it per month. We finally chose Weibo and Douban as our research sites. The procedure of our data sampling is as follows. First, we randomly sampled 7,593 users who exhibited their Weibo usernames or homepage URLs in Douban. Second, we attempted to find these users corresponding profiles in Weibo and removed 388 individuals whose Weibo usernames or URLs were non-existent. The users who had no posts in Weibo or Douban were also filtered out. Finally, we sampled the rest 6,229 users' attributes in Weibo and Douban. Through the above procedure, we totally obtained these users' 20,899,266 posts in Weibo and 469,295 posts in Douban.

4 Model and Variables

4.1 Model

We chose the random-effects negative binomial regression model which combined Regression Discontinuity Design (RDD) [3] with Panel Data Analysis (PDA) [10] as our analysis method. To build the model, the first step is to construct an observation time period to observe people's content sharing behaviors. As prior research [24] has exhibited the evolution pattern of user behavior in

SNS month-to-month, we constructed a time period of six months to observe each user's content sharing behavior. The six observation windows are (M_{-3}, M_{-2}), (M_{-2}, M_{-1}), (M_{-1}, M_0), (M_0, M_1), (M_1, M_2) and (M_2, M_3), where M_0 is the time point when the user begins sharing content in the new SNS, M_{-1} and M_1 are respectively one month before and after it. The other symbols are defined in the same manner. In our research, we originally utilized different time periods including six months, ten months and twelve months to observe users' content sharing behavior. As the results are similar, and RDD requires a minor observation window, we finally set six months as the observation period.

4.2 Dependent Variables

Variables for content quantity measurement: *Content sharing frequency in the existing SNS.* It represents the number of posts the user publishes in the existing SNS in each observation window.

Variables for content quality measurement: *Like receiving frequency in the existing SNS.* It represents the average number of likes the user's posts receive in the existing SNS in each observation window, which is evaluated by dividing the number of post likes by the number of posts in the existing SNS in each observation window.

4.3 Independent Variables

For RQ 1, an independent variable *having shared content in the new SNS* was considered, and for RQ 2, we introduced 6 independent variables which we believe are key ingredients according to previous findings. In the following, we explain the rationale and extraction process for each variable.

having shared content in the new SNS. It represents whether the user has begun sharing content in the new SNS. It is set to 0 before the user begins sharing content in the new SNS and 1 after that.

having shared content in the new SNS × user tenure in the existing SNS. user tenure in the existing SNS means the number of months since a user registered in the existing SNS. The interaction term is utilized to test whether the effects of sharing content in a new SNS are different between longer-time users and shorter-time users. sites [1,2,16].

having shared content in the new SNS × follower size in the existing SNS. follower size in the existing SNS means the number of followers a user has in the existing SNS. The interaction term is utilized to test whether the effects of sharing content in a new SNS are different depending on users' follower sizes.

having shared content in the new SNS × interaction partner size in the existing SNS and *having shared content in the new SNS × interaction message size in the existing SNS. interaction partner size in the existing SNS* means the number of individuals the user communicates with in the existing SNS in each observation window, and *interaction message size in the existing SNS* counts the messages through such interactions. The two interaction terms are utilized to test whether the effects of sharing content in a new SNS are different depending on users' interactions.

having shared content in the new SNS × *content interest diversity in the existing SNS. content interest diversity in the existing SNS* is the topic diversity of a user's content posted in the existing SNS in each observation window. The interaction term is utilized to test whether the effects of sharing content in a new SNS are different depending on users' content interest diversity. We treated a user's content in each time slot as a document and leveraged Latent Dirichlet Allocation (LDA) [5] model with default parameter setting to extract topics from it. Shannon Entropy was used to analyze the topic distribution to obtain the topic diversity.

having shared content in the new SNS ×*group size in the existing SNS. group size in the existing SNS* means the number of groups the user has joined in the existing SNS. The interaction term is utilized to test whether the effects of sharing content in a new SNS are different depending on users' group sizes.

4.4 Control Variables

user tenure in the existing SNS, follower size in the existing SNS, interaction partner size in the existing SNS, interaction message size in the existing SNS, content interest diversity in the existing SNS and *group size in the existing SNS* are set as control variables in order to control their effects on the dependent variable.

We found that variables *follower size in the existing SNS, interaction partner size in the existing SNS, interaction message size in the existing SNS* and *group size in the existing SNS* were skewed. So base-*e* logarithmic transformations were conducted for them, and then all independent variables were standardized for ease of interpretation. In the data set, there are no values in some users' one or more time slots as their time intervals between content sharing in the existing SNS and that in the new SNS are less than three months. The random-effects negative binomial regression model can handle such imbalanced data. So we do not need to process the missing values additionally. We also analyzed correlations between all independent variables before analysis. The results indicate there is no strong correlation between these variables.

5 Results

5.1 Discontinuity of Content Quantity and Content Quality

RDD is initialized by observing whether there is a discontinuity right before and after the intervention. Thus we first analyzed the evolving trend of users' content quality and content quality in the existing SNS right before and after starting content sharing practice in a new site. We conducted analysis from two directions by setting Douban as an existing SNS and Weibo as the new site as well as Weibo as the existing SNS and Douban as the new one. The results are shown in Figs. 1 and 2. X-axis lists the six time slots, and Y-axis shows the

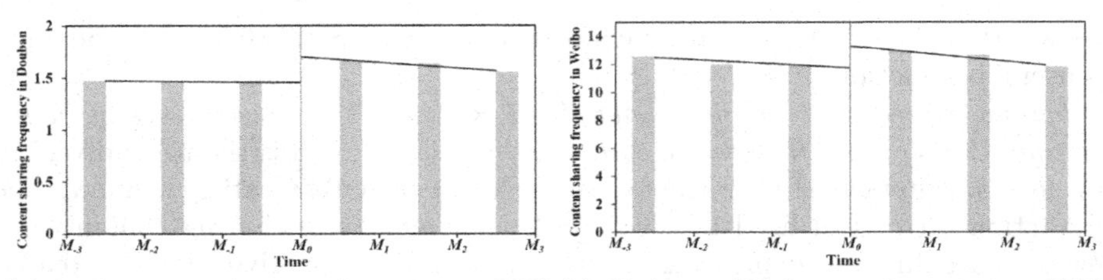

(a) Utilizing Douban as the existing SNS and Weibo as the new SNS.
(b) Utilizing Weibo as the existing SNS and Douban as the new SNS.

Fig. 1. Temporal evolution of *content sharing frequency in the existing SNS.*

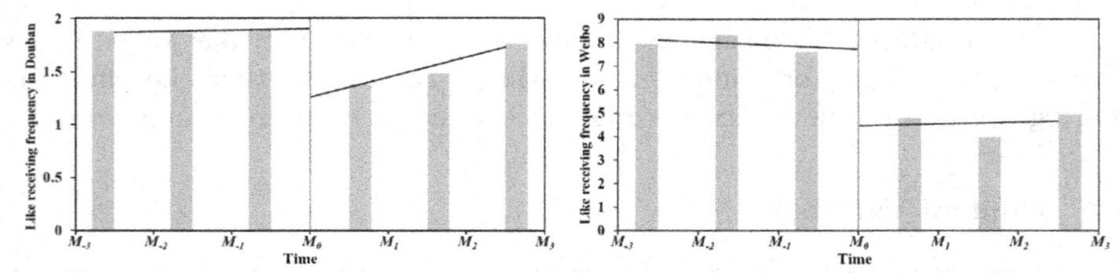

(a) Utilizing Douban as the existing SNS and Weibo as the new SNS.
(b) Utilizing Weibo as the existing SNS and Douban as the new SNS.

Fig. 2. Temporal evolution of *like receiving frequency in the existing SNS.*

mean values of *content sharing frequency in the existing SNS* and *like receiving frequency in the existing SNS* in each time period. As can be seen from these two figures, the mean values of *content sharing frequency in the existing SNS* and *like receiving frequency in the existing SNS* are smooth from time point M_{-3} to M_0, which is in accordance with RDD's assumption - human behavior is typically orderly or smooth on the aggregate or average level. However, the two variables all experience a discontinuity right before and after M_0. Specifically, in Fig. 1(a), 75.81% of users show increase on content sharing frequency right after M_0, and in Fig. 1(b), such a percentage is 65.68%, which suggests that most users' content quantity in the existing SNS increases right after they begin sharing content in the new one. However, most users experience content quality decrease around M_0. The percentage values are 89.06% in Fig. 2(a) and 65.28% in Fig. 2(b) respectively.

5.2 Results of Random-Effects Negative Binomial Regression Model

Results of Content Quantity Analysis. The results by setting Douban as the existing SNS and Weibo as the new one as well as Weibo as the existing SNS and Douban as the new site are shown in Tables 1 and 2 respectively (***: $p < 0.001$, **: $p < 0.01$, *: $p < 0.05$, Tables 3 and 4 have same settings). In each table, Model 1 only considers control variables in order to control their effects on the dependent variable. Model 2 introduces the variable *having shared content in the new SNS*. In Model 3, the other variables are added.

As can be seen from the results of Model 1 in each table, there are significant correlations between *content sharing frequency in the existing SNS* and most of the variables. *user tenure in the existing SNS* is negatively associated with *content sharing frequency in the existing SNS*, which suggests that longer-time SNS users tend to post less contents. The other variables including *follower size in the existing SNS*, *interaction partner size in the existing SNS*, *interaction message size in the existing SNS* and *group size in the existing SNS* are all positively related to the dependent variable. It indicates the SNS users who have more audiences, more interactions and more groups are inclined to share content more frequently. However, the variable *content interest diversity in the existing SNS* is not significant in the model. One SNS user's interest diversity is not essentially predictive for her/his content sharing frequency.

Table 1. Content quantity analysis based on random-effects negative binomial regression model by utilizing Douban as the existing SNS.

Variable	Model 1		Model 2		Model 3	
	Coef.	p	Coef.	p	Coef.	p
user tenure in the existing SNS	−0.37	*	−0.55	**	−0.47	**
follower size in the existing SNS	3.88	***	3.92	***	3.81	***
interaction partner size in the existing SNS	1.40	***	1.39	***	1.56	***
interaction message size in the existing SNS	1.33	***	1.36	***	1.14	***
content interest diversity in the existing SNS	0.01		0.01		0.00	
group size in the existing SNS	0.09	*	0.09	*	0.13	*
having shared content in the new SNS			0.06	***	0.09	*
having shared content in the new SNS × *User tenure in the existing SNS*					−0.67	**
having shared content in the new SNS × *Follower size in the existing SNS*					0.44	**
having shared content in the new SNS × *Interaction partner size in the existing SNS*					−0.36	***
having shared content in the new SNS × *Interaction message size in the existing SNS*					0.39	**
having shared content in the new SNS × *Content interest diversity in the existing SNS*					0.04	
having shared content in the new SNS × *Group size in the existing SNS*					0.09	*

In Model 2, *having shared content in the new SNS* has a significant positive effect on *content sharing frequency in the existing SNS*. It suggests that users' newly conducted content sharing practice in Weibo (Douban) can promote the content sharing frequency in Douban (Weibo), which is consistent with the results of temporal analysis. Based on the results of Model 1 and Model 2, we can see SNS users' content sharing frequency in the existing SNS declines with their tenure, while as an intervention, adopting a new site to share content can make it to decline less or even increase.

In Model 3, first, *having shared content in the new SNS* × *user tenure in the existing SNS* is negatively associated with *content sharing frequency in the existing SNS*, which suggests that longer-time users tend to experience less improvement on content sharing frequency in the existing SNS after they adopt a new site to distribute content. Second, *having shared content in the new SNS* × *follower size in the existing SNS* is positive in the model. It means users who have a larger amount of audiences in the existing SNS are more likely to intensify their content sharing frequency in that after they begin sharing content in a new site. Third, *having shared content in the new SNS* × *group size in the existing SNS* is also positively associated with the dependent variable. Content sharing frequency of users who have joined more groups in the existing SNS tend to have more content sharing frequency improvement after they utilize a new site to share content. The last, *having shared content in the new SNS* × *interaction message*

Table 2. Content quantity analysis based on random-effects negative binomial regression model by utilizing Weibo as the existing SNS.

Variable	Model 1		Model 2		Model 3	
	Coef.	p	Coef.	p	Coef	p
user tenure in the existing SNS	−1.49	***	−1.70	***	−1.16	***
follower size in the existing SNS	1.96	***	1.89	***	2.32	***
interaction partner size in the existing SNS	3.65	***	3.47	***	4.74	***
interaction message size in the existing SNS	0.45	*	0.64	*	0.54	*
content interest diversity in the existing SNS	−0.10		−0.11		0.17	
group size in the existing SNS	0.10	*	0.13		0.10	*
having shared content in the new SNS			0.20	***	0.77	***
having shared content in the new SNS × *User tenure in the existing SNS*					−1.19	***
having shared content in the new SNS × *Follower size in the existing SNS*					0.34	*
having shared content in the new SNS × *Interaction partner size in the existing SNS*					−1.33	**
having shared content in the new SNS × *Interaction message size in the existing SNS*					0.48	*
having shared content in the new SNS × *Content interest diversity in the existing SNS*					−0.53	
having shared content in the new SNS × *Group size in the existing SNS*					0.36	*

Table 3. Content quality analysis based on random-effects negative binomial regression model by utilizing Douban as the existing SNS.

Variable	Model 1		Model 2		Model 3	
	Coef.	*p*	Coef.	*p*	Coef.	*p*
user tenure in the existing SNS	1.12	***	1.15	**	1.05	***
follower size in the existing SNS	3.84	***	3.86	***	3.82	***
interaction partner size in the existing SNS	−1.19	***	−1.20	***	−1.38	***
interaction message size in the existing SNS	7.78	***	7.78	***	7.84	***
content interest diversity in the existing SNS	0.25		0.25		0.24	
group size in the existing SNS	1.45	***	1.45	***	1.40	***
having shared content in the new SNS			−0.07	*	−0.12	*
having shared content in the new SNS × *User tenure in the existing SNS*					0.19	*
having shared content in the new SNS × *Follower size in the existing SNS*					0.10	*
having shared content in the new SNS × *Interaction partner size in the existing SNS*					−0.34	*
having shared content in the new SNS × *Interaction message size in the existing SNS*					0.14	*
having shared content in the new SNS × *Content interest diversity in the existing SNS*					−0.01	
having shared content in the new SNS × *Group size in the existing SNS*					0.03	*

size in the existing SNS is positively related to *content sharing frequency in the existing SNS*, while *having shared content in the new SNS* × *interaction partner size in the existing SNS* is negative in the model. It suggests that the improvement degree of content sharing frequency in the existing site increases with users' interaction message size but decreases with the number of interaction partners. However, *having shared content in the new SNS* × *content interest diversity in the existing SNS* is not significant in the model, which implies the effects of *having shared content in the new SNS* on users' content sharing frequency in the existing site are not different significantly depending on their content interest diversity.

Results of Content Quality Analysis. The results of content quality analysis by setting Douban as the existing SNS and Weibo as the new site as well as Weibo as the existing SNS and Douban as the new one are given in Table 3 and Table 4 respectively. Similar to content quantity analysis, there are also three models wherein Model 1 only considers the control variables, Model 2 introduces the variable *having shared content in the new SNS* and Model 3 considers the interaction terms.

In Model 1, we can see most of the variables are significant. It can obtain results that the contents of users with longer tenure, more followers, more groups and more interactions concentrating on a more restrict span of partners tend to receive more likes from others. In Mode 2, the variable *having shared content in the new SNS* is negatively associated with the dependent variable, which implies users tend to experience a like receiving frequency setback in the existing SNS after they begin sharing content in a new site.

Table 4. Content quality analysis based on random-effects negative binomial regression model by utilizing Weibo as the existing SNS.

Variable	Model 1		Model 2		Model 3	
	Coef.	p	Coef.	p	Coef.	p
user tenure in the existing SNS	0.17	*	0.37	**	0.06	*
follower size in the existing SNS	1.55	***	1.59	***	1.75	***
interaction partner size in the existing SNS	−1.59	***	−1.51	***	−2.01	***
interaction message size in the existing SNS	2.77	***	2.83	***	2.58	*
content interest diversity in the existing SNS	−0.11		−0.12		0.07	
group size in the existing SNS	0.20	*	0.18	*	0.07	
having shared content in the new SNS			−0.16	***	−0.69	**
having shared content in the new SNS × User tenure in the existing SNS					1.08	***
having shared content in the new SNS × Follower size in the existing SNS					0.11	*
having shared content in the new SNS × Interaction partner size in the existing SNS					−0.56	**
having shared content in the new SNS × Interaction message size in the existing SNS					0.12	*
having shared content in the new SNS × Content interest diversity in the existing SNS					−0.39	
having shared content in the new SNS × Group size in the existing SNS					0.40	*

In Model 3, we can see the newly introduced variables are significant except *having shared content in the new SNS × content interest diversity in the existing SNS*. First, *having shared content in the new SNS × user tenure in the existing SNS* and *having shared content in the new SNS × follower size in the existing SNS* are all positively associated with *like receiving frequency in the existing SNS*, which means longer-time users and users with a broader audience in the existing SNS correspond to less decline of content quality after the new site use. Second, *having shared content in the new SNS × interaction partner size in the existing SNS* is negative, while *having shared content in the new SNS × interaction message size in the existing SNS* is positive. It suggests that the setback degree of content quality in the existing site increases with users' interaction partner size but decreases with the number of interaction messages. Third, *having shared content in the new SNS* × *group size in the existing SNS* is positively related to the dependent variable. The content quality of users with more groups in the existing SNS is likely to decline less after they utilize a new site to share content. The last, similar to the results of content frequency analysis, the variable *having shared content in the new SNS* × *content interest diversity in the existing SNS* is also not significant in the model. There is no significant difference of content quality declining trend when considering content diversity as a metric.

6 Discussion

Our results indicate that users' content quantity in the existing SNS experiences an improvement right before and after they begin sharing content in a new site. It complies with previous research which suggests that SNS users' tendency for self-disclosing and information sharing increases with the number of sites they use. However, content quality in the existing SNS experiences a setback right after users begin sharing content in a new site. We thought as people' time and efforts for SNS use are generally limited [24], when they are involved into the content sharing practice in a new SNS, it can take their time and efforts for content crafting and censoring in the existing site, which weakens the content quality in that. Considering user characteristics, we find that the changes of users' content quality and content quality in the existing SNS right after a newly conducted content sharing practice are associated with some user features like user tenure, audience, interactions, etc.

Longer-time users tend to correspond to less improvement on content quantity but less decline on content quality. The tendency for content sharing declines with user tenure in general, which implies the longer-time users correspond to weaker desires to distribute content compared with shorter-time users. It can limit the improvement degree of content sharing frequency of longer-time users in the existing SNS. However, longer-time users in the existing SNS are generally characterized with stronger commitment to the platform [23], and such users are likely to have more abundant knowledge and extensive experiences [4], which can make their content quality in that more robust (to decline less) when they start content sharing practice in other mediums.

The degree of content quantity improvement increases with follower size, while the extent of content quality decline decreases with that. The SNS users with more audiences tend to have stronger motivation to publish content as they perceive that the newly shared content can permeate more people. In the meanwhile, users generally craft each post to be published to ensure the content can be perceived as appropriate by the audience [20]. The existence of larger number of followers in the existing SNS can simulate users to censor the content to be published more carefully, which is helpful for the content quality robustness in that.

Social interaction cannot be always helpful for a user's content sharing. To conclude, our results indicate more interactions concentrating on a more restricted span of interaction relationships can be more helpful for stimulating improvement of content quantity and alleviating setback of content quality in the existing SNS. As users' time and efforts for SNS use are limited, to concentrate interaction efforts on a more restricted span of friends highlights the stronger significance and strength of such relationships compared with the other ties.

Content interest diversity is not a significant factor associated with the change of content sharing in the existing SNS. Users with more diverse interests are difficult to generate sense of common social identity in the existing SNS, which can weaken their content quantity and content quality. However, diverse interests and needs are significant intrinsic factors that motivate users to be involved into multiple SNSs [20]. The users with more diverse interests trend to combine different platforms to share content as one site cannot meet their goals. Thus it is less likely that such users abandon participating in the existing sites, which is helpful for their sustained content sharing in that. These two contradictory aspects can interpret why interest diversity is not significant in our analysis.

Users participating in more groups in the existing SNS correspond to more improvement on content quantity and less decline on content quality after beginning shared content in a new SNS. By organizing users into groups in SNSs, it can generate stronger sense of social identity among them and intensify their commitment to the platform, which is helpful for motivating users to contribute more and high-quality contents and efficient for stimulating content quantity improvement and alleviating content quality setback shown in this paper.

6.1 Implications

Our findings provide valuable insights into both alleviating the content quality setback as well as prompting further improvement of content quantity. First, SNS should not just consider to recruit newcomers to expand the user base, but also lay emphasis on longer-time users as such users tend to contribute high-quality but less contents both before and after they are involved into content sharing practice in other platforms. We also suggest practitioners to pay attention to users with less followers, less groups and less interactions especially through strong ties as their content quantity and content quality are more sensitive to sharing content in a new site. Second, based on our results, some actions can be taken to maintain these users' content quantity and content quality. The major suggestion is making it easy for users to conduct interactions through strong ties. For content quantity improvement promotion and content quality setback alleviation, communicating several times through one strong relationship is more meaningful than communicating one time through each of several common relationships. SNSs should not simply emphasize on the number of communication partners, but pay attention to strategies that can stimulate communication between relationship parties characterized with strong strength. Moreover, SNSs are better to recommend more groups to users as it can not only promote users' content quantity and content quality in that but also stimulate their sustained content contribution when they are involved into content sharing practice in other platforms.

7 Conclusion

In this paper, we utilized Weibo and Douban as research sites to see how content sharing in Weibo (Douban) could affect that in Douban (Weibo), and what kinds of users' content sharing in Weibo or Douban was more or less likely to be affected. According to the results, one user's newly conducted content sharing practice in Weibo (Douban) can promote her/his content quantity in Douban (Weibo) in a short time period, while the content quality (number of likes) tends to decline right after the new content sharing practice. We also uncovered many mechanisms that can be utilized to stimulate the improvement of content quantity and alleviate the decline of content quality, including improving interaction experience to guide the users to communicate more through strong relationships and encouraging them to connect with more followers and participate in more groups. As the difficulty to sample data from multiple SNSs, only two sites are utilized for cross-SNS user content analysis in our work. In the future, such cross-SNS analysis can be migrated to other popular SNSs to obtain more general and plenteous results.

Acknowledgements. This work is supported by the National Natural Science Foundation of China (NSFC) under the Grants nos. 61902075 and 61932007.

References

1. Nov, O., Naaman, M., Ye, C.: Motivational, structural and tenure factors that impact online community photo sharing. In: Proceedings of the Third International ICWSM Conference, pp. 138–145. AAAI (2009)
2. Kairam, S., Kaye, J., Guerra-Gomez, J.A., Shamma, D.A.: Snap decisions?: how users, content, and aesthetics interact to shape photo sharing behaviors. In: Proceedings of the 2016 SIGCHI Conference on Human Factors in Computing Systems, pp. 113–124. ACM (2016)
3. Angrist, J.D., Krueger, A.B.: Empirical strategies in labor economics. Handb. Labor Econ. **3**, 1277–1366 (1999)
4. Arguello, J., et al.: Talk to me: foundations for successful individual-group interactions in online communities. In: Proceedings of the SIGCHI Conference on Human Factors in Computing Systems, pp. 959–968. ACM (2006)
5. Blei, D.M., Ng, A.Y., Jordan, M.I.: Latent dirichlet allocation. J. Mach. Learn. Res. **3**, 993–1022 (2003)
6. Fang, B., Liu, X.: Do money-based incentives improve user effort and UGC quality? Evidence from a travel blog platform. In: Proceedings of the Twenty-Second Pacific Asia Conference on Information Systems, p. 132 (2018)

7. Gao, R., Hao, B., Li, H., Gao, Y., Zhu, T.: Developing simplified Chinese psychological linguistic analysis dictionary for microblog. In: Imamura, K., Usui, S., Shirao, T., Kasamatsu, T., Schwabe, L., Zhong, N. (eds.) BHI 2013. LNCS (LNAI), vol. 8211, pp. 359–368. Springer, Cham (2013). https://doi.org/10.1007/978-3-319-02753-1_36
8. Gilbert, E.: Predicting tie strength in a new medium. In: Proceedings of the 2012 ACM Conference on Computer Supported Cooperative Work, pp. 1047–1056. ACM (2012)
9. Gilbert, E., Karahalios, K.: Predicting tie strength with social media. In: Proceedings of the 2009 SIGCHI Conference on Human Factors in Computing Systems, pp. 211–220. ACM (2009)
10. Hsiao, C.: Analysis of Panel Data, no. 54. Cambridge University Press (2014)
11. Huberman, B., Romero, D.M., Wu, F.: Social networks that matter: Twitter under the microscope. First Monday **14**, 1 (2008)
12. Kittur, A., Chi, E., Pendleton, B.A., Suh, B., Mytkowicz, T.: Power of the few vs. wisdom of the crowd: Wikipedia and the rise of the bourgeoisie. World Wide Web **1**(2), 19 (2007)
13. Lampe, C., Wash, R., Velasquez, A., Ozkaya, E.: Motivations to participate in online communities. In: Proceedings of the 2010 SIGCHI Conference on Human Factors in Computing Systems, pp. 1927–1936. ACM (2010)
14. Lim, B.H., Lu, D., Chen, T., Kan, M.-Y.: # mytweet via instagram: exploring user behaviour across multiple social networks. In: Proceedings of the 2015 IEEE/ACM International Conference on Advances in Social Networks Analysis and Mining, pp. 113–120. ACM (2015)
15. Manikonda, L., Meduri, V.V., Kambhampati, S.: Tweeting the mind and instagramming the heart: exploring differentiated content sharing on social media. In: Proceedings of the Tenth International AAAI Conference on Web and Social Media, pp. 639–642. AAAI (2016)
16. Manikonda, L., Meduri, V.V., Kambhampati, S.: Analysis of participation in an online photo-sharing community: a multidimensional perspective. J. Am. Soc. Inf. Sci. Technol. **61**(3), 555–566 (2010)
17. Ottoni, R., et al.: Of pins and tweets: investigating how users behave across image- and text-based social networks. In: Proceedings of the Eighth International AAAI Conference on Weblogs and Social Media, pp. 386–395. AAAI (2014)
18. Panciera, K., Priedhorsky, R., Erickson, T., Terveen, L.: Lurking? cyclopaths?: a quantitative lifecycle analysis of user behavior in a Geowiki. In: Proceedings of the 2010 SIGCHI Conference on Human Factors in Computing Systems, pp. 1917–1926. ACM (2010)
19. Sleeper, M., Balebako, R., Das, S., McConahy, A.L., Wiese, J., Cranor, L.F.: The post that wasn't: exploring self-censorship on facebook. In: Proceedings of the 2013 ACM Conference on Computer Supported Cooperative Work, pp. 793–802. ACM (2013)
20. Sleeper, M., Melicher, W., Habib, H., Bauer, L., Cranor, L.F., Mazurek, M.L.: Sharing personal content online: exploring channel choice and multi-channel behaviors. In: Proceedings of the 2016 SIGCHI Conference on Human Factors in Computing Systems, pp. 101–112. ACM (2016)
21. Smith, A., Anderson, M.: Social media use in 2018 (2018). https://www.pewresearch.org/internet/2018/03/01/social-media-use-in-2018/

22. Wang, Y.-C., Burke, M., Kraut, R.E.: Gender, topic, and audience response: an analysis of user-generated content on Facebook. In: Proceedings of the 2013 SIGCHI Conference on Human Factors in Computing Systems, pp. 31–34. ACM (2013)
23. Yang, D., Kraut, R., Levine, J.M.: Commitment of newcomers and old-timers to online health support communities. In: Proceedings of the 2017 CHI Conference on Human Factors in Computing Systems, pp. 6363–6375. ACM (2017)
24. Zhang, P., Zhu, H., Lu, T., Gu, H., Huang, W., Gu, N.: Understanding relationship overlapping on social network sites: a case study of Weibo and Douban. In: PACMHCI, vol. 1, no. CSCW, pp. 120–121 (2017)

An Improved Label Propagation Algorithm for Community Detection Fusing Temporal Attributes

Wenjing Gu, Chengjie Mao(✉), Ronghua Lin, Wande Chen, and Yong Tang

South China Normal University, Guangzhou 510631, China
{gwen,rhlin,chenwande,ytang}@m.scnu.edu.cn, maochj@qq.com

Abstract. Community detection algorithms are used to analyze the community structure. However, existing algorithms are based on undirected or undimensional network structure, which cannot make full use of the user's attributes. Based on the academic social network, we propose an algorithm that integrates multidimensional network structures and user attributes, which can divide community by using an improved Label Propagation Algorithm (LPA). We construct a real social network by integrating multidimensional network, and use the similarity of research fields of scholars to generate edges between nodes. The directed edges are weighted by the four-dimensional network and temporal attributes, while the nodes are weighted by the social status and influences in the network structure of users. Then, the propagation probability of label is defined by the weighted indicators, and the community is divided via the directed weighted academic network. The experimental results show that the proposed algorithm LPA-STN can effectively improve not only the accuracy of the community detection results, but also the compactness within the communities.

Keywords: Academic social network · Heterogeneous network · Community detection · LPA

1 Introduction

With the development of social network [1], users are pursuing faster and more efficient interactions. Meanwhile, the academic social network between scholars are also developing rapidly, such as SCHOLAT, Academia, ResearchGate, etc. There is a large amount of academic information every day. Therefore, analyzing the complex social network of scholars cannot only help us study the network structure of real communities and mine scholars' academic information, but also make scholars communicate efficiently and promote the development of academic social network.

This work was supported in part by the National Natural Science Foundation of China under Grant 61772211 and Grant U1811263.

Y. Sun et al. (Eds.): ChineseCSCW 2021, CCIS 1492, pp. 312–325, 2022.
https://doi.org/10.1007/978-981-19-4549-6_24

Many researchers construct an academic social network with multi-dimensional heterogeneous characteristics, and detect community by integrating the dimensions of social relationships. However, they ignore one characteristic of a real community, which can affect the results of community divisions. This characteristic is the time-sensitive [2,3], and the relationships between users change over time, which affects the results of community division.

In this paper, we propose a community detection algorithm for directed weighted academic social network embedded with temporal attributes, which can derive hidden edges by using the similarity of academic research interests of scholars [4,5] and building a real social network via multi-dimensional social connections. For the weight of the network, the local influences of nodes are calculated by combining the interactions between users and the similarity of research interests of users, which is defined as the weights of directed edges. The topology of the network and the social status of users are used to calculate the global influences, and we define it as the weights of the nodes.

In summary, we select LPA as the baseline since it has original teams and courses in SCHOLAT, which is beneficial to our experiments. There are some approaches to generate edges with the similarity of research fields such as TF-IDF , LDA topic model and word2vec [6,7]. Considering the algorithm cost, we use LDA topic model to mine users' research fields, calculate the similarity of users' research fields via cosine similarity, and set a threshold to decide whether edges be derived. We calculate the time weights of edges via time decay function. We consider both the influences and social status of users to weight node to improve the results of community detection. For the influences, we use PageRank to analyze the influences of the nodes in the network. PageRank judges the rank of one node by using it's in-degrees and adjacent nodes. For the social status, both the influences of personal and social attributes of the user are calculated as factors. And then, the multidimensional directed heterogeneous network with edges and weighted nodes is constructed. The improved label propagation algorithm makes full use of the team and course information to define the initial subgroup labels. For directed network, users tend to accept the labels of users they trust rather than whom trust them [8], so we define users to accept the labels of users pointing to themselves. In real network, users do not belong to only one community. Therefore, we construct the set of accepted labels of users by using the propagation probability of labels. We achieve the community discovery of academic social network based on label propagation by leveraging the directed weighted network, the improved label initialization, and label update rules.

The main contributions of this paper are as follows:

1. We generate edges between nodes with no interactions, add the time attributes, and weight the user nodes to construct a four-dimensional heterogeneous network.
2. We use the weighting nodes to improve the rule of label propagation, and label the nodes by the original information of datasets.

3. We propose an algorithm LPA-STN that combines temporal attributes and node weights. We not only construct a network with directed edges, and weight both edges and nodes, initial core group labels, but also define the propagation probability of the labels by considering the two weighted metrics.

2 Related Work

There are many researches about the structure of complex social network. For mining the network structure of complex network and partitioning it, researchers have studied in various aspects.

The community detection methods are mainly as follows: (1) Methods based on graph segmentation, such as Kernighan-Lin algorithm, spectrum halving method, etc.; (2) Methods based on hierarchical clustering, such as GN algorithm, Newman fast algorithm, etc.; (3) Methods based on modularity optimization, such as greedy algorithm, simulated annealing algorithm, Memetic algorithm, PSO algorithm, evolutionary multi-objective optimization algorithm, etc. The traditional cluster percolation-based algorithm [9] detects the initial communities by using CPM algorithm. And Zhou [10] et al. proposed an improved CPMK-Means algorithm that combines depth-first and breadth-first search to generate the maximum number of communities to determine the number of clustering centers and iterations are repeatedly executed until the centers become stable.

Most of such algorithms need to be detected from the global analysis of the network, which need some data such as the structural information of the whole network and the scale of the community division. The complexity of the algorithms will become higher, and much more researchers start to study the network structure based on the local features [11] to mine the network structure. Firstly, we take the sub-network with some certain features as the seed community, and then expand it to the adjacent nodes via merging, so as to obtain the most optimized community we want. Shi et al. [12] proposed a local partial spectrum approximation (LBSA) algorithm based on the traditional local expansion algorithm, which samples smaller subgraphs via fast random wandering, personalized PageRank [13] and heat kernel diffusion, and use a few seed numbers to identify other potential members. Currently researchers mostly use semi-supervised machine learning via LPA [14] to detect community, but the community is not stable because of the randomness in label selection.

3 Algorithm Design and Implementation

In this section, we will analyze the structural features of the datasets in scholar network. Then we introduce how to abstract the social behaviors of real network into nodes and edges, and build a connected network. Moreover, we use the interaction of users to weight nodes and edges, and divide communities by our proposed algorithm.

3.1 Building a Four-Dimensional Network with Embedded Temporal Properties

There are different interactions between scholars, which constitute a complex heterogeneous network. Heterogeneous network can be classified as multimodal and multidimensional. The nodes in our network are all scholar users, and the interactions between users are various. One connection relationship represents a dimension, so it belongs to single-mode, namely multidimensional network. There are three kinds of links among users, such as academic cooperation, interaction, and follow. And we construct a scholar network by building the connections between nodes. Considering that some nodes have no social links, we build a connection between two nodes with the similarity of research fields. In real network, the community structure changes over time. Therefore, the dynamic division can affect the real community structure by considering temporal attributes. The shorter the time interval between nodes' interactions is, the bigger the probability that belong to the same community in this period is. At present, the decay functions proposed for time attribute are mainly exponential, linear and Ebbinghaus type, etc. Combining the academic characteristics of scholars, we joint exponential decay with linear variation to generate a logical time decay function. The equation is defined as:

$$W(t_i, t) = \frac{1}{1 + e^{\mu(t - t_i)}}, \tag{1}$$

where t is the current time and t_i is the initial time.

Academic Cooperation Network. If there is academic cooperation between users i and j, then we consider they belong to the same community and have double connected edge, m is the proportion of time attributes. The weight value between user i and j can be obtain in Eq. (2):

$$AW(i, j) = m(\frac{coauthor(i, j)}{max\{coauthor(i, k)\}}) + (1 - m)W(t_i, t), \tag{2}$$

where $coauthor(i, j)$ is the number of academic cooperation between users, and k is the scholar who cooperates with i.

User Interaction Network. In SCHOLAT, if scholar i comments ($commemt$ (i, j)), repost ($transpond(i, j)$), and likes ($likes(i, j)$) scholar j's dynamics and other academic information, we consider scholar i may be interested in j, and it is highly likely that they belong to the same community. Thus, we build a directed edge from scholar i to j, the equation as follow:

$$sum(i, j) = comment(i, j) + transpond(i, j) + like(i, j) \tag{3}$$

$$IW(i, j) = m\frac{sum(i, j)}{max\{sum(i, k)\}} + (1 - m)\frac{1}{t - (t_i)} \tag{4}$$

User Follow Network. If scholar i follows scholar j, it indicates that scholar i is interested in j. Therefore they may come from the same community, and there is a directed edge from i to j. If scholar i follows j, the (i, j)-entry of adjacency matrix A will be assigned a value of 1. The equation is shown as follows:

$$F(i,j) = \begin{cases} 0, A_{ij} = 0 \\ 1, A_{ij} = 1 \end{cases} \tag{5}$$

$$FW(i,j) = mF(i,j) + (1-m)\frac{1}{t-t_i} \tag{6}$$

User Similarity Network. For scholars who have no interactions temporarily, if the research fields of them are highly similar, we also consider that they may be from the same community and have a double directed edge. The similarity is defined as:

$$S(i,j) = S(j,i) = cos(D_i, D_j) = \frac{\Sigma_{k=1}^{n}(w_{ik} \times w_{jk})}{\sqrt{\Sigma_{k=1}^{n}(w_{ik})^2}\sqrt{\Sigma_{k=1}^{n}(w_{jk})^2}}, \tag{7}$$

$$SW(i,j) = m \cdot S(i,j) + (1-m) \cdot W(t_i, t), \tag{8}$$

where the vector $D_i = \{w_{i1}, w_{i2}, \cdots, w_{ik}\}$ is used to represent the research fields that user i is interested in, w_{ik} is the probability that user i is interested in research topic k, and m is the number of research fields that user i is interested in.

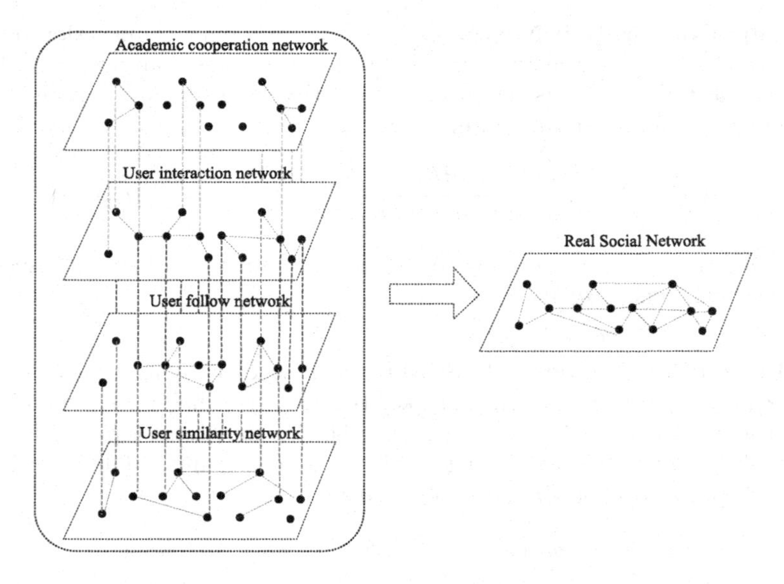

Fig. 1. The integration of the four-dimensional network

The network fusing those four dimensions is showed as Fig. 1 We need to consider the four heterogeneous graphs to determine whether it has edge between node i and j. The fusion equation is defined as:

$$RW(i,j) = \alpha AW(i,j) + \beta FW(i,j) + \gamma IW(i,j) + (1-\alpha-\beta-\gamma)SW(I,J) \quad (9)$$

3.2 Node Weighting

The nodes with different status have different influences in community structure. For example, the quality of a professor's academic information is significantly higher than an undergraduate. Combining the influences of the nodes and the intensity of user interactions helps us analyze the propagation pattern of labels in the network more clearly. So we use both the influences of users based on the tightness and the status of individuals in the network to measure the global influences of nodes.

Node Influences Based on Network Structure. To analyze the importance of the user in the network, we use the PageRank algorithm to analyze the nodes and calculate the importance of the nodes in the whole network. The PageRank(PR) is defined as:

$$PR(i) = (1-\alpha)\frac{1}{N} + \alpha\Sigma_{j\in M_i}\frac{PR(j)}{L(j)} \quad (10)$$

where N denotes the total number of network nodes, M_i is the set of all nodes pointing to i, and $L(j)$ denotes the sum of the out-degree of node j.

Social Status of Users in the Network. In academic social network, the social status of scholars can be measured from two aspects: personal attributes (such as title rank, academic achievements). And social attributes(such as courses, teams and academic dynamics). $AcademicAchievements(i)$ denotes the sum of academic achievements of user i and $max\{AcademicAchievements\}$ is the maximum value of the academic achievements of the users in the network. And then we then digitize the title rank of the user as $Title(i)$ where Professor is defined as 10, Associate Professor as 8, Doctor as 5, Master as 3, Undergraduate as 1, and other as 0. The personal attribute of a user is defined as:

$$Personal(i) = \frac{AcademicAchievements(i)}{max\{AcademicAchievements\}} + \frac{AcademicTitle(i)}{10} \quad (11)$$

The number of academic socialization in which scholars participate is $course(i)$, $team(i)$, $dynamic(i)$, respectively, and the sum of those and the social attribute of a user is defined as follows:

$$sum(i) = course(i) + team(i) + dynamic(i) \quad (12)$$

$$Social(i) = \frac{sum(i)}{max\{sum(k)\}} \quad (13)$$

The social status of the user nodes is obtained by the users' personal and social attributes:

$$SP(i) = \eta Personal(i) + (1 - \eta)Social(i) \tag{14}$$

In summary, we calculate the influences of scholars in the network by combining the network structure with the social status of scholars, and thus we define the weighted node value as:

$$NW(i) = \theta(i) + (1 - \theta)PR(i) \tag{15}$$

3.3 Label Propagation Algorithm Based on Node Influence

The user nodes in SCHOLAT have different feature attributes, so we assign corresponding weights to the nodes to measure their influences in the network, and improve the label initialization rules and label propagation rules. The nodes we selected from the SCHOLAT datasets include labels of team or course, so we can analyse the relatedness of nodes labeled with team or course and get the initial group during label initialization, and each initial group is given a unique label l. At this time, all nodes in this group label l, and the probability that a node accepts label l is 1, which improves the speed and quality of community structure mining, by using label initialization with a semi-manual approach.

In the label propagation phase, the values of local influences and global influences of nodes are used as the impact factors to measure the probability of label propagation. And then we define the propagation probability of labels, which can avoid the instability of the traditional algorithm that randomly selects neighbor nodes. At the same time, we believe that in a directed network, users are more inclined to receive labels from users they trust than from users trust them. Therefore, we just consider the outgoing edges of users. Taking Fig. 2 as an example, if the node p trust q, then it accepts the label T1 of q. Similarity, if node i and j, since node j already has the label T2, it only accepts label T1. However, node m does not belong to any community, so it accepts both T1 and T2 from the node n it trusts.

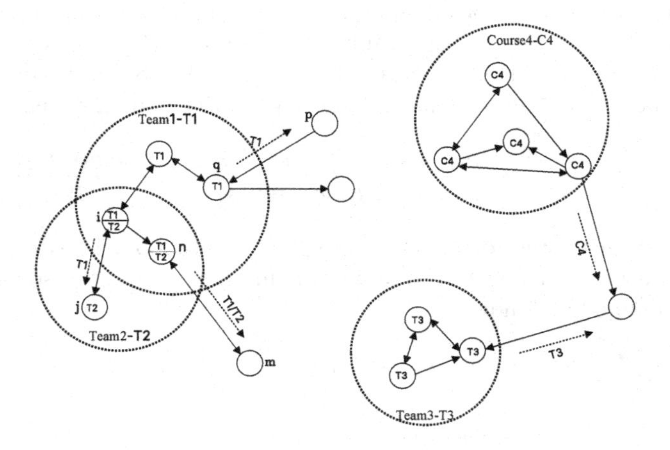

Fig. 2. The integration of the four-dimensional network

We denote the pair of label owned by a node as $L_n = (l_n, p_n), (n = 1, 2, \cdots, n)$, where l is a label of node i and p is the probability that node i owns label l. Initially, the probabilities of all the labels owned by nodes are 1. By continuously iterating the labels until the label probabilities converge, the label set of node i tends to be stable, and the label set of node i is denoted as $L(i) = label(i) = (L_1, L2, \cdots, L_n)$, in descending order of label probabilities. When the label is updated, we perform two operations as follows: 1)we remove labels with the probability less than the threshold value, 2)we scan its adjacent node j to add a new label pair, and denote V_{ij} as the set of neighbor nodes trusted by node i. In the end, the probability of label l is updated as p_1'.

4 Experiment

In this section, we will introduce our experimental settings, including tested datasets, compared baselines, evaluation metrics (Modularity, Similarity, Normalized Mutual Information), and so on. In addition, we concisely illustrate some primary settings of this work for making this work self-completed.

4.1 Data

This paper uses the real datasets from SCHOLAT (www.scholat.com), an online academic platform that provides academic information management, team management, and scholar exchange services. At present, the number of visits to SCHOLAT exceeded 3 million, and there is rich semantic information.

We select the cooperation, interaction and follow information, as well as the achievements and title of users, constructing a network via those. And we use the similarity of research fields to generate the hidden edges among similar users by mining the achievements of users. The datasets are shown in Table 1.

Table 1. Network data set

Type	Node	Edge
Academic cooperation network	868	4372
User interaction network	2135	21536
User follow network	34471	108472
User similarity network	2154	199396
Four-dimensional real network	35125	134380

4.2 Evaluation Metrics

The results of community division are measured by modularity, average similarity [15], Normalized Mutual Information(NMI) and Adjusted Rand Index(ARI)[16]. The value of modularity Q is between [0,1], the value of Q is computing as:

$$Q = \Sigma_i (e_{ij} - a_i^2), \tag{16}$$

where e_{ij} denotes the ratio of the sum of the number of edges inside community i and community j to the total number of edges in the network, and a_i denotes the ratio of the number of all edges associated inside community i to the total number of edges.

The average similarity is an indicator of the similarity of the nodes in the divided network, which is used to measure the degree of clustering of the nodes in the community:

$$Sim = \frac{\Sigma S_i}{N}, \tag{17}$$

where S_i is the average of the similarity between all nodes within a subcommunity two by two, and N is the number of communities. The higher average similarity indicates the closer the connection within the community.

The value of NMI is based on the contingency values of true classes $Y = (Y1, Y2, \cdots)$ and output classes $X = (X1, X2, \cdots)$, $p(x, y)$ is the joint distribution of random variables (X, Y), $p(x), p(y)$ is the marginal distribution. $MI(X, Y)$ is the relative entropy of $p(x, y)$ and $p(x)(y)$, which is defined as:

$$MI(X, Y) = \Sigma_{i=1}^{Y} \Sigma_{i=1}^{X} p(x, y) log \frac{p(x, y)}{p(x)p(y)} \tag{18}$$

And NMI can be computed as follows:

$$NMI(X, Y) = 2\frac{MI(X, Y)}{H(X) + H(Y)}, \tag{19}$$

where $H(X) = -\Sigma_i p(x_i) logp(x_i)$ and $H(Y) = -\Sigma_i p(y_i) logp(y_i)$ is the quantitative measurement of information, respectively.

4.3 Result

Our proposed algorithm will be compared with the following algorithms, in order to verify whether it makes full use of the labeling attributes and can get a better and more realistic structure of community via applying directed weighted network graphs with embedded temporal attributes.

(1) Label propagation algorithm with similarity (SLPA) for based on user similarity. It deals with nodes having no interaction but being similar for research fields, and then forms a directed weighted network by connecting all nodes in the network.
(2) Label propagation algorithm based on network structure and user information (LPA-NU): this algorithm just considers the network structure and the node weighting of users.
(3) LPA-TNU: Based on the LPA-NU algorithm, it considers both the local influence of users to improve the quality of community division.
(4) LPA-NNU: Based on the LPA-NU algorithm, the temporality of the network structure is considered, and temporal attributes are added to the user relationships.

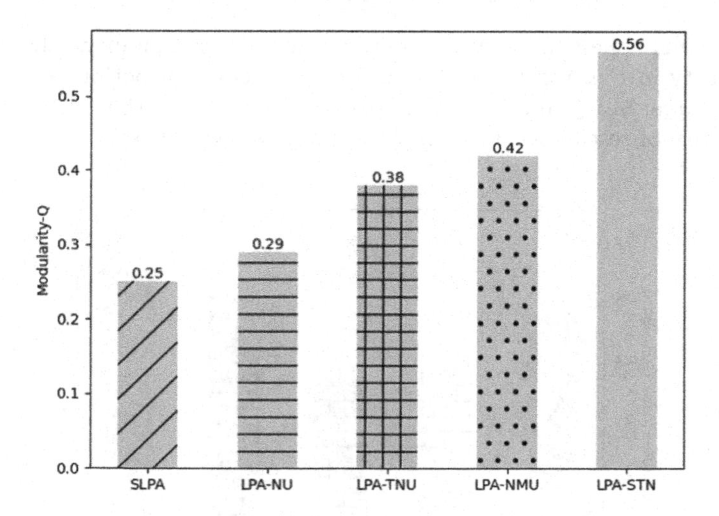

Fig. 3. The modularity comparisons

The results showed in Fig. 3 is value of modularity Q on the five algorithms. The values of Q are 0.25 and 0.29 when we just consider the similarity between users and the network structure respectively. If we add the temporal attributes or the influences of user nodes, the Q values are 0.38 and 0.42, which is better. The Q value of LPA-STN is 0.56 indicates that the combination of the user similarity, the influences of nodes and the temporal attributes perform a better result.

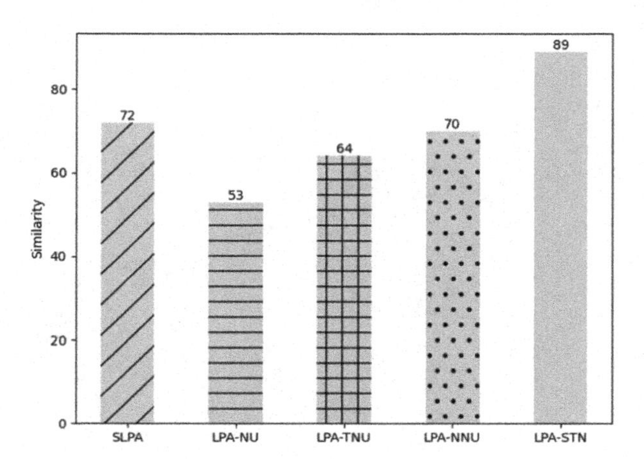

Fig. 4. The similarity comparisons

We calculate the similarity of nodes within each community. The results of the similarity in Fig. 4 indicate that LPA-STN algorithm performs better than the other three. SLPA and LPA-STN perform better than the others show that the similarity of research fields is key to detect communities.

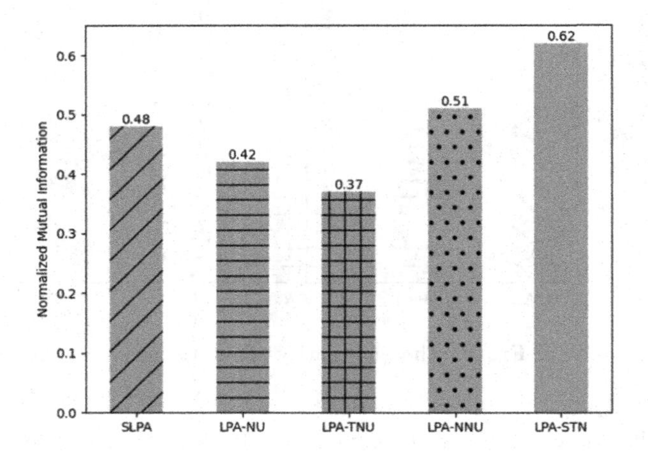

Fig. 5. The Normalized Mutual Information comparisons

The result showed in Fig. 5 indicates that when we consider the influence of nodes, we can get better performances on the metric of NMI. The similarity of user nodes and the network structure are also beneficial for our experiment, the time attributes is not the crucial factor.

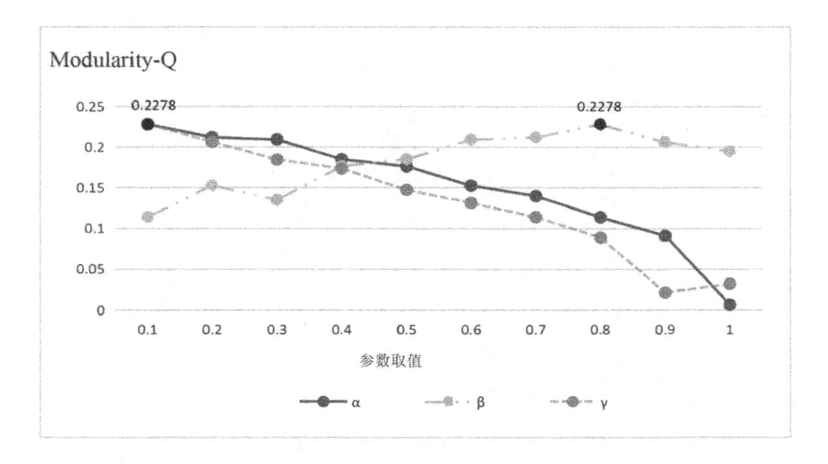

Fig. 6. The influences of α-β-γ

We also discuss the effects of different parameters on the results when weighting edges and nodes, respectively. In weighting edges, we integrate four network to form the real social network, and their weights were α, β, γ, and $1-\alpha-\beta-\gamma$, respectively. The results of modularity Q in our proposed LPA-STN with different parameters in Fig. 6 We not only know that the maximum modularity Q is 0.2278, but also find that the four weights of the real social network are 0.5, 0.1, 0.1, and 0.3, respectively, which present that academic cooperation is a key factor for detecting the community. In addtion, we can know that the similarity of research fileds is a key factor.

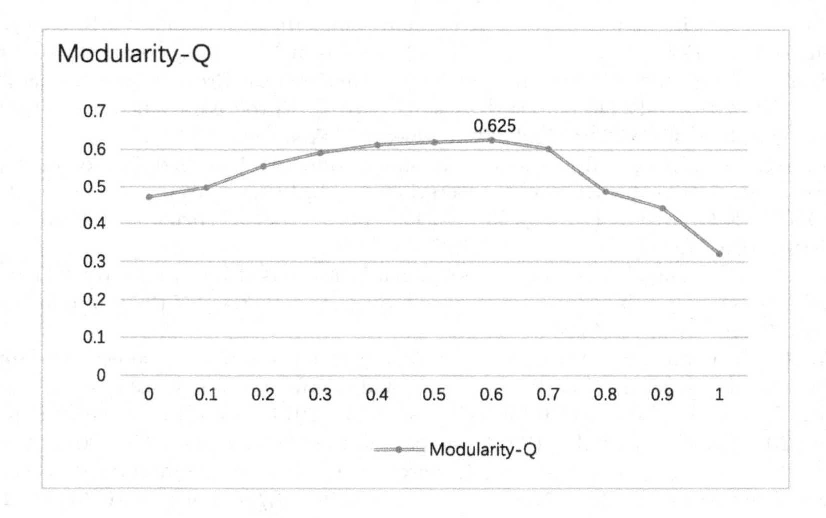

Fig. 7. The influences of α-β-γ

The influence of parameter θ is showed in Fig. 7 It presents the proportion of weighting nodes. The value of Q is the highest when θ is 0.58, which indicates that it performs better considering the influences of nodes than the structure of network.

5 Conclusion

We propose the LPA-STN, which combines the similarity of researches fields, temporal attributes and the influences of user nodes. The improved algorithm solved the problems of nodes with no interactions, added the time attributes to simulate the dynamic detection and the influences of nodes, these all make the results better. And the improvement of label initialization and label propagation rules optimizes the time efficiency of the algorithm and improves the quality of the community. However, we just consider the discrete time attributes, and ignore the correlation between the current time and the previous time. We cannot guarantee the smooth evolution of the community, so that we need to evaluate the structure of the network at different times with terrible time complexity. So how to smooth the long-term evolution of dynamic community with time balance algorithm, it is what we need to challenge.

References

1. Li, N., et al.: A review of the research progress of social network structure. Complex **2021**, 6 692 210:1–6 692 210:14 (2021). https://doi.org/10.1155/2021/6692210
2. Dall'Amico, L., Couillet, R., Tremblay, N.: Community detection in sparse time-evolving graphs with a dynamical bethe-hessian. In: Larochelle, H., Ranzato, M., Hadsell, R., Balcan, M., Lin, H., (eds.) Advances in Neural Information Processing Systems 33: Annual Conference on Neural Information Processing Systems 2020, NeurIPS 2020, 6–12 December 2020 (2020). https://proceedings.neurips.cc/paper/2020/hash/54391c872fe1c8b4f98095c5d6ec7ec7-Abstract.html
3. Ji, Y., et al.: Temporal heterogeneous interaction graph embedding for next-item recommendation. In: Hutter, F., Kersting, K., Lijffijt, J., Valera, I. (eds.) ECML PKDD 2020. LNCS (LNAI), vol. 12459, pp. 314–329. Springer, Cham (2021). https://doi.org/10.1007/978-3-030-67664-3_19
4. Meng, Q., Xiong, H.: A doctor recommendation based on graph computing and LDA topic model. Int. J. Comput. Intell. Syst. **14**(1), 808–817 (2021). https://doi.org/10.2991/ijcis.d.210205.002
5. Maity, S., Rath, S.K.: Extended clique percolation method to detect overlapping community structure. In: 2014 International Conference on Advances in Computing, Communications and Informatics, ICACCI 2014, Delhi, India, 24–27 September 2014, pp. 31–37. IEEE (2014). https://doi.org/10.1109/ICACCI.2014.6968420
6. He, K., Li, Y., Soundarajan, S., Hopcroft, J.E.: Hidden community detection in social networks. Inf. Sci. **425**, 92–106 (2018). https://doi.org/10.1016/j.ins.2017.10.019
7. Pang, S., Ban, T., Kadobayashi, Y., Kasabov, N.K.: LDA merging and splitting with applications to multiagent cooperative learning and system alteration. IEEE Trans. Syst. Man Cybern. Part B **42**(2), 552–564 (2012). https://doi.org/10.1109/TSMCB.2011.2169056
8. Zhang, J., Zhu, Y., Chen, Z.: Evolutionary game dynamics of multiagent systems on multiple community networks. IEEE Trans. Syst. Man Cybern. Syst. **50**(11), 4513–4529 (2020). https://doi.org/10.1109/TSMC.2018.2854294
9. Li, P., Chien, I.E., Milenkovic, O.: Optimizing generalized pagerank methods for seed-expansion community detection. In: Wallach, M., Larochelle, H., Beygelzimer, A., d'Alché-Buc, F., Fox, E.B., Garnett, R.H., (eds.) Advances in Neural Information Processing Systems 32: Annual Conference on Neural Information Processing Systems 2019, NeurIPS 2019, 8–14 December 2019, Vancouver, BC, Canada, pp. 11 705–11 716 (2019). https://proceedings.neurips.cc/paper/2019/hash/9ac1382fd8fc4b631594aa135d16ad75-Abstract.html
10. Shi, P., He, K., Bindel, D., Hopcroft, J.E.: Locally-biased spectral approximation for community detection. Knowl. Based Syst. **164**, 459–472 (2019). https://doi.org/10.1016/j.knosys.2018.11.012
11. Deng, X., Wen, Y., Chen, Y.: Highly efficient epidemic spreading model based LPA threshold community detection method. Neurocomputing **210**, 3–12 (2016). https://doi.org/10.1016/j.neucom.2015.10.142

12. Dey, A., Jenamani, M., Thakkar, J.J.: Lexical TF-IDF: an n-gram feature space for cross-domain classification of sentiment reviews. In: Shankar, B.U., Ghosh, K., Mandal, D.P., Ray, S.S., Zhang, D., Pal, S.K. (eds.) PReMI 2017. LNCS, vol. 10597, pp. 380–386. Springer, Cham (2017). https://doi.org/10.1007/978-3-319-69900-4_48
13. Dupuy, C., Bach, F., Diot, C.: Qualitative and descriptive topic extraction from movie reviews using LDA. In: Perner, P. (ed.) MLDM 2017. LNCS (LNAI), vol. 10358, pp. 91–106. Springer, Cham (2017). https://doi.org/10.1007/978-3-319-62416-7_7
14. Polpinij, J., Srikanjanapert, N., Sopon, P.: Word2Vec approach for sentiment classification relating to hotel reviews. In: Meesad, P., Sodsee, S., Unger, H. (eds.) IC2IT 2017. AISC, vol. 566, pp. 308–316. Springer, Cham (2018). https://doi.org/10.1007/978-3-319-60663-7_29
15. Kirkley, A., Newman, M.E.J.: Representative community divisions of networks, CoRR, vol. abs/2105.04612 (2021). https://arxiv.org/abs/2105.04612
16. Singh, J., Singh, A.K.: NSLPCD: topic based tweets clustering using node significance based label propagation community detection algorithm. Ann. Math. Artif. Intell. **89**(3–4), 371–407 (2021). https://doi.org/10.1007/s10472-020-09709-z

Understanding Scholar Social Networks: Taking SCHOLAT as an Example

Min Gao[1,2], Yang Chen[1,2](✉), Qingyuan Gong[1,2], Xin Wang[1,2], and Pan Hui[3,4]

[1] School of Computer Science, Fudan University, Shanghai, China
mgao21@m.fudan.edu.cn, {chenyang,gongqingyuan,xinw}@fudan.edu.cn
[2] Shanghai Key Lab of Intelligent Information Processing, Fudan University, Shanghai, China
[3] Department of Computer Science, University of Helsinki, Helsinki, Finland
panhui@cs.helsinki.fi
[4] Department of Computer Science and Engineering, Hong Kong University of Science and Technology, Clear Water Bay, Hong Kong

Abstract. Scholar social networks are composed of scholars and social connections among them. Studying such social networks can help promote academic exchanges and cooperation, and predict future trends in research. In this paper, we analyze SCHOLAT, a representative scholar social network in China, from three perspectives. First, we explore SCHOLAT's social graph, and we find this graph has a smaller average shortest-path length and a higher clustering coefficient than other social networks, for example, the collaboration network of Google Scholar and the Flickr social network. Moreover, we leverage the structural hole theory to identify important users on SCHOLAT. By comparing the top-500 structural hole spanners with 500 randomly selected users, we have found that the former have the higher values of several graph-based metrics, and they also connect more communities. Finally, we also undertake user group-based analysis, and we discover that the users belonging to Guangdong province, and the users from the top universities in China are well-connected and occupy important positions in the network.

Keywords: SCHOLAT social network · Social graph analysis · Structural hole theory · User group-based analysis

1 Introduction

Scholar social networks [1–6] are a kind of social networks made up of scholars and their social connections. Such social networks have two main functions. On one hand, these networks allow researchers to create their homepages, upload their papers, maintain their information, and advertise their research outcomes to the public. On the other hand, these networks provide a platform for scholars to communicate and collaborate on scientific research topics, and build social connections. Plenty of scholar social networks are emerging, such as HASTAC [7],

Y. Sun et al. (Eds.): ChineseCSCW 2021, CCIS 1492, pp. 326–339, 2022.
https://doi.org/10.1007/978-981-19-4549-6_25

Academia [8], ResearchGate [10,11], and SCHOLAT [12]. SCHOLAT social network is a representative scholar social network, serving a large number of Chinese scholars. It provides services such as scholar information management and scholarly interaction, through which one can easily build her scholar homepage, and find scholars with the same research interests. And it also supports various types of interactions, such as making comments, posting "likes", and reposting articles.

Scholar social networks are a particular type of social networks used for the purpose of scholarly communication. Researchers and scholars act as nodes and their relationships serve as edges, while for the general online social networks, the registered users and social connections between them correspond to the nodes and edges in the network, respectively. For example, Flickr [13] is a photo-sharing social network with millions of users, where users and their friendships can be modeled to nodes and edges. There are also certain differences between scholar social networks and collaboration networks [14,15]. Although both have researchers acting as nodes, the edges of the former are generated by online interactions (making friends, following or other forms of interactions) between scholars, while the edges of the latter are generated by joint publications between scholars. These differences in network composition motivate us to further explore scholar social networks. We aim to figure out whether there are differences among them and whether there are particularities of the user groups [17] in scholar social networks. However, existing work [18] have not carried these differences or give an analysis of the characteristics of scholar social networks. In this paper, we start with SCHOLAT social network to explore these differences and carry out a comprehensive analysis.

First, we conduct an informative analysis on the social graph of SCHOLAT. Specifically, we use social graph to model the users and their social connections and analyze this graph with representative graph metrics. We also compare the differences between SCHOLAT's social graph and other social graphs, for example, the collaboration social graph of Google Scholar and Flickr's social graph. We have found lots of unique features of SCHOLAT's social graph, such as a lower average shortest-path length and higher clustering coefficient, which reflects the feature of small-world networks [19]. By examining the distributions of some graph metrics, we discover that SCHOLAT's social graph is significantly different from the collaboration social graph of Google Scholar and Flickr's social graph.

Second, we identify important users [20] based on the structural hole theory [14,22,37] on SCHOLAT. Important users refer to the scholars bridging scholar communities and occupying key positions in the network. Due to the differences in academic impact of scholars, mining and tracking influential scholars in interdisciplinary research areas can help to identify research hotspots, and predict future trends. However, existing studies do not carry out the mining of structural hole spanners (SH spanners) in scholar social networks. To fill this gap, we dive into this problem by abundant structural hole theory-based analysis [21,22] with pivotal metrics, for example, effective size. We first use the Louvain algorithm [16] to detect communities and study the sizes of the top-10 communities. Through comparative analysis between the top-500 SH spanners and the randomly selected 500 users, we can identify important users by effective size, who occupy more critical positions and resources than other users within the network.

Finally, we pay attention to the attributes of users and conduct a user group-based study on SCHOLAT. The same or similar user attributes often promote the establishment of direct or indirect associations among scholars, which are embodied in the forms of friendships, followings, collaborations, and other interactions. First, we extract users with attributes of research interests and affiliations. Then, we analyze user groups according to these two attributes. We have found the research interest with the largest number of users is "computer science". By comparing the distribution of user groups with different affiliations, we find that both geographic location and the ranking of universities in China can be used to divide users, and the former has a greater impact. We also discover that the users in Guangdong province or from top universities have a significant structural advantage on SCHOLAT, which indicates these users are more well-connected and span over important positions in the social graph.

The paper structure is illustrated as follows. In Sect. 2, we provide some related works. Then we analyze the SCHOLAT social network from three perspectives and uncover some characteristics of SCHOLAT social network in Sect. 3. Conclusion and future work are discussed in Sect. 4.

2 Related Work

HASTAC [7] was commented by a National Science Foundation review as the world's first and oldest scholar social network in 2014[1]. Users in this network include researchers, scholars, and the general public with various interests. HASTAC contains more than 14,000 members, and offers them a free and open access community to teach and learn.

AMiner [23,24], is a big data mining and service system platform for scientific and technological information. It has collected more than 130,000,000 scholar profiles and 100,000,000 papers from different publication datasets. It offers some useful services, such as profile extraction, scholar ranking, and name disambiguation.

Google Scholar offers a wealth of information about authors and papers. Chen et al. [14] built a collaboration network of Google Scholar by referring to author profiles. They conducted a demographic analysis and provided an informative overview of scholars. In addition, they used social network analysis methods to analyze the collaboration network and explored several citation indicators.

Academia [8] is a typical scholar social network, which offers access to the users who tend to share their works and videos with others. Users can edit their personal information and upload papers to their profiles. In addition, it supports message sending between users. In the social graph of Academia, nodes are researchers, and edges are friendships among them. Their friendships are formed by paper sharing. Niyazov et al. [9] found that papers shared to Academia receive a 69% boost in citation over five years.

ResearchGate [10,11] is a scientific social networking website that was launched in May 2008. The site is designed to promote scientific collaboration

[1] https://www.hastac.org/about/history.

on a global scale. Users can maintain their profiles, and connect with colleagues, share research methods, and exchange ideas. In ResearchGate's social graph, nodes are researchers, and edges are their social connections, such as sharing publications, collaborating with scholars, and communicating with researchers.

3 Data Analysis

In this section, we analyze the SCHOLAT social network from several perspectives. Firstly, we conduct a social graph-based analysis to observe the features of SCHOLAT's social graph (Subsect. 3.1). Then, we employ the structural hole theory to identify important users within this network (Subsect. 3.2). Finally, we carry out an analysis of the attribute-based user groups. We discover that the number of users with research interest "computer science" is the largest within this network, and we conclude that scholars from high-ranking universities are influential and occupy important positions within the network (Subsect. 3.3).

SCHOLAT offers an open-access dataset[2] of a scholar social network, which includes 16,007 nodes and 202,248 edges. Nodes in this network refer to the scholars within this network, and the edges correspond to the friendships between scholars.

3.1 Social Graph Analysis

We do social graph analysis by *igraph* package[3] for SCHOLAT and a series of findings are presented in this subsection. Specifically, we examine the difference between SCHOLAT's social graph and social graphs of two other networks, i.e., the collaboration network of Google Scholar [14] and the friendship network crawled from Flickr [36]. The information of these graphs is listed as Table 1.

Table 1. The description of three social graphs

Information	SCHOLAT	Google Scholar	Flickr
Nodes	16,007	402,392	80,513
Edges	202,248	1,234,019	5,899,882
Type	Indirect	Indirect	Indirect

We select four representative metrics to conduct social graph analysis as follows.

Degree [28]: in a network, the degree of a node refers to the number of its neighbors. In SCHOLAT's social graph, a scholar's degree equals the number of her connected scholars.

Eigenvector Centrality [29]: this metric of a node holds that its centrality depends on the centrality of its neighbors. Bihari et al. [30] believed that the

[2] https://www.scholat.com/research/opendata/#social_network.
[3] https://igraph.org/python/.

eigenvector centrality is more suitable for the discovery of influential researchers, and they used this metric to measure the influences of researchers. Similarly, in SCHOLAT's social graph, a scholar's influence can be calculated based on the influences of her friends.

Clustering Coefficient [31]**:** for a node in a network, the definition of this indicator is the ratio of the number of edges between its neighbors divided by the number of possible edges between them. This metric quantifies the extent to which its neighbors form a cluster. In SCHOLAT's social graph, this metric reflects that the extent to which a scholar's friends are also friends.

Shortest-Path Length [31]**:** this indicator refers to the length (or distance) of the shortest path between two nodes in a network. In SCHOLAT's social graph, the value of this metric can be used to indicate the smallest hops between two scholars establishing a connection.

We have made comparisons mainly on the following representative graph-based metrics including average degree, average clustering coefficient, network diameter of the largest connected component (LCC), and average shortest-path length of the LCC. The results are listed as Table 2.

Table 2. Comparison between social graphs of three social networks

Metrics	SCHOLAT	Google Scholar	Flickr
Average degree	25.27	6.13	146.56
Average clustering coefficient	0.55	0.20	0.17
Network diameter (LCC)	10	24	27
Average shortest-path length (LCC)	4.31	5.96	2.90

The average degree of SCHOLAT's social graph is 25.27, which is much larger than that of the Google Scholar's collaboration social graph. The reason for the large difference is that in SCHOLAT's social graph, edges can be generated as long as there is a friendship, while in the collaboration social graph of Google Scholar, edges are generated on the condition that there is a co-authorship. Therefore, forming an edge in SCHOLAT's social graph is much convenient than that in the collaboration social graph of Google Scholar. The average degree of SCHOLAT's social graph is less than that in the Flickr's social graph. This is because the users interact more intensively on Flickr. The average clustering coefficient of SCHOLAT's social graph is more than twice as large as that of the collaboration social graph of Google Scholar. This indicator of a scholar quantifies the extent to which her neighbors aggregate to form clusters with each other. The average clustering coefficient of SCHOLAT's social graph is also greater than that of Flickr. The average shortest-path length (LCC) of SCHOLAT's social graph is 4.31, which is smaller than that of LCC of the collaboration social graph of Google Scholar and greater than that of the LCC of Flickr's social graph. The SCHOLAT's social graph has the feature of small average shortest-path length and high clustering coefficient, which conforms to the criteria of small-world networks [19].

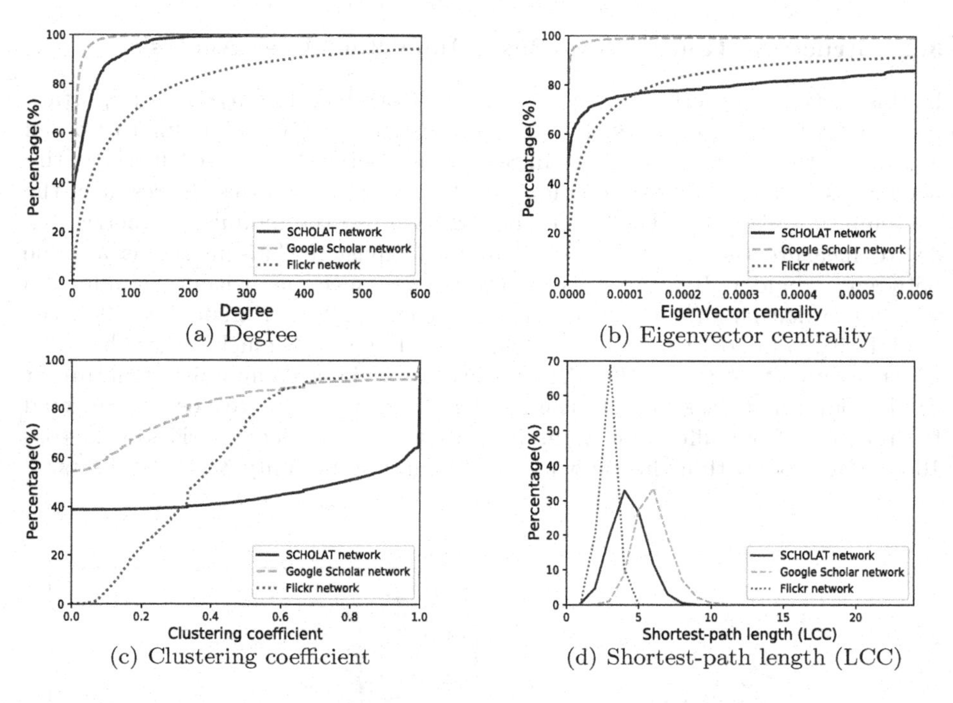

(a) Degree

(b) Eigenvector centrality

(c) Clustering coefficient

(d) Shortest-path length (LCC)

Fig. 1. Comparison on representative graph metrics among three social graphs

To analyze the SCHOLAT social graph, we compare the differences between this graph with the other two social graphs of the collaboration network of Google Scholar and Flickr network. Figure 1(a)–Fig. 1(d) depicts the cumulative distribution function (CDF) of the metric of degree, eigenvector centrality, clustering coefficient, and the LCC's shortest-path length of these three graphs. In Fig. 1(a), for the SCHOLAT's social graph, the median value of the degrees is 11, which indicates that half of the users in this graph have a degree value greater than 11. While for the Flickr social graph, more than half of the users have over 150 neighbors. This is because friendships among users are relatively denser on Flickr than the relationship of connections between scholars. The distribution of the metric of eigenvector centrality is revealed in Fig. 1(b). Apparently, in SCHOLAT's social graph, the high values of this metric are noticeably more than that in the other two social graphs. That is to say, in SCHOLAT's social graph, there are high-influence scholars with influential friends. Figure 1(c) corresponds to the CDF of the clustering coefficient. We find that the users in SCHOLAT's social graph have larger values of clustering coefficient than those in the social graphs of Google Scholar and Flickr. Figure 1(d) plots the distribution of the shortest-path lengths (LCC) in each of the three social graphs. It can be observed that the average shortest-path length between users in SCHOLAT's social graph is smaller than that in the social graph of Google Scholar.

3.2 Structural Hole Theory-Based Important User Analysis

In this section, we perform structural hole theory-based analysis on important users in SCHOLAT network. Firstly, we use Louvain [16], a modularity-based algorithm, to detect communities in SCHOLAT network. As a result, the entire network is divided into 295 communities. Figure 2(a) visualizes the result of the community division of SCHOLAT network. We use the modularity metric [32, 33] to measure the result of the community division. This metric is a value between −1 and 1. It measures the density of connections within communities versus the density of connections between communities. Pursuant to [34], the modularity value of the division on SCHOLAT is 0.856, much larger than 0.3, which signifies that the SCHOLAT network has a viable community structure. It can be found that the scholars within each community are more closely connected than scholars from different communities. Figure 2(b) depicts the sizes of the top-10 communities within this network. The largest community has 1,310 users.

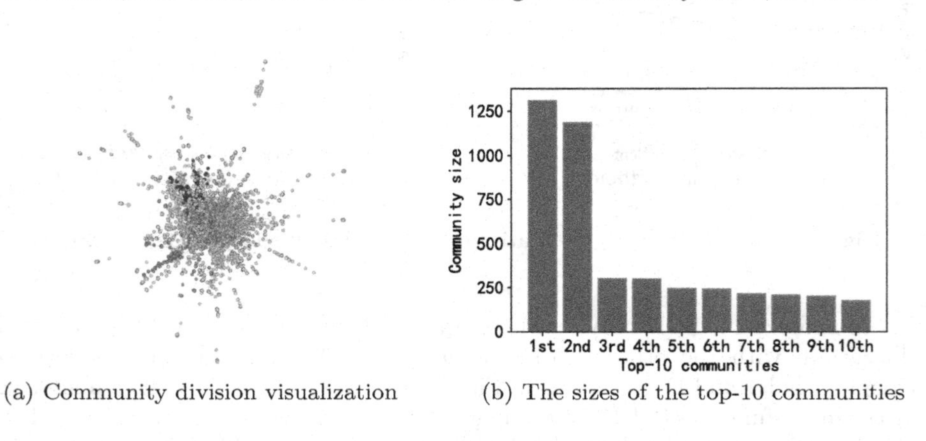

(a) Community division visualization (b) The sizes of the top-10 communities

Fig. 2. Analysis of communities within SCHOLAT social network

Based on the data contained in this network, we investigate the problem of important user identification by a structural hole theory-based analysis. In a social network, if there is no direct connection between several communities, it seems that there is a hole in the network structure. Structural hole (SH) refers to the gap between user communities. The users, occupying these special positions and playing a bridging role between different communities, are known as SH spanners. Burt [37] put forward the structural hole theory, which studies the structure formation of interpersonal networks and analyzes occupying what kind of positions in the network can bring more benefits or rewards to the nodes. The structural hole theory emphasizes that users occupying the positions of structural holes in networks can bring advantages in information dissemination and other resources to organizations and individuals [14,22,37]. As an important part of the structural hole theory, a series of SH metrics, including effective size, constraint, efficiency and hierarchy, are proposed to evaluate whether one node is probably serving as an SH spanner [22,37]. There is a wide range of applications of the structural hole theory, such as identification of high potential talents from newly-enrolled employees of a company [38], influence maximization in social networks [39], and proposing new metrics for measuring the importance of nodes [40]. Efforts have also been made to solve the problem of effectively mining the top-k SH spanners in social networks [41,42]. We utilize SH theory to study the top-k important users within SCHOLAT network. Specifically, we use effective size [43,44], a representative metric measuring the non-redundant links, to discover the SH spanners. The effective size of the ego network of node i is denoted as $e(i)$, and its definition is

$$e(i) = n - \frac{2t}{n} \tag{1}$$

where n and t are the number of nodes and edges within the ego network of node i, respectively.

On the basis of this metric, we study the feature of the top-500 SH spanners from several perspectives, including the degrees, the betweenness centrality values, the PageRank values and the number of communities (N_{com}) they bridge. In order to illustrate the difference between the top-500 SH spanners and the randomly selected 500 users, we use the CDF to characterize the distribution of the values on the above metrics of the two user groups. The results are shown in Fig. 3(a) – Fig. 3(d).

According to Fig. 3, we find some significant differences between the top-500 SH spanners (represented as the solid lines in the figure) and the randomly selected 500 users (represented as the dotted lines in the figure). In Fig. 3(a), we can find that half of the top-500 SH spanners have more than 200 neighbors, while about 50% of the 500 randomly selected users have degree values lower than 100. Similarly, compared with the 500 randomly selected users, the top-500 SH spanners whose betweenness centrality values are more than 0.0023 account for nearly 50%, which means that these SH spanners play a decisive role in the process of information dissemination on the entire network. Figure 3(c) shows the difference on the metric of PageRank, and we can find that half of the top-500 SH spanners have PageRank values larger than 0.00025, which are much higher than those of the 500 randomly selected users. As shown in Fig. 3(d), we discover that the top-500 SH spanners bridge more communities than other users. And we also find these important nodes with higher values of graph-based metrics, such as the degrees, betweenness centrality values, which further validates that the top-500 SH spanners are more important in the whole network. For example, node #4809, a scholar from South China Normal University, has the largest value of effective size and connects 22 communities. What is more, it is the user with the greatest values of degree, betweenness centrality in the network. This means that user #4809 is able to play a critical role in the information propagation and resource control on the whole network.

3.3 User Group-Based Analysis

This subsection mainly analyzes different groups of users according to two important attributes, i.e., research interest and affiliation. We first conduct the data preprocessing and the extraction on users providing the attributes of research interest and affiliation. As a result, we get 11,996 users providing the research interest information and 8,970 users providing the affiliation information. Then, through statistical calculation and comparative analysis, we have explored the differences between groups of users with different research interests and affiliation information, respectively.

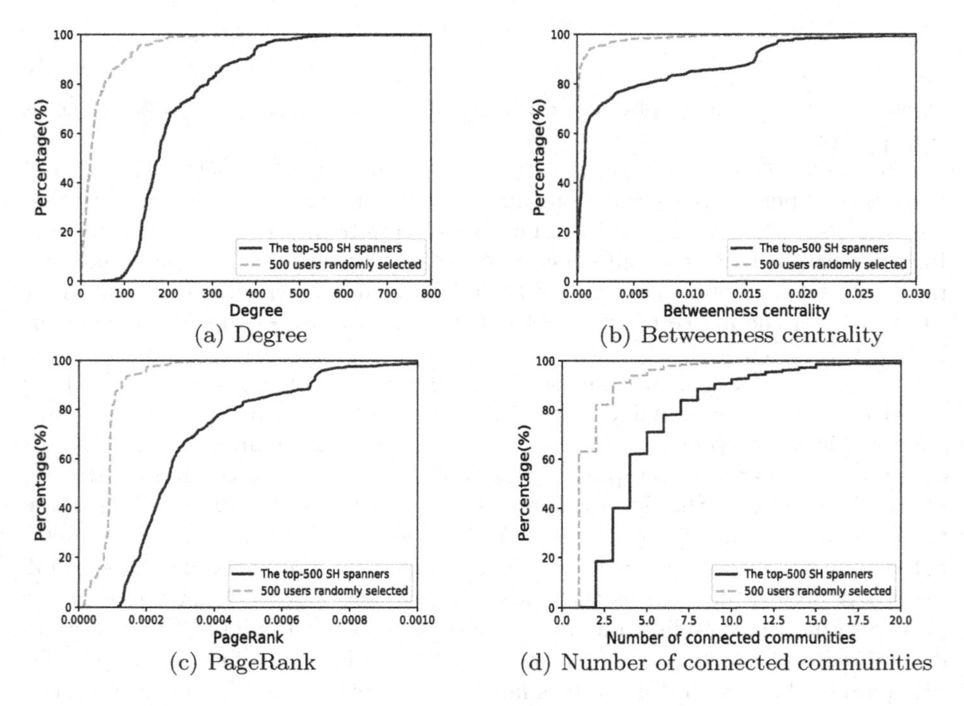

Fig. 3. Difference between the top-500 SH spanners and the randomly selected 500 users

Due to the unbalanced distribution of the attributes of users, which may lead to inaccurate analysis, we need to make some data preprocessing. First, we filter out incorrect values and null values that violate common sense. Second, by observing the dataset, we find that some attributes of users are ambiguous. One possible reason is that users are not willing to let others know the precise information about themselves. Such as, "university of science and technology" is a vague expression. So we remove such data entries for user group-based analysis.

We first divide the users according to their research interests. Table 3 lists the top-10 keywords of research interests of users. The keyword of "computer science" has the highest proportion of users, indicating that there are about 6,084 (50.7%) scholars interested in computer science among all users of SCHOLAT. And the average degree (Avg. Degree) of users with 7 out of the top-10 keywords is higher than the average degree of the entire graph. In general, it can be recognized that the top-10 keywords include most of the research hot-spots in computer science and engineering. There are also research directions in other disciplines, such as culture industry and law.

Table 3. The top-10 keywords of research interests of SCHOLAT users

Research interest	#Users	Avg. Degree
Computer science	6,084	70.5
Artificial intelligence	1,284	33.7
Automation	1,116	30.3
Big data	636	36.2
Machine learning	636	26.9
Network security	516	38.5
Privacy preservation	478	25.0
Internet of things	504	26.7
Culture industry	422	24.9
Law	320	23.6

In addition, we investigate the users' affiliation based on the geographic locations and the university ranking. Since the SCHOLAT website is located in Guangdong province, we count the number of users whose affiliations belong to Guangdong province or not. We find that the number of users in Guangdong province is 5,808, higher than that out of Guangdong province, which is 3,912. This reveals the impact of geographic location on attracting users. We compare the users from Guangdong province and out of Guangdong province in terms of the average degree (Avg. Degree) and average effective size (Avg. ES), as shown in Fig. 4(a). It can be spotted that the Avg. Degree and the Avg. ES of users from Guangdong province are also higher than those out of Guangdong province. This also indicates that these users are more well-connected and occupy more important positions.

In order to further analyze user groups according to university ranking, we exclude the users with the affiliations of universities outside China and vague meanings, and the final number of domestic universities is 208. We consider one widely adopted standard in China, i.e., we refer to the "Project 211[4]", which covers 116 top universities in China. In this paper, we define the universities

[4] Project 211, commonly known as 211 universities, is a project of the Education Ministry of China to build about 100 key disciplines universities for the 21st century, including 116 universities.

belonging to "Project 211" as top universities in China. We compare the users with different ranked universities in terms of the Avg. Degree and Avg. ES, and the results are depicted in Fig. 4(b). The numbers of users belonging to the top universities and other universities are 6,118 and 2,852, respectively. In other words, users from top-ranked universities account for a larger proportion of SCHOLAT users. Similarly, the Avg. Degree and the Avg. ES of users from the top universities are higher than those from other universities. This is because the top universities have more educational resources and scientific research advantages than other universities, which have a greater influence on the academic ecosystem in China.

By comparing the two subgraphs in Fig. 4, the differences reflect that geographic location and university ranking are important to distinguish users, and the former takes a more significant role. Moreover, compared with other users, the users in Guangdong province or from the top-ranked universities occupy more structural advantages and are important in SCHOLAT.

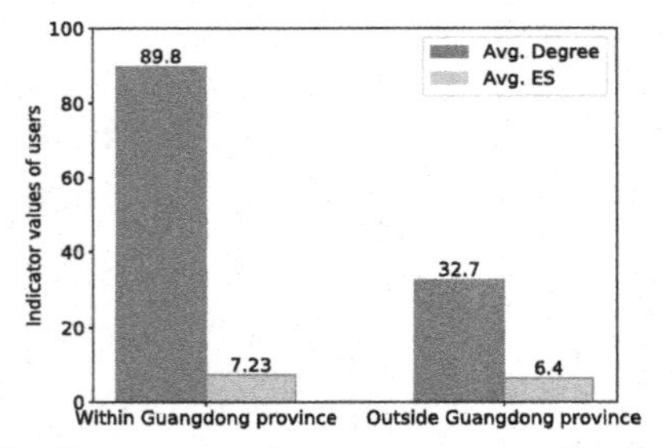

(a) User groups based on geographic location

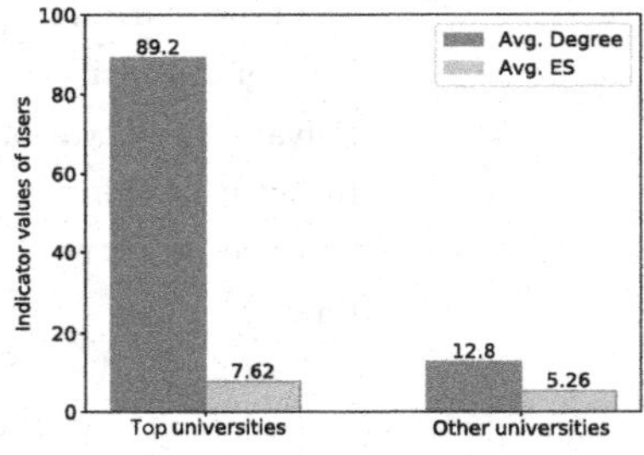

(b) User groups based on university ranking

Fig. 4. Differences between user groups on two metrics

4 Conclusion and Future Work

In this paper, we conduct a series of analysis on SCHOLAT social network, including social graph analysis, structural hole theory-based important user analysis, and user group-based analysis. We have found that SCHOLAT's social graph has obvious small-world features with a large average clustering coefficient and a small average shortest-path length, which are different from other social graphs of social networks, i.e., the collaboration network of Google Scholar (collaboration network) and the Flickr network (online social network). Based on the structural hole theory, we have observed the top-500 SH spanners are more important with the network structure than the randomly selected 500 users, and the SH spanners have a higher level of connectivity, and span over critical positions within the network. We further study the research interests of the scholars and find that a large portion of the scholars are working in the computer science and engineering-related field of research. We also find that the users from top universities in China, and the users from Guangdong province, play a more important role in SCHOLAT network.

We plan to further enhance our work in scholar social networks from the following three aspects. First, we would like to explore the composition and characteristics of user communities. Second, we plan to study the dynamics and evolution of the social connections. Last but not least, we would like to investigate the identification of potential fake accounts [45–47].

Acknowledgement. This work is sponsored by National Natural Science Foundation of China (No. 62072115, No. 71731004, No. 61602122, No. 61971145), China Postdoctoral Science Foundation (No. 2021M690667), the Research Grants Council of Hong Kong (No. 16214817), the 5GEAR project and FIT project from the Academy of Finland. Yang Chen is the corresponding author.

References

1. Giglia, E.: Academic Social Networks: it's time to change the way we do research. Eur. J. Phys. Rehabil. Med. **47**(2), 345–349 (2011)
2. Kong, X., Shi, Y., Yu, S., Liu, J., Xia, F.: Academic social networks: Modeling, analysis, mining and applications. J. Netw. Comput. Appl. **132**, 86–103 (2019)
3. Mason, S.: Adoption and usage of Academic Social Networks: a Japan case study. Scientometrics **122**(3), 1751–1767 (2020). https://doi.org/10.1007/s11192-020-03345-4
4. Huang, C.: Social network site use and academic achievement: a meta-analysis. Comput. Educ. **119**, 76–83 (2018)
5. Subelj, L., Fiala, D., Bajec, M.: Network-based statistical comparison of citation topology of bibliographic databases. Scientific Reports **4**(1), 6496:1–6496:10 (2014)
6. Tang, Y.: Scholar Social Networks and Big Data Research. In: Proceedings of the 24th International Conference on Computer Supported Cooperative Work in Design (CSCWD) (2021)
7. Puschmann, C., Bastos, M.: How digital are the digital humanities? An analysis of two scholarly blogging platforms. PLoS ONE **10**(2), e0115035 (2015)
8. Sugimoto, C., Work, S., Lariviere, V., Haustein, S.: Scholarly use of social media and altmetrics: a review of the literature. J. Am. Soc. Inf. Sci. **68**(9), 2037–2062 (2017)
9. Niyazov, Y., et al.: Open access meets discoverability: citations to articles posted to Academia. edu. PLOS ONE **11**(2), e0148257 (2016)
10. Yu, M., Wu, Y., Alhalabi, W., Kao, H., Wu, W.: ResearchGate: an effective altmetric indicator for active researchers? Comput. Hum. Behav. **55**, 1001–1006 (2016)
11. Thelwall, M., Kousha, K.: ResearchGate articles: age, discipline, audience size, and impact. J. Am. Soc. Inf. Sci. **68**(2), 468–479 (2017)
12. Xu, Q., Qiu, L., Lin, R., Tong, Y., He, C., Yuan, C.: An improved community detection algorithm via fusing topology and attribute information. In: Proceedings of the 24th International Conference on Computer Supported Cooperative Work in Design (CSCWD), pp. 1069–1074 (2021)
13. Cha, M., Mislove, A, Gummadi, K. A measurement-driven analysis of information propagation in the Flickr social network. In: Proceedings of the 18th International Conference on World Wide Web (WWW), pp. 721–730 (2009)

14. Chen, Y., Ding, C., Hu, J., Chen, R., Hui, P., Fu, X.: Building and analyzing a global co-authorship network using google scholar data. In Proceedings of the 26th International Conference on World Wide Web Companion (WWW Companion), pp. 1219–1224 (2017)
15. Abbasi, A., Chung, K., Hossain, L.: Egocentric analysis of co-authorship network structure, position and performance. Inf. Proc. Manag. **48**(4), 671–679 (2012)
16. Blondel, V., Guillaume, J., Lambiotte, R., Lefebvre, E.: Fast unfolding of communities in large networks. J. Stat. Mech: Theory Exp. **2008**(10), P1000 (2008)
17. Peleshchyshyn, A., Vus, V., Markovets, O., Albota, S.: Identifying specific roles of users of social networks and their influence methods. In: Proceedings of the 13th International Scientific and Technical Conference on Computer Sciences and Information Technologies (CSIT), vol. 2, pp. 39–42 (2018)
18. Wu, X., Zhang, C.: Finding high-impact interdisciplinary users based on friend discipline distribution in academic social networking sites. Scientometrics **119**(2), 1017–1035 (2019). https://doi.org/10.1007/s11192-019-03067-2
19. Watts, D.J.: Networks, dynamics, and the small-world phenomenon. Am. J. Sociol. **105**(2), 493–527 (1999)
20. Gong, Q., et al.: Cross-site Prediction on Social Influence for Cold-start Users in Online Social Networks. In ACM Transactions on the Web **15**(2), 6:1–6:23(2021)
21. Rezvani, M., Liang, W., Xu, W., Xu, C.: Identifying top-k structural hole spanners in large-scale social networks. In: Proceedings of the 24th ACM International on Conference on Information and Knowledge Management (CIKM), pp. 263–272 (2015)
22. Lin, Z., Zhang, Y., Gong, Q., Chen, Y., Oksanen, A., Ding, A.: Structural hole theory in social network analysis a review. IEEE Trans. Comput. Soc. Syst. **9**, 1–16 (2021). https://doi.org/10.1109/TCSS.2021.3070321
23. Tang, J.: Aminer: toward understanding big scholar data. In: Proceedings of the 9th ACM International Conference on Web Search and Data Mining (WSDM), pp. 467–467 (2016)
24. Wan, H., Zhang, Y., Zhang, J.: AMiner: search and mining of academic social networks. Data Intelligence **1**(1), 58–76 (2019)
25. Biryukov, M., Dong, C.: Analysis of computer science communities based on DBLP. In: Proceedings of International Conference on Theory and Practice of Digital Libraries (TPDL), pp. 228–235 (2010)
26. Moreira, C., Calado, P., Martins, B.: Learning to rank academic experts in the DBLP dataset. Expert. Syst. **32**(4), 477–493 (2015)
27. Aggrawal, N., Arora, A.: Visualization, analysis and structural pattern infusion of DBLP co-authorship network using Gephi. In: Proceedings of the 2nd International Conference on Next Generation Computing Technologies (NGCT), pp. 494–500 (2016)
28. Freeman, L.C.: Centrality in social networks conceptual clarification. Social Networks **1**(3), 215–239 (1978)
29. Ruhnau, B.: Eigenvector-centrality-a node-centrality? Social Networks **22**(4), 357–365 (2000)
30. Bihari, A., Pandia, K.: Eigenvector centrality and its application in research professionals' relationship network. In: Proceedings of the 2015 International Conference on Futuristic Trends on Computational Analysis and Knowledge Management (ABLAZE), pp. 510–514 (2015)
31. Zhou, T., Yan, G., Wang, B.: Maximal planar networks with large clustering coefficient and power-law degree distribution. Phys. Rev. E **71**(4), 046141 (2005)

32. Newman, M.-E.: Modularity and community structure in networks. Proc. Natl. Acad. Sci. U.S.A. **103**(23), 8577–8582 (2006)
33. Chen, Y., Hu, J., Xiao, Y., Li, X., Hui, P.: Understanding the user behavior of foursquare: a data-driven study on a global scale. IEEE Trans. Comput. Social Syst. **7**(4), 1019–1032 (2020)
34. Kwak, H., Choi, Y., Eom, Y.-H., Jenog, H., Moon, S.: Mining communities in networks: a solution for consistency and its evaluation. In: Proceedings of the 9th ACM SIGCOMM Conference on Internet Measurement Conference (IMC), pp. 301–314 (2009)
35. Saccenti, E., Hendriks, M., Smilde, A.: Corruption of the Pearson correlation coefficient by measurement error and its estimation, bias, and correction under different error models. Scientif. Reports **10**(1), 438:1–438:19 (2020)
36. Tang, L., Liu, H.: Scalable Learning of Collective Behavior based on Sparse Social Dimensions. In: Proceedings of the 18th ACM Conference on Information and Knowledge Management (CIKM) (2009)
37. Burt, R. S.: Structural Holes: The Social Structure of Competition. Harvard University Press (1992)
38. Ye, Y., Zhu, H., Xu, T., Zhuang, F., Yu, R., Xiong, H.: Identifying high potential talent: a neural network based dynamic social profiling approach. In: Proceedings of the 2019 IEEE International Conference on Data Mining (ICDM), pp. 718–727 (2019)
39. Zhu, J., Liu, Y., Yin, X.: A new structure-hole-based algorithm for influence maximization in large online social networks. IEEE Access **5**, 23405–23412 (2017)
40. Lu, M.: Node importance evaluation based on neighborhood structure hole and improved topsis. Comput. Netw. **178**, 107336 (2020)
41. Lou, T., Tang, J.: Mining structural hole spanners through in-formation diffusion in social networks. In: Proceedings of the 22nd International Conference on World Wide Web (WWW), pp. 825–836 (2013)
42. Xu, W., Rezvani, M., Liang, W., Yu, J., Liu, C.: Efficient algorithms for the identification of top-k structural hole spanners in large social networks. IEEE Trans. Knowl. Data Eng. **29**(5), 1017–1030 (2017)
43. Kwon, Y.D., Mogavi, R.H., Haq, E.U., Kwon, Y., Ma, X., Hui, P.: Effects of ego networks and communities on self-disclosure in an online social network. In: Proceedings of International Conference on Advances in Social Networks Analysis and Mining (ASONAM) (2019)
44. Ying, Q., Chiu, D., Zhang, X.: Diversity of a user's friend circle in OSNs and its use for profiling. In: Proceedings of the 10th International Conference on Social Informatics (SocInfo) (2018)
45. Gong, Q., et al.: DeepScan: Exploiting deep learning for malicious account detection in location-based social networks. IEEE Communications Magazine **56**(11), 21–27 (2018)
46. Gong, Q., et al.: Detecting malicious accounts in online developer communities using deep learning. In: Proceedings of the 28th ACM International Conference on Information and Knowledge Management (CIKM), pp. 1251–1260 (2019)
47. He, X., Gong, Q., Chen, Y., Zhang, Y., Wang, X., Fu, X.: DatingSec: detecting malicious accounts in dating Apps using a content-based attention network. IEEE Trans. Dependable Secure Comput. **18**(5), 2193–2208 (2021)

SCHOLAT Link Prediction: A Link Prediction Dataset Fusing Topology and Attribute Information

Ronghua Lin[1], Yong Tang[1(✉)], Chengzhe Yuan[2], Chaobo He[3], and Weisheng Li[1]

[1] School of Computer Science, South China Normal University, Guangzhou, China
{rhlin,ytang,weishengli}@m.scnu.edu.cn
[2] School of Electronics and Information, Guangdong Polytechnic Normal University, Guangzhou, China
ycz@gpnu.edu.cn
[3] School of Information Science and Technology, Zhongkai University of Agriculture and Engineering, Guangzhou, China

Abstract. Link prediction is an important research field on social network analysis. However, most existing link prediction datasets have not taken text attribute information into account. In this paper, we propose a novel link prediction dataset named SCHOLAT Link Prediction based on the academic social network SCHOLAT, which contains multitype relationships and user attribute information. We conduct extensive experiments on our dataset and the results show the potential of our dataset on link prediction. Our proposed dataset is available at https://www.scholat.com/research/opendata/#link_prediction.

Keywords: Link prediction · SCHOLAT · Topology · User attribute information

1 Introduction

Due to the rapid development of social networks, massive user interaction data will be generated on social networks every day. The social networks can be treated as the topology structures or graphs where the nodes are enrolled users in social networks and the edges represent different interactions or relationships (such as follow, friending, etc.) between different users. Moreover, the relationships in social networks are substantially dynamic since the new nodes and edges will be added over time [9,12]. In this case, social network analysis has attracted more and more attentions in recent years, such as link prediction, community detection, personalized recommendation, and so on.

Link prediction is an important research problem in social network analysis [16]. The link prediction problem can be described as to infer which new interactions will be occurred in the future according to the existing network structure. An effective link prediction method can mine the potential relationships among users and further support many personalized applications including friend recommendation, relationship completion/prediction, etc. [9,28].

Y. Sun et al. (Eds.): ChineseCSCW 2021, CCIS 1492, pp. 340–351, 2022.
https://doi.org/10.1007/978-981-19-4549-6_26

Since traditional social networks such as Twitter, Facebook, etc. are with genericity and non-specialty, some domain-specific or vertical social networks have emerged and gained more and more popularity. Academic social networks are generally designed for scholars or in certain disciplines, which provide effective research exchanges and sharing. Existing academic social networking websites include ResearchGate,[1] Academia,[2] SCHOLAT,[3] and so on. By taking comprehensive investigations, we list the statistics among ResearchGate, Academia and SCHOLAT in Table 1.

Table 1. Statistics among ResearchGate, Academia and SCHOLAT.

	ResearchGate	Academia	SCHOLAT
Personal homepage	✓	✓	✓
Posts	✓	✓	✓
Calendar/schedule			✓
Publication management	✓	✓	✓
Onsite chat/email	✓	✓	✓
Research institute			✓
Online course		✓	✓
Free/premium	Free	Free/Premium*	Free

* Some features are offered for free while some are offered for premium.

Link prediction on academic social networks can provide more accurate recommendations and research collaborations by mining the potential user interactions in the future. In the last decades, more and more researchers have paid attention to link prediction on different social networks and further applications [3,6,29]. However, to the best of our knowledge, most existing datasets for link prediction only contains the user nodes and the relationships among them while neglecting the user attribute information. The user attribute information on academic social networks can provide extra text information such as research interests by extracting from the user biographies, user posts, academic publications or other long texts. It is promising to utilize the user attribute information to improve the accuracy of link prediction.

In this paper, we propose a novel link prediction dataset named SCHOLAT Link Prediction based on the academic social networking website SCHOLAT. It contains not only the user multitype relationships but also the user attribute information. We also list the statistics and differences among our dataset and other existing link prediction datasets. We utilize supervised methods and unsupervised methods respectively with several machine learning models and deep learning models for evaluations on our proposed dataset. Experimental results

[1] https://www.researchgate.net/.
[2] https://www.academia.edu/.
[3] https://www.scholat.com/.

demonstrate that our dataset can achieve satisfactory performance on link prediction by using most machine learning models and deep learning models.

The contributions of this paper are listed as follows. (1) We propose a novel link prediction dataset based on the academic social networking website SCHOLAT. The proposed dataset contains user multitype relationships and user attribute information. (2) To show the differences, we compare our dataset with several other existing link prediction datasets. (3) Extensive experiments on several supervised learning models and unsupervised learning models demonstrate the potential of our proposed dataset on link prediction.

2 Related Work

The literature about link prediction has lasted for decades. Traditional researches on link prediction can be divided into two categories which are supervised methods and unsupervised methods. Pecli et al. [23] proposed three different automatic feature selection strategies for link prediction with supervised learning. Aghabozorgi et al. [1] developed a novel similarity measure for link prediction and used supervised learning for evaluations. Ahmed et al. [2] utilized the machine learning methods and treated the link prediction as a classification problem. Muniz et al. [19] proposed an unsupervised link prediction in social networks by using three weighting criteria that combine contextual, temporal and topological information. Sherkat et al. [25] introduced a novel unsupervised link prediction approach based on ant colony algorithm. Zhang et al. [31] designed two incremental dynamic link prediction algorithms which can improve prediction accuracy and incur low running time respectively.

Recently, more and more researchers have paid attention to link prediction in knowledge graphs or knowledge bases. Zhang et al. [30] proposed a novel knowledge graph embedding model which can effectively improve the accuracy of link prediction. Wang et al. [27] proposed a predictive network representation learning for link prediction task. Tay et al. [26] proposed Parallel Universe TransE (puTransE) which is a robust knowledge graph embedding model for link prediction on dynamic knowledge graphs. Ostapuk et al. [22] designed a novel deep active learning framework and applied for neural link predictor in knowledge graphs.

However, the link prediction datasets used by aforementioned works mainly only contain the nodes/entities and edges/relationships while have less text attribute information. In social networks, the text attributes can provide extra features or information that may be beneficial to improve the accuracy of link prediction.

3 Data Processing

Since the SCHOLAT Link Prediction dataset mainly contains user multitype relationships and user attribute information, we divide the data processing into two parts, which are relationship extraction and attribute extraction.

3.1 Relationship Extraction

The relationships among SCHOLAT enrolled users can be considered a heterogeneous networks. Specifically, SCHOLAT Teamwork and Online Course platforms are two special features on SCHOLAT. Users can create or join their own research labs, institutes, or organizations (collectively referred as "teams" in SCHOLAT) and share or exchange their latest research news or publications. Besides that, users can take the online courses and manage their course space (uploading/downloading course resources, uploading/checking out homeworks, etc.) on SCHOLAT. Hence, there are multiple types of relationships on SCHOLAT, which are in detail friend relationship, team member relationship, course classmate relationship and so on.

We mainly extract three types of relationships (friend relationship, team member relationship and course classmate relationship) as the edges in our dataset. Actually, there are other types of relationships such as academic paper co-author relationship in SCHOLAT, which will be considered as our future work. It is obvious that these edges are undirected. We subsequently conduct the deduplication to remove the duplicate edges. For example, if user u_A and user u_B are friends, then there is a undirected edge between u_A and u_B. In this case, if there is another edges between u_A and u_B since they are in the same team or course, then this edge will be removed because it is duplicate. As a result, we obtain the multitype relationship undirected edges among SCHOLAT users. The procedure of relationship extraction is shown in Fig. 1.

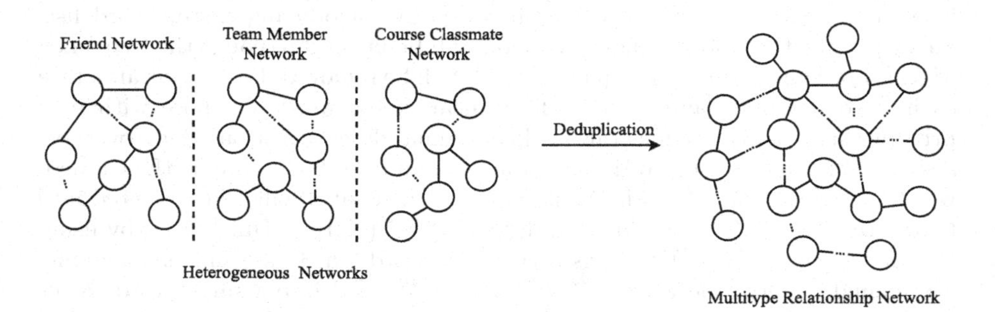

Fig. 1. Procedure of relationship extraction.

3.2 Attribute Extraction

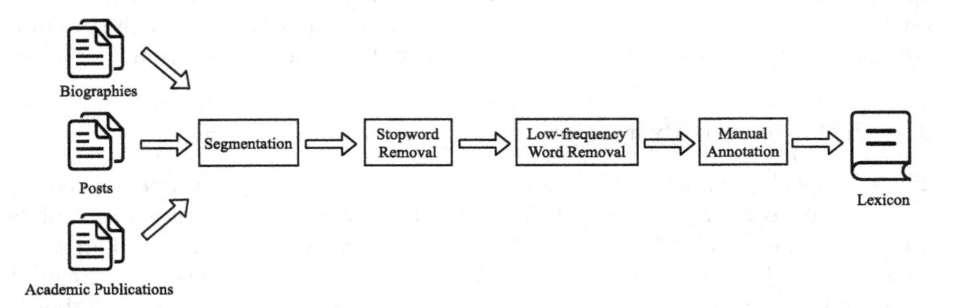

Fig. 2. Overview of lexicon processing.

Users in SCHOLAT have massive text information. Our goal is to extract the keywords from the text data as user attribute information. First of all, we need to construct the lexicon that to decide which words will be regarded as user attribute information. Figure 2 shows how we build our exclusive lexicon. We extract three different types text data from SCHOLAT, which are user biographies, user posts, and user academic publications. In a nutshell, we get the text data formatted as $text = \{t_1, t_2, t_3, ..., t_n\}$ where t_i is the text data of user u_i which contains u_i's biography, posts and academic publications. User biographies usually contain users' affiliated organizations, education experiences, research interests, and etc. User posts can reflect what are user doing recently such as some academic news, some learning experiences and so on. User academic publications mainly contain academic papers, books, projects, and patents, from which can extract rich user attribute information such as research fields.

We use HanLP[4] to segment the aforementioned three types of text data into words. We then remove the stop words by introducing a stop word list. Subsequently, the word frequency counter will be utilized to count the frequency of each word and those low-frequency words will be removed. Finally, we annotate each word manually and remove those meaningless words or those with user privacies such as user names (also called manual desensitization). Moreover, we also correct the errors in word segmentation manually. For example, the Chinese word "华南师范大学" (South China Normal University) will be segmented into two words "华南" (South China) and "师范大学" (Normal University) by using HanLP. Since "华南师范大学" is a complete word which is a university name, we manually merge the words "华南" and "师范大学" into a single word. After conducting above processing, we obtain the exclusive SCHOLAT lexicon that contains all the words which are appeared in users' attributes.

The last step is that we use the above lexicon to re-segment the text data of users individually. As a result, we get the user attributes formatted as $attribute =$

[4] https://github.com/hankcs/HanLP.

Table 2. Statistics of existing link prediction datasets.

Name	Nodes	Edges	Attributes
HTC	7,610	15,751	
NSC	1,461	2,742	
GRQ	5,241	14,484	
FBK	4,024	87,887	
EML	1,133	5,451	
INF	410	2,765	
Cora*	11,881 papers	34,648 citations relations	
	16,114 authors	27,020 authorship relations	
FB15K	14,951	592,213	
WN18RR	40,943	93,003	
YAGO3-10	123,182	1,089,040	Some attributes (citizenship, gender, profession, etc.)
SCHOLAT link prediction	10,755	202,248	23,527

* The **Cora** dataset contains two types of nodes and two types of relationships.

$\{a_1, a_2, a_3, ..., a_n\}$ where a_i is the attributes of user u_i that are the words by re-segmenting the text data t_i with the above exclusive SCHOLAT lexicon.

4 Dataset Overview

After data processing mentioned in Sect. 3, we get user relationships and user attribute information respectively. In this case, we manual choose a subset of users and only remain their relationships and attributes. We anonymize the user names into user IDs and shuffle these IDs.

We compare our dataset with several existing link prediction datasets which can be mainly categorized into two types. The first type is mainly based on social networks or relationship networks. Among this type of datasets, we mainly choose the following datasets for comparison which are **HTC** [21], **NSC** [20], **GRQ** [15], **FBK** [18], **EML** [5], **INF** [13], **Cora** [24]. Specifically, **HTC**, **NSC** and **GRQ** are co-authorship networks. **FBK** is a social network. **EML** is a personal email shared network. **INF** is a face-to-face contact network in an exhibition. **Cora** is a paper citation network, which specifically contains not only the citation relations between papers but also the authorship relations between papers and authors.

The second type of link prediction datasets is mainly based on knowledge graph triples. Among this type of datasets, we choose three widely used datasets which are **FB15K** [4], **WN18RR** [8] and **YAGO3-10** [17]. **FB15K** is a knowledge base completion dataset which is also applied for link prediction problem. It contains the knowledge base relation triples of Freebase. **WN18RR** is a link prediction dataset created from WN18 which is a subset of WordNet. **YAGO3-10** is a subset of YAGO3, which contains massive triples used for knowledge base completion and link prediction.

The statistics of the above datasets and our proposed dataset are overviewed in Table 2. The main difference is that our link prediction dataset contains 23,527 user attributes. Although **YAGO3-10** also contains some user attributes such as citizenship, gender and profession, our dataset can provide richer information such as user degrees, research interests, education experiences and so on. Actually, our dataset belongs to the first type of link prediction datasets. By comparing with **HTC**, **NSC**, **GRQ**, **FBK**, **EML**, **INF**, and **Cora**, our dataset contains more user nodes and edges.

5 Evaluation

In this section, we try to conduct evaluations on our dataset. We use supervised learning methods and unsupervised learning methods respectively for experiments.

5.1 Supervised Methods

In the first round experiment, we treat link prediction as a classification problem and use some classic machine learning classification models for evaluations. Specifically, we use four machine learning models which are Logistic Regression (LR), Support Vector Machine (SVM), Random Forest (RF), and XGBoost. Since the original dataset only contains the real existing edges (i.e. positive edges), we need to generate some negative edges which are not present in SCHOLAT for supervised learning. In detail, assuming that the number of positive edges is e_{num}, we will generate e_{num} negative edges whose shortest path is greater than 2. We map the nodes and relationships in our dataset into an undirected graph $\mathcal{G}$. The single generated negative edge $e = (u_i, u_j)$ should meet the criteria that $\{(u_i, u_j) | (u_i, u_j) \notin \mathcal{G} \wedge \neg \exists u_k (u_k \in \mathcal{G} \wedge u_k \neq u_i \wedge u_k \neq u_j \wedge (u_i, u_k) \in \mathcal{G} \wedge (u_k, u_j) \in \mathcal{G})\}$.

We annotate the positive edges as label 1 and the negative edges as label 0. As a result, we get total 404,496 (2 $\times$ 202,248) samples where 202,248 samples labelled as 1 and 202,248 samples labelled as 0. We shuffle these samples and randomly split into training set and test set with the ratio 9:1. For the user attribute information, we simply use the Bag-of-Words (BOW) model to convert the user attributes into vectors and further reduce dimensionality by Principal Component Analysis (PCA). We subsequently use the cosine distance as

$$cos(e_i, e_j) = \frac{\sum_{k=1}^{m} (e_{i\_k} \times e_{j\_k})}{\sqrt{\sum_{k=1}^{m} (e_{i\_k})^2} \times \sqrt{\sum_{k=1}^{m} (e_{j\_k})^2}} \tag{1}$$

to measure the user similarity where e_i and e_j are the attribute information embedding of u_i and uj, e_{i_k} and e_{j_k} are the k-th element of e_i and e_j. Both user nodes between each edge and user similarity will be feed into supervised learning models as features.

We use precision, recall, F1-score and AUC (Area Under Curve) as the metrics to measure the performance [7,10]. The experimental results are shown in Table 3. In the above four supervised models, LR performs poorly since the data features are nonlinear, which causes the model easily underfitting or overfitting. In contrast, RF reaches up to the highest results on all the selected metrics due to the capacity on processing high-dimensional features and no feature selection required since the features are selected randomly. Moreover, XGBoost achieves the performance close to that of RF because XGBoost has in-built L1 and L2 regularization which can effectively avoid overfitting. SVM outperforms LR since SVM is effective in both linearly separable data and nonlinearly separable data as well as in high dimensional spaces.

In our experiments, we only utilize the cosine distance as user similarity. However, many other features (such as common friends, etc.) can be considered to further improve the classification performance.

Table 3. Experimental results by using supervised learning models.

	Precision	Recall	F1-score	AUC
LR	0.5733	0.6606	0.6139	0.5845
SVM	0.6741	0.7772	0.7220	0.7007
RF	**0.8933**	**0.8871**	**0.8902**	**0.8906**
XGBoost	0.8635	0.8458	0.8545	0.8560

5.2 Unsupervised Methods

In the second round experiment, we use some unsupervised learning methods for evaluations on our dataset. The data preprocessing is the same as we conduct in supervised learning experiments. We use three deep learning models which are Graph Convolutional Network (GCN) [14], GraphSAGE [11] and HinSAGE (a variant of GraphSAGE that used for heterogeneous networks) respectively for experiments. In this round experiment, we use accuracy, MAE (Mean Absolute

Error) and MSE (Mean Squared Error) as the metrics to measure the performance. The equations of MAE and MSE are shown in Eq. (2) and Eq. (3) respectively, which are all used to measure the errors between true values and predictive values. Generally, the lower MAE and MSE indicate the better performance.

$$MAE = \frac{1}{n}\sum_{i=1}^{n}|(y_i - \hat{y}_i)| \tag{2}$$

$$MSE = \frac{1}{n}\sum_{i=1}^{n}(y_i - \hat{y}_i)^2 \tag{3}$$

The experimental results on unsupervised learning methods are shown in Table 4. Similar with supervised learning methods, other features can be introduced to improve the performance of link prediction.

Table 4. Experimental results by using unsupervised learning models.

	Accuracy	MAE	MSE
GCN	0.6188	0.9669	4.8673
GraphSAGE	0.5882	0.4197	0.3063
HinSAGE	**0.8969**	**0.2283**	**0.0613**

As shown in Table 4, GCN and GraphSAGE achieve close performance on the selected metrics. GCN is a transductive learning method which cannot get the embedding of new nodes, while GraphSAGE is an inductive learning method which can address the above problem. However, both GCN and GraphSAGE cannot effectively process heterogeneous networks. HinSAGE gets the best performance since it is designed for heterogeneous networks, which is in accordance with the feature of our dataset.

6 Conclusion

Link prediction on social networks has attracted more and more popularity nowadays. However, most existing link prediction datasets only contain nodes or entities and relationships, while neglecting the text attribute information. In this paper, we propose a novel link prediction dataset named SCHOLAT Link Prediction based on the academic social network SCHOLAT. Our proposed dataset contains multitype relationships among users and user attribute information. We conduct extensive evaluations on our dataset with supervised and unsupervised methods. The experimental results show the potential of our dataset on link prediction.

Acknowledgments. This work was supported in part by the National Natural Science Foundation of China under Grant 61772211, Grant U1811263 and Grant 62077045, in part by the Humanity and Social Science Youth Foundation of Ministry of Education of China under Grant 19YJCZH049 and in part by the Natural Science Foundation of Guangdong Province of China under Grant 2019A1515011292.

References

1. Aghabozorgi, F., Khayyambashi, M.R.: A new similarity measure for link prediction based on local structures in social networks. Physica A **501**, 12–23 (2018). https://doi.org/https://doi.org/10.1016/j.physa.2018.02.010, https://www.sciencedirect.com/science/article/pii/S0378437118300864
2. Ahmed, C., ElKorany, A., Bahgat, R.: A supervised learning approach to link prediction in twitter. Soc. Netw. Anal. Min. **6**(1), 24:1–24:11 (2016). https://doi.org/10.1007/s13278-016-0333-1
3. Benchettara, N., Kanawati, R., Rouveirol, C.: A supervised machine learning link prediction approach for academic collaboration recommendation. In: Amatriain, X., Torrens, M., Resnick, P., Zanker, M. (eds.) Proceedings of the 2010 ACM Conference on Recommender Systems, RecSys 2010, Barcelona, Spain, 26–30 September 2010, pp. 253–256. ACM (2010). https://doi.org/10.1145/1864708.1864760
4. Bordes, A., Usunier, N., García-Durán, A., Weston, J., Yakhnenko, O.: Translating embeddings for modeling multi-relational data. In: Burges, C.J.C., Bottou, L., Ghahramani, Z., Weinberger, K.Q. (eds.) Advances in Neural Information Processing Systems 26: 27th Annual Conference on Neural Information Processing Systems 2013. Proceedings of a Meeting held 5–8 December 2013, Lake Tahoe, Nevada, United States, pp. 2787–2795 (2013)
5. Bu, D., et al.: Topological structure analysis of the protein-protein interaction network in budding yeast. Nucleic Acids Res. **31**(9), 2443–2450 (2003)
6. Bu, Z., Wang, Y., Li, H., Jiang, J., Wu, Z., Cao, J.: Link prediction in temporal networks: integrating survival analysis and game theory. Inf. Sci. **498**, 41–61 (2019). https://doi.org/10.1016/j.ins.2019.05.050
7. Davis, J., Goadrich, M.: The relationship between precision-recall and ROC curves. In: Cohen, W.W., Moore, A.W. (eds.) Machine Learning, Proceedings of the Twenty-Third International Conference (ICML 2006), Pittsburgh, Pennsylvania, USA, 25–29 June 2006. ACM International Conference Proceeding Series, vol. 148, pp. 233–240. ACM (2006). https://doi.org/10.1145/1143844.1143874
8. Dettmers, T., Minervini, P., Stenetorp, P., Riedel, S.: Convolutional 2D knowledge graph embeddings. In: McIlraith, S.A., Weinberger, K.Q. (eds.) Proceedings of the Thirty-Second AAAI Conference on Artificial Intelligence, (AAAI-18), the 30th innovative Applications of Artificial Intelligence (IAAI-18), and the 8th AAAI Symposium on Educational Advances in Artificial Intelligence (EAAI-18), New Orleans, Louisiana, USA, 2–7 February 2018, pp. 1811–1818. AAAI Press (2018). https://www.aaai.org/ocs/index.php/AAAI/AAAI18/paper/view/17366
9. Dong, Y., et al.: Link prediction and recommendation across heterogeneous social networks. In: Zaki, M.J., Siebes, A., Yu, J.X., Goethals, B., Webb, G.I., Wu, X. (eds.) 12th IEEE International Conference on Data Mining, ICDM 2012, Brussels, Belgium, 10–13 December 2012, pp. 181–190. IEEE Computer Society (2012). https://doi.org/10.1109/ICDM.2012.140

10. Goutte, C., Gaussier, E.: A probabilistic interpretation of precision, recall and F-score, with implication for evaluation. In: Losada, D.E., Fernández-Luna, J.M. (eds.) ECIR 2005. LNCS, vol. 3408, pp. 345–359. Springer, Heidelberg (2005). https://doi.org/10.1007/978-3-540-31865-1_25
11. Hamilton, W.L., Ying, Z., Leskovec, J.: Inductive representation learning on large graphs. In: Guyon, I., et al. (eds.) Advances in Neural Information Processing Systems 30: Annual Conference on Neural Information Processing Systems 2017, Long Beach, CA, USA, 4–9 December 2017, pp. 1024–1034 (2017)
12. Hopcroft, J.E., Lou, T., Tang, J.: Who will follow you back?: reciprocal relationship prediction. In: Macdonald, C., Ounis, I., Ruthven, I. (eds.) Proceedings of the 20th ACM Conference on Information and Knowledge Management, CIKM 2011, Glasgow, United Kingdom, 24–28 October 2011, pp. 1137–1146. ACM (2011). https://doi.org/10.1145/2063576.2063740
13. Isella, L., Stehlé, J., Barrat, A., Cattuto, C., Pinton, J.F., Van den Broeck, W.: What's in a crowd? Analysis of face-to-face behavioral networks. J. Theor. Biol. **271**(1), 166–180 (2011)
14. Kipf, T.N., Welling, M.: Semi-supervised classification with graph convolutional networks. In: 5th International Conference on Learning Representations, ICLR 2017, Conference Track Proceedings, Toulon, France, 24–26 April 2017. OpenReview.net (2017). https://openreview.net/forum?id=SJU4ayYgl
15. Leskovec, J., Kleinberg, J.M., Faloutsos, C.: Graph evolution: densification and shrinking diameters. ACM Trans. Knowl. Discov. Data **1**(1), 2 (2007). https://doi.org/10.1145/1217299.1217301
16. Liben-Nowell, D., Kleinberg, J.M.: The link-prediction problem for social networks. J. Assoc. Inf. Sci. Technol. **58**(7), 1019–1031 (2007). https://doi.org/10.1002/asi.20591
17. Mahdisoltani, F., Biega, J., Suchanek, F.M.: YAGO3: a knowledge base from multilingual Wikipedias. In: Seventh Biennial Conference on Innovative Data Systems Research, CIDR 2015, Online Proceedings, Asilomar, CA, USA, 4–7 January 2015 (2015). https://www.cidrdb.org, http://cidrdb.org/cidr2015/Papers/CIDR15_Paper1.pdf
18. McAuley, J.J., Leskovec, J.: Learning to discover social circles in ego networks. In: Bartlett, P.L., Pereira, F.C.N., Burges, C.J.C., Bottou, L., Weinberger, K.Q. (eds.) Advances in Neural Information Processing Systems 25: 26th Annual Conference on Neural Information Processing Systems 2012. Proceedings of a meeting held 3–6 December 2012, Lake Tahoe, Nevada, United States, pp. 548–556 (2012)
19. Muniz, C.P.M.T., Goldschmidt, R.R., Choren, R.: Combining contextual, temporal and topological information for unsupervised link prediction in social networks. Knowl. Based Syst. **156**, 129–137 (2018). https://doi.org/10.1016/j.knosys.2018.05.027
20. Newman, M.E.: Finding community structure in networks using the eigenvectors of matrices. Phys. Rev. E **74**(3), 036104 (2006)
21. Newman, M.E.: The structure of scientific collaboration networks. In: The Structure and Dynamics of Networks, pp. 221–226. Princeton University Press (2011)
22. Ostapuk, N., Yang, J., Cudré-Mauroux, P.: ActiveLink: deep active learning for link prediction in knowledge graphs. In: Liu, L., et al. (eds.) The World Wide Web Conference, WWW 2019, San Francisco, CA, USA, 13–17 May 2019, pp. 1398–1408. ACM (2019). https://doi.org/10.1145/3308558.3313620
23. Pecli, A., Cavalcanti, M.C., Goldschmidt, R.R.: Automatic feature selection for supervised learning in link prediction applications: a comparative study. Knowl. Inf. Syst. **56**(1), 85–121 (2018). https://doi.org/10.1007/s10115-017-1121-6

24. Sen, P., Namata, G., Bilgic, M., Getoor, L., Gallagher, B., Eliassi-Rad, T.: Collective classification in network data. AI Mag. **29**(3), 93–106 (2008). https://doi.org/10.1609/aimag.v29i3.2157
25. Sherkat, E., Rahgozar, M., Asadpour, M.: Structural link prediction based on ant colony approach in social networks. Physica A **419**, 80–94 (2015). https://doi.org/10.1016/j.physa.2014.10.011, https://www.sciencedirect.com/science/article/pii/S0378437114008498
26. Tay, Y., Luu, A.T., Hui, S.C.: Non-parametric estimation of multiple embeddings for link prediction on dynamic knowledge graphs. In: Singh, S.P., Markovitch, S. (eds.) Proceedings of the Thirty-First AAAI Conference on Artificial Intelligence, San Francisco, California, USA, 4–9 February 2017, pp. 1243–1249. AAAI Press (2017). http://aaai.org/ocs/index.php/AAAI/AAAI17/paper/view/14524
27. Wang, Z., Chen, C., Li, W.: Predictive network representation learning for link prediction. In: Kando, N., Sakai, T., Joho, H., Li, H., de Vries, A.P., White, R.W. (eds.) Proceedings of the 40th International ACM SIGIR Conference on Research and Development in Information Retrieval, Shinjuku, Tokyo, Japan, 7–11 August 2017, pp. 969–972. ACM (2017). https://doi.org/10.1145/3077136.3080692
28. Xie, F., Chen, Z., Shang, J., Feng, X., Li, J.: A link prediction approach for item recommendation with complex number. Knowl. Based Syst. **81**, 148–158 (2015). https://doi.org/10.1016/j.knosys.2015.02.013
29. Zhang, M., Chen, Y.: Link prediction based on graph neural networks. In: Bengio, S., Wallach, H.M., Larochelle, H., Grauman, K., Cesa-Bianchi, N., Garnett, R. (eds.) Advances in Neural Information Processing Systems 31: Annual Conference on Neural Information Processing Systems 2018, NeurIPS 2018, Montréal, Canada, 3–8 December 2018, pp. 5171–5181 (2018)
30. Zhang, Z., Cai, J., Zhang, Y., Wang, J.: Learning hierarchy-aware knowledge graph embeddings for link prediction. In: The Thirty-Fourth AAAI Conference on Artificial Intelligence, AAAI 2020, The Thirty-Second Innovative Applications of Artificial Intelligence Conference, IAAI 2020, The Tenth AAAI Symposium on Educational Advances in Artificial Intelligence, EAAI 2020, New York, NY, USA, 7–12 February 2020, pp. 3065–3072. AAAI Press (2020). https://aaai.org/ojs/index.php/AAAI/article/view/5701
31. Zhang, Z., Wen, J., Sun, L., Deng, Q., Su, S., Yao, P.: Efficient incremental dynamic link prediction algorithms in social network. Knowl. Based Syst. **132**, 226–235 (2017). https://doi.org/10.1016/j.knosys.2017.06.035

A Graph Neural Network-Based Approach for Predicting Second Rise of Information Diffusion on Social Networks

Jiaxing Shang[1,2](✉), Yijie Wang[1,2], Yuxin Gong[1,2], Yanli Zou[1,2], and Xinjun Cai[1,2]

[1] College of Computer Science, Chongqing University, Chongqing 400044, China
shangjx@cqu.edu.cn
[2] Key Laboratory of Dependable Service Computing in Cyber Physical Society, Ministry of Education, Chongqing University, Chongqing 400044, China

Abstract. Recently, with the rise of social media, the research of information diffusion prediction has drawn much attention from scholars. The problem has important applications in public opinion monitoring, social advertising, etc. Through an in-depth diffusion analysis, we observed an interesting phenomenon where a piece of message may endure a "second rise" after it peaked at the maximum popularity for a while. However, this phenomenon has not yet been investigated by existing works. Moreover, the valuable information contained in the repost comments were not fully utilized. To fill this gap, this paper proposes a graph neural network-based model for predicting second rise of information diffusion. Specifically, we first design a simple but efficient algorithm to determine whether a message has second rise. Then, text analysis is carried out on the repost comments, and different message-text bipartite graphs are constructed according to different types of textual information (topics, comments, @users, emojis). After that, we use random walk to generate node sequences and apply skip-gram model to learn node representations, followed by a PCA-based dimension deduction. The compressed textual features are combined with embeddings learned from repost network, before they are finally fed to downstream machine learning models to generate predictions. Experimental results on the Weibo dataset show that the overall prediction accuracy can be improved significantly by incorporating the textual features.

Keywords: Information diffusion · Second rise · Graph neural network · Text analysis · Representation learning

1 Introduction

In the era of big data, with the rapid development of the Internet, communication has become more convenient and frequent, resulting in the prosperity of social networks. Today's social platforms consist of a huge number of users, such a large and complex network has greatly changed the way users access information. Through social platforms, it becomes easier for people to get information. At the same time, social platforms are

Y. Sun et al. (Eds.): ChineseCSCW 2021, CCIS 1492, pp. 352–363, 2022.
https://doi.org/10.1007/978-981-19-4549-6_27

also affected by negative factors such as malicious users, false marketing, fake news, etc.

Therefore, it is of great significance to study information propagation on social platforms. For example, in social advertising, through social diffusion and influence analysis, social users with higher influence can be selected as seed users, so as to maximize the effect of advertising. For another example, analysis and popularity prediction of social contents can help government departments to understand the latest hot topics, which is beneficial to public opinion monitoring. Therefore, the prediction of information propagation on social networks has great economic and social significance. Based on an in-depth analysis of real diffusion data, we observed a general phenomenon of second rise in information propagation. The second rise means that after the information propagation reaches a plateau, it will suddenly endure a rapid second rise. However, the existing research has paid little attention to this phenomenon. Therefore, this paper mainly focuses on this phenomenon, and utilizes the latest graph neural network model to predict the second rise.

Although there have been a lot of studies on the prediction of social network information propagation [1], there are still some deficiencies in the existing research methods. For the methods based on manual feature [2–4], due to the need to handcrafting features, on the one hand, it takes time and effort, on the other hand, the performance of the algorithms is often unsatisfactory. Methods based on probabilistic generation models [5–7] usually need to make strong assumptions on the model, which cannot provide good scalability. In recent years, with the popularity of graph neural network (GNN) model [8], a number of diffusion prediction methods based on network representation learning have emerged [9–11]. The main idea is to learn users' low-dimensional representation vectors from the social network through network representation learning, and feed the features to downstream machine learning model for prediction. However, the existing methods generally have the following deficiencies: First, they have not considered the aforementioned second rise phenomenon; Second, existing methods did not make full use of the rich textual information such as repost comments, leaving rooms for further improvement.

In view of the above problems, this paper studies how to predict the second rise phenomenon based on the multi-modal structural and textual information in the forwarding network during the initial stage of information propagation. In order to solve this problem, a graph neural network-based model is proposed. To be specific, we first design a simple but efficient algorithm to determine whether the information has second rise, so as to reduce the cost of manual annotation. Then, text analysis is carried out on the repost comments, and different message-text bipartite graphs are constructed according to different types of textual information (topics, comments, @users, emojis). After that, we use random walk to generate node sequences and apply skip-gram model to learn node representations, followed by a PCA-based dimension deduction. The compressed textual features are combined with embeddings learned from repost network, before they are finally fed to downstream machine learning models to generate predictions. Experimental results on the Weibo dataset show that the overall prediction accuracy can be improved significantly by combining the text features.

2 Related Work

This paper mainly focuses the prediction of information propagation, so the related works are mainly divided into three categories: feature engineering-based studies, methods based on probabilistic generative models and network representation learning-based works.

The first category of methods is usually based on node importance metrics, which manually extracts the relevant features based on network structure, and combines with the traditional machine learning models to predict the diffusion. The commonly used centrality measures include degree centrality [2], closeness and betweenness centrality [3], PageRank centrality [12], HITS centrality [13], MADA-based centrality [14], etc. The key disadvantage of these methods is that these features are highly dependent on human experience and cannot fully reflect the structural information related to information diffusion, so they often result in unsatisfactory results. Moreover, some centrality metrics, such as the betweenness centrality, is too expensive to be calculated on large-scale networks.

The second category of methods is based on probabilistic generative models, which predicts the information diffusion by fitting the data with the underlying theoretical models. Shen et al. [5] used an enhanced Poisson process to fit the information propagation process. Bao et al. [6] proposed a probabilistic model based on self-excited Hawkes process (SEHP). Mishra et al. [7] proposed to use labeled Hawkes self-excitation point process to model the information propagation process. Gao et al. [15] proposed a probabilistic model based on mixed processes, Wang et al. [16] defined an information diffusion model to predict the propagation process. These methods usually need to make strong hypothesis to the models, so they usually exhibit poor scalability across different datasets.

The third type of methods is based on user relationship network and historical propagation data, which uses network representation learning method to extract user features. In recent years, with the popularity of graph neural network model, it is gradually favored by researchers. Typical examples include: representation learning method based on thermal diffusion process [9], representation learning prediction method based on Independent Cascade (IC) propagation model [10], forwarding network representation learning method RNe2Vec [11], role-based network representation learning method [17], representation learning method based on COSINE probability generation model [18], Deep Collaborative Embedding (DCE) Model [19], etc. The advantage of such methods is their high prediction accuracy. However, existing methods have not yet considered the second rise phenomenon, and have not fully utilized the rich textual information in repost comments, so there is still a large space for exploration.

3 Methodology

3.1 Identification of Cascades with Second Rise

This paper aims to predict whether there will be a second rise in the diffusion process of a message on Weibo[1], which is rarely studied by previous studies. To this end, we need to define the phenomenon of second rise and identify the cascades with this phenomenon.

Figure 1 shows an example, where the left subfigure shows a message with a second rise while the right one not. To automatically identify whether a certain message has second rise, calculation rules are defined as follows:

1) Divide the whole reposts of a message into 10 equal segments, each of which is defined as R_i, $i = 0, 1, \cdots, 9$.
2) Respectively calculate the corresponding temporal length T_i for each section R_i, $i = 0, 1, \cdots, 9$.
3) Calculate the forwarding trend K, which is defined as $K = \sum_i T_i / \sum_i R_i$.
4) Define the state of each segment R_i as S_i:

$$S_i = \begin{cases} 1, \ if \ (T_i/R_i) < K \ and \ (K - (T_i/R_i)) > p \\ 0, \ else \end{cases} \tag{1}$$

   where p is a tunable parameter. After small scale experiments, we choose $p = K/2$, since it yields to best overall identification effect of second rise phenomenon.
5) Based on the state sequence $S_0S_2 \ldots S_9$, if a subsequence '11001' is observed, then the message identified to endure the phenomenon of second rise.

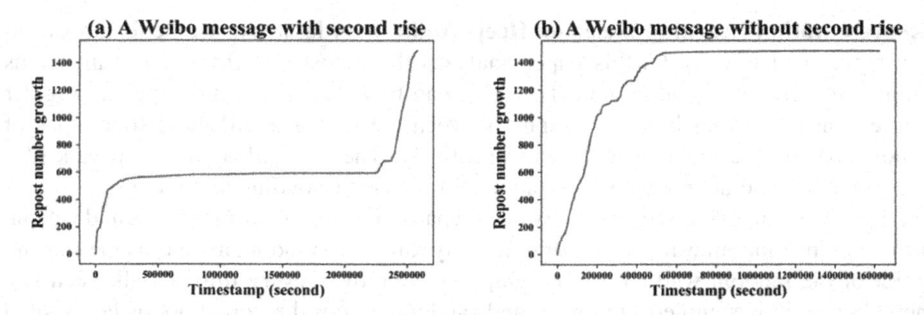

Fig. 1. Repost number growth of Weibo messages with and without second rise.

3.2 Text Analysis

Repost Text Analysis: Considering that the text associated with reposts is likely to be an important factor to distinguish whether the message will have a second rise, we first conduct a textual analysis of repost text. Figure 2 shows the word cloud figures

[1] https://weibo.com/.

of Weibo messages with (left subfigure) and without (right subfigure) second rise. A significance in the important words can be observed between the two types of messages, which conforms our assumption. Therefore, it can be concluded that textual analysis is beneficial to the problem.

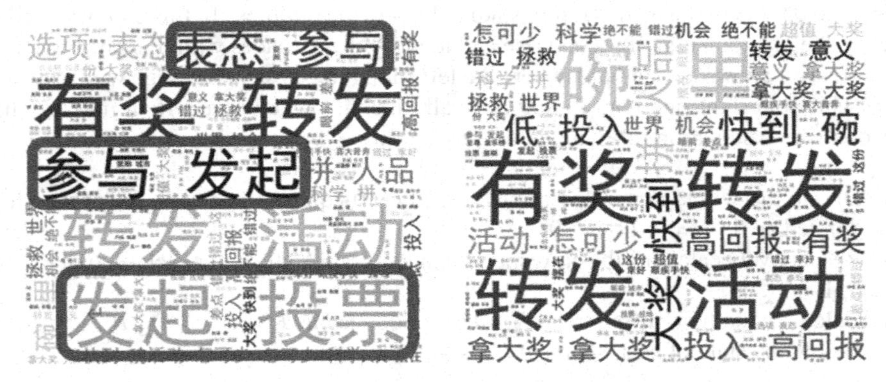

Fig. 2. Word cloud with (left) and without (right) second rise.

After removing the stop words, we find that for messages with second rise, the most frequently used term in the repost texts is "initiate voting". At the same time, the most frequently referred account (through @account) is "City Supreme Music List". Actually, these words are all prominent characteristics that differentiate two types of messages. The difference can reflect the significant influence of idol culture on social media platforms. Specifically, the support of fans to their idols is a significant contributor for the second rise of a message.

Representation Learning Based on Deep Walk: In order to extract the features from the repost text effectively, this paper analyzes the repost text from four dimensions (topic, @user, emoji and comment), which can be extracted through specific regular expressions; The four bipartite graphs between the message and these four types of repost textual objects are established respectively; Then we utilize the DeepWalk [20] algorithm to generate node representations for the corresponding text features.

DeepWalk algorithm consists of two main parts: Random Walk Generator and Update Process. The random walk generator uniformly samples a random node v_i as the starting point of the random walk W_i in the graph G. Moreover, every time a walk occurs, a neighbor node is sampled uniformly and randomly from the neighbors of last visited node until the walk reaches the maximum length t. The pseudo-code of the DeepWalk algorithm is shown in **Algorithm 1**.

For each constructed bipartite graph, we first regard it as a homogeneous graph, and the apply Random Walk sampling from each node to obtain node sequences. Then SkipGram (as shown in Algorithm 2) model is applied to train these node sequences to obtain representation vectors. Meanwhile, the Hierarchical Softmax technique is utilized to improve the model training efficiency. Through these steps, we are able to obtain the textual representation vectors for the four types of repost text, i.e., topics, @ users, emojis, and forwarding words.

Algorithm 1 DeepWalk (G, ω, d, γ, t)

Input: graph $G(V, E)$

window size ω

embedding size d

walks per vertex γ

Output: matrix of vertex representations $\Phi \in \mathbb{R}^{|V|\times d}$

1: Initialization: Sample Φ from $u^{|V|\times d}$

2: Build a binary Tree T from V

3: **for** i = 0 to γ **do**

4: O = Shuffle(V)

5: **for each** $v_i \in O$ **do**

6: $W_{v_i} = RandomWalk(G, v_i, t)$

7: SkipGram (Φ, W_{v_i}, ω)

8: **end for**

9: **end for**

Algorithm 2 SkipGram (Φ, W_{v_i}, ω)

1: **for each** $v_j \in W_{v_i}$ **do**

2: **for each** $u_k \in W_{v_i}[j - \omega : j + \omega]$ **do**

3: $J(\Phi) = -\log \Pr(u_k \mid \Phi(v_j))$

4: $\Phi = \Phi - \alpha * (\partial J / \partial \Phi)$

5: **end for**

6: **end for**

4 Experimental Evaluation

4.1 Dataset

The dataset used in this paper is the Sina Weibo dataset, which contain a total of 27,000 original Weibo messages and 15.97 million corresponding repost records, provided by the Data Castle competition platform[2]. Among them, each repost record includes Weibo ID, repost user ID, source user ID, repost text, and the repost timestamp which is the relative time interval between the repost behavior and the original message publication.

[2] https://js.dclab.run/v2/cmptDetail.html?id=166.

After appropriate preprocessing of the dataset, we finally get 1058 samples consisting of 500+ positive (with second rise) and 500+ negative messages (without second rise) from the Weibo messages whose final cascade size is more than 300. The ratio between positive and negative sample is close to 1:1 and the total number of associated repost records is 792,110. Figures 3 and 4 show examples of original Weibo messages and the repost records.

	weibo_id	author_id	time	content
0	3794305741726764	2724513	20:47:26	看到这条微博的人，羊年都会走大运！
1	3794545218812248	7460165	12:39:02	2015你最想获得下面那一项?
2	3794726233244929	7387806	00:38:19	从剧组偷了个闲出来认真的看过沙果"贱"入佳境了，幽默精彩，点赞支持，票房小黑马马力十足祝...
3	3794986804085757	8091415	17:53:44	Hey,从事教师这一职业的我不甘于现状，心中总有股劲儿想要挑战自己！一次偶然的机会让我接触了...
4	3795261795455671	8189779	12:06:27	我操你们妈了隔壁[呵呵]
5	3795438593411753	7298946	23:48:59	[讨论]周杰伦的歌曲和正能量影响着几代人！当然还有现在爆红的00后组合TFBOYS的队长王俊...
6	3795659825938762	7900610	14:28:05	#2015我的开年心愿#圣诞已过，除夕未至。新年伊始，大家一定有新的心愿或者目标：收获一份甜...

Fig. 3. Original Weibo messages from the dataset.

15730378	3976605364067619	5257388	113	3542382	看哭了
15730379	3974264695126996	3018880	106	3583050	NaN
15730380	3869617355730958	3021266	104	28171843	NaN
15730381	3969150235298748	3092539	87	5232788	NaN
15730382	3938415609472208	7579011	84	12209189	NaN
15730383	3975243367813460	7996665	75	3423037	NaN
15730384	3946499828206442	5687616	47	10376457	NaN
15730385	3937831519031094	8210321	40	12554562	NaN
15730386	3831471334364039	5222540	29	37787115	NaN
15730387	3954122480659216	2418468	23	8698115	1
15730388	3968615998437099	974639	11	5403789	NaN

15730389 rows × 5 columns

Fig. 4. Repost records from Weibo dataset.

On the basis of the aforementioned data, we aim to predict whether a Weibo message will endure a second rise in the future based on the available data before the point when the Weibo message reaches its first plateau, as shown in Fig. 5 (which is marked with dotted line).

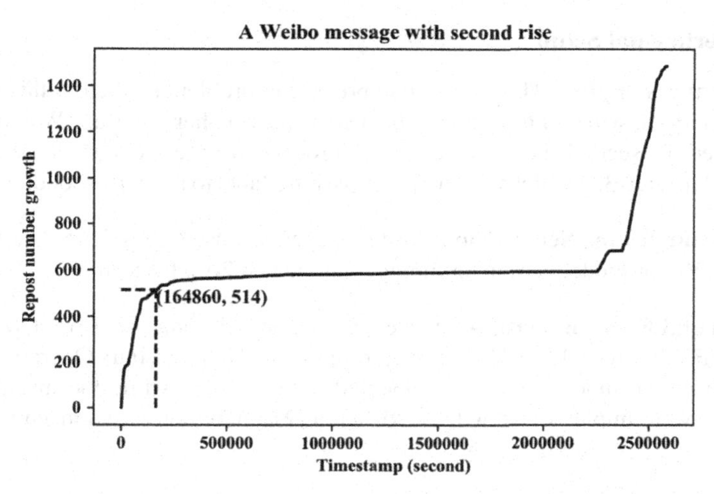

Fig. 5. The time point when a Weibo message reaches its first plateau (marked with dotted line)

4.2 Comparing Methods

In this paper, we feed the learned features to several downstream machine learning models, and compare their performance under different textual features, so as to see how these features would benefit the prediction performance. The machine learning models considered in this paper include: Logistic Regression (**LR**), Decision Tree (**DT**), Multi-Layer Perceptron (**MLP**), K-Nearest Neighbors (**KNN**), and Random Forest (**RF**).

4.3 Evaluation Metrics

In this paper, the performance of the machine learning models is evaluated by four metrics: *Accuracy*, *Precision*, *Recall* and *F1-Score*, the calculation formulas of which are as follows:

$$Accuracy = \frac{TP + TN}{TP + TP + TN + FN} \tag{2}$$

$$Precision = \frac{TP}{TP + FP} \tag{3}$$

$$Recall = \frac{TP}{TP + FN} \tag{4}$$

$$F1 - score = \frac{2Precision \times Recall}{Precision + Recall} \tag{5}$$

where *TP* (True Positive) is the number of cases correctly identified as second rise, *FP* (False Positive) is the number of cases incorrectly identified as second rise, *TN* (True Negative) is the number of cases correctly identified as non-second rise, and *FN* (False Negative) is the number of cases incorrectly identified as non-second rise.

4.4 Experimental Setup

Labelling the Samples: The second rise prediction problem studied in this paper is a typical binary classification problem. To determine whether a certain Weibo message has endured the second rise in the diffusion process, we use the identification method introduced in Sect. 3.1, which automatically assign a label to the corresponding message.

Training and Testing Set Division: In this paper, we use 60% of the data for model training, 20% for model parameter tuning, and the rest 20% for testing.

Experimental Environment: All the experiments were conduced on a workstation with Intel E5-2630v4 CPU, 64G Memory, running the Ubuntu Linux Operating System. The graph neural models were implemented with PyTorch while the machine learning models were implemented with Scikit-learn [21], a python-based machine learning package.

4.5 Results

Results Under Different Textual Features: We first compare the results of adding any single textual feature (topic, @user, emoji, and repost comment) and the combination of these features on the prediction performance of the learning models. The results of *accuracy* and *F1-score* are shown in Table 1 and Table 2 respectively. By comparing the experimental results of different machine learning models, we can see that the @user textual feature exhibit the best prediction performance in single feature experiments, where the highest accuracy value reaches up to 64%. However, when we consider the combination of different textual features, the multi-layer perceptron (MLP) model gives the best overall performance. When using MLP to train combined text features, its *Accuracy* and *F1-score* both can exceed 61%. In addition, among the four textual features, due to the relatively poor training performance of the topic feature, we choose the other three text features (@user, emoji and repost comment) to combine and reduce the dimensionality to 20 as the final text features. The 2-D visualization of dimension reduction result for the combined text features is shown in Fig. 6.

Improvements with Textual Features: Table 3 shows the improvement of machine learning models with textual features over the basic models without textual features, evaluated by *accuracy*, *precision*, *recall*, *F1-score*. For the basic models, we first learn the node features from the repost network using the DeepWalk method, and the aggregate node features to obtain the message features. It can be seen from the results that most of the evaluation metrics have been significantly improved with textual features. Specifically, the most significant improvement is observed on the precision measure with the absolution percentage increase of 15%, which shows the effectiveness of the method proposed in this paper.

Table 1. Results of accuracy with text features

Accuracy	Topic	@User	Emoji	Words	Combination
LR	0.5330	**0.5896**	0.5613	0.5755	0.5802
DT	0.5425	**0.6415**	0.5896	0.6179	0.5613
MLP	0.5047	**0.6321**	0.5613	0.5708	0.6179
KNN	0.5047	**0.5613**	0.5472	0.5236	0.5472
RF	0.5425	**0.6415**	0.5896	0.6179	0.5366

Table 2. Results of F1-score with Text features

F1-score	Topic	@User	Emoji	Words	Combination
LR	0.5325	**0.5892**	0.5541	0.5724	0.5800
DT	0.5416	**0.6320**	0.5847	0.6106	0.5366
MLP	0.5022	**0.6318**	0.5570	0.5691	0.6177
KNN	0.4717	**0.5517**	0.5337	0.4512	0.4717
RF	0.5416	**0.6320**	0.5847	0.6106	0.5613

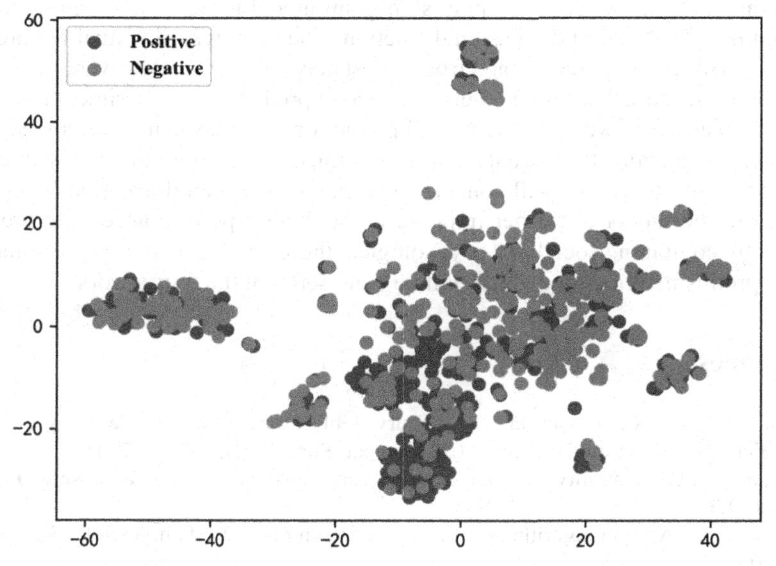

Fig. 6. Visualization of dimension reduction results of combined text features

Table 3. Improvement over basic models with textual features

Evaluation	With text features	Without text features	Improvement (%)
Accuracy	57%	48%	+9%
Precision	63%	48%	+15%
Recall	77%	83%	−6%
F1-score	69%	63%	+6%

5 Conclusion

In this paper, we investigated the second rise phenomenon of information diffusion in social networks, which has rarely been investigated in previous studies. In order to predict the second rise of information diffusion in social media, this paper proposed a graph neural network-based method by incorporating rich information from repost texts. Specifically, we first designed a simple but efficient algorithm to determine whether a message has second rise. Then, text analysis was carried out on the repost comments, and different message-text bipartite graphs were constructed according to different types of textual information (topics, comments, @users, emojis). After that, we used random walk to generate node sequences and apply skip-gram model to learn node representations, followed by a PCA-based dimension deduction. The compressed textual features were combined with embeddings learned from repost network, before they were finally fed to downstream machine learning models to generate predictions. Experimental results on the Weibo dataset showed that the overall prediction accuracy can be improved significantly by incorporating the textual features. In sum, this paper investigated an interesting problem. In the future, we will consider incorporating more information (e.g., image, url, etc.) in the model to further improve its prediction performance. Interpretability research by combining social and psychological theories to explain this phenomenon is another promising direction, which will be considered in the future work.

References

1. Zhou, F., Xu, X., Trajcevski, G.: A survey of information cascade analysis: models, predictions, and recent advances. ACM Comput. Surv. **54**(2), 1–36 (2021)
2. Freeman, L.C.: Centrality in social networks conceptual clarification. Soc. Netw. **1**(3), 215–239 (1978)
3. Brandes, U.: A faster algorithm for betweenness centrality. J. Math. Sociol. **25**(2), 163–177 (2001)
4. Zhong, L., Gao, C.: Identifying influential nodes in complex networks. In: International Conference on Active Media Technology, pp. 11–22 (2014)
5. Shen, H., Wang, D.: Modeling and predicting popularity dynamics via reinforced poisson processes. In: Twenty-Eighth AAAI Conference on Artificial Intelligence (2014)

6. Bao, P., Shen, H.W.: Modeling and predicting popularity dynamics of Weibos using self-excited Hawkes processes. In: Proceedings of the 24th International Conference on World Wide Web, pp. 9–10 (2015)
7. Mishra, S., Rizoiu, M.A.: Feature driven and point process approaches for popularity prediction. In: Proceedings of the 25th ACM International on Conference on Information and Knowledge Management, pp. 1069–1078 (2016)
8. Wu, Z., Pan, S., Chen, F., Long, G., Zhang, C.: A comprehensive survey on graph neural networks. IEEE Trans. Neural Netw. Learn. Syst. **32**(1), 4–24 (2020)
9. Bourigault, S., Lagnier, C.: Learning social network embeddings for predicting information diffusion. In: Proceedings of the 7th ACM International Conference on Web Search and Data Mining, pp. 393–402 (2014)
10. Bourigault, S., Lamprier, S.: Representation learning for information diffusion through social networks: an embedded cascade model. In: Proceedings of the Ninth ACM International Conference on Web Search and Data Mining, pp. 573–582 (2016)
11. Shang, J., et al.: RNe2Vec: information diffusion popularity prediction based on repost network embedding. Computing **103**(2), 271–289 (2020)
12. Page, L., Brin, S., Motwani, R.: The pagerank citation ranking: bringing order to the web. Stanford InfoLab (1999)
13. Kleinberg, J.M.: Authoritative sources in a hyperlinked environment. J. ACM **46**(5), 604–632 (1999)
14. Du, Y., Gao, C., Hu, Y.: A new method of identifying influential nodes in complex networks based on TOPSIS. Phys. A **399**, 57–69 (2014)
15. Gao, J., Shen, H., Liu, S.: Modeling and predicting retweeting dynamics via a mixture process. In: Proceedings of the 25th International Conference Companion on World Wide Web, pp. 33–34 (2016)
16. Wang, Y., Zhang, Z.-M., Peng, Z.-S., Duan, Y.-Y., Gao, Z.-Q.: A cascading diffusion prediction model in micro-blog based on multi-dimensional features. In: Barolli, L., Zhang, M., Wang, X.A. (eds.) EIDWT 2017. LNDECT, vol. 6, pp. 734–746. Springer, Cham (2018). https://doi.org/10.1007/978-3-319-59463-7_73
17. Wang, Z., Chen, C., Li, W.: Information diffusion prediction with network regularized role-based user representation learning. ACM Trans. Knowl. Discov. Data **13**(3), 1–23 (2019)
18. Zhang, Y., Lyu, T., Zhang, Y.: Community-preserving social network embedding from information diffusion cascades. In: Thirty-Second AAAI Conference on Artificial Intelligence (2018)
19. Zhao, Y., Yang, N., Lin, T.: Deep collaborative embedding for information cascade prediction. Knowl. Based Syst. **193**, 105502 (2020)
20. Perozzi, B., Rami, A., Steven, S.: Deepwalk: online learning of social representations. In: Proceedings of the 20th ACM SIGKDD International Conference on Knowledge Discovery and Data Mining (KDD), pp. 701–710 (2020)
21. Pedregosa, F., et al.: Scikit-learn: machine learning in python. J. Mach. Learn. Res. **12**, 2825–2830 (2011)

HPEMed: Heterogeneous Network Pair Embedding for Medical Diagnosis

Mengxi Li[1], Jing Zhang[1(✉)], Lixia Chen[2], Yu Fu[1], and Cangqi Zhou[1]

[1] School of Computer Science and Engineering, Nanjing University of Science and Technology, 200 Xiaolingwei Street, Nanjing 210094, China
{mengxili,jzhang,cqzhou}@njust.edu.cn

[2] The Second Affiliated Hospital of Nanjing University of Chinese Medicine, 23 Nanhu Road, Nanjing 210017, China

Abstract. Capturing a high-quality representation of clinical events in Electronic Health Record (EHR) is critical for upgrading public medical applications such as intelligent auxiliary diagnosis systems. However, the relationships among clinical events have different semantics and different contributions to disease diagnosis. This paper proposes a novel heterogeneous information network (HIN) based model named HPEMed for disease diagnosis tasks. HPEMed takes advantage of high-dimensional EHR data to model the nodes (with features) and edges of a graph. It exploits meta paths to higher-level semantic relations among EHR data and employs a pair-node embedding scheme that considers patient nodes with rich features and diagnosis nodes together, which achieves a more reasonable clinical event representation. The experimental results show that the performance of HPEMed in diagnosis tasks is better than that of some advanced baseline methods.

Keywords: Network embedding · Heterogeneous Information Networks · Representation learning · Electronic Health Record

1 Introduction

The embedding of medical concepts in data sets can make it possible to learn the correlation of clinical events, such as calculating the similarity between diagnosis and symptoms [3]. These medical concepts can be utilized as features in the network to predict other medical tasks [2] in the future. Among the numerous medical information, EHR [11] is full of adequate provision for detailed information on a variety of clinical records that occurred during the patient's hospitalization. For instance, microbiology tests, procedures, and symptoms are all examples of different types of events [5]. In the previous years, clinical records are often utilized to extract medical concepts and predict diseases by traditional feature engineering methods [10,14], which may result in missing values in EHR.

This work has been partially sponsored by the National Natural Science Foundation of China (No. 62076130 and No. 61902186) and the internal programs of the Second Affiliated Hospital of Nanjing University of Chinese Medicine (No. SEZ202121).

Y. Sun et al. (Eds.): ChineseCSCW 2021, CCIS 1492, pp. 364–375, 2022.
https://doi.org/10.1007/978-981-19-4549-6_28

In order to overcome the above problem and successfully capture the structure of EHR, previous methods concentrate on the use of admission history and diagnosis history in EHR data to predict future diseases by the way of homogeneous networks [7]. In usual, a clinical event isn't targeted for a specific disease, the co-occurrence of multiple clinical events may indicate this disease. Generally, in the single diagnosis process, it is more important to carefully consider the relationship between the clinical events (such as medical record and diagnosis) [5].

In the real world, many tasks can be modeled as two different types of nodes in HIN, the pair-wise relationship between them can be analyzed by heterogeneous information network-based embedding methods [9]. For instance, the pair-wise relationship for disease diagnosis tasks can be defined as the patient-diagnosis pairwise relationship. More specifically, the model can be constructed according to the patient's features and his/her diagnosis results. When constructing the task-guided embedding models, especially in the field of disease diagnosis, it makes sense to concentrate on modeling pair-wise relationships embedding instead of the traditional node-centric embedding. To this end, we use heterogeneous networks for modeling medical data and implement pair-node embeddings for tackling the potential semantics in pairwise relationships. In summary, we list the research contributions of this paper:

- We put forward a HIN-based medical diagnosis model named HPEMed which condenses the rich semantic relationships in EHR data into patients' representation features and tailors the patient-diagnosis node embeddings method for the diagnosis task. This design can adapt to efficient diagnosis tasks and enhance the performance of the diagnosis results.
- The experimental results quantitatively illustrate that HPEMed provides important improvements on mainstream baselines in predicting disease codes and lifting the proximity of diagnosis results.

2 Problem Definition

In this segment, we first provide some problem definitions and clinical terminology that appear in our research. Secondly, we describe how to model the EHR into a HIN. In the end, we formalize the disease diagnosis task.

2.1 Clinical Events and Patient Features

For each patient's record, we summarize a series of basic information, admission data, and treatment results of the patient into a clinical event. In consideration of the different compositions of nodes in HIN, we formulize a clinical event c as $c = (n, t, v)$, where n, t and v represent the type of events, clinical terminology, and value, respectively. For example, the blood glucose level of 4.0 can be viewed as $type = laboratory\ test$, $terminology = blood\ glucose$, and $value = 4.0$.

From the perspective of the division of medical events, a clinical event C shown in Fig. 1 consists of a patient event P and a diagnosis event D, which can be formalized as $P = \{laboratory\ test, symptom, age, gender, ethnicity,$

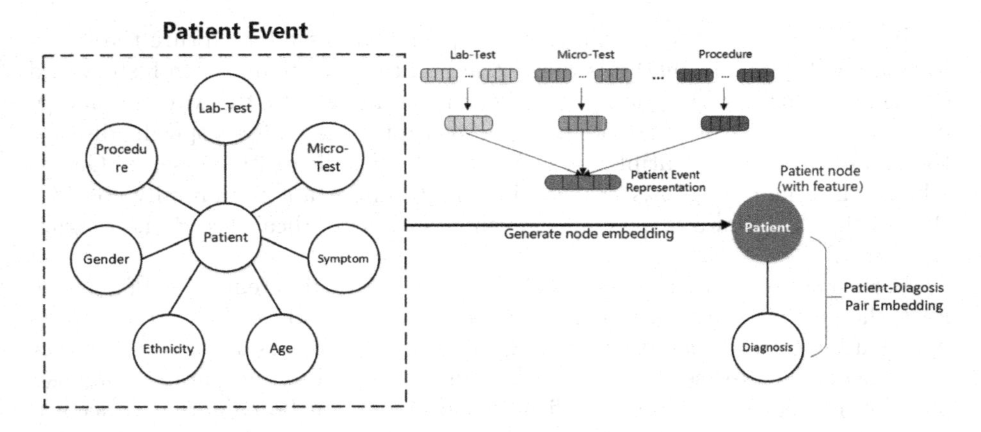

Fig. 1. The structure of patient-diagnosis pair

microbiology test} and $D = \{diagnosis\}$. Patient events are the source of information for patients and there are treated as the features of patient nodes.

2.2 Concepts Related to Heterogeneous Network

Heterogeneous Network. A HIN can be defined as a graph $G = (V, E, T_v, T_e, \psi, \phi)$, where node sets V and edge sets E in graph G have different types. Then, we map nodes $v \in V$ to their type by the defined node mapping function $\psi : V \rightarrow T_v$, where T_v is the union of different types of nodes. Meanwhile, we also map each edge $e \in E$ to edge type mapping function $\phi : E \rightarrow T_e$, where T_e is the union of various types of edges. Due to E is the edge of heterogeneous network, we define $|T_v > 1|$ or $|T_e > 1|$.

Meta path. Set $N_i \in T_v$ and $Z_i \in T_e$ constitute a meta-path schema $N_1 \xrightarrow{Z_1} N_2 \xrightarrow{Z_2} \dots \xrightarrow{Z_m} N_{m+1}$, which shows a composite relation Z between objects N_i and N_j. That is, $Z = Z_1 \circ Z_2 \circ ... \circ Z_l$, where $\circ$ denotes the composition operator on relations in a given meta-path.

Context path. We presume two nodes $v_i, v_j \in V$, a union of context paths for all random walks $w \in W^{\mathcal{P}}$ from v_i to v_j, where $W^{\mathcal{P}}$ is a union of collected random walks. $\mathcal{P}$ represents a specific meta-path $\mathcal{P}$ represented by $C^{\mathcal{P}}_{i \rightarrow j}$.

The diagnosis task can be defined as follows:

Input: Given node sets which consist of patient-diagnosis pairs defined as (v_i, v_j), and context path sets which node sets related to. (Particularly, the single context path is defined as $C^{\mathcal{P}}_{i \rightarrow j}$, and context path sets are distilled from different random walks directed by a meta-path $\mathcal{P}$ in meta-path scheme $S(\mathcal{P})$).

Output: Calculate the possibility of the defined patient-diagnosis pairwise relationship between a specific pair of nodes (v_i, v_j) in V.

3 Construction of Clinical Events for HIN

3.1 Structure of Clinical Events and Data Preprocessing

In this segment, we present a modeling method for building HIN from the EHR dataset and the technology of extracting some clinical events from the original text. For diagnosis nodes, we map diagnosis events established on the ICD-9 coding system [15]. We utilize Autophrase [12] to extract the common symptoms in the original clinical records and provide clinical phrases for Autophrase, which come from Medical Subject Headings (MeSH) and the (ICD-9) database.

3.2 Feature Representation of Patient Nodes from Clinical Events

In order to use heterogeneous network technique to acquire the latent embedding of the patients from the context of words in a corpus, we apply metapath2vec [4] to accomplish the meta-path guided random walk in conjunction with the heterogeneous skip-gram model [1]. We define a HIN $G = (V, E, T_v, T_e, \psi, \phi)$ and a graph embedding function $\boldsymbol{\xi} : V \to R^d$ that projects each node to a d-dimensional vector. Finally, the purpose of $\boldsymbol{\xi}(v)$ is to maximize the possibility of visiting the neighborhood $N(v)$ of a node v:

$$\underset{\boldsymbol{\xi}}{\operatorname{argmax}} \prod_{v \in V} \prod_{s \in N(v)} \Pr(s \mid \boldsymbol{\xi}(v)) \tag{1}$$

we denote the possibility $\Pr(s \mid \boldsymbol{\xi}(v))$ as a softmax function, and normalize the representation of all network nodes. More precisely, the probability of visiting a neighbor s of a node v under path z with route $V_1 \to ... \to V_l$, is represented as:

$$\Pr(s \mid \boldsymbol{\xi}(v), z) = \frac{\exp(\boldsymbol{\xi}(s) \cdot \boldsymbol{\xi}(v))}{\sum_{y \in V_l} \exp(\boldsymbol{\xi}(y) \cdot \boldsymbol{\xi}(v))} \tag{2}$$

$$\Pr(s \mid v, z) = \log \sigma(\boldsymbol{\xi}(s) \cdot \boldsymbol{\xi}(v)) + \sum_{1}^{k} \mathbb{E}_{y_l \sim P_l(y_l)} \log \sigma\left(-\boldsymbol{\xi}\left(y_l\right) \cdot \boldsymbol{\xi}(v)\right) \tag{3}$$

We apply negative sampling [8] to attain the following objective function in Eq. (3). Among them, one training step stochastically samples k negative nodes, path z, two nodes v, and s connected under z. Then, we use Stochastic Gradient Descent to update the embeddings and construct the clinical events of patients:

$$\boldsymbol{f}(p) = \sum_{t} \boldsymbol{w}_t \boldsymbol{f}_t(p), \quad \boldsymbol{f}_t(p) = \sum_{n \in N_t(p)} \frac{\boldsymbol{f}(n)}{|N_t(p)|} \tag{4}$$

where $\boldsymbol{w}_t$ denotes different types of weights. We calculate the similarity $s(p, d) = \boldsymbol{f}(d) \cdot \boldsymbol{f}(p)$ between patient event $\boldsymbol{f}(p)$ and diagnosis event $\boldsymbol{f}(d)$ to learn $\boldsymbol{w}_t$. Finally, we utilize hinge loss $\max(0, -s(d, p) + s(\sim d, p) + \sigma)$ to update $\boldsymbol{f}$ and $\boldsymbol{w}_t$, where $\sim d$ denotes negative sample, σ denotes hinge margin.

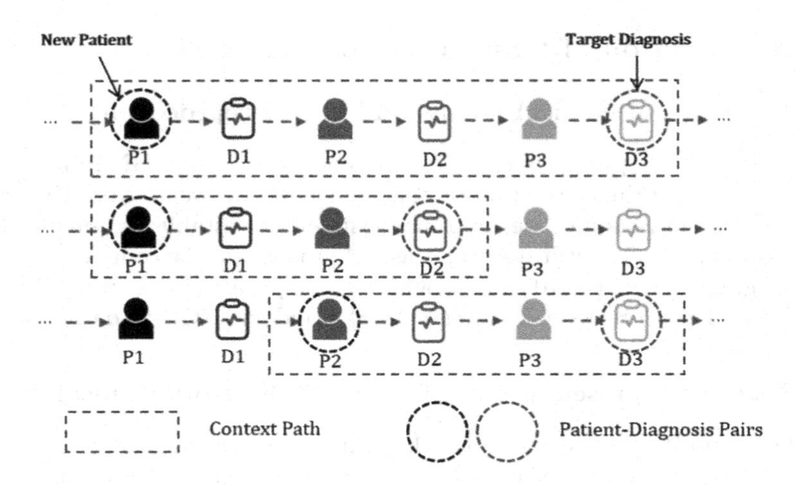

Fig. 2. Possible patient-diagnosis pairs and the related context path

4 The Proposed HPEMed Model

4.1 Method

In this segment, we propose a patient-diagnosis pairwise embedding method based on HIN, whose model is shown in Fig. 3. First, we utilize the measure mentioned in Sect. 3.2 to represent the patient's features. Second, we implement a random walk directed by several different meta-paths in HIN composed of EHR data. Third, we extract the patient-diagnosis pairs and the related context path of each pair from the specific meta-paths, after capturing patient-diagnosis pairs and related context paths. Finally, we need to design a classifier to verify whether pair embeddings and context path embeddings are effective for disease diagnosis tasks.

Task: Disease Diagnosis. There are some patients with their diagnostic records, each of which is related with more detailed information in MIMIC [6] records (the processed format is shown in Fig. 2). Our task is to rank all potential diagnosis that existed in the diagnostic system, so that, the highest-ranked diagnosis is the correct diagnosis. In order to better present the form of context path, Fig. 2 presents how patient-diagnosis pairs and their related context paths from the part of a random walk are directed by the specific path.

4.2 Pair Nodes of Patient-Diagnosis Embedding

We define a pair embedding of patient embedding $\boldsymbol{p}_a \in R^k$ and diagnosis embedding $\boldsymbol{d}_b \in R^k$ as $\boldsymbol{Group}(\boldsymbol{p}_a \in R^k, \boldsymbol{d}_b \in R^k) \in R^{4k}$. The pair embedder $\boldsymbol{pairEM}$: $R^{4k} \rightarrow R^d$ denotes an n-layer multi-layer perceptron that generates a d-dimensional embedding for a patient-diagnosis $pair(a, b)$. We have

$$h^{(l)} = \begin{cases} \text{ReLU}\left(W^{(l)} \text{ Dropout } \left(h^{(l-1)}\right) + b^{(l)}\right), & 0 < l < n \\ W^{(l)}\text{Dropout}\left(h^{(l-1)}\right) + b^{(l)}, & l = n \end{cases} \tag{5}$$

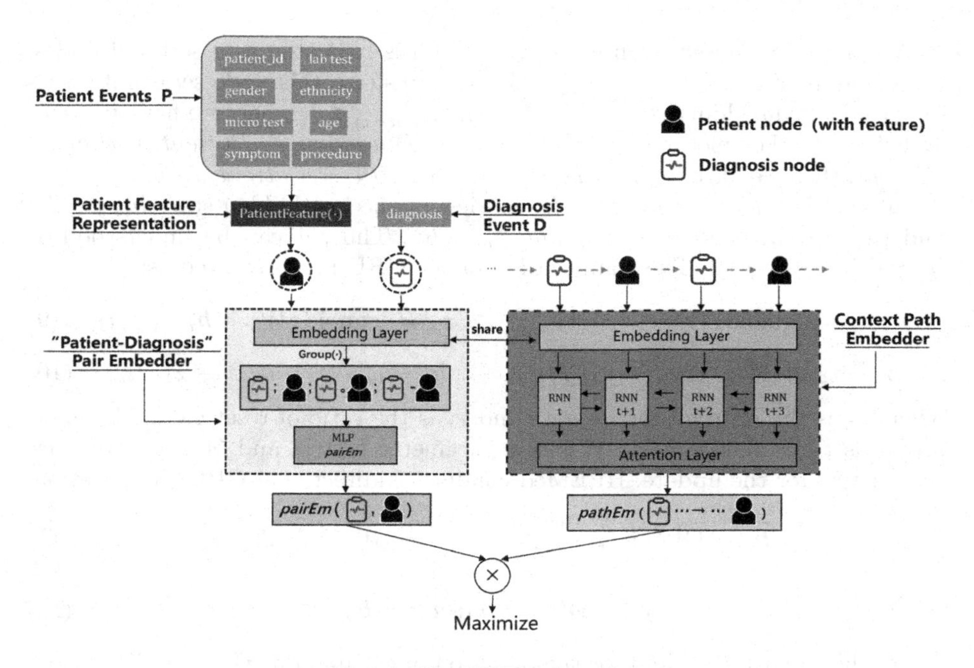

Fig. 3. The HPEMed model

$$h^{(0)} = \boldsymbol{Group}\,(\boldsymbol{p}_a, \boldsymbol{d}_b) = [\boldsymbol{p}_a; \boldsymbol{d}_b; \boldsymbol{p}_a \circ \boldsymbol{d}_b; \boldsymbol{p}_a - \boldsymbol{d}_b] \in \mathbb{R}^{4K} \tag{6}$$

$$\boldsymbol{pairEM}(a, b) = h^{(n)} \in R^d \tag{7}$$

In Eqs. (5), (6), and (7), ReLU is denoted as $\text{ReLU}(x) = max(0, x)$, $\boldsymbol{Group}(\cdot)$ is an MLP with two layers where each layer has 100 hidden units which $\circ$ means element-wise vector multiplication. In addition, we use the Dropout [13] in the hidden layers. Afterward, we will get the pair embedding from the last layer output of MLP, where $Input_a$ is the index of patient a, and PatientFeaEncoder$(\cdot)$ is a heterogeneous network-based feature generator presented in Sect. 3.2 that encodes a patient node with his/her records into an embedding. We adopt PatientFeaEncoder$(\cdot)$ to generate the feature of patients, and it exploits meta paths to extract important semantic relationships in HIN, to improve the preciseness of disease diagnosis results, i.e.,

$$\boldsymbol{p}_a = \text{PatientFeaEncoder}(Input_a) \in R^k \tag{8}$$

4.3 Context Path of Pair Nodes Embedding

We define a set of context paths for patient-diagnosis $pair(a, b)$ as $C^{\mathcal{P}}_{a \to b}$. We adopt a bi-directional Gated Recurrent Unit (GRU) [16] model to embed a context path consisting of a series of nodes.

We define a context path $c \in C_{a\to b}^{\mathcal{P}}$, which is indicated as a series of nodes. For example, $c = \{a, c_1, c_2, ..., c_{n-1}, b\}$. We transform this node series into a K-dimensional embedding vector $\{\boldsymbol{p}_a, \boldsymbol{f}_2, \boldsymbol{f}_3, ..., \boldsymbol{f}_{n-1}, \boldsymbol{d}_b\}$, where we have $\boldsymbol{p}_a = \boldsymbol{f}_1$ and $\boldsymbol{d}_b = \boldsymbol{f}_n$. For metapath $P = Patient \to Diagnosis \to Patient$, if $\psi(c_1) = Patient$, then we have $\psi(c_2) = Diagnosis$ and $\psi(c_3) = Patient$. At unit t, GRU calculates the hidden state $\boldsymbol{h}_t \in R^d$ in view of the last hidden state $\boldsymbol{h}_{t-1} \in R^d$ and the current context path input $\boldsymbol{f}_t \in R^k$. Thus, it can be understood as $\boldsymbol{h}_t = GRU(\boldsymbol{h}_{t-1}, \boldsymbol{f}_t)$. The computation of the GRU is represented as:

$$\boldsymbol{z}_t = \text{sigmoid}\left(\boldsymbol{W}_z \cdot (\boldsymbol{h}_{t-1} \| \boldsymbol{f}_t)\right), \qquad \boldsymbol{r}_t = \text{sigmoid}\left(\boldsymbol{W}_r \cdot (\boldsymbol{h}_{t-1} \| \boldsymbol{f}_t)\right) \tag{9}$$

$$\tilde{\boldsymbol{h}}_t = \tanh\left(\boldsymbol{W}_h \cdot (\boldsymbol{r}_t \cdot (\boldsymbol{h}_{t-1} \| \boldsymbol{f}_t))\right), \qquad \boldsymbol{h}_t = (1 - \boldsymbol{z}_t) \cdot \boldsymbol{h}_{t-1} + \boldsymbol{z}_t \cdot \tilde{\boldsymbol{h}}_t \tag{10}$$

where $\boldsymbol{z}_t$ is the ratio of update gate and $\boldsymbol{r}_t$ is the ratio of reset gate. The operator $\|$ is the concatenation. $\boldsymbol{W}_h$ is a parameter matrix and $\boldsymbol{W}_z$ and $\boldsymbol{W}_r$ are parameters for the update. HPEMed conducts bidirectional GRU computation:

$$\overrightarrow{\boldsymbol{h}}_t = \overrightarrow{\text{GRU}}\left(\boldsymbol{f}_t, \overrightarrow{\boldsymbol{h}}_{t-1}\right), \quad \overleftarrow{\boldsymbol{h}}_t = \overleftarrow{\text{GRU}}\left(\boldsymbol{f}_t, \overrightarrow{\boldsymbol{h}}_{t-1}\right) \tag{11}$$

$$\boldsymbol{h}_t = \boldsymbol{W}_{\text{proj}}\left[\overrightarrow{\boldsymbol{h}}_t; \overleftarrow{\boldsymbol{h}}_t\right] + \boldsymbol{b}_{\text{proj}} \tag{12}$$

where the top arrows stand for the direction of calculation. $\boldsymbol{W}_{proj} \in R^{d\times 2d}$ and $\boldsymbol{b}_{proj} \in R^d$ stand for the parameters on the projection layer. The operator $[\cdot;\cdot]$ stands for vector concatenation.

The purpose of the above is to embed nodes along with the context path $c \in C_{a\to b}^{\mathcal{P}}$ into the matrix representation $\boldsymbol{h}$. To better measure the importance of different nodes in the $c \in C_{a\to b}^{\mathcal{P}}$ and represent the whole context path, we use attentive pooling [17] to define the context path embedder as $\boldsymbol{pathEM}$:

$$a_t = \text{softmax}\left(\boldsymbol{g}\boldsymbol{h}_t\right) = \frac{\exp\left(\boldsymbol{g}\boldsymbol{h}_t\right)}{\sum_{i=1}^{n} \exp\left(\boldsymbol{g}\boldsymbol{h}_i\right)} \tag{13}$$

$$\boldsymbol{pathEM}(c) = \sum_t a_t \boldsymbol{W}_{att} \boldsymbol{h}_t \tag{14}$$

The adoption of the attention mechanism allows the focus to be concentrated on the context path whose contribution is much higher. Refer to Eq. (14), where $\boldsymbol{W}_{att} \in R^{d\times d}$ and $\boldsymbol{g} \in R^d$. We hope that $\boldsymbol{pathEM}(c) \in R^d$ encodes the disease field related to the context path c.

4.4 Enrich the Context Details of Pair Nodes

Now we have designed a patient-diagnosis $pair(a, b)$. We hope to maximize the probability of the $pair(a, b)$ within the context path c. We transform the objective into calculating the minimum of the negative log probability:

$$\mathcal{L}_{\text{ctx}}(a, b) = \sum_{c \in C_{a\to b}^{\mathcal{P}}} -\log p((a, b) \mid c, \mathcal{P}). \tag{15}$$

Table 1. The MIMIC III dataset utilized in our experiment

Table name	Fields	Explanations
Patients	hadm_id, dob, dod, gender, ethnicity	Fundamental information about patients
Admissions	admit_time, disch_time	Admission information about patients
Laboratory	itemid, value, flag	Laboratory test results
Note events	Text, Category	Initial explanation of patients' symptoms
Microbiology	org_itemid,interpretation	Microbiological experiments and records
Diagnosis icd	ICD-9 code	Prescribed diagnoses codes
Procedure icd	ICD-9 code	Procedures records represented by ICD-9-CM

More precisely, we acquire the context paths $C_*^{\mathcal{P}}$ from $W^{\mathcal{P}}$ as much as possible and the probability of $p((a,b) \mid c, P)$ is denoted as:

$$p((a,b) \mid c, \mathcal{P}) = \frac{\exp[(\boldsymbol{pairEM}(a,b) \cdot \boldsymbol{pathEM}(c))]}{\sum_{c' \in C_*^{\mathcal{P}}} \exp\left[(\boldsymbol{pairEM}(a,b) \cdot \boldsymbol{pathEM}\left(c'\right))\right]}. \tag{16}$$

In the same way, as directly calculating $p((a,b) \mid c, \mathcal{P})$ in large-scale HIN is high-cost, we denote k random sampled contents and employ negative sampling to achieve further optimization as follows:

$$\begin{aligned} \log p((a,b) \mid c, \mathcal{P}) \approx \log \sigma(\boldsymbol{pairEM}(a,b) \cdot \boldsymbol{pathEM}(c)) + \\ \sum_{j=1}^{k} \log \sigma \left(-\boldsymbol{pairEM}(a,b) \cdot \boldsymbol{pathEM}\left(c_{\text{rand}}^{j}\right)\right). \end{aligned} \tag{17}$$

The purpose of HPEMed is to let $\boldsymbol{pairEM}(a,b)$ close to $\boldsymbol{pairEM}(c)$. However, when c is currently the context path between a and b, $\boldsymbol{pairEM}(a,b)$ should retain itself disclose to the embedding of $\boldsymbol{pairEM}(c_{rand})$, which stands for random context paths. We can maximize Eq. (16) to achieve this purpose. At the meantime, it promotes $\boldsymbol{pairEM}(a,b)$ to encode its associated disease field.

4.5 Diagnosis Prediction

We calculate the prediction result of HPEMed by outputting the pair-effective classifier $\gamma(\cdot)$, which is defined as $\gamma : R^d \rightarrow R$ and represented by a binary cross-entropy loss as below:

$$\mathcal{L}_{\text{pv}}(a,b) = j_{a,b}\sigma(\gamma(\mathbf{pairEM}(a,b))) + (1 - j_{a,b})\left(1 - \sigma(\gamma(\mathbf{pairEM}(a,b)))\right). \tag{18}$$

To be exact, we determine the true diagnosis of patient a and rank $\sigma\left(\gamma(\boldsymbol{pairEM}(a,b))\right)$ for $b \in B$, where B indicates the union of diagnosis results, to observe the number of positive results in the top-ranked diagnosis results.

Table 2. Node statistics of HPEMed extracted from the MIMIC dataset

Node type	Train	Test
Patient	36610	10000
Microbiology	209	61
Laboratory	1867	1046
Symptom	1598	436
Age Group	3	3
Gender	2	2
Ethnicity	5	5
Diagnosis	5603	2739

Table 3. Comparison of predicting disease codes

Method Name	MAP@3	MAP@4	MAP@5	MAP@10
Med2vec	0.75	0.75	0.78	0.79
Skipgram	0.73	0.74	0.76	0.77
HeteroMed-embedded	0.78	0.78	0.79	0.80
HPEMed	**0.82**	**0.84**	**0.87**	**0.86**

5 Experiment

5.1 Dataset and Evaluation Details

Our experiments operate on the MIMIC III dataset [6]. MIMIC III is divided into 26 tables containing medical records for admitted patients and other general information. By processing the dataset, we total apply 46,610 admitted patients sample collection, meanwhile, we separate sample collection into a 10,000-instance test set and a 36,610-instance training set. Table 1 lists the data tables used in the MIMIC dataset, the node types, and a brief description of each table. Table 2 lists the number of different types of nodes in the network.

We define the evaluation process of disease diagnosis as a ranking problem where the object is the diagnosis code prediction, which is evaluated with MAP@k. The prediction model exports a sorting diagnosis list where AP@k indicates the average precision of all the contents in the list. The diagnosis list requires that the diagnosis result is right and its index is less than k. The unsupervised method extracts 200 negative diagnosis samples for every patient according to the diagnosis node degree. In order to train the model and learn embedding representations better, each training step uses a batch of 1000 patients and sets the embedding vector size to 128.

5.2 Comparison and Predict Performance

To better evaluate the HPEMed model, we compare it with three state-of-the-art models and put forward a question when we need to predict accurate disease codes? The models in comparison are as follows:

Table 4. Similarity search results for cold.

HPEMed	HeteroMed	Skipgram
Nasal edema	**General pain**	**Fever**
Runny nose	**Fever**	**Sick contact**
RTI	**Chill**	Constipation
Cough	**Sore throat**	**Muscle pain**
Respiration disturbance	Swelling	**Recent travel**
Pharyngitis	**Allergy**	Limb pain
Fever	**Tightness**	Urinary changes
Nasal obstruction	**Sinus congestion**	**Cough**
Nasosinusitis	**Cough**	Stiff neck
Sore throat	Blurred vision	**Runny nose**

– **Med2Vec** [3]: A multi-layer embedding neural network that uses a method inspired by word2vec to learn the embedding of medical events and visits.
– **Skipgram** [1]: In this model, all clinical events related to admission are treated as words and connected into sentences.
– **HeteroMed** [5]: This method uses the unsupervised representation learning method introduced to learn node embeddings as the former method, and then applies a supervised diagnosis method to rank the diagnosis code.

Predict Disease Codes. Table 3 shows the evaluation results in code prediction, which shows that HPEMed is superior to all baseline models in terms of accuracy. Meanwhile, HPEMed is better than the skip-gram embedding model and other relation-aware embedding methods like HeteroMed.

Case Analysis. HPEMed needs to verify the sensitivity of the medical representation after modeling EHR data. Table 4 lists the results of three models according to **cold** related symptoms and observations which are examined by medical specialists. We implement a similarity search for the related symptoms(marked in bold) of the common disease. The rank of related symptoms by HPEMed is higher than that of HeteroMed and Skipgram model. HPEMed is also the most powerful in understanding the relations of symptoms to diseases. Even for words not marked in bold, HPEMed is closer to **cold** than other models.

6 Conclusion

In this paper, we put forward a new framework named HPEMed for disease diagnosis tasks by heterogeneous information network embedding. HPEMed extracts detailed information from the EHR data as the features of the patient nodes to enrich the content of the node and simultaneously retains the construction and relation of HIN when capturing the representation of medical records. The objective of HPEMed is to calculate the similarity of pairing between patient nodes and diagnosis nodes and learn to embed node pairs between context paths considering their association. The output of our experiment indicates that HPEMed is better than some advanced methods in the prediction of diagnosis codes.

References

1. Cao, Y., Peng, H., Yu, P.S.: Multi-information source Hin for medical concept embedding. Adv. Knowl. Discov. Data Mining **12085**, 396–408 (2020)
2. Chen, T., Sun, Y.: Task-guided and path-augmented heterogeneous network embedding for author identification. In: The 10th ACM International Conference on Web Search and Data Mining, pp. 295–304 (2017)
3. Choi, E., et al.: Multi-layer representation learning for medical concepts. In: The 22nd ACM SIGKDD International Conference on Knowledge Discovery and Data Mining, pp. 1495–1504 (2016)
4. Dong, Y., Chawla, N., Swami, A.: Metapath2vec: scalable representation learning for heterogeneous networks. In: The 23rd ACM SIGKDD International Conference on Knowledge Discovery and Data Mining, pp. 135–144 (2017)
5. Hosseini, A., Chen, T., Wu, W., Sun, Y., Sarrafzadeh, M.: Heteromed: heterogeneous information network for medical diagnosis. In: The 27th ACM International Conference on Information and Knowledge Management, pp. 763–772 (2018)
6. Johnson, A., et al.: Mimic-III, a freely accessible critical care database. Sci. Data **3**(1), 1–9 (2016)
7. Lei, Y., Zhang, J.: Capsule graph neural networks with em routing. In: The 30th ACM International Conference on Information and Knowledge Management (2021)
8. Mikolov, T., Sutskever, I., Chen, K., Corrado, G.S., Dean, J.: Distributed representations of words and phrases and their compositionality. In: Advances in Neural Information Processing Systems, pp. 3111–3119 (2013)
9. Park, C., Kim, D., Zhu, Q., Han, J., Yu, H.: Task-guided pair embedding in heterogeneous network. In: The 28th ACM International Conference on Information and Knowledge Management, pp. 489–498 (2019)
10. Perozzi, B., Al-Rfou, R., Skiena, S.: Deepwalk: Online learning of social representations. In: The 20th ACM SIGKDD International Conference on Knowledge Discovery and Data Mining, pp. 701–710 (2014)
11. Rajkomar, A., et al.: Scalable and accurate deep learning for electronic health records. NPJ Digit. Med. **1**, 18 (2018)
12. Shang, J., Liu, J., Jiang, M., Ren, X., Voss, C.R., Han, J.: Automated phrase mining from massive text corpora. IEEE Trans. Knowl. Data Eng. **30**(10), 1825–1837 (2018)
13. Srivastava, N., Hinton, G., Krizhevsky, A., Sutskever, I., Salakhutdinov, R.: Dropout: a simple way to prevent neural networks from overfitting. J. Mach. Learn. Res. **15**(1), 1929–1958 (2014)

14. Tang, J., Qu, M., Wang, M., Zhang, M., Yan, J., Mei, Q.: Line: Large-scale information network embedding. In: The 24th International Conference on World Wide Web, pp. 1067–1077 (2015)
15. Trott, P.: International classification of diseases for oncology. J. Clin. Pathol. **30**, 782–782 (1977)
16. Zhang, J., Li, M., Gao, K., Meng, S., Zhou, C.: Word and graph attention networks for semi-supervised classification. Knowl. Inf. Syst. **63**(11), 2841–2859 (2021). https://doi.org/10.1007/s10115-021-01610-3
17. Zhou, P., et al.: Attention-based bidirectional long short-term memory networks for relation classification. In: The 54th Annual Meeting of the ACL, vol. 2, pp. 207–212 (2016)

MR-LGC: A Mobile Application Recommendation Based on Light Graph Convolution Networks

Weishi Zhong, Buqing Cao(✉), Mi Peng, Jianxun Liu, and Zhenlian Peng

School of Computer Science and Engineering, Hunan Provincial Key Laboratory for Services Computing and Novel Software Technology, Hunan University of Science and Technology, Xiangtan, China
buqingcao@gmail.com, zlpeng@hnust.edu.cn

Abstract. With the rapid growth of the number and types of mobile applications, how to recommend mobile applications to users accurately has become a new challenge. Graph convolution neural networks is a typical technique to facilitate mobile application recommendation. However, non-linear activation, feature transformation, and other operations in the existing mobile application recommendation based on graph convolution neural networks, which are used to model and characterize high-order interaction relationships between users and mobile applications, increase the difficulty of model training and lead to over-smoothing effects, reducing recommendation performance. To solve this problem, this paper proposes a mobile application recommendation method based on light graph convolution networks. In this method, firstly, a bipartite graph is used to model the interaction between users and mobile applications. Then, light graph convolution networks is utilized to smooth the features on the graph and extract the high-order connection between users and mobile applications, and three convolution layers are exploited to generate the feature representations of users and mobile applications. Finally, the inner product is used to predict the user's preference for different mobile applications and complete the recommendation task. "Shopify app store", a real dataset of Kaggle, is used to perform many groups of comparative experiments, and the experimental results show that the proposed method is superior to other methods.

Keywords: Mobile applications · Recommendation system · Light graph convolution networks · High-order connectivity

1 Introduction

According to statistics, by the end of 2020, the number of mobile apps in China closes to 3.45 million, ranking the first one globally. E-government, games, short videos, and other rich applications will affect people's necessities of life in an all-around way. Recently, the number of mobile applications is growing exponentially. Aiming to these massive mobile applications, it is difficult for users to choose suitable mobile applications for

Y. Sun et al. (Eds.): ChineseCSCW 2021, CCIS 1492, pp. 376–390, 2022.
https://doi.org/10.1007/978-981-19-4549-6_29

their personalized requirements. Therefore, it is necessary to provide a high-quality recommendation to achieve a good experience for user.

Traditional mobile application recommendation methods, such as collaborative filtering [1], matrix factorization [2], usually transform the mobile application recommendation problem into a supervised learning problem. In essence, the user and the app are initially embedded, respectively, and then the interaction information between them is used to optimize the model and perform the recommendation. Although these methods have good performance, the disadvantage is that they regard users and apps as independent instances, ignoring the possible interaction and the original implied information between them. The reason is that the interaction information is not embedded into the initial input feature in the process of initial embedding. In recent years, mobile application recommendation methods based on deep learning have become popular [3], such as graph convolution networks (GCN) [4] and neural graph collaborative filtering (NGCF) [5]. They explicitly encode the structural information between users and mobile applications, such as user-item bipartite graph, user-item high-order connectivity graph, etc., into the initial embedding of collaborative filtering task and complete recommendation. However, their design is quite complicated and redundant, that is to say, many operations are inherited directly from GCN without modification or deletion. The two most common designs in GCN feature transformation and nonlinear activation make little contribution to collaborative filtering, and these operations in the model will increase the difficulty of training and reduce the recommendation performance. To solve this problem, He et al. [6] proposed a new graph convolution neural network model, i.e., light graph convolution networks (LightGCN), which only retains the essential component (neighborhood aggregation) in GCN for collaborative filtering. This simple, linear and concise model is more comfortable in the process of implement and training. Inspired by this work, this paper considers introducing LightGCN into mobile application recommendations and proposes a mobile application recommendation method based on light graph convolution networks. As we all know, vector embedding of users and items is the core of current recommendation systems, but nowadays, many algorithms only use the characteristics of users or items to embed, and other related information is rarely used. In particular, LightGCN can represent users and mobile applications as a bipartite graph for initial vector, and propagate them on the graph through algorithms to utilize interactive information between users and apps.

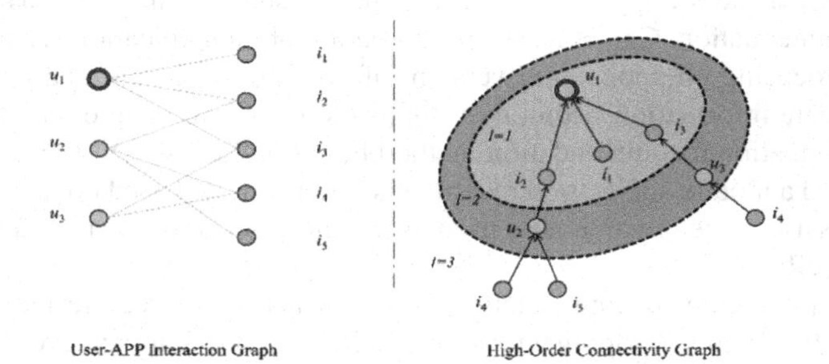

Fig. 1. Interaction graph between user and mobile application, and high-order connectivity diagram.

As shown in Fig. 1 above, LightGCN displays and encodes the interaction information between users and mobile applications into the user-item bipartite graph structure (on the left side of Fig. 1), updates and learns the high-order interaction features between users and mobile applications in the propagation process, so as to achieve the effect of the high-order connectivity graph on the right side of Fig. 1. LightGCN only retains the most critical component, i.e., neighborhood aggregation for collaborative filtering, learns the embeddings of users and mobile applications by linearly propagating them on the user-item interaction graph, and uses the weighted sum of the embedding learned from all layers as the final embedding. Suppose a mobile application preferred by user u_1 needs to predicted, the right part of Fig. 1 is a tree structure expanded with u_I as the root node, and the path length (or hop number) $l = 1$ indicates the app's path that can be reached in one step, including $i_1 \rightarrow u_1, i_2 \rightarrow u_1, i_3 \rightarrow u_1$. At this time, when $l =$ 3, it can be seen that the user's interest in i_4 is higher than that of i_5 when the number of hops in the outermost layer of i_4 is the same, because two paths are starting from i_4, *i.e.*, $i_4 \rightarrow u_2 \rightarrow i_2 \rightarrow u_1, i_4 \rightarrow u_3 \rightarrow i_3 \rightarrow u_1$. Therefore, through the tree structure, we can observe the user's preference for different mobile applications, and we can also see which users may use a mobile application. This reflects the connectivity between users and mobile applications, i.e., high-order connectivity. This kind of high-order connectivity contains rich interaction semantics between users and mobile applications, which can embed better and characterize mobile applications and users to improve the recommendation results.

2 Related Work

Mobile applications recommendation method mainly includes content-based, network structure-based and user interaction relationships-based mobile applications recommendation.

The content-based mobile applications recommendation method recommends mobile applications to users via exploiting the text description, tags and user information. For example, according to Twitter followers' characteristics of various mobile applications, Lin et al. [7] used the LDA topic model to generate potential groups to predict the possibility of target users' preferences for applications. Chen et al. [8] analyzed the content similarity of mobile applications according to the meaning relationship and category between mobile applications and then performed content-oriented mobile application recommendation. Cao et al. [9] proposed a mobile application recommendation framework focusing on application version information based on dual heterogeneous data and update information within user application version. Peng et al. [10] designed a mobile application recommendation method based on license and function. Xu et al. [11] presented a mobile application recommendation method based on neural networks, which makes use of the semantic way used by mobile applications to recommend mobile applications effectively.

Mobile application recommendation based on network structure mainly uses the similarity between mobile applications and builds a network to achieve mobile application recommendation. Among this, Woerndl et al. [12] introduced social network and context information into recommendation tasks at the same time and proposed a hybrid mobile service recommendation method. Xie et al. [13] acquired the relationship structure and semantic information of applications and users through different information networks and utilized weighted meta path and heterogeneous information networks to recommend mobile applications. Upasna et al. [14] obtained users' preferences by using mobile applications initially installed on mobile phones and built similarity charts of mobile applications to provide personalized recommendations. Guo et al. [15] proposed a mobile application recommendation model based on depth factor decomposition machine by using classification information and text information of applications. Liu et al. [16] designed a new structure selection model, which uses the hierarchical structure of the application tree to capture the competition between mobile applications and learn fine-grained user preferences to perform recommendations.

In addition, some studies infer user preferences for mobile applications through user-item interaction graphs and achieve mobile application recommendation based on user interaction relationships. The HopRec model proposed by Yang et al. [17] in the early stage, by combining graph-based and embedding-based methods, alleviates the problem of model parameters lacking optimization objective function to a certain extent. Firstly, the random walk algorithm is used to enrich the interaction between users and multi-hop connectors. Based on the abundant user-item interaction data, the BPR function objective training of matrix factorization is carried out, and the recommendation model is established. Later, NMF proposed by He et al. [18] is an improved factorization model, which seamlessly combines FM and neural networks to model the potential interaction between users and mobile applications. Recently, Chen et al. [19] presented an interaction-based mobile application recommendation method (MR-UI) based on the neural graph collaborative filtering model. MR-UI integrates collaborative signals into model-based CF embedding function and uses the high-order correlation in the integration graph of users and mobile applications to achieve the recommendation task of mobile applications. However, it is not easy to train the model with nonlinear activation function in the high-order interaction graph of users and mobile applications. Moreover, due to the excessive smoothing effect of graph convolution, most GCN-based models cannot model more in-depth features and interactions, resulting in the final recommendation performance is not so well. Therefore, this paper introduces light graph convolution networks to solve this problem to achieve more accurate mobile application recommendation.

3 Proposed Method

The proposed mobile application recommendation method based on light graph convolution network is named as MR-LGC, and its framework is shown in Fig. 2. In the MR-LGC model, firstly, the nodes of user and mobile application are initialized and embedded in the initial embedding layer. Secondly, three convolution layers (layer1/2/3) are used to generate each layer's embeddings in the lightweight graph convolution layer. The high-order connection relationship between the user and the mobile application is extracted by smoothing the graph's features. The embeddings of each layer are further combined to form the final embedding representation of the user or mobile application. Finally, in the mobile application recommendation layer, the inner product is calculated through the final generation representation of the user and mobile application, and the prediction scores are generated and sorted to complete the mobile application recommendation.

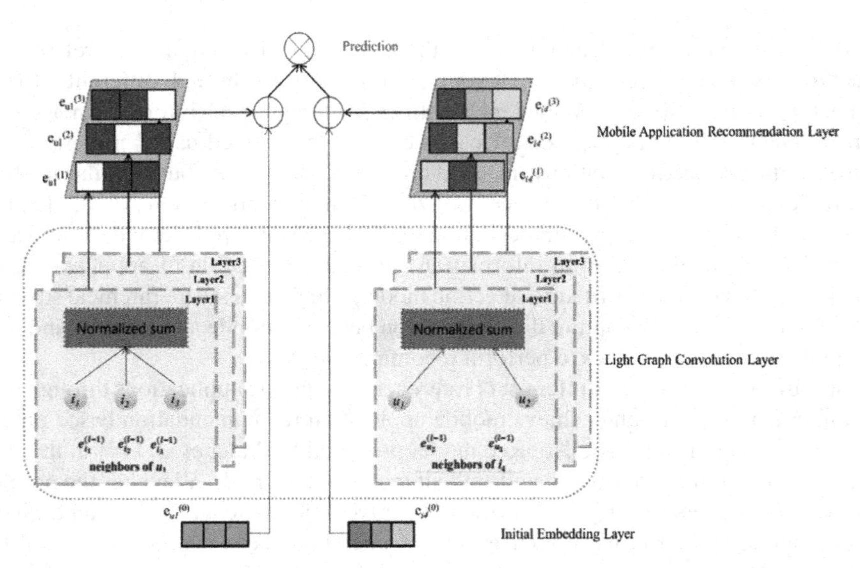

Fig. 2. The MR-LGC model.

3.1 Initial Embedding Layer

The embedding vector $e_u \in \mathbb{R}^d$ represents the embedding matrix of user u and $e_i \in \mathbb{R}^d$ represents the embedding matrix of the mobile application i, where d is the embedding dimension. Because the embedding dimension of the user or mobile application is the same, it can be considered that the parameter matrixes are integrated into an embedded large matrix, as shown in Eqs. (1) and (2):

$$E_u = [e_{u_1}, e_{u_2}, \cdots, e_{u_n}] \tag{1}$$

$$E_i = [e_{i_1}, e_{i_2}, \cdots, e_{i_n}] \tag{2}$$

3.2 Light Graph Convolution Layer

In the LGC layer, only the normalized neighborhood embedding summation is performed on the next layer, and the operations such as self-connection, feature transformation, and nonlinear activation are removed, which significantly simplifies the GCN. In the combination process of the LGC layer, each layer's embedding is summed to obtain the final vector representation. The basic idea of GCN is to learn the vector representation of new nodes by aggregating the characteristics of mobile application nodes and user nodes on the graph. To achieve this goal, the model iteratively performs graph convolution, that is, aggregating the features of neighbors into new vector representations of target nodes. This neighborhood aggregation can be abstracted as:

$$e_u^{(k+1)} = AGG(e_u^{(k)}, \{e_u^{(k)} : i \in N_u\}) \tag{3}$$

Among them, AGG represents the linear aggregation function, and LGC only propagates information from neighbors, eliminating the ordinary self-join operation. Similarly, the representation of a mobile application can be obtained by propagating information from its connected users. Intuitively speaking, the interaction between users and mobile applications can directly reflect users' preferences. At the same time, users who have used mobile applications can be regarded as the characteristics of corresponding mobile applications and utilized to measure the collaborative similarity between two mobile applications. In LGC, a simple weighted sum aggregator is used instead of feature transformation and nonlinear activation function. The propagation rules in LGC are as follows:

$$e_u^{(k+1)} = \sum_{i \in N_u} [\frac{1}{\sqrt{N_u}\sqrt{N_i}} e_u^{(k)}] \tag{4}$$

$$e_u^{(k+1)} = \sum_{i \in N_u} [\frac{1}{\sqrt{N_u}\sqrt{N_i}} e_u^{(k)}] \tag{5}$$

According to the formulas (4) and (5), taking the second-order propagation embedding as an example, it is known that:

$$\begin{aligned} e_u^{(2)} &= \sum_{i \in N_i} [\frac{1}{\sqrt{N_i}\sqrt{N_u}} e_i^{(1)}] \\ &= \sum_{i \in N_i} \frac{1}{N_i} \sum_{v \in N_i} [\frac{1}{\sqrt{|N_u|}\sqrt{|N_v|}} e_v^{(0)}] \end{aligned} \tag{6}$$

From the above Eq. (6), it can be found:

- the more the number of common interaction items between the mobile application node and the user node, the more significant the impact on feature updating.
- the lower the popularity of the interactions between the mobile application node and the user node (that is, the lower the popularity of the interactive mobile application can better reflect the user's personalized preference), the more significant the impact on feature updating.
- the lower the activity degree of v, the more significant the impact on feature updating.

In the LGC layer, the embedding of different layers captures different semantics. For example, the first layer smoothes users and mobile applications with interaction. The second layer smoothes users or mobile applications with overlapping interactive applications or user and combines embedding and the weighted sum of different layers to capture the effect of the graph convolution and self-connection. The simplified network can alleviate the problem of excessive smoothing in graph convolution aggregation operation by aggregating the embedding of different layers. The specific propagation process is shown in Fig. 3 below.

For example, as for user u_1, its three-hop neighbors i_2, i_4, and i_5 begin to perform the above weighting and then aggregate its two-hop neighbors u_1 and u_2 for weighting (at this time, u_1 completes the update) and update i_2. Similarly, other mobile applications i_1 and

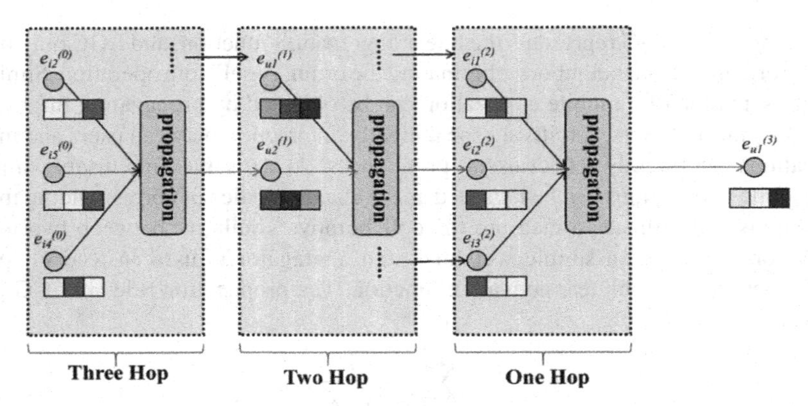

Fig. 3. An example of the LGC propagation

i_3 will be updated accordingly. Finally, u_1 is updated by i_1, i_2 and i_3. Thus, through the mutual updating and iteration, the high-order interactive feature learning and embedding propagation of the LGC layer are completed. In LGC, the only model parameter that can be trained is the embedding of layer 0, that is, $e_u^{(0)}$ (for all users) and $e_i^{(0)}$ (for all mobile applications).

When the embedding of the layer 0 is given, the higher layers' embedding can be calculated by LGC propagation rules defined in formulas (4) and (5). By stacking two embedded propagation layers, users (or mobile applications) can receive messages propagated from their hop neighbors to form the final vector representation of users and mobile application nodes:

$$e_u = \sum_{k=0}^{K} \alpha k e_u^{(k)} \tag{7}$$

$$e_i = \sum_{k=0}^{K} \alpha k e_i^{(k)} \tag{8}$$

In the above formula, αk is the weight coefficient that can be learned and adjusted, and optimized by neural network training. After many experiments, it is found that when αk is equal to $1/(K+1)$, the experimental performance is the best.

Next, matrix propagation rules are defined. To provide an overall view of embedded propagation and facilitate batch implementation, the user-mobile application interaction matrix is set to $R \in \Re^{M \times N}$, where M and N represent the number of users and mobile applications, respectively. If u interacts with i, R_{ui} is 1; otherwise, it is 0. Therefore, the adjacency matrix of user-mobile application interaction is as follows:

$$A = \begin{pmatrix} 0 & R \\ R^T & 0 \end{pmatrix} \tag{9}$$

where T is the embedded size. Given the embedding matrix of the layer 0, the equivalent form of LGC matrix can be obtained:

$$E^{(k+1)} = (D^{-\frac{1}{2}} A D^{-\frac{1}{2}}) E^{(k)} \tag{10}$$

where D is a diagonally positive definite matrix of $(M + N) \times (M + N)$, also known as the degree matrix of adjacency matrix A, and $D^{-\frac{1}{2}}AD^{-\frac{1}{2}}$ is a symmetric normalized matrix.

3.3 Recommendation Layer of Mobile Application

After LGC layers' embedding propagation, multiple vector representations of users u and mobile applications i can be obtained. According to the generated node vector representation, the inner product calculation can predict the user's preference score for mobile applications based on the below formula (11):

$$\hat{y}LGC_{(u,i)} = e_u^T e_i \tag{11}$$

4 Experimental Evaluation and Result Analysis

4.1 Dataset Description

The experimental data came from the real dataset "Shopify app store" made public by Kaggle (https://www.kaggle.com/usernam3/shopify-app-store). The dataset includes 2,831 mobile applications, 191,781 users and 292,029 pieces of information about the interaction between users and mobile applications. Its details are shown in Table 1. From the dataset, we observe that most users have few interaction records with mobile applications, and there are usually only one or two records. To facilitate the verification of the test set for the experimental process, the interaction records without user information are eliminated. The data with four or more interaction records between users and mobile applications is selected as the final data set from the cleaned data set, and 75% of them are selected as the training set and 25% as the testing set.

Table 1. DataSet information.

	User	Interaction	Sparsity
Original data set	191,781	292,029	0.00053
After processing	5,969	44,530	0.00264
Training set	5,969	30,320	0.00179
Testing set	5,969	14,210	0.00084

4.2 Evaluation Metrics

To evaluate the effectiveness of *Top-K* recommendation and user preference ranking, three widely used evaluation metrics are used in the experiment, that are *NDCG@K*, *Precision@K*, and *Recall@K*. The details of them are as follows:

- Precision: the proportion of correctly related mobile applications *Recom(A_i)* in all related mobile applications *Real(A_i)*. Its formula is as follows:

$$Precision = \frac{|Real(A_i) \cap Recom(A_i)|}{Real(A_i)} \tag{12}$$

- Recall: the proportion of correct and relevant mobile applications *Recom(A_i)* in the total recommended mobile applications *Real(A_i)*. Its formula is as follows:

$$Recall = \frac{|Real(A_i) \cap Recom(A_i)|}{Recom(A_i)} \tag{13}$$

- Normalized Discounted Cumulative Gain (*NDCG*): an indicator to measure the ranking quality of recommended mobile applications in the recommendation list. The formula is as follows:

$$DCG@N = \sum_{1}^{N} \frac{2^{(rel_i)} - 1}{\log_2(i+1)} \tag{14}$$

$$DCG@N = \sum_{1}^{N} \frac{2^{(rel_i)} - 1}{\log_2(i+1)} \tag{15}$$

Among them, rel_i is the user's rating on mobile application *i*, *DCG* makes the results that rank higher affect the final results, and *IDCG* is the biggest value of *DCG* under ideal circumstances.

4.3 Baseline Methods

To verify the experimental performance, the following methods are employed as comparative methods:

- Matrix Factorization (*MF*) [2]: firstly the matrix is built by the historical interaction information between user and mobile application, then user or mobile application's ID is embedded directly as a vector through decomposition strategies such as SVD, finally the interaction between users and mobile applications is modeled by using the inner product.
- Neural Matrix Factorization (*NeuMF*) [18]: the potential features between users and mobile applications are extracted, and then a more complicated operation is used in place of the simple vector inner product to complete feature interaction to make up for the defect that the target cannot be fully fit. This model can be used in combination with the linear relationship of MF and the nonlinear relationship of DNNS to build a model for the potential interaction between users and mobile applications.
- Graph Convolution Networks (*GCN*) [4]: integrates the node information and topology structure naturally, expresses as user-user social network and user-mobile application graph, and learns the potential characteristics of users and mobile applications. The local first-order approximation of spectral convolution is used to determine the vector representation of users and mobile applications.

- Graph Convolution Matrix Completion (*GC-MC*) [20]: from the perspective of link prediction, considering matrix complement recommended by mobile applications, a graph automatic encoder framework based on distinguishable messages transmitted on a bidirectional interaction graph is adopted.
- Mobile Application Recommendation Based on User Interaction (*MR-UI*) [19]: the high-order connectivity of user-item is introduced into specific mobile application recommendation, and an interaction function is used to make up for the shortcomings of the embedded representation of mobile applications and users. The *MR-UI* will be used as the main contrast method for experimentation and analysis.

4.4 Experimental Performance

In this section, we will compare the performance of all methods and analyze the experimental results. In this method, the number of embedded propagation layers is 3, and *K* is equal to 20. The experimental results of various methods are shown in Table 2, and it can be found that:

- The performance of *MF* is the worst in all the comparison methods. Matrix factorization is an early model, which directly projects the user's single ID to the embedded system. Its inner product cannot fully fit the complex interaction between mobile applications and users, limiting the performance of the model to a certain extent. *NeuMF* is always better than *MF*, which shows that the neural attention mechanism distinguishes the importance of mobile applications in the interaction history during the similarity calculation process of mobile applications. However, from the perspective of user-mobile application interaction graph, this improvement originated from the user's sub-graph structure. More specifically, its one-hop neighbor improves embedded learning, but it does not explicitly model the connections in the process of embedded learning.
- In most cases, *GC-MC* usually has better performance than *GCN*, which may be attributed to the fact that the self-coding framework with matrix completion can generate a more accurate embedded representation of users and mobile applications.
- On the whole, *MR-LGC* maintains preferable performance. In particular, when *K* is 20, *MR-LGC* increases by 166.55%, 50.77%, 116.75%, and 113.6% on *Recall@K* compared with *MF*, *NeuMF*, *GCN*, and *GC-MC*. Also, it rises by 74.32%, 12.12%, 46.48%, and 43.90% on *NDCG@K*. The most significant improvement is that *MR-LGC* increases by 127.46%, 44.51%, 127.05%, and 93.67% on *Precision@K*. By stacking multiple embedding propagation layers, *MR-LGC* can explicitly explore higher-order connectivity, while *GCN* and *GC-MC* only use the first-order neighborhood to obtain the representations of mobile applications and users. This proves the significance of capturing collaborative signals in embedding functions.

Compared to MR-UI, the experimental performance of *MR-LGC* is tested in detail when K = 20, 40 and 60 and the number of embedded propagation layers, L, is 1, 2 and 3, respectively. The specific experimental results are shown in Tables 3, 4, and 5. From these tables and figures, it can be found that:

- As mentioned in Section III.B, LGC can capture higher-order connectivity by stacking convolution layers. The experimental performance of different layers in LGC also verifies this point. When the number of convolution layer is set from one to three, the experimental performance of each metric is continuously improved. At the same time, it also proves that it is necessary to extract the semantics of overlapping items and high-order connectivity in mobile application recommendation tasks.
- Compared with *MR-UI*, *MR-LGC* performs better in all cases using different embedded propagation layers in terms of *Recall@K*, *Precision@K* and *NDCG@K*. *MR-LGC* performs better than *MR-UI* in any metric, which verifies that *MR-LGC* can not only extract the semantics of interaction and even overlapping items through multi-layer embedding propagation, but also verify that the simplified *NGCF* model can bring better performance in mobile application recommendation task.

Table 2. Performance comparison of various methods.

	Recall	Precision	NDCG
MF	0.03839	0.00498	0.03093
NeuMF	0.06787	0.00784	0.04810
GCN	0.04721	0.00499	0.03681
GC-MC	0.04790	0.00585	0.03747
MR-UI	0.07850	0.00989	0.05278
MR-LGC	**0.10233**	**0.01133**	**0.05392**

Table 3. Recall@K.

	$K = 20$	$K = 40$	$K = 60$
MR-UI(L1)	*0.08347*	*0.13324*	*0.17106*
MR-LGC(L1)	**0.09260**	**0.14157**	**0.18562**
MR-UI(L2)	0.08659	0.14192	0.18520
MR-LGC(L2)	**0.09675**	**0.15142**	**0.19273**
MR-UI(L3)	0.07850	0.13335	0.18066
MR-LGC(L3)	**0.10233**	**0.15496**	**0.19661**

Table 4. NDCG@K.

	$K = 20$	$K = 40$	$K = 60$
MR-UI(L1)	0.00929	0.00749	0.00652
MR-LGC(L1)	**0.01028**	**0.00796**	**0.00701**
MR-UI(L2)	0.00956	0.00790	0.00692
MR-LGC(L2)	**0.01071**	**0.00846**	**0.00723**
MR-UI(L3)	0.00989	0.00792	0.00698
MR-LGC(L3)	**0.01133**	**0.00868**	**0.00737**

Table 5. Precision@K.

	$K = 20$	$K = 40$	$K = 60$
MR-UI(L1)	0.00929	0.00749	0.00652
MR-LGC(L1)	**0.01028**	**0.00796**	**0.00701**
MR-UI(L2)	0.00956	0.00790	0.00692
MR-LGC(L2)	**0.01071**	**0.00846**	**0.00723**
MR-UI(L3)	0.00989	0.00792	0.00698
MR-LGC(L3)	**0.01133**	**0.00868**	**0.00737**

4.5 Model Training

In the mobile application recommendation scenario, the recommendation system needs to recommend single-digit mobile applications to users in tens of millions of mobile applications. To learn the parameters in the model, we need a loss function, which can sort all the mobile applications corresponding to each user according to their preferences, and can better reflect the user preferences. The interaction should be given a higher predictive value. Therefore, this paper selects the BPR loss function [21], which has been widely used in mobile application recommendation, and use Adam [22] optimizer as mini batch. The specific loss function is as follows:

$$L_{BPR} = -\sum_{u=1}^{M} \sum_{i \in Nu} \sum_{j \notin Nu} \ln \sigma\left(\hat{y}_{ui} - \hat{y}_{uj}\right) + \lambda \left\| E^{(0)} \right\|^2 \tag{16}$$

where λ controls the weight of L2 regularization to prevent over fitting. The trainable parameters of LGC are only embedded in layer 0, that is $E^{(0)}$. This makes the complexity of the model is same as that of matrix factorization, i.e. $O(|\Omega|K)$ for one pass of all observed entries.

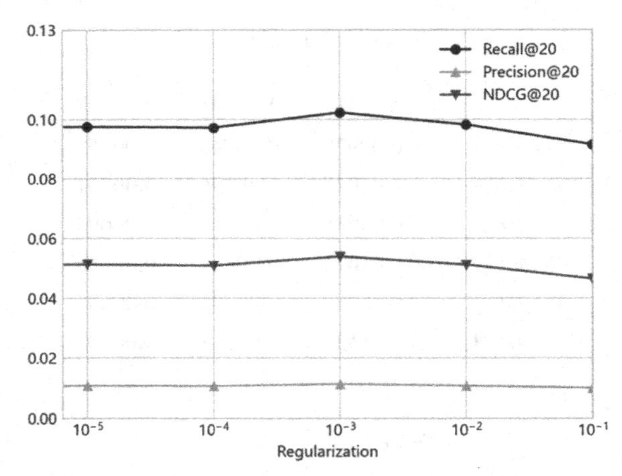

Fig. 4. Performance of 3-layer LGC w.r.t. Different Regularization Coefficient λ on Shopify App Store.

To analyze the influence of super parameters λ on model accuracy. Experiments will be carried out for different values of λ, and the results are shown in the Fig. 4. It can be known from the above figure that when λ is set to *1e−3*, the model has the highest accuracy, when λ is set from 0 to *1e*−3, the training effect is improved steadily. However, when λ is set from *1e*−*2* to *1e*−*1*, the training effect decreased significantly, which may be caused by the overfitting of the model.

5 Conclusion and Future Work

This paper proposes a mobile application recommendation method based on light graph convolutional networks (*MR-LGC*). In this method, firstly, a bipartite graph is utilized to model the interaction between mobile applications and users, and high-order connectivity (tree graph) is used for initial embedding. Then, in the lightweight graph convolutional layer, embedding propagation is used to capture collaborative filtering signals along the graph structure to further refine the embedding between mobile applications and users. Finally, the preference of users for different mobile applications is predicted by inner product, completing the recommendation task. The rationality and effectiveness of the *MR-LGC* are proved by the comparative experiments on the real data set of Kaggle. In the next work, we will consider introducing the association graph attention network to mine the deep interaction between users and mobile applications, and use the self-attention mechanism to calculate the weight and influence of different neighbor nodes, so as to achieve more accurate mobile application recommendation.

Acknowledgement. The work is supported by the National Natural Science Foundation of China (No. 61873316, 61872139, 61832014 and 61702181), the National Key R&D Program of China (No. 2018YFB1402800), the Natural Science Foundation of Hunan Province (No. 2021JJ30274), and the Key Projects of Hunan Provincial Department of Education(No. 20A175). Buqing Cao is the corresponding author of this paper.

References

1. Wang, M.-J., Han, J.-T.: Collaborative filtering recommendation based on item rating and characteristic information prediction. In: Proceedings of the 2nd International Conference on Consumer Electronics, Communications and Networks (CECNet), pp. 214–217. IEEE (2012)
2. Koren, Y., Bell, R., Volinsky, C.: Matrix factorization techniques for recommender systems. Computer **42**(8), 30–37 (2009)
3. Wang, H., Wang, N., Yeung, D.-Y.: Collaborative deep learning for recommender systems. In: Proceedings of the 21th ACM SIGKDD International Conference on Knowledge Discovery and Data Mining, pp. 1235–1244 (2015)
4. Kipf, T.N., Welling, M.: Semi-supervised classification with graph convolutional networks. arXiv preprint arXiv:1609.02907 (2016)
5. Wang, X., He, X., Wang, M., Feng, F., Chua, T.-S.: Neural graph collaborative filtering. In: Proceedings of the 42nd International ACM SIGIR Conference on Research and Development in Information Retrieval, pp. 165–174 (2019)
6. He, X., Deng, K., Wang, X., Li, Y., Zhang, Y., Wang, M.: Lightgcn: simplifying and powering graph convolution network for recommendation. In: Proceedings of the 43rd International ACM SIGIR Conference on Research and Development in Information Retrieval, pp. 639–648 (2020)
7. Lin, J., Sugiyama, K., Kan, M.-Y., Chua, T.-S.: Addressing cold- start in app recommendation: latent user models constructed from twitter followers. In: Proceedings of the 36th International ACM SIGIR Conference on Research and Development in Information Retrieval, pp. 283–292 (2013)
8. Chen, N., Hoi, S., Li, S., Xiao, X.: Simapp: a framework for detecting similar mobile applications by online kernel learning. In: The 8th ACM International Conference on Web Search and Data Mining (WSDM 2015) (2015)
9. Cao, D., et al.: Cross-platform app recommendation by jointly modeling ratings and texts. ACM Trans. Inf. Syst. **35**(4), 1–27 (2017)
10. Peng, M., Zeng, G., Sun, Z., Huang, J., Wang, H., Tian, G.: Personalized app recommendation based on app permissions. World Wide Web J. **21**(1), 89–104 (2018)
11. Xu, X., Dutta, K., Datta, A.: Functionality-based mobile app recommendation by identifying aspects from user reviews. J. Assoc. Inf. Sci. Technol. **69**, 2 (2014)
12. Woerndl, W., Schueller, C., Wojtech, R.: A hybrid recommender system for context-aware recommendations of mobile applications. In: 2007 IEEE 23rd International Conference on Data Engineering Workshop, pp. 871–878. IEEE (2007)
13. Xie, F., Chen, L., Ye, Y., Liu, Y., Zheng, Z., Lin, X.: A weighted meta-graph based approach for mobile application recommendation on heterogeneous information networks. In: Pahl, C., Vukovic, M., Yin, J., Yu, Q. (eds.) ICSOC 2018. LNCS, vol. 11236, pp. 404–420. Springer, Cham (2018). https://doi.org/10.1007/978-3-030-03596-9_29
14. Bhandari, U., Sugiyama, K., Datta, A., Jindal, R.: Serendipitous recommendation for mobile apps using item-item similarity graph. In: Banchs, R.E., Silvestri, F., Liu, T.-Y., Zhang, M., Gao, S., Lang, J. (eds.) AIRS 2013. LNCS, vol. 8281, pp. 440–451. Springer, Heidelberg (2013). https://doi.org/10.1007/978-3-642-45068-6_38
15. Guo, C., Xu, Y., Hou, X., Dong, N., Xu, J., Ye, Q.: Deep attentive factorization machine for app recommendation service. In: 2019 IEEE International Conference on Web Services (ICWS), pp. 134–138. IEEE (2019)
16. Liu, B., Yao, W., Gong, N.Z., Wu, J., Xiong, H., Ester, M.: Structural analysis of user choices for mobile app recommendation. ACM Trans. Knowl. Discov. Data **11**(2), 1–23 (2016). https://doi.org/10.1145/2983533

17. Yang, J.-H., Chen, C.-M., Wang, C.-J., Tsai, M.-F.: Hop-rec: high-order proximity for implicit recommendation. In: Proceedings of the 12th ACM Conference on Recommender Systems, pp. 140–144 (2018)
18. He, X., Liao, L., Zhang, H., Nie, L., Hu, X., Chua, T.-S.: Neural collaborative filtering. In: Proceedings of the 26th International Conference on World Wide Web, pp. 173–182 (2017)
19. Chen, J., Cao, B., Liu, J., Li, B.: MR-UI: a mobile application recommendation based on user interaction. In: 2020 IEEE International Conference on Web Services (ICWS), pp. 134–141. IEEE (2020)
20. Berg, R.v.d., Kipf, T.N., Welling, M.: Graph convolutional matrix completion. KDD2018
21. Rendle, S., Freudenthaler, C., Gantner, Z., Schmidt-Thieme, L.: BPR: Bayesian personalized ranking from implicit feedback. UAI (2009)
22. Kingma, D.P., Ba, J.: Adam: a method for stochastic optimization. Comput. Sci. (2014)

Neural Matrix Decomposition Model Based on Scholars' Influence

Ying Li[1], Chenzhe Yuan[2(✉)], Yibo Lu[1], Chao Chang[1], and Yong Tang[1]

[1] School of Computer Science and Technology, South China Normal University, Guangzhou, Guangdong, China

[2] School of Electronic and Information, Guangdong Polytechnic Normal University, Guangzhou, Guangdong, China

260412876@qq.com

Abstract. Recommending conferences for scholars in academic social networks (ASNs) not only can enhance communication between scholars, but also have a power to promote the integration and development of the Internet and other fields. This paper propose to design a neural matrix decomposition model based on scholar influence to recommend conferences in ASNs, we define the scholar influence by measuring the local influence and global influence of scholars on ASNs based on their papers, books and projects, as well as their followers, tweets. We evaluate the effectiveness of our models on the SCHOLAT data set. The promising experimental results demonstrate the effectiveness of the proposed model.

Keywords: Academic social network · Academic influence · Social influence · Neural matrix decomposition

1 Introduction

With the rapid development of Internet, social networks have also developed from the early stage of conceptualization to becoming more and more complex and mature. Nowadays, the amount of information is growing explosively at an exponential rate, and the efficiency of information use is not increasing but decreasing. Users cannot get what they need, which leads to the problem of "information overload" [1]. How to obtain high-quality information from such a huge information cluster is a common challenge in various research fields. Therefore, recommendation system came into being. The recommendation system recommends information and products that users may be interested in by mining users' needs and preferences [2].

The initial academic social network was mainly an academic search engine. The well-known Google scholar was founded in 2004 and covered most academic journals when it was first published. Until 2008, research gate went online. Users can share academic achievements, understand academic trends in their own field, and participate in some interest groups. CiteULike is an academic knowledge base, which is usually used to store, manage and share users' academic papers. In 2009, the website SCHOLAT came out. Scholar users can build their own academic home page, manage their own academic

Y. Sun et al. (Eds.): ChineseCSCW 2021, CCIS 1492, pp. 391–402, 2022.
https://doi.org/10.1007/978-981-19-4549-6_30

information and create their own scientific research team; At the same time, it has the function of recommending scholars, recommending scholars with similar aspirations for users, and sharing your academic achievements with them by publishing academic trends. In addition, users have their own attention and fans, so as to establish their own unique social circle.

Academic conferences can quickly spread advanced technology, facilitate scientific researchers to exchange scientific achievements, and help to create a good education and scientific research environment. However, in practical applications, researchers often face one problem or another, as a result, users cannot obtain the conferences information in time and accurately. Therefore, it is necessary to study the processing and calculation of conference information in academic social networks and make personalized conference recommendation for scholars.

2 Related Work

At present, domestic and foreign recommendation algorithms mainly include content-based recommendation algorithm (CBF), collaborative filtering algorithm (CF), graph based community discovery algorithm, knowledge-based recommendation, group recommendation, and hybrid recommendation algorithms combining two or more to overcome some defects of a single algorithm [3].

The traditional collaborative filtering system considers that the behaviors of users are independent of each other, and does not consider the influence relationship between users [4]. However, as a node in the recommendation system, the impact of its various interactive behaviors on other users can not be underestimated. Indraneil Paul [5] and others studied the authenticated users of Twitter. They found that the activity of authenticated users was not only high and stable, while that of ordinary users was the opposite, so they predicted the authentication rules of Twitter and proved that authenticated users had greater influence. Li [6] et al. modeled and estimated the influence of users in social networks. They introduce the concepts of individual influence and type influence, and found that only considering local neighbors and outdated information can get better approximate values of individual impact, subtype impact and type impact. Shi [7] et al. proposed a new rating prediction model- RTRM to study the impact of internal factors on users, including users' credibility and popularity. The results show that the accuracy is greatly improved by using trust data mining to mine the internal factors affecting the relationship between different users.

In recent years, deep neural network has achieved some results in recommendation system. He [8] et al. combined the two and proposed a neural collaborative filtering (NCF) model. Bai [9] et al. introduced the neighbor information of users in social networks and proposed a neural collaborative filtering recommendation method based on community interaction.

With the continuous improvement of social network, the of recommendation system and recommendation algorithm are also innovating, and its application field is more and more extensive. However, the research on conference recommendation is difficult. Wongchokprasitti [10] and others designed a conference navigator. By processing the data of the user's community, the system finds out the interests of the participating scholars, and then adds the meeting that the user may be most interested in to the schedule and exports it to the calendar program for recommendation. Zhang [11] and others designed and developed a conference recommendation system - Confer, which aims to help participants find interesting papers and speeches, even meet and communicate with people with common interests, and use personalized conference schedule management.

Most of the above meeting recommendation systems or algorithms are for the schedule of a single meeting. Recommending interested meetings for them cannot be extended to academic social networks, the recommendation results will inevitably appear one-sided and not convincing.

3 Model Design

The traditional method regards all users as independent individuals without considering the interaction between users. Academic social networks generally have attention and fan functions, such as Research Gate and SCHOLAT. Therefore, the more friends and fans a scholar user has, the greater the user's influence. The research of social science [12] literature also shows that an authenticated user can increase the credibility of his tweets and other behaviors in social networks.

3.1 Scholar Influence Mechanism

In social networks, influence is like a business card showing users' confidence, similarly, influence in academic social networks is also an important reference to measure the academic value of a scholar. According to the coverage of influence, this paper divides scholar user influence into academic influence and social influence.

Academic Influence. The behavior characteristics of academic users in academic social networks can be divided into academic characteristics and active characteristics. Among them, academic features include papers, works, patents, scientific research projects, etc. Active features refer to the dynamic behavior of scholars in academic social networks, such as publishing dynamics, attention, fans, etc.

To measure the academic value of a scholar in this field, we must use the academic characteristics of scholar users, this paper extracts the papers, academic works, projects and other information of scholars and users. Define the number of papers P, the number of works B and the scientific research project Pj, where u represents the current scholar user. Considering the timeliness value of papers, works and scientific research projects, that is, with the change of time, their authority and value will gradually decrease or even disappear; Therefore, in order to better and more accurately measure the influence of scholars and users, this paper selects three time intervals, 1 year, 3 years and 5 years, that is, the academic achievements in recent 1 year, 3 years and 5 years are counted separately for selection. The specific calculation can be expressed as follows:

$$A_u = \frac{SUM(P_u + B_u + Pj_u)}{t}. \tag{1}$$

In formula (1), A_u is the academic credibility of the current scholar user. When A_u is lower than a certain threshold, the academic influence of the scholar user can be regarded as 0. Academic credibility A_u is defined as the academic influence of academic users.

Social Influence. Social influence can be understood as the credibility of academic users in the whole academic social network. Tang [13] defined the global influence as the influence gained through statistical indicators of the network, and defined a unified model framework to describe it. In this paper, the global influence of scholar users will be measured and calculated from three perspectives: the number of fans concerned by scholar users, the amount of dynamic likes and forwards, and whether they are authenticated users.

Attention, Fans

The social network of scholar users in the SCHOLAR is mainly built by paying attention and being concerned. The calculation formula of the number of concerns and fans defined in this paper is shown in (2):

$$\overline{F}_u = \frac{SUM(Fw + Fr)}{\Delta t} \tag{2}$$

where Fw, Fr represent the number of concerns and fans respectively, Δt is the time difference, that is, the time interval from the registration of the scholar user's account to the present, $\overline{F}_u$ is the average growth of the number of followers and followers.

Dynamic

The dynamics published by users in the SCHOLAR is similar to Microblog. The likes and forwarding volume of a microblog can reflect its degree of attention [14]. The dynamic number published by a scholar user in a certain period of time can reflect the activity of the scholar user to a certain extent. When a scholar publishes a dynamic, its likes and forwards can be used to measure the influence of the scholar's users. Therefore, the praise amount is defined as Tc and the forwarding amount is defined as Tr.

$$\overline{T}_w = \frac{\sum_{i=1}^{n} (T_{ic}, T_{ir})}{t} \tag{3}$$

Formula (3) is the dynamic calculation method published by scholars in their personal space,$\overline{T}_w$ is the dynamic weight value, t is the time interval, which is the same as the calculation method of academic influence, $t \in [1, 3, 5]$, that is, the value of t is 1 year, 3 years and 5 years, and the most appropriate one is selected. n is the total number of dynamics published by the scholar user in the current time interval of T_{ic}, T_{ir} respectively represent the praise and forwarding of the i scholar user to the dynamics. When $T_{ic} = 1$, it indicates that the scholar user likes this dynamic. On the contrary, when $T_{ic} = 0$, the scholar user does not like this dynamic. Similarly, when $T_{ir} = 1$, it indicates that the scholar user forwarded this dynamic message; If $T_{ir} = 0$, there is no forwarding.

Certified User

Whether a user is an authenticated user or not, its credibility (communicability) is different. Define whether it is an authenticated scholar user, and the influence calculation is shown in formula (4).

$$Ver = \frac{V_u}{N} \tag{4}$$

where, V_u refers to the total number of authenticated users in the SCHOLAR, and N refers to the total number of scholar users. Therefore, Ver is a fixed value.

3.2 Construction of Analytic Hierarchy Process Model of Scholars' Influence

Analytic hierarchy process [15] (AHP) is a subjective weighting method based on the idea of linear algebra. Generally speaking, it decomposes the influencing factors related to decision-making into three parts: influencing factors, criteria and objectives, and then carries out calculation, analysis and decision-making on this basis. It has a very good effect in dealing with complex decision-making problems.

Construction of Scholars' Influence. Based on the particularity of scholars' influence in academic social networks, combined with analytic hierarchy process, this paper divides scholars' influence measurement into target layer, benchmark layer and feature layer, as shown in Fig. 1. Among them, feature layer F is all the influencing factors mentioned above in this paper. Academic influence includes the number of papers, works and projects, and social influence includes the number of attention and fans, the number of dynamic users and whether they are authenticated users.

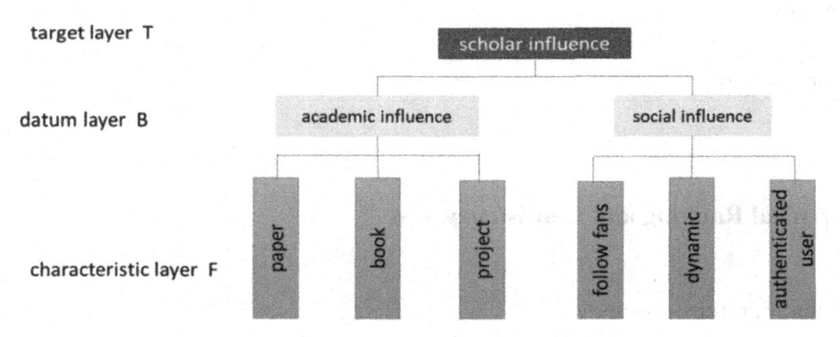

Fig. 1. AHP model of scholar influence

Build a Judgment Matrix. In this paper, scholars' influence has been divided into academic influence and social influence, and their influencing factors have been refined. In order to more accurately measure the influence of scholars' users, according to the characteristics of scholars' data in the SCHOLAR, first, artificially score each influencing factor - expert score. Compare any two influencing factors with each other, and set the scoring scale X_{ij} of influencing factors as 1–9 according to the importance of influencing factors i and j. when the importance of influencing factors i and j is the same, $X_{ij} = 1$, the more important the influencing factor i is compared with j, the higher the value of X_{ij}; When i is compared with j to get X_{ij}, define j and i to get $X_{ji} = \frac{1}{X_{ij}}$.

Then, through the three-layer AHP structure model of scholar influence constructed above, combined with the above scale X_{ij} value method, the influencing factors are compared and the judgment matrix is constructed. Considering the privacy and confidentiality of the scholar data, in order to obtain more accurate expert scores while preventing the disclosure of scholar user information, this paper investigates and fills in the scoring questionnaire for the research and development personnel of the SCHOLAR laboratory. After statistical analysis, most people believe that "project" is the most important, followed by "paper", The third is "dynamic", "pay attention to fans", "certified users" and "works". Since the premise of AHP is to have a full understanding of the concepts in the real situation, the expert scores of various influencing factors are shown in Table 1. The scoring is based on filling the upper right corner and the lower left corner is the reciprocal. Therefore, through the pairwise comparison of the above six factors, we get the judgment matrix $X_{m\times n}$ of scholars' influence. It can be seen from $m = n$ that this matrix is a square matrix $X_{n\times n}$. Therefore, in this matrix $x_{ij} > 0$, $x_{ij} = \frac{1}{x_{ij}}$, $x_{ij} = 1$.

Table 1. Expert score of scholar influence factors

Influential factors	Paper	Book	Project	Follow fans	Dynamic	Authenticated user
Paper	1	3	1/5	4	5	1/3
Book		1	1/9	1/3	1/5	1/2
Project			1	5	3	7
Follow fans				1	1/2	2
Dynamic					1	3
Authenticated user						1

Hierarchical Ranking and Consistency Test

(1) Total hierarchy sorting:

After the previous two steps, we have obtained the judgment matrix $X_{n\times n}$ of scholars' influence factors. Normalize it by column, as shown in formula (5):

$$\overline{x}_{ij} = \frac{x_{ij}}{\sum_{i=1}^{n} x_{ij}} \tag{5}$$

Normalization is to convert the sum of elements in each column vector into 1, and then calculate the row sum, as shown in formula (6):

$$W_i = \sum\nolimits_{i=1}^{n} \overline{x}_{ij} \tag{6}$$

Formula (7) normalizes it:

$$\overline{W}_t = \frac{W_i}{\sum_{i=1}^{n} W_i} \tag{7}$$

The weight matrix $W_{n\times 1}$ is obtained. Finally, maximum characteristic value of the matrix $X_{n\times n}$ is calculated by formula (8):

$$\lambda_{max} = \frac{1}{n} \sum\nolimits_{i=1}^{n} \frac{X_{n\times n} W_{n\times 1}}{\overline{W}_t} \tag{8}$$

where, $X_{n\times n}$, $W_{n\times 1}$ is the multiplication of two matrices.

(2) Conformance Inspection:

Imagine, what if the judgment matrix constructed above is unreasonable? The answer is consistency test. if and only if When $\lambda_{max} = n$ is the consistency matrix. Therefore, define the inspection index CI:

$$CI = \frac{\lambda_{max} - n}{n - 1} \tag{9}$$

In formula (9), λ_{max} is the maximum eigenvalue of the matrix and n is the order of the matrix. When $CI = 0$, it indicates that the matrix has perfect consistency; The farther the CI value is from 0, the more inconsistent the matrix is. By introducing the random consistency index RI (the value range is shown in Table 2), the size of CI can be better measured. At the same time, define the consistency ratio CR.

Table 2. Random consistency index

N	1	2	3	4	5	6	7	8	9
RI	0	0	0.58	0.9	1.12	1.24	1.32	1.41	1.45

$$CR = \frac{CI}{RI} \tag{10}$$

Judgment matrix $X_{n \times n}$ is generally considered when and only when $CR < 0.1$ passed the consistency test. Then, the calculated normalized eigenvector can be used as the weight vector.

Hierarchical Total Ranking and Consistency Test

(1) Total hierarchy sorting:

The total hierarchy sorting is to calculate the weight of the relative importance of feature layer F and reference layer B relative to target layer t from top to bottom. The above has obtained the hierarchical single ranking f_{ij} of the six factors in the feature layer F relative to the two factors in the reference layer B, where $I \in [1, 6]$, $J \in [1, 2]$, and the hierarchical single ranking b1, b2 of the target layer t by the reference layer B. Therefore, the final total ranking of levels is obtained, that is, the weight of the six factors of feature layer f on the scholar influence of target layer t is:

$$Fi = \sum\nolimits_{j=1}^{2} b_j f_{ij} \tag{11}$$

(2) Conformance inspection:

It is the same as the consistency inspection in hierarchical single sorting. Similarly, if and only if CR < 0.1, the consistency inspection is qualified. So far, the weight values of the six influencing factors of scholars' influence are obtained.

Result Calculation

According to Table 1, our judgment matrix is $\begin{bmatrix} 1 & 3 & 1/3 & 2 & 2 & 1/3 \\ 1/3 & 1 & 1/5 & 3 & 3 & 1/3 \\ 3 & 5 & 1 & 7 & 7 & 1/3 \\ 1/2 & 1/3 & 1/7 & 1 & 1/2 & 1/5 \\ 1/2 & 1/3 & 1/7 & 2 & 1 & 1/5 \\ 3 & 3 & 3 & 5 & 5 & 1 \end{bmatrix}$,

After calculationed, the maximum eigenvalues of feature layer F of academic influence and feature layer F of social influence are 3.038 and 3.054 respectively; CR values were 0.021 and 0.03 respectively, which passed the inspection; The weights obtained are $[0.258\ 0.105\ 0.637]^T$, $[0.113\ 0.179\ 0.709]^T$. Further, the maximum characteristic value of reference layer B is 6.519 and CR value is 0.079, which passes the test. Finally, the weight value of feature layer F relative to target layer T is obtained, that is, the weight value of six influencing factors of scholar influence is $[0.128\ 0.095\ 0.303\ 0.044\ 0.055\ 0.375]^T$.

3.3 Neural Matrix Decomposition

In this paper, the neural network layer is used to model the interaction between users and projects, so as to deal with collaborative filtering completely, as shown in Fig. 2.

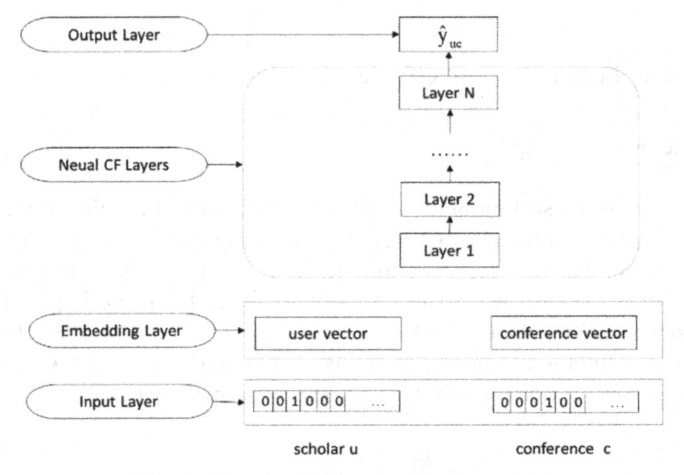

Fig. 2. Neural collaborative filtering model

The input layer is the feature vector of scholars and conferences, which is represented by u and c respectively. Here, the unique identification of scholars and conferences is introduced u _ id and c_ id is used as the feature input, and then it is transformed into a binary sparse vector and coded. Because this is a general representation of input characteristics, users and meetings can be represented according to it, so as to solve the cold start problem of data. The second layer is the embedded layer, which is a fully connected layer, which represents the sparse vector projection of the input layer as a dense vector. In the latent factor model, the embedding of scholars and conferences can be regarded as the latent vector of scholars and conferences. The third layer contains multi-layer neural networks, each of which can be customized, so as to explore the potential characteristics between scholars and conferences more widely, that is, neural collaborative filtering layer.

To sum up, the whole process can be described as formula (12):

$$\hat{y}_{uc} = f(P^T V_u^U, QV_c^C) = \Phi_{output}(\Phi_N(\ldots \Phi_2(\Phi_1(P^T V_u^U, QV_c^C))\ldots)) \tag{12}$$

where P, Q ∈ R are the potential factor matrices of scholars and conferences respectively, and V_u, V_c are the eigenvectors of scholars and conferences respectively, Φ_{output}, Φ_N represents the mapping function of the output layer and the N layer respectively, the final predicted value is $\hat{y}_{uc}$, and there are N neural collaborative filtering layers in total.

Let $m_u = P^T V_u^U$, $n_c = QV_c^C$, then the mapping function of the first neural collaborative filter layer is formula (13):

$$\Phi_1(m_u, n_c) = m_u \odot n_c \tag{13}$$

where, $\odot$ is the symbol of vector inner product operation. Then, map it to the output layer to get:

$$\hat{y}_{uc} = a_s(W^T(m_u \odot n_c)) \tag{14}$$

In formula (14), a_s is the activation function and W^T is the weight. When $W^T = 1$, it is the matrix decomposition model; Since the matrix decomposition is linear modeling, Sigmoid is selected as the activation function. Next, multiply the scholar influence weight obtained above to obtain the output result of the final model. Its calculation is shown in formula (15):

$$\hat{y}_{uc} = a_s((I_w^T m_u \odot n_c)) \tag{15}$$

Among them, I_w^T is the scholar influence weight vector obtained above.

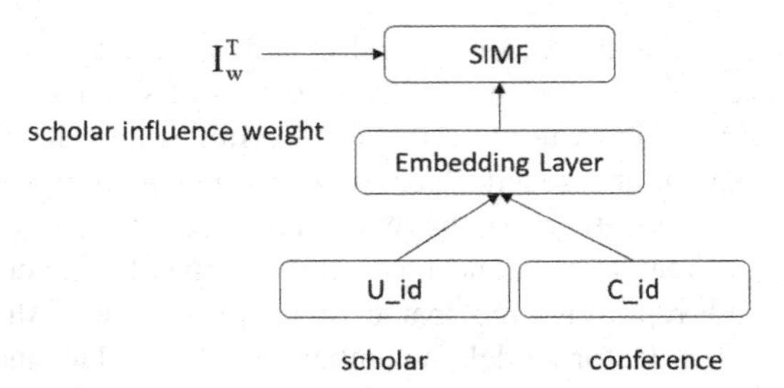

Fig. 3. SIMF model

Thus, the neural matrix decomposition model based on the influence of scholars shown in Fig. 3 is obtained. The SIMF (academic influence matrix factorization) layer is a single-layer neural network, that is, neural matrix decomposition based on the influence of scholars.

3.4 Experimental Results and Analysis

In order to verify the effectiveness of the model proposed in this paper, the leave one method is used for evaluation. For each scholar user, the latest interaction behavior is selected as the test data, and other interaction behaviors are used as the training data; At the same time, because the data set is too large, 100 samples without interaction with scholar users are randomly selected by random sampling to rank the test data. Finally, the conference recommendation results are measured by calculating the hit rate HR, the average reciprocal ranking MRR and the formula normalized impairment cumulative return NDCG.

This paper selects the following data of SCHOLAR for experiment:

(1) Basic data of scholar users, including ID and whether they are unauthenticated users;
(2) Academic achievement data of scholars and users, including published papers, conferences, works and projects under research;
(3) Social space data of scholar users, including paying attention to fans and dynamics.

By processing the conference paper data displayed by scholars in the SCHOLAR personal space, we get the scholar conference list of each scholar user, and get the one-to-one correspondence of 3847 scholars conferences. After manual de filtering, we finally get the correspondence of 3064 scholars conferences.

As can be seen from Fig. 4, HR, MRR and NDCG have reached more than 0.45, which verifies the effectiveness of the neural matrix decomposition model based on the influence of scholars proposed in this paper and can achieve satisfactory conference recommendation results.

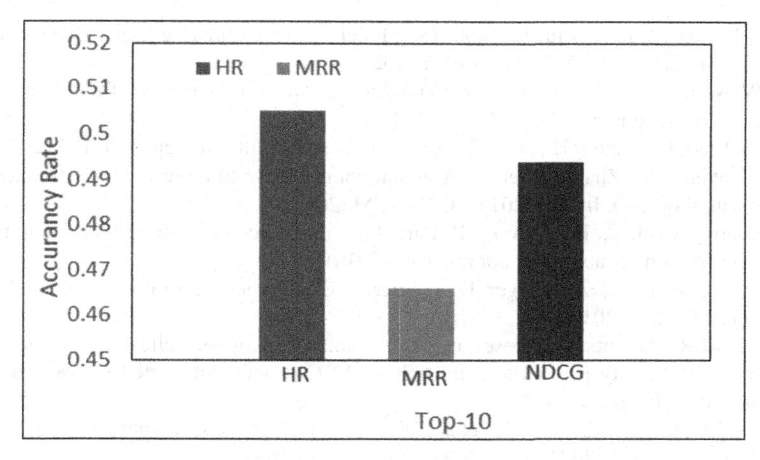

Fig. 4. HR@10, MRR@10, NDCG@10

4 Summary

Starting from the academic social network, this paper points out the necessity of conference recommendation research. We propose to design a neural matrix decomposition model based on scholar influence to recommend conferences in ASNs. We used the data set of academic social network - scholar to carry out the experiment. Firstly, we define the local influence and global influence of scholars, and calculate their weight by AHP. Secondly, the traditional MF is extended to the neural network and multiplied by the influence weight of scholars. Finally, an improved neural matrix decomposition model based on scholar influence is obtained. The promising experimental results demonstrate the effectiveness of the proposed model.

Acknowledgment. This work was supported in part by the National Natural Science Foundation of China under Grant 61772211 and Grant U1811263.

References

1. Zhang, X.S., Zhang, X., Kaparthi, P.: Combat information overload problem in social networks with intelligent information-sharing and response mechanisms. IEEE Trans. Comput. Soc. Syst. **7**(4), 924–939 (2020)
2. Xiang, L.: Recommended System Practice. Posts & Telecom Press (2012)
3. Shi, H.Y., Sun, T.H., Li, S.Q., et al.: A matrix factorization recommendation algorithm with time and type weight. J. Chongqing Univ. **42**(01), 79–87 (2019)
4. Zhang, Y.P., Zhang, S., Qian, F.L., et al.: Local and global user influence combined social recommendation algorithms. J. Nanjing Univ. (Nat. Sci.) **51**, 858–865 (2015)
5. Paul, I., Khattar, A., Kumaraguru, P., et al.: Elites tweet? Characterizing the twitter verified user network. In: 2019 IEEE 35th International Conference on Data Engineering Workshops (ICDEW). IEEE (2019)

6. Li, P., Nie, H., Yin, F., Liu, J., Zhou, D.: Modeling and estimating user influence in social networks. IEEE Access **8**, 21943–21952 (2020)
7. Shi, W., Wang, L., Qin, J.: Extracting user influence from ratings and trust for rating prediction in recommendations. Sci. Rep. **10**(1) (2020)
8. He, X., Liao, L., Zhang, H., et al.: Neural Collaborative Filtering, pp. 173–182 (2017)
9. Bai, T., Wen, J.R., Zhang, J., et al.: A neural collaborative filtering model with interaction-based neighborhood. In: The 2017 ACM. ACM (2017)
10. Wongchokprasitti, C., Brusilovsky, P., Para, D.: Conference navigator 2.0:community-based recommendation for academic conferences (2010)
11. Zhang, A., Bhardwaj, A., Karger, D.: Confer: A Conference Recommendation and Meetup Tool, pp. 118–121 (2016)
12. Morris, M.R., Counts, S., Roseway, A., et al.: Tweeting is believing?: Understanding microblog credibility perceptions. In: CSCW'12 Computer Supported Cooperative Work, Seattle, 11–15 February 2012 (2012)
13. Tan, Q., Zhang, F.L., Zhang, Z.Y., et al.: Modeling methods of social network user influence. Comp. Sci. **2**, 76–86 (2021)
14. Zhang, C., Tang, K., Peng, Y.B.: Fuzzy comprehensive evaluation of social network user's influence. Comp. Syst. Appl. (2017)
15. Saaty, T.L.: The Analytic Hierarchy Process (2001)

DOCEM: A Domain-Embedding-Based Open-Source Community Event Monitoring Model

Hong Huang, Jian Cao(✉), Qing Qi, and Boxuan Zhao

Shanghai Jiao Tong University, Shanghai, China
{hh1172622140,qi_ng616,b-x-zhao}@sjtu.edu.cn, cao-jian@cs.sjtu.edu.cn

Abstract. Open source software (OSS) development has become a trend and many popular software systems are developed and maintained by open source communities. In open source communities, developers are loosely organized which brings difficulties to the management of OSS projects. The events happening in the open source community may affect the development of the OSS project, which should be taken care of by the project organizers. Unfortunately, to identify these events from large amounts of messages generated in the community, especially from multiple sources, is still a big challenge. In this paper, a Domain-embedding-based Open-source Community Event Monitoring Model (DOCEM) is proposed to identify events from multiple sources. Specifically, DOCEM is based on DualGAN and StarGAN. An event dataset for Tensorflow project is constructed to train and test DOCEM. The experiment results show that DOCEM has much better performance than counterparts.

Keywords: Open source community · Event monitor · Domain embedding

1 Introduction

Internet is changing the way of working including software development. Open source software (OSS) development on the Internet is becoming more and more popular. Many successful software packages are developed and maintained by open source communities.

In an open source community, developers coming from different organizations, different places even different countries work together in a self-organized way. They submit codes, raise issues, answer questions, post review comments and organize discussions to push the project forward. Unfortunately, most OSS projects fail in practice and how to run OSS projects is still a big challenge.

It is unavoidable that events happen from time to time in open source communities. Here, the event is defined as the reaction and discussion of users on the Internet due to specific reasons within a given time [3]. These events should be taken care of because they affect the development of the community, such as the discovery of potential bugs, user complaints, and requests to new features.

Y. Sun et al. (Eds.): ChineseCSCW 2021, CCIS 1492, pp. 403–417, 2022.
https://doi.org/10.1007/978-981-19-4549-6_31

OSS project managers should notice the happening of these events and take measures to handle these events in time to reduce the risk in the community as early as possible. This has to bring a huge burden to managers. Some open source platforms, such as GitHub, Gitee and Bitbucket, have implemented the notification function, which only provides updates about the activity on platforms that users subscribed to. In this paper, we try to discover important events that often corresponds to multiple messages and it's difficult to identify events by reading messages and summarizing them by managers manually. Therefore, we try to develop a monitoring model to identify the events happening in the open source community automatically.

At the same time, some important events also appear outside of the open source platform. For instance, members of an OSS project often form discussion groups on Twitter, Stack Overflow or other technical forums. They exchange ideas and discuss problems about an OSS project. Therefore, to identify events needs making use of data from multiple sources. However, each technical forum has it's language style so that a model working well on a platform often perform bad on another platform.

Therefore, we proposed a Domain-embedding-based Open-source Community Event Monitoring Model (DOCEM) to detect events from GitHub, Stack Overflow and Twitter. DOCEM can identify events in a timely manner, which helps community manager take measures to reduce the risk of community operations and capture the common needs of community users.

The remaining of this paper is organized as follows. Section 2 reviews the related work on event detection and open source community. In Sect. 3, we introduce the framework of DOCEM and its three components: event detection module, event ranking module and event summarization module. Section 4 introduces the three models in event detection module, especially the domain embedding model. Section 5 introduces the event ranking module and event summarization module. Section 6 presents the experiment in detail. Conclusions are given in Sect. 7.

2 Related Work

Event monitoring technology can be classified according to event type (specific or non-specific), monitoring task (retrospective or new event type) and monitoring method (supervised or unsupervised) [1]. The types of events in the open source community are not known in advance and most of the data is not labelled. For this task, a common framework can be adopted [8], which includes an event detection module, an event ranking module and an event summarization module.

The core module is the event detection module, which can be classified into three main classes: term interestingness, topic modeling and incremental clustering [8]. The method of event detection based on term interestingness is dependent on the choice of keywords. This type of method often discovers potential keywords from the text, and then uses search or clustering methods to detect events [12,14,28]. Event detection based on topic modeling refers to the identification of possible events in the text information flow through possible topics [2,27]. The event detection algorithm based on incremental clustering can grow the cluster by itself as the data grows, which is the most suitable for the task of event detection [7,15,19].

The event ranking module is used to find the most popular events. The event can be ranked based on novelty score [12], information entropy [9,19] or user diversity [9]. And the event summarization module can give a legible summary to user, which can be constructed by TF-IDF [11], Phrase reinforcement algorithm [23] or TextRank [13,16].

For the event processing of the open source community, Wahyudin et al. [25] proposed an event-based OSS project monitoring system to find the risks in open source community, but they only focus on subscription and other simple trigger information, instead of the more important text information.

For the anomaly detection of the open source community, Goyal et al. [6] can identify abnormal git commit, for example, new user, new language, too long code and so on. However, it can only detect commit events.

3 A Domain-Embedding-Based Open-Source Community Event Monitoring Model: DOCEM

In an open source community, project members not only discuss on the OSS platforms, but also discuss on other platforms such as Twitter and Stack Overflow. Specifically, we find people are more likely to express their opinions towards the project on social media, and describe their problems on technical forums. Moreover, event detection based on the data from forums and social medias can help community to find common requirements, discover potential bugs, recognize product weakness and even identify the gap between product competitors. Therefore, the data from open source platforms and other sources should be used for event detection together.

We propose the Domain-embedding-based Open-source Community Event Monitoring Model (DOCEM) based on multi-source data. As shown in Fig. 1, DOCEM also consists of event detection module, event ranking module and event summarization module. Event detection module discovers instance clusters from multi-source instances as events; event ranking module is responsible for ranking the events based on their information and popularity and event summarization module extracts legible text summaries from events. Moreover, event detection module consists of Embedding models, Domain Embedding model and Group model. Embedding model maps the text from each domain to text vector; domain embedding model is used to map text vectors from different domains to a specific domain that can eliminate domain transferring cost; and group model clusters instances based on text vector similarity.

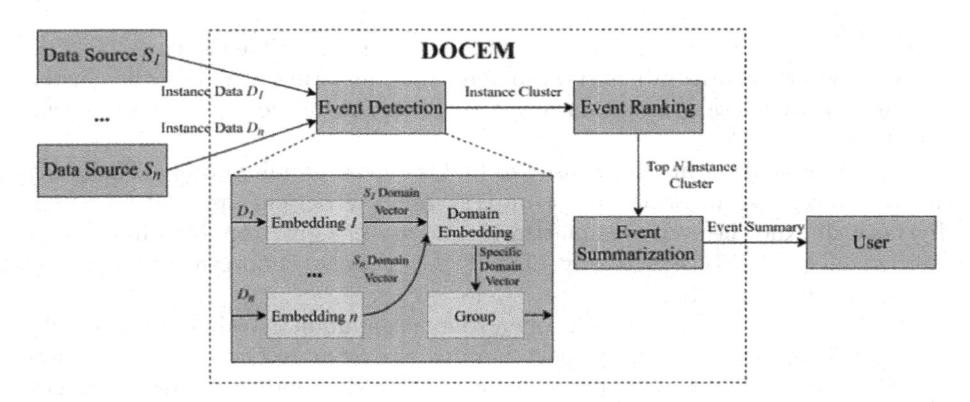

Fig. 1. The framework of DOCEM

4 Event Detection Module

4.1 Embedding Model

An embedding model can map the text into a fixed-length vector so that the distance of the vectors mapped from semantically similar texts is close. In this paper, we choose Doc2Vec [10] and SentBERT [21] as embedding models.

Doc2Vec [10] is a sentence embedding model, which represents each sentence as a vector. Doc2Vec is further classified into two kinds of model including PV-DM and PV-DBOW. We choose PV-DM in this paper, which uses frozen pre-trained Word2Vec [17] to train paragraph matrix which represents the sentence. The negative sampling [18] is used in the output stage instead of calculating the loss from the total vocabulary.

The BERT [5] model has achieved State-of-the-Art (SOTA) in many NLP fields, including machine translation, semantic similarity, etc. But because the BERT model needs to pair sentences when performing semantic similarity calculations, the time complexity is very high. The SentBERT [21] model uses the Siamese network model to input sentences into two BERT models with shared parameters, and then obtains the representation vector of each sentence. There are three ways for SentBERT to obtain fixed-length sentence vectors. One is to add [CLS] to the beginning of each sentence as a vector representation of the entire sentence like BERT, the next one is to take the mean pooling of the word vector of each word, the last one is to take the maximum pooling of the word vector of each word. And three different structures and objective functions are formulated for three different task settings. We only use mean pooling to train SentBERT.

4.2 Domain Embedding Model

As shown in Table 1, the text data from different sources differs in style, length distribution, vocabulary and information distribution, which lead to the domain loss between vectors from different sources. Take text in Table 1 as an example, due to domain loss, although the first GitHub text and the first Twitter tweet talk about the same topic (DLL load error while importing TensorFlow), their cosine similarity under SentBERT embedding is 0.6782, compared to 0.7348 which is the cosine similarity between two different topic Twitter tweets in Table 1.

Table 1. Table of different source text data about Tensorflow community

GitHub text	Stack Overflow text	Twitter tweet
Win10: ImportError: DLL load failed while importing tensorflow: A dynamic link library (DLL) initialization routine failed.	Keras model training memory leak	Hi @Tensorflow i install you but when i import you as "import tensorflow as tf" you will give a dll load error of this plz help my Whole Final Year Project is on you
Cannot use keras estimator_from_model() in distributed cluster	Tensorflow importerror: dllload failed while importing pywrap_tensorflow_internal	I met a module missing error when I install @Tensorflow. pls help a friend out

Therefore, to eliminate loss between different domains, we propose the double domain embedding model (DoubleDEM) based on DualGAN [26] and the multiple domain embedding model (MultiDEM) based on StarGAN [4], which exploit unsupervised methods to embed vectors from different domains into the specific domain.

In this section, we will introduce DoubleDEM and MultiDEM in detail.

Double Domain Embedding Model (DoubleDEM). Inspired by DualGAN used in image translation, we propose DoubleDEM which can transfer vectors in two domain to each other. We trained DoubleDEM in an unsupervised adversarial method. As shown in Fig. 2, the goal is training generator G_{A2B} and G_{B2A} which can generate the target vector from source vector. The training goals are following:

1. The Domain discriminators D_A and D_B need to distinguish the real vector and fake vector.
2. The generators G_{A2B} and G_{B2A} can generate a fake vector in target domain with a vector in source domain as input.
3. The generators G_{A2B} and G_{B2A} can reconstruct vectors by generating twice.
4. The generators G_{A2B} and G_{B2A} try to generate vectors that have similar distribution with the target domain, so as to confuse the domain discriminators D_A and D_B.

As shown in Fig. 2, before the training, it is required to pre-train one or two embedding models and froze them. Firstly, using an embedding model to embed

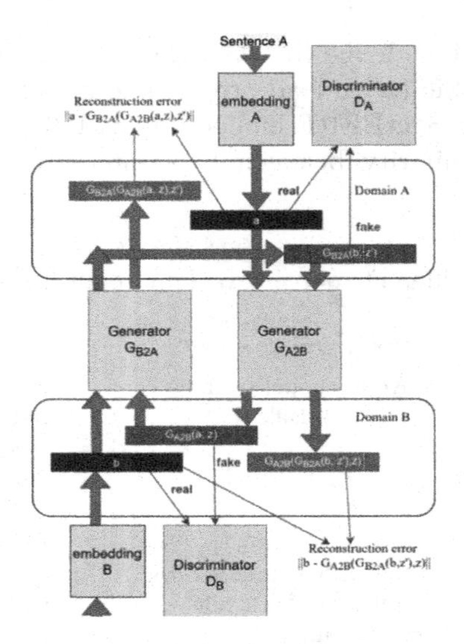

Fig. 2. DoubleDEM training processing

two sentences from different sources into vectors in two domains. These vectors are real vectors so they are denoted by green a and b. Then red fake vector $G_{A2B}(a, z)$ and $G_{B2A}(b, z')$ are generated by generators G_{A2B} and G_{B2A}, where z and z' are random noises. Meanwhile, domain discriminators D_A and D_B distinguish the real vectors a and b, fake vectors $G_{A2B}(a, z)$ and $G_{B2A}(b, z')$ respectively. Finally, orange reconstructed vectors $G_{B2A}(G_{A2B}(a, z), z')$ and $G_{A2B}(G_{B2A}(b, z')$ are generated by using generators G_{A2B} and G_{B2A} again to obtain the reconstruction loss.

Training loss consists of two parts,i.e., adversarial loss and reconstruction loss. Generator loss function and discriminator loss function are calculated correspondingly. The two discriminator loss functions L_A^d and L_B^d in discriminator D_A and D_B respectively, are defined by Eqs. 1 and 2.

$$L_A^d(a, b) = D_A(G_{B2A}(b, z')) - D_A(a) \tag{1}$$

$$L_B^d(a, b) = D_B(G_{A2B}(a, z)) - D_B(b) \tag{2}$$

Due to the same goal of generators G_{A2B} and G_{B2A}, both generators share the same generator loss function L^g. Generator loss function L^g is calculated from the reconstruction loss L_{Re}^g and the adversarial discrimination loss L_{Dis}^g according to Eq. 3.

$$L^g(a, b) = L_{Re}^g(a, b) + L_{Dis}^g(a, b) \tag{3}$$

The discrimination loss l^g_{Dis} is obtained by calculating the discriminator error according to Eq. 4. The reconstruction loss L^g_{Re} is obtained according to the L1 error according to Eq. 5, where λ_A and λ_B are two constants.

$$L^g_{Dis}(a,b) = -D_A(G_{B2A}(b,z')) - D_B(G_{A2B}(a,z)) \tag{4}$$

$$L^g_{Re}(a,b) = \lambda_A||a - G_{B2A}(G_{A2B}(a,z),z')||_1 + \lambda_B||b - G_{A2B}(G_{B2A}(b,z'),z)||_1 \tag{5}$$

After training, we choose a specific domain, take domain A for example, then exploit the trained generator G_{B2A} as the domain embedding model. For embedding vector in domain B, we use the domain embedding model to transfer it to domain A. By doubleDEM, we can transfer the vectors from different domains to one specific domain, which can eliminate the gap between two domains.

Multiple Domain Embedding Model (MultiDEM). DoubleDEM is not suitable for multiple domain transferring task. When the number of domains is n, it needs n^2 generators to achieve domain transferring task. As shown in Fig. 3, each double-headed arrow represents two generators. It will make the model really complex.

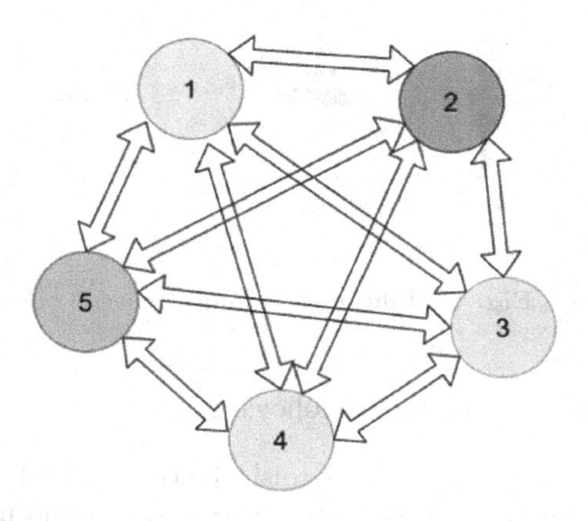

Fig. 3. Multiple domain transferring with DoubleDEM

Therefore, we propose the Multiple Domain Embedding Model(MultiDEM) based on StarGAN [4], which contains only one generator no matter of the number of domains.

MultiDEM has only one generator G. Unlike ordinary generators, this generator takes text vector and target domain label as input, and outputs the target domain vector. In addition, there is only one discriminator D to discriminate real and fake vector in training. This discriminator not only needs to distinguish the real and fake vector, but also needs to classify the input vector into its domain.

Before training, it is still necessary to pre-train and freeze one or more embedding models. As shown in Fig. 4, the MultiDEM is trained in sequence along the blue, brown and pink paths. Firstly, the text are mapped to the green real vector v through the embedding module, corresponding to the domain S, and the domain label is s. After that, the red fake vector $G(v,t)$ in domain T is generated by the generator G with vector v and target domain label t as input. In the next stage, orange reconstruction vector $G(G(v,t),s)$ is generated by the source field label s and the generated fake vector $G(v,t)$. In this stage, reconstruction loss L_{Re} is calculated. Finally, discriminator D distinguishes and classifies the fake vector $G(v,t)$ and the real vector v. In this stage, adversarial loss L_{Ad} and classification loss L_{Cl} are calculated.

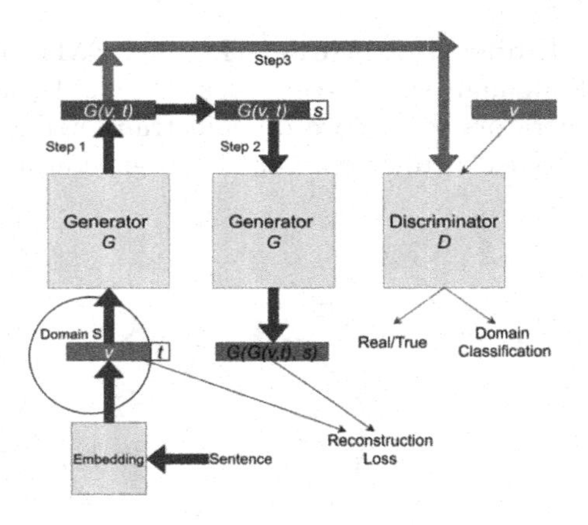

Fig. 4. MulitDEM training process

In short, the goals of training are following:

1. The discriminator D needs to distinguish the true and false of the text vector and be able to classify the text vector to the corresponding domain.
2. The generator G takes a certain domain vector and the target domain label as input, and generates a fake vector in target domain.
3. The generator G can reconstruct the vector based on the source domain label.
4. The generator G tries to generate vectors that have similar distribution to the target domain, in order to confuse the discriminator D.

As described in above, The loss of the MultiDEM training includes three aspects. The first one is the adversarial loss L_{Ad}, which is determined by the ability of discriminator D to distinguish the real and fake vector; the second one is the classification loss L_{Cl}, which is determined by the ability of discriminator

D to classify the vector domain; the last one is reconstruction loss L_{Re}, which is determined by the ability of generator G to reconstruct source vector.

For the adversarial loss L_{Ad}, it can be calculated by Eq. 6, where v is the real source vector in domain S, $G(v,t)$ is the fake generated vector in domain T, and D_{src} is the function of the discriminator D to distinguish real and fake vector, whose output is 0 or 1.

$$L_{Ad} = \mathbf{E}_v[\log D_{src}(v)] + \mathbf{E}_{v,t}[\log(1 - D_{src}(G(v,t)))] \tag{6}$$

For the classification loss L_{Cl}, we divide it into discriminator classification loss L^d_{Cl} and generator classification loss L^g_{Cl}, which are obtained by Eqs. 7 and 8, where s is the source domain label of real source vector v and D_{Cl} is the function of the discriminator D to classify the vector domain, whose output is the possibility of specific domain.

$$L^d_{Cl} = \mathbf{E}_{v,s}[-\log D_{Cl}(s|v)] \tag{7}$$

$$L^g_{Cl} = \mathbf{E}_{v,t}[-\log D_{Cl}(t|G(v,t))] \tag{8}$$

For the reconstruction loss L_{Re}, it is similar with DoubleDEM's reconstruction loss. With real vector v in domain S and the target domain label t, L_{Re} is calculated by Eq. 9.

$$L_{Re} = \mathbf{E}_{v,s,t}[||v - G(G(v,t),s)||_1] \tag{9}$$

According to the adversarial loss L_{Ad}, classification loss L_{Cl} and reconstruction loss L_{Re}, we can get the discriminator loss L_D and generator loss L_G by Eqs. 10 and 11 respectively, where λ_1 and λ_2 are both constants.

$$L_D = -L_{Ad} + \lambda_1 L^d_{Cl} \tag{10}$$

$$L_G = L_{Ad} + \lambda_1 L^g_{Cl} + \lambda_2 L_{Re} \tag{11}$$

Since the input of the domain embedding model is the vector, the generator G is constructed by a fully connected layers. The discriminator D needs two outputs, so there are two output layers, which are the n-Softmax layer for domain classification and the Sigmoid layer for true and false discrimination. The hidden layers are built with fully connected layers.

After training, we exploit the generator G as MultiDEM. After selecting a specific domain S, it is able to transfer the other domain vectors into domain S with the domain label s as input.

4.3 Group Model

Group Model is used to group the similar semantic texts into same cluster by the embedding vectors. Based on First Story Detection (FSD) [19,22] algorithm, we propose the Time-based First Story Detection (TFSD) algorithm, whose main difference with FSD [19,22] is that it uses the time span as the time window threshold ws instead of time window length. In other words, when the time span of the window exceeds the threshold ws, the older instance will be pushed out of the time window, while the traditional FSD algorithm will push out the oldest instance out only when the time window length exceeds the threshold ws. Therefore, the length of the time window W in the TFSD algorithm is unlimited. The reason is that we hope the DOCEM can detect intensive events in a certain period, without being bound by the length of the time window.

Algorithm 1: Time-based First Story Detection(TFSD)

Data: similarity threshold t, time window threshold ws, list of text vectors sorted by time L

Result: Text vector cluster dictionary *classification*

```
1  initialization;
2  queue W := [], int c := 0;
3  dict classification := {} ;
4  for text vector v in L do
5  |   if W is empty then
6  |   |   classification[v] := c ;
7  |   |   c := c + 1 ;
8  |   else
9  |   |   v_nearest := nearest vector of d in W ;
10 |   |   if cosine_sim(v, v_nearest) < t then
11 |   |   |   classification[v] := classification[v_nearest] ;
12 |   |   else
13 |   |   |   classification[v] := c ;
14 |   |   |   c := c + 1 ;
15 |   |   end
16 |   end
17 |   if time interval of W > ws then
18 |   |   remove oldest vectors from W until time interval of W ≤ ws ;
19 |   else
20 |   end
21 |   push v to W;
22 end
23 return classification;
```

Through the TFSD Algorithm 1, each instance vector will first perform similarity calculation with each instance in the time window, and the neighbor instance with the biggest similarity will be recorded. If the maximum similarity exceeds the similarity threshold t, then the current instance will be divided into the event cluster that this neighbor instance belongs. If the maximum similarity does not exceed the similarity threshold t, then the current instance will be treated as a new event cluster. If the time interval between the newest instance and the oldest instance in the time window exceeds the threshold ws, the oldest instance will be pushed out of the window W until the time interval of $W \leq ws$.

5 Event Ranking and Event Summarization

5.1 Event Ranking

A lot of instance clusters will be generated by event detection. An Event Ranking module is required to find the most important event. We choose average attention score C and average information entropy E as ranking metrics. Average information entropy E is the geometric mean of the information entropy from each source, calculated by Eq. 12. Average attention score C is calculated by the sum of the total number of comments on GitHub issue data, the total number of votes and comments on Stack Overflow data, and the total number of likes and retweets on Twitter data, then divide by the total number of instances in the event cluster.

$$E = -\sum_{w} \frac{n_w}{N} \log \frac{n_w}{N} \tag{12}$$

5.2 Event Summarization

This paper uses the TF-IDF algorithm [11] as Event Summarization module to summarize the keywords of the event cluster. After removing the stop words, calculate TF_w and IDF_w of each word w separately according to the formulas 13 and 14, where n_{wi} represents the number of times the word w appears in the text i, $\sum_w n_{wi}$ represents the total number of words in the text i, $|c|$ represents the number of instance texts in the event cluster, and $|i : w \in i|$ represents the total number of instance texts in the event cluster that contain the word w. Use the product of TF_w and IDF_w as the importance of the word w, and select top six as the keyword summary of the event.

$$TF_w = \frac{n_{wi}}{\sum_w n_{wi}} \tag{13}$$

$$IDF_w = \frac{|c|}{|i : w \in i, i \in c| + 1} \tag{14}$$

6 Experiments

6.1 Datasets

We evaluate the embedding model and domain embedding model on a dataset. Because there is no existing dataset for open source community event monitoring task, we construct the dataset by ourselves.

First, we constructed the TensorFlow GitHub event dataset based on connections between issue comments. In GitHub, as shown in Fig. 5, if a new issue is similar to a previous one, the precedent comments might mention the previous one. It is called mentioned event and is very common in open source community. Therefore, we use PyGitHub to obtain the issue data from the TensorFlow community, then we utilize each issue as a node and connect two nodes if they have a mentioned event. A GitHub event dataset is constructed by choosing some large issue sub-graphs that are disconnected from each other and each issue sub-graph contains similar issues.

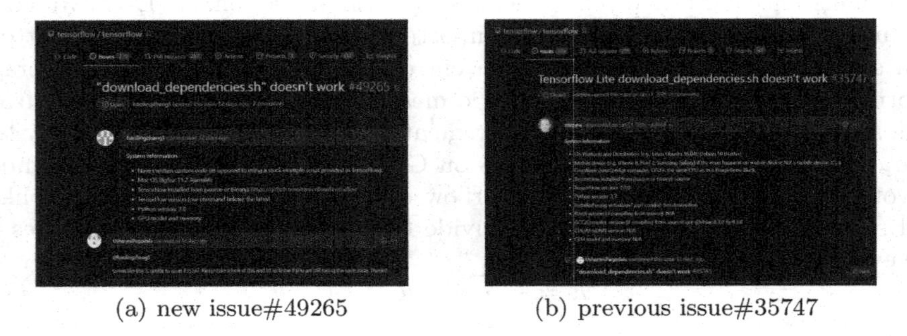

(a) new issue#49265 (b) previous issue#35747

Fig. 5. Similar issues in Tensorflow

Then according to the topic keywords extracted from GitHub events by TF-IDF [11], Tensorflow Twitter dataset is obtained through Twint and Stack Overflow dataset is constructed through Google BigQuery. Some keywords are shown in the Table 2. In addition, for some common errors reported by Stack Overflow, such as "runtime error", and some common complaints on Twitter, such as "hard to use", keyword search and filtering is performed to get the unique events.

Table 2. Some keywords extracted from GitHub Dataset

Event label	Keywords
1	rnn, transformed, keras, error, executed, bug
2	tf.function-decorated, decorated, variables, valueerror, non-first call, error
3	gpu, cuda, set_visible_devices, setting up virtual devices, visible device, device
4	build, libtensorflow, libtensorflow_cc.so, library, c++, config
5	bazel, cuda, gpu, build, error, dependency

Relying on the above method, the GitHub event dataset contains 40 events and 834 instances; the Twitter event dataset contains 24 events and 523 instances; and the Stack Overflow event dataset contains 31 events and 723 instances. Among them, the Stack Overflow event dataset and GitHub dataset overlap with 23 events, and the Twitter event dataset and GitHub event dataset overlap with 21 events. The text in GitHub event dataset consists of issue title, issue body after removing irrelevant text and the last line of the code; the text in Stack Overflow event dataset consists of question title; and the text in Twitter event dataset consists of tweets.

6.2 Experiment Results

TFSD is used to group instances and Best Adjusted Rand Index(Best-ARI) is selected as the main evaluation metrics. Best Adjusted Rand Index is defined as the largest Adjusted Rand Index (ARI) [20] score under different similarity threshold t in TFSD. Meanwhile, time window threshold ws is set to positive infinity in evaluation. The purpose of all the above methods is to eliminate the effect of the parameters of the TFSD algorithm.

We choose two embedding models in our experiment. Doc2Vec [17] is trained on GitHub, Stack Overflow and Twitter data, denoted as Doc2Vec Source. Another model is paraphrase-mpnet-base-v2 in the SentBERT [21] model which uses mpnet-base [24] as the base model, denoted as SentBERT. Both of them can choose DoubleDEM, MultiDEM or None.

The experiment results are shown in the Table 3. It can be seen that using the domain embedding model is generally better than that of the single domain embedding model. Especially for the Doc2Vec Source model that is trained by domain dataset, the performance is more obvious.

Table 3. Best-ARI evaluation results of multi-source event detection

	Github+Stack Overflow	Twitter+Github+Stack Overflow
Doc2Vec Source	0.3589	0.3248
Doc2Vec Source+DoubleDEM	0.4828	\
Doc2Vec Source+MultiDEM	\	0.4862
SentBERT	0.4916	0.5176
SentBERT+DoubleDEM	**0.5146**	\
SentBERT+MultiDEM	\	**0.5368**

Table 4. V-Measure evaluation results of multi-source event detection

	Github+Stack Overflow			Github+Stack Overflow+Twitter		
	Homogeneity	Completeness	v-measure	Homogeneity	Completeness	v-measure
Doc2Vec Source	**0.8176**	0.4707	0.5974	0.6741	0.4502	0.5399
Doc2Vec Source+DoubleDEM	0.6341	0.5236	0.5735	\	\	\
Doc2Vec Source+MultiDEM	\	\	\	0.6223	0.5469	0.5821
SentBERT	0.7142	0.5051	0.5917	**0.7284**	0.5291	0.6129
SentBERT+DoubleDEM	0.6349	**0.5814**	**0.6069**	\	\	\
SentBERT+MultiDEM	\	\	\	0.6849	**0.6224**	**0.6521**

In order to show the results more prominently, prediction results of above models are also evaluated by V-Measure, which consists of homogeneity and completeness. The result is shown in Table 4. After adding DoubleDEM and MultiDEM, generally, homogeneity will decrease and completeness will increase. The reason is that domain embedding can map different sources data belonging to the same event into a closer position, which reduces the number of classes to be divided.

7 Conclusions

In this paper, we propose a DOCEM to monitor events from multiple sources in the open source community. We design the domain embedding model to map text vectors from different domains to a specific domain, which can estimate the domain transferring cost. Besides, we established an event dataset to evaluate event detection module and found domain embedding model performing better in merging multi-source events.

Acknowledgement. This work is partially supported by National Key Research and Development Plan (No. 2018YFB1003800) and China National Science Foundation (Granted Number 62072301).

References

1. Atefeh, F., Khreich, W.: A survey of techniques for event detection in twitter. Comput. Intell. **31**(1), 132–164 (2015)
2. Cai, H., Yang, Y., Li, X., Huang, Z.: What are popular: exploring Twitter features for event detection, tracking and visualization. In: Proceedings of the 23rd ACM International Conference on Multimedia, pp. 89–98 (2015)
3. Chen, X., Li, Q.: Event modeling and mining: a long journey toward explainable events. VLDB J. **29**(1), 459–482 (2019). https://doi.org/10.1007/s00778-019-00545-0
4. Choi, Y., Choi, M., Kim, M., Ha, J.W., Kim, S., Choo, J.: StarGAN: unified generative adversarial networks for multi-domain image-to-image translation. In: Proceedings of the IEEE Conference on Computer Vision and Pattern Recognition, pp. 8789–8797 (2018)
5. Devlin, J., Chang, M.W., Lee, K., Toutanova, K.: BERT: Ppre-training of deep bidirectional transformers for language understanding. In: NAACL (2019)
6. Goyal, R., Ferreira, G., Kästner, C., Herbsleb, J.: Identifying unusual commits on GitHub. J. Softw. Evol. Process **30**(1), e1893 (2018)
7. Hasan, M., Orgun, M.A., Schwitter, R.: TwitterNews+: a framework for real time event detection from the Twitter data stream. In: Spiro, E., Ahn, Y.-Y. (eds.) SocInfo 2016. LNCS, vol. 10046, pp. 224–239. Springer, Cham (2016). https://doi.org/10.1007/978-3-319-47880-7_14
8. Hasan, M., Orgun, M.A., Schwitter, R.: A survey on real-time event detection from the twitter data stream. J. Inf. Sci. **44**(4), 443–463 (2018)
9. Kumar, S., Liu, H., Mehta, S., Subramaniam, L.V.: From tweets to events: exploring a scalable solution for twitter streams. arXiv preprint arXiv:1405.1392 (2014)
10. Le, Q., Mikolov, T.: Distributed representations of sentences and documents. In: International Conference on Machine Learning, pp. 1188–1196. PMLR (2014)
11. Lee, S., Kim, H.J.: News keyword extraction for topic tracking. In: 2008 Fourth International Conference on Networked Computing and Advanced Information Management, vol. 2, pp. 554–559. IEEE (2008)
12. Li, C., Sun, A., Datta, A.: Twevent: segment-based event detection from tweets. In: Proceedings of the 21st ACM International Conference on Information and Knowledge Management, pp. 155–164 (2012)
13. Madani, A., Boussaid, O., Zegour, D.E.: Real-time trending topics detection and description from twitter content. Soc. Netw. Anal. Min. **5**(1), 1–13 (2015)

14. Mathioudakis, M., Koudas, N.: TwitterMonitor: trend detection over the Twitter stream. In: Proceedings of the 2010 ACM SIGMOD International Conference on Management of Data, pp. 1155–1158 (2010)
15. Mazoyer, B., Cagé, J., Hervé, N., Hudelot, C.: A French corpus for event detection on Twitter. In: Proceedings of the 12th Language Resources and Evaluation Conference, pp. 6220–6227 (2020)
16. Mihalcea, R., Tarau, P.: TextRank: bringing order into text. In: Proceedings of the 2004 Conference on Empirical Methods in Natural Language Processing, pp. 404–411 (2004)
17. Mikolov, T., Chen, K., Corrado, G., Dean, J.: Efficient estimation of word representations in vector space. arXiv preprint arXiv:1301.3781 (2013)
18. Mikolov, T., Sutskever, I., Chen, K., Corrado, G.S., Dean, J.: Distributed representations of words and phrases and their compositionality. In: Advances in Neural Information Processing Systems, pp. 3111–3119 (2013)
19. Petrović, S., Osborne, M., Lavrenko, V.: Streaming first story detection with application to Twitter. In: Human Language Technologies: The 2010 Annual Conference of the North American Chapter of the Association For Computational Linguistics, pp. 181–189 (2010)
20. Rand, W.M.: Objective criteria for the evaluation of clustering methods. J. Am. Stat. Assoc. **66**(336), 846–850 (1971)
21. Reimers, N., et al.: Sentence-BERT: sentence embeddings using Siamese BERT-networks. In: Proceedings of the 2019 Conference on Empirical Methods in Natural Language Processing. Association for Computational Linguistics (2019)
22. Repp, Ø., Ramampiaro, H.: Extracting news events from microblogs. J. Stat. Manag. Syst. **21**(4), 695–723 (2018)
23. Sharifi, B., Hutton, M.A., Kalita, J.: Summarizing microblogs automatically. In: Human Language Technologies: The 2010 Annual Conference of the North American Chapter of the Association for Computational Linguistics, pp. 685–688 (2010)
24. Song, K., Tan, X., Qin, T., Lu, J., Liu, T.Y.: MPNet: masked and permuted pre-training for language understanding. arXiv preprint arXiv:2004.09297 (2020)
25. Wahyudin, D., Tjoa, A.M.: Event-based monitoring of open source software projects. In: The Second International Conference on Availability, Reliability and Security (ARES 2007), pp. 1108–1115. IEEE (2007)
26. Yi, Z., Zhang, H., Tan, P., Gong, M.: DualGAN: unsupervised dual learning for image-to-image translation. In: Proceedings of the IEEE International Conference on Computer Vision, pp. 2849–2857 (2017)
27. You, Y., et al.: GEAM: a general and event-related aspects model for Twitter event detection. In: Lin, X., Manolopoulos, Y., Srivastava, D., Huang, G. (eds.) WISE 2013. LNCS, vol. 8181, pp. 319–332. Springer, Heidelberg (2013). https://doi.org/10.1007/978-3-642-41154-0_24
28. Zhang, X., Chen, X., Chen, Y., Wang, S., Li, Z., Xia, J.: Event detection and popularity prediction in microblogging. Neurocomputing **149**, 1469–1480 (2015)

Diversified Concept Attention Method for Knowledge Tracing

Hao Wu(✉) and Yuekang Cai

Computer and Information Engineering, Zhejiang Gongshang University, Hangzhou 310018, China
19020100036@pop.zjgsu.edu.cn

Abstract. Through the learner's historical learning to study the knowledge state over time, knowledge tracing can predict the learner's future learning performance. It is an important issue in personalized tutoring. Although there are many related works on knowledge tracing, the existing methods still have some problems. For example, the tracing exercises are limited to a single historical exercise, the concept to be tested is limited to a single knowledge, and the semantic information on multiple concepts is rarely explored. To address this issue, a Diversified Concept Attention model for Knowledge Tracing (DCAKT) is proposed in this paper. Our method makes up for the shortcoming of the single concept of tracing, which applies natural language processing technology to get multiple concepts from the attention layer of emotional consciousness over time. Then we use an attention mechanism that evaluates the correlation between the exercises to be tested and historical concepts. This article uses the real datasets of the ASSISTments intelligent tutoring platform and a college engineering course on statics to assess the performance of DCAKT. In addition, our method can independently learn meaningful exercise sequences containing correct concepts.

Keywords: Personalized tutoring · Knowledge tracing · Diversified concept · Attention layer of emotional consciousness · Attention mechanism

1 Introduction

A large-scale online learning courses and intelligent learning systems have emerged on the Internet recently, learners obtain the relevant knowledge and appropriate guidance necessary to complete the exercises from a variety of ways. Faced with a new question, learners can guess the answer to the test exercises depended on a lot of concepts that has appeared in many exercises. For instance, the learner answers the exercise "$5 + 2 + 7$", and then the learner masters the concepts of decimal and addition. When encountering an exercise "$8 + 2$", learners can answer this exercise based on the concepts they have learned. In the learning process, how to systematically and accurately trace the knowledge state of learners has become very important. Knowledge tracing is an important link in the realization of personalized teaching. Therefore, the study of the knowledge forgetting and updating is particularly important.

Y. Sun et al. (Eds.): ChineseCSCW 2021, CCIS 1492, pp. 418–430, 2022.
https://doi.org/10.1007/978-981-19-4549-6_32

The knowledge tracing could predict the current learner's concept mastery status based on the learner's historical exercise performance. The problem of knowledge tracing is transformed into: the exercise label contains exercise key $\mathrm{K} = \{k_1, \mathrm{k}_2, \ldots, \mathrm{k}_{t-1}\}$ and knowledge concept $\mathrm{C} = \{\mathrm{c}_1, \mathrm{c}_2, \ldots, \mathrm{c}_{t-1}\}$. The interactive sequence of historical exercise answers and knowledge components are combined into a sequence containing knowledge state [2]. The current time is marked as t, and there are $t - 1$ historical times in the time axis $1, 2, \ldots\ldots, t - 1$. Record the sequence of historical exercise answers as $X = \{x_1, x_2, \ldots, x_{t-1}\}$, and the sequence of exercise concepts as $C = \{c_1, c_2, \ldots, c_{t-1}\}$. The tuple in the interaction sequence is defined as $x_\mathrm{t} = (e_\mathrm{t}, r_\mathrm{t}, \mathrm{t})$, where $e_\mathrm{t} \in \{1, 2, \ldots, E\}$ is the sequence of exercises [7]. The learner's answer vector to the exercise key k_t is denoted as r_t. Finally, the model predicts the probability that learners answer the exercises correctly, which is $\mathrm{p} = (r_t = 1, k_t|\mathrm{X})$. At present, rich knowledge tracing methods have provided a basic solution to the exercise of a single knowledge concept. However, this field also faces many challenges: Firstly, the knowledge concepts contained in the exercises exist in various situations. Secondly, the state of knowledge acquired by learners is dynamically changing [5]. Furthermore, how to deal with the changing, dynamic and different learning ability of learners in time is also a difficult problem. The development of knowledge tracing methods and related technologies has become important.

In order to overcome the above obstacles and deal with the exercises of multiple and complex concepts in all aspects, we transfer the problems of multiple fields to the newly emerging exercises. Use the transformed exercise concepts to detect the cognitive state of the test questions and predict the learner's knowledge mastery. This paper proposes a new method called Diversified Concept Attention Model for Knowledge Tracing (DCAKT). This method introduces the emotional awareness attention layer to capture the multiple knowledge concepts in multiple domain problems, and first obtains the data of several attention modules from the multiple concept attention modules. Then, in the diversified semantic module, the content of the context and the semantic relationship of the content are used to calculate the data of the context chunks. After that, the attention mechanism is used for calculating the weight of the concept distribution of exercise keys and multiply it with the vector of historical exercise answers to obtain the characteristic scores of the learners' answers to the test exercises. In order to trace related concepts of historical exercises, this paper proposes a new method of calculating attention scores. Finally, we apply four real datasets of the ASSISTments intelligent tutoring platforms and international university engineering courses in statics to assess performance of our algorithm.

The article is set up as follows: the related work of deep learning and multiple semantic attention algorithms is discussed in Sect. 2, and the DCAKT is present in Sect. 3, which including diversified concept module, diversified semantic module, and attention classification layer. In Sect. 4, the comparative experimental analysis of the DCAKT method is carried out. Finally, there are conclusions and prospects.

2 Related Work

According to the learner's historical performance, knowledge tracing infers mastery of the concept. Then it predicts the learner's answer to the newly emerging exercises, so as to provide corresponding data support for future learning tutoring [13]. In teaching, some assumptions are the premise of knowledge tracing. The exercises will contain some similar and unique concepts. One topic can transform to multiple concepts, etc. [4]. Due to the establishment of these conditions, researchers can better study the effect of knowledge on learners' performance in answering exercises. The general knowledge tracing model contains the two steps. Firstly, we get the concepts depended on deep learning. Secondly, we calculate the similarity between learning concepts based on the content of the exercise context.

2.1 Concept Extraction Method Based on Deep Learning

Existing deep learning concept extraction models such as Deep Knowledge Tracing (DKT) [3] and Dynamic Key-Value Memory Networks (DKVMN) [1].

Deep Knowledge Tracing uses recurrent neural networks for learning. Without the explicit coding knowledge, the ability to express more complex knowledge states can be obtained. DKT uses the Long Short-Term Memory [11]. For the interaction between the DKT model and students, the input is $x_t = \{q_t, a_t\}$. Where a_t is the answer result, q_t is the exercise query.

Dynamic Key-Value Memory Networks are composed of concept correlation and each traced concept state. When a new exercise q_t arises, the q_t needs to find the concepts c_t^i and c_t^j. Then we seek the corresponding knowledge states s_{t-1}^i and s_{t-1}^j to predict the mastery c_t^i and c_t^j under corresponding t timestamp, and the question answering situation. After students solve the problems, our model will update the two concepts. All concepts under this timestamp will be merged.

Deep learning [8] has become more and more popular. The nature of the traditional exercise document leads to the knowledge information with few contextual links in the description of exercises. This means that the concept extraction method based on deep learning does not work very well in the semantics. Such as "soccer player" and "football player" have the similar meaning, but the grammar does not associate together very well. Thus, how to conduct more knowledge mining on semantics has become a major challenge.

2.2 Diversified Semantic Algorithm

The diversified semantic algorithm divides all exercises into a number of semantic combinations through semantic division. For the exercises to be queried, the attention feature is calculated in each of the divided semantic combinations. After that, the conceptual information of all semantic combinations is integrated. At present, the scholars have done much research for multiple semantic algorithms, which can contain semantics and grammar.

Semantic-based attention method adopts an ontological approach to enhance semantic descriptiveness. Combining the way of sub-concept matching, the concept matching problem is transformed into a bipartite graph expansion matching problem. In the cases, the semantic-level attention algorithms are harmful to the effectiveness of knowledge tracing algorithms.

Except the above shortcomings, most of the existing research ignores the diversity of the characteristics of the exercises, which is very beneficial to obtain the difference of conceptual information.

According to the above analysis, we apply the mention-aware attention layer to capture diversified concepts in multiple domains. Then, in the diversified semantic module, the model uses the context content and content semantics to calculate the data of the context chunks. Finally, the attention mechanism is used to calculate the relationship of exercise concepts. This paper proposes an attention calculation method for exercises combined with multiple semantics to discover global historical concepts and individual nuances, which enhances entity type recognition. Our method has the global characteristics of the concept and the overall representation of exercise concept, focusing on the diverse and distracting characteristics of exercise descriptions. So as to achieve the expected effect of knowledge state calculation.

3 The DCAKT Method

The DCAKT method takes the learner's answering time at the entrance to the requirements. It analyzes the relationship between the statement descriptions of the exercises and uses natural language processing technology to extract the knowledge in the exercises needed for knowledge tracing. At the same time, the model extracts similar elements from the exercise description text and defines them as conceptual attributes, and then measures the attention score between the knowledge requirements for the exercises to be tested and the knowledge attributes contained in the learners' historical exercises. Finally, the knowledge of the exercises is sorted according to the attention score, and the concepts with large attention scores are selected and returned to the learner [14] for their choice. The following chapters are based on the extraction diversified conceptual attention needs and the integration of multiple semantic attention features. Figure 1 shows the framework of the knowledge tracing in the article. The model contains three parts:

(1) Information required for diversified based on the learners' historical answers.
(2) Diversified Semantic Module.
(3) Attention classification layer.

The data in the first part of the figure come from the ASSISTments intelligent guidance platform and the engineering statics courses of international universities.

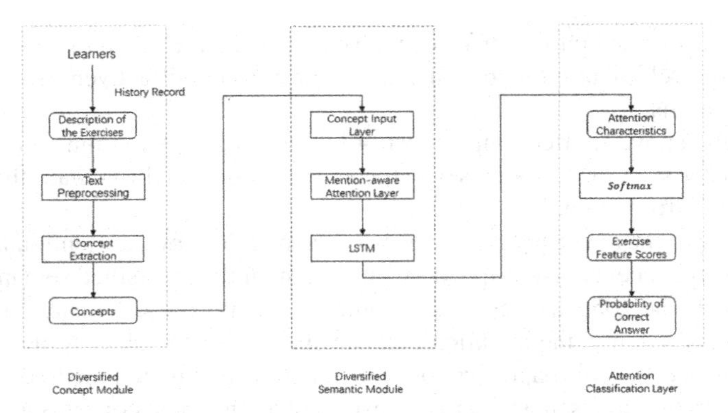

Fig. 1. A framework of diversified concept attention model for knowledge tracing

3.1 Concepts Required Knowledge Tracing

The concepts needed for knowledge tracing are stored in the text describing the exercises, which is the most direct and specific form that learners can obtain by reading the exercises. The concepts required for traditional knowledge tracing are described in natural language [10–12]. In the text, the elements required for multiple knowledge tracing are described based on the learners' historical answer sentences [13]. For example, sum-of-interior-angles-more-than-three-sides.

The knowledge element of history exercises refers to the key elements that can reflect the needs of knowledge concepts starting from the learners in the smart learning platform, including finding the role of the learner, the demand for solving the problem, and the expression of the calculation, as shown below:

The learner is the role of doing exercises. For example, "music player". The specific manifestations can be roughly divided into four types [11, 12]: single nouns, such as learner, etc.; nouns modify nouns, such as chess learner; adjectives modify nouns, such as good learner; mixed modified nouns, such as good chess learner. Obviously, the role is only composed of two parts of speech, noun and adjective.

Requirement to ask for knowledge. For example, "music in website". The specific manifestations can be roughly divided into: transitive verb + single noun, such as search music; intransitive verb + preposition + noun, such as: sit on a chair; there are also roles described in the role of adding adjectives or before nouns modifies the form of nouns; The specific scenario used is: preposition + noun, such as: in the website. It can be seen that the function of the question is composed of verbs, nouns, and adjectives, but due to the distinction between main functions and usage scenarios, the nouns in the function are refined into direct objects and indirect objects.

Operation is the mapping of the acquired new collection. For example, "five and seven". It is specifically related to form and knowledge.

3.2 Concept Extraction

The characteristics of the exercises are diversified. In order to better trace the learning process of learners, this article needs to acquire diversified concepts. According to the learner's exercise key k, the interactive sequence X of historical answers is divided into M attention segments, and each attention segment has multiple different concepts. In this way, the interactive sequence of answering the long historical exercises is divided into several small attention segments. The size of each attention segment is different, and the larger attention segment contains more knowledge and concept information [4]. In each attention segment, the knowledge concept c and the exercise key k are used for attention mapping, and the exercise key of the data unit is obtained. In this model, each embedding vector has a key, query and learner answer, we map the vector to a large-scale output key, query and value, the dimensions of the vector are D_k, D_q, D_v, where $D_k = D_q$. D_{h} is the dimension of the hidden state.

The attention segment provides a wealth of exercise description elements, which.

is conducive to achieve better multiple knowledge description. The multi-concept attention model can focus on more knowledge component information, we need to combine the attention score of each data unit in the attention module. Then the model obtains the output data of each attention module, and the calculation formula of the output data S_m of the $m-$th attention segment is:

$$S_m = \sum_{i=1}^{m^*} \alpha_i^m k^{m,i} \tag{1}$$

wherein m^* represents the total number of an attention segment, $k^{m,i} \in R^{D_k \times 1}$ is the exercise key in the m-th attention segment, $\alpha_i^m \in R^{D_v \times 1}$ represents the attention score of data unit of the m-th attention segment.

$$\alpha_i^m = \text{Softmax}(e_i^m) = \frac{exp(e_i^m)}{\sum_{j=1}^{m^*} exp(e_j^m)} \tag{2}$$

where e_i^m is the interaction between the weight matrix W^m of the m-th attention segment and the exercise key $k^{m,i}$ in the m-th attention segment, which is expressed as the following calculation formula:

$$e_i^m = \text{Tanh}(W^m k^{m,i}) \tag{3}$$

where $W^m \in R^{D_{\mathrm{h}} \times D_k}$ is the calculated weight matrix of key $k^{m,i}$.

3.3 Diversified Semantic Module

Existing attention methods mostly focus on the distinction of conceptual features of exercises, while ignoring the diversity of exercises in the process of solving the problems. At the same time, we find that when learning entities and knowledge representative features containing multiple semantic distinguishing words, the diversity of exercise concepts is very important. For solving the above problems, the diversified semantic module that includes an input layer, a mention-aware attention, and a long short-term memory network integration. The LSTM integration layer as shown in Fig. 2.

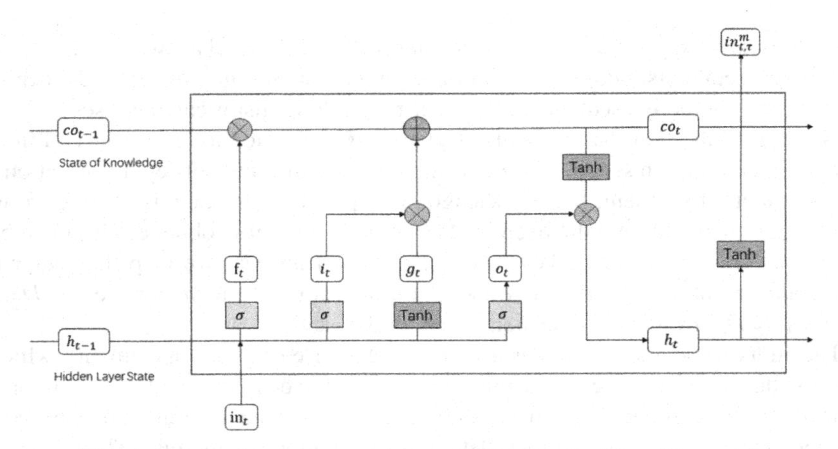

Fig. 2. A long shot-term memory network integration layer framework

3.4 Concept Input Layer

At each timestamp t, the output of each diversified concept segment $S_t = \{s_{t,1}, s_{t,2}, \ldots, s_{t,m}\} \in R^{D_h \times D_k}$ reaches the level of emotional awareness and attention. The $K_t = \{k_{t,1}, k_{t,2}, \ldots, k_{t,m}\} \in R^{D_k \times 1}$ obtained at the same time represents the passage of emotion in the diversified concept segment. The concept of contextualization after the mention-aware attention layer. Where $k_{t,i}$ is the contextualized concept at the time t.

3.5 Mention-Aware Attention Layer

In the layer, we pay more attention to knowledge-containing exercises [4]. We apply the content in the context and the semantic relationship of the content to calculate the context chunk data co_t. To better integrate multiple information, the context chunk data co_t, which is expressed as a calculation formula as follows:

$$co_t = \sum\nolimits_{i=1}^{N} \alpha_{t,i} k_{t,i} \tag{4}$$

where N represents the number of context chunks, $k_{t,i} \in R^{D_k \times 1}$ represents the i-th contextualized concept at t, $\alpha_{t,i\in} \in R^{D_v \times 1}$ is the attention score of the context chunk at t:

$$\alpha_{t,i} = \text{Softmax}(W_i^h h_{t-1} + W_i(k_{t,i} \oplus S_m)) = \frac{exp(w_i^h h_{t-1} + W_i(k_{t,i} \oplus S_m))}{\sum_{j=1}^{N} exp(w_j^h h_{t-1} + W_j(k_{t,j} \oplus S_m))} \tag{5}$$

where h_{t-1} is the hidden state of the LSTM, W_i^h is the weight vector connecting the h_{t-1} and the i-th attention score in the LSTM, W_i represent the weight for calculating an attention ability of $k_{t,i}$, and $\oplus$ represents the operation of matrix splicing in a row.

Finally, the attention feature vector in_t of the m-th attention module is formed by diversified concept segment $S_{t,m}$ of the multivariate concept segment at time t, and the context chunk data co_t in a matrix row splicing manner, The calculation formula is as follows:

$$int_t = co_t \oplus S_{t,m} \tag{6}$$

3.6 LSTM Integration Layer

The LSTM can handle events with a long time, so we apply it to integrate the conceptual features in historical exercises [8]. The in_t is the input unit in LSTM, and then the similarity between the context blocks is integrated in the LSTM. The LSTM in this paper is composed of 1 memory tuple, 1 activation function and 3 gating units. The gating unit is used to update the context chunk data co_t, so as to trace the learner's knowledge states, which is expressed as:

$$co_t = f_t \odot co_{t-1} + i_t \odot g_t \tag{7}$$

$$g_t = \mathrm{Tanh}(W^g \odot (h_{t-1} \oplus in_t)) \tag{8}$$

$$h_t = o_t \odot \mathrm{Tanh}(co_t) \tag{9}$$

where f_t is the forgetting gate, which controls the forgetting situation of the concept; i_t is the input gate, which controls the memory of the concept; h_t is the hidden state in LSTM. g_t represents the effective data saved in the attention feature vector in_t, which represents the concept retained by the mention-aware attention layer; W^g is the weight matrix of in_t of the LSTM and the attention feature vector in_t spliced by rows; o_t is the output gate, which controls the output of the effective concept of the memory tuple. The updated co_t is marked as $\overline{co_{t,\tau}}$ for the prediction of feature information on the diversified semantic attention model and the mapping of the next moment of attention. $\odot$ represents the basic productive operation of the matrix.

For the historical timestamp τ, splicing the updated $\overline{co_{t,\tau}}$ and $S_{t,m}$ to form the attention feature vector $in_{t,\tau}^m$ of the $m-$th attention segment, which is expressed as:

$$in_{t,\tau}^m = \overline{co_{t,\tau}} \oplus S_{t,m} \tag{10}$$

3.7 Attention Characteristics

This section mainly introduces in detail how to obtain the probability of the learner's correct answer to the exercises. Finally, in order to optimize the objective, this model adopts the cross entropy loss.

The attention layer proposed in this paper could evaluate the relative weights between the questions and the knowledge of historical acquisition. The input of the attention layer is the attention feature vector $in_{t,\tau}^m$ of the output of the LSTM integration layer. The attention feature vector $in_{t,\tau}^m$ needs to interact with the exercises to be tested at time t. In this article, the query q_t represents the information of the exercises to be tested.

Firstly, the historical attention feature vector $in^{m}_{t,\tau}$ and the query q_t are used as the inner product, and then taking the Softmax activation to obtain the relevant weights and store them in the vector $RW_{t,\tau}$, the vector $RW_{t,\tau}$ represents the relevant weights between the exercises to be tested and the multiple concepts contained in the exercises, the formula is as follows:

$$RW_{t,\tau} = Softmax(q_t^T in^{m}_{t,\tau}) \tag{11}$$

where $Softmax(y_i) = \frac{e^{y_i}}{\sum_j e^{y_j}}$.

3.8 Exercise Feature Scores

In the attention classification layer, multiply the relevant weight $RW_{t,\tau}$ between the diversified concepts [9] contained in the exercises to be tested and the historical exercises by the v_τ of the corresponding exercises for the current learner's capability:

$$scores = RW_{\mathrm{t},\tau} \times v_\tau \tag{12}$$

where $v_\tau \in R^{D_v \times 1}$ represents the embedded vector of the historical timestamp τ learner's answer.

3.9 Optimization Goal

Due to predict the performance of learners in answering the exercise x_t correctly. Our model uses a sigmoid [6] to predict students' capability to a series of exercises.

$$p_t = \sigma(w^T scores + b) \tag{13}$$

where $\sigma(x) = \frac{1}{e^{-x}}$ and the dimension of the calculation matrix w is $D_v \times D_v$, $p_t \in (0, 1)$ represents the probability that the learner answers the test questions correctly.

Where the diversified semantic attentive is trained, the cross entropy loss [16] could evaluate the model convergence, and the back-propagation is applied for the parameters converging. According to the learner's answer at time t, the probability of the correct answer to the current exercise uses the cross-entropy loss function as:

$$L = -\sum_t (r_t logp_t + (1 - r_t)log(1 - p_t)) \tag{14}$$

where r_t represents the real result of the answer. p_t represents the probability that the students answer the test exercise correctly.

4 Experiment

All experiments are implemented in Python. The models are applied in Windows10, AMD Ryzen 5 4600H CPU, 16G memory, 3.0 GHz and a single NVIDIA 1650 GPU. The Python version is 3.7.0. The python framework is mainly pytorch 1.5.1, numpy 1.16.6 and pandas 1.0.5.

The methods apply 4 public datasets: Statics2011, ASSISTments2009, ASSISTments2015 and ASSISTments2017. They are all collected from the answer data of learners in real teaching. It can be found in Table 1 for details.

Table 1. Dataset contents

Dataset	Learners	Exercises	Responses
ASSISTments2009	4,151	110	325,637
ASSISTments2015	19,840	100	683,801
ASSISTments2017	1,709	102	942,816
Statics2011	333	1,223	189,297

4.1 ASSISTments2009

This dataset is collected on the ASSISTment intelligent tutoring platform. Due to the duplication of tags in the published original dataset. Therefore, our experiment uses the updated "concept builder" dataset in the paper. In the preprocessing, the data without the knowledge name is discarded, while the tags that only appear once are retained. After processing, this dataset contains 4,151 learners and 110 exercises, with a total of 325,637 answers to the exercises.

4.2 ASSISTments2015

In the dataset, we deleted all the sequences whose "is Correct" field is neither 0 nor 1. Specifically, it is 683,801 problem-solving feedback messages from 19,840 learners to 100 exercises.

4.3 ASSISTments2017

The ASSISTments2017 dataset contains the learner's answer value $v \in \{0, 1\}$, and the total number of exercises answered is 942,816. At the same time, it collected 1,709 learners and 102 exercises. Each exercise contains one or more concepts. A few of learners provided the most feedback on answering exercises, which shows that the learners of the dataset interact well with the exercises, and the learners have the strongest willingness to answer the exercises.

4.4 Statics2011

This is a dataset widely used in knowledge research. It originated from the engineering statics courses of international universities. Specifically, this dataset contains 189,927 answer feedbacks, 333 learners and 1,223 exercises. We connect the problem name with the step as a concept. Therefore, it is rare for learners to answer the same exercise multiple times.

4.5 Evaluation Metric

This article uses the terms of Area Under Curve (AUC) to evaluate the performance of the model. AUC [2] is the area under the ROC curve enclosed by the coordinate axis. And the ROC curve is usually above the line $y = x$. The evaluation metric is 0.5 to 1. The excellent performance of a model is demonstrated by a large AUC.

4.6 Parameters of Knowledge Tracing Algorithm

In our experiment, we use the Adam optimizer [1], in which each batch size is set to 24 to ensure that each datum can be completely transmitted to our models. There will be no loss of time caused by too many transmission batches. The initial learning rate is 1×10^{-5}. A small initial learning rate shows that the learner's initial learning ability is weak, which is conducive to verifying the model's strong answer prediction ability to the new exercises. The initial dimensions of the exercise key, the embedding query and the learner's answer are all set to 50. In the initial state, we embed each variable into more vectors, and then pass them into the model. We set the dropout rate in the network to 0.05 to reduce overfitting. It is obtained empirically during our experiment. In some classical knowledge tracing algorithms using neural networks, the hidden layers are all set to 512, so we also choose the same number to facilitate algorithm comparison. In all baseline methods, we apply 300 epochs as the maximum number of learning iterations for experiment [2]. In this way, we can obtain the best performance of each method.

4.7 Baseline Methods

This section compares the DCAKT with DKVMN and AKT to verify the effectiveness of the DCAKT.

DKVMN uses two storage forms: one is to store concepts with keys, and the other uses a dynamic matrix to represent concepts that need to be updated. This method digs out the mastery of the concept $M_t^v(i)$, the formula as follows:

$$e_t = \text{Sigmoid}(E^T v_t + b_e) \tag{15}$$

$$\widetilde{M_{t-1}^{v}(i)} = M_{t-1}^{v}(i)[1 - \omega_t(i)e_t] \tag{16}$$

$$M_t^v(i) = \widetilde{M_{t-1}^{v}(i)} + \omega_t(i)\alpha_t \tag{17}$$

where the shape of matrix E is $D_v \times D_v$, v_t is the original mastery, and b_e is the conversion deviation. $\widetilde{M_{t-1}^{v}(i)}$is the modification amount of the concept mastery at the $t - 1$, and $M_t^v(i)$ is the concept mastery at the t. $\omega_t(i)$ is the correlation matrix, α_t is used for updating in memory.

AKT uses the attention mechanism to evaluate learners' answers to exercises. The method obtains the key, query and value embedding layer of the exercises from the encoder and knowledge retriever. Then the model adds a time decay variable to calculate the learner's mastery of the test exercises, the formula as follows:

$$s_{t,\tau} = \frac{exp(-\phi \times c(t,\tau)) \times q_t^T k_{t-\tau}}{\sqrt{D_k}} \tag{18}$$

$$\alpha_{t,\tau} = Softmax(s_{t,\tau}) = \frac{e^{s_{t,\tau}}}{\sum_{\alpha_{t,\tau}} e^{s_{t,\tau}}} \tag{19}$$

where $k_{t-\tau} \in R^{D_k \times 1}$ represents the exercise key of historical time $t - \tau$, $q_t \in R^{D_k \times 1}$ represents the query for exercises at t, ϕ represents the attenuation parameter, $c(t, \tau)$ represents the time distance between t and $t - \tau$.

In all methods, we apply 60% of the data for training the model, 20% for the validation dataset, and the remaining 20% for the test. The evaluation results of the four knowledge tracing methods are as follows (Table 2):

Table 2. Testing AUC of the comparison methods

Datasets	Metric	DKVMN	AKT	DCAKT
ASSISTments2009	AUC	0.8060	0.8154	0.8188
ASSISTments2015	AUC	0.7271	0.7652	0.8890
ASSISTments2017	AUC	0.7110	0.7238	0.8150
Statics2011	AUC	0.8103	0.8221	0.8433

Figure 3 shows the AUC of each knowledge tracing method. As shown in the figure, DCAKT's AUC is the best one on all datasets. The AUC of the DCAKT method reaches 0.889 on the Assistments2015. On the Assistants2017, our model'AUC is 0.815, which is better than the DKVMN and AKT methods. DCAKT achieved an AUC in 0.8433 from the Statics2011, which is more advantageous than DKVMN and AKT.

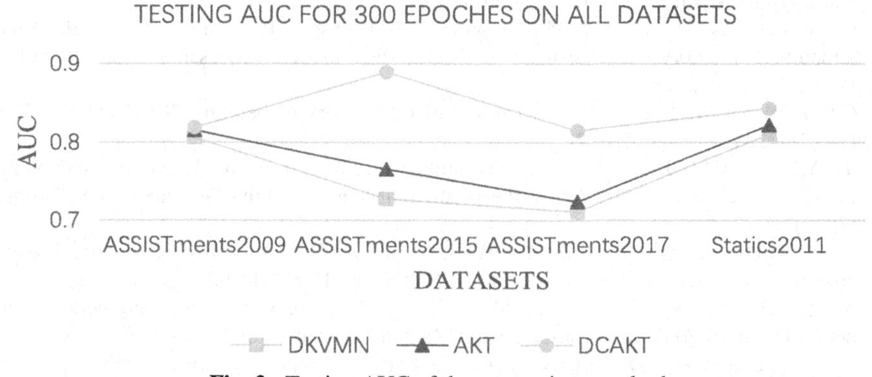

Fig. 3. Testing AUC of the comparison methods

5 Conclusions and Prospects

In this article, we reveal a new DCAKT method to deal with knowledge tracing issues, which applies natural language processing methods to get the context of exercise concepts. At the same time, we set up a mention-aware attention layer to extract multiple semantics and knowledge, and then evaluate the similarity between the historical exercises the learners have answered and the exercises to be tested. Then the model calculates the characteristic scores of the answers to the exercises. The experimental data all come from the ASSISTments intelligent tutoring platform and the engineering statics courses of international universities. Extensive experiments have verified our method. Experiments demonstrate that our model has excellent performance over other classic for classification and prediction.

For future tasks, combining time variables, we will evaluate whether our method can reduce the impact of predictive changes brought about by memory decline. We will research whether there is a dynamic relationship between the feature information on the multiple semantic method and the concept mastery state.

References

1. Zhang, J.N., Shi, X.J., King, I., Yeung, D.Y.: Dynamic key-value memory networks for knowledge tracing. In: Proceedings of the International Conference on World Wide (2017). https://arxiv.org/abs/1611.08108
2. Ghosh, A., Heffernan, N., Lan, A.S.: Context-aware attentive knowledge tracing. In: Proceedings of the 26th ACM SIGKDD Conference on Knowledge Discovery and Data Mining USB Stick (KDD 2020), pp. 2330–2339 (2020)
3. Piech, C., et al.: Deep knowledge tracing. In: Proceedings of the Conference on Advances in Neural Information Processing Systems, pp. 505–513 (2015). https://arxiv.org/abs/1506.05908
4. Hu, Y.F., Qiao, X., Luo, X., Peng, C.: Diversified semantic attention model for fine-grained entity typing. IEEE Access **9**, 2251–2265 (2020)
5. Grefenstette, E., Hermann, K.M., Suleyman, M., Blunsom, P.: Learning to transduce with unbounded memory. In: Advances in Neural Information Processing Systems, vol. 2, pp. 1828–1836 (2015)
6. Cui, P., Wang, X., Pei, J., Zhu, W.: A survey on network embedding. IEEE Trans. Knowl. Data Eng. **31**, 833–852 (2017)
7. Ye, Y.W., Li, F.M., Liu, Q.Q., et al.: The influence of forgetting and datavolume factors into the knowledge tracking model on the predictionaccuracy. China Distance Educ. **8**, 20–26 (2019)
8. Chen, Y., Dai, X., Liu, M., Chen, D., Yuan, L., Liu, Z.: Dynamicconvolution: attention over convolution kernels. In: Proceedings of the CVPR, pp. 11030–11039 (2020)
9. Santoro, A., Bartunov, S., Botvinick, M., et al.: Meta-learning with memory-augmented neural networks. In: International Conference on Machine Learning, pp. 1842–1850 (2016)
10. Thai-Nghe, N., Lucas, D., Artus, K.G., Lars, S.T.: Recommender system for predicting student performance. Procedia Comput. Sci. **1**, 2811–2819 (2010)
11. Liu, D., Yuan, Y., Zhu, H., Teng, S., Huang, C.: Balance preferences with performance in group role assignment. IEEE Trans. cybern (6), 1800–1813 (2017). https://ieeexplore.ieee.org/document/8012442
12. Aureli, S., Giampaoli, D., Ciambotti, M., et al.: Key factors that improve knowledgeintensive business processes which lead to competitive advantage. Bus. Process Manag. **1**, 126–143 (2019)
13. Tan, Z., Wang, M., Xie, J., Chen, Y., Shi, X.: Deep semantic role labeling with self-attention. In: Thirty-Second AAAI Conference on Artificial Intelligence (2018)
14. Zhang, L., Zhan, J., Xu, Z.: Covering-based generalized IF rough sets with applications to multi-attribute decision-making. Inf. Sci. **478**, 275–302 (2019)

Self-auxiliary Hashing for Unsupervised Cross Modal Retrieval

Jingnan Xu, Tieying Li, Chong Xi, and Xiaochun Yang(✉)

Northeastern University, Shenyang 110169, China
{1901816,1910624}@stumail.neu.edu.cn, yangxc@mail.neu.edu.cn

Abstract. Recently, cross modality hashing has attracted significant attention for large scale cross-modal retrieval owing to its low storage overhead and fast retrieval speed. However, heterogeneous gap still exist between different modalities. Supervised methods always need additional information, such as labels, to supervise the learning of hash codes, while it is laborious to obtain these information in daily life. In this paper, we propose a novel self-auxiliary hashing for unsupervised cross modal retrieval (SAH), which makes sufficient use of image and text data. SAH uses multi-scale features of pairwise image-text data and fuses them with the uniform feature to facilitate the preservation of intra-modal semantic, which is generated from Alexnet and MLP. Multi-scale feature similarity matrices of intra-modality preserve semantic information better. For inter-modality, the accuracy of the generated hash codes is guaranteed by the collaboration of multiple inter-modal similarity matrices, which are calculated by uniform features of both modalities. Extensive experiments carried out on two benchmark datasets show the competitive performance of our SAH than the baselines.

Keywords: Cross-modal retrieval · Multi-scale fusion · Cross-modal hashing

1 Introduction

With the development of science and technology, more and more multimedia data, such as images and texts, appear on the Internet. Owing to the explosive increase of these data, the requirement of cross-modal retrieval increases sharply. Cross-modal retrieval aims to search semantically related images (texts) with text (image) query and vice versa. Image retrieval hashing is a long-established research task to retrieve images with similar contents [17], it is common for us to process images with VGG [19] or some other neural networks. For text, Word2Vec technology is widely used, which also try to exploit latent semantic [23]. One of the biggest challenges of cross-modal retrieval is how to bridge the heterogeneous gap between two different modalities. The cause of the heterogeneous gap is the difference distribution between the feature from different modalities. Data from intra-modality also have heterogeneous information, which can be tackled from multiple views [5]. To tackle the problem of the heterogeneous gap between modalities, many cross-modal hashing methods have been

Y. Sun et al. (Eds.): ChineseCSCW 2021, CCIS 1492, pp. 431–443, 2022.
https://doi.org/10.1007/978-981-19-4549-6_33

proposed because of the advantages of low storage cost and high query speed by mapping data into binary codes.

The development of cross-modal retrieval can be divided into two phases: shallow cross-modal hashing and deep learning-based cross-modal hashing. Shallow cross-modal hashing is based on hand-crafted features and learns the hash codes by linear functions. The advantage of these methods is easily implemented, while they cannot fully explore the semantic information of two modalities. Recently, with the development of deep learning, the deep neural network(DNN) has been deployed to cross-modal hashing. DNN-based cross-modal hashing can be divided into two categories: supervised hashing and unsupervised hashing. Supervised methods, with label information, such as tags, always perform remarkably. While in the real-life, it is a waste of time for us to obtain labels of the image-text pairs. Unsupervised methods that do not use label information in the training phase have also shown remarkable performance in recent years. Unsupervised methods focus more on the information of raw features. As a result, the quality of the hash codes that used in retrieval task is dramatically concerned with the feature learning stage.

However, there are still some issues that should be tackled. Firstly, at the feature extraction phase, these methods only focus on the single source feature, neglecting the rich semantic information gained from multiple views. Secondly, the general similarity matrix of features can not bridge the heterogeneous gap well, because the distribution information or similarities of different scales are not considered. In this paper, we propose a novel self-auxiliary hashing (SAH) method for unsupervised cross-modal retrieval. SAH provides a two-branch network for each modality, including the uniform branch and the auxiliary branch. Each branch will generate specific features and hash codes. Moreover, based on the features and hash codes of two branches, we construct multiple similarity matrices for inter-modality and intra-modality. These similarity matrices will be calculated to preserve more semantic and similarity information. Extensive experiments demonstrate the superior performance of our method.

2 Related Work

Cross modality hashing can be roughly divided into supervised cross modality hashing and unsupervised cross modality hashing. The task of cross modality retrieval is to retrieve images (or texts) with similar semantics to the input text (or image). Shallow cross-modal hashing methods [12,13,15,16] and deep cross-modal hashing methods [1,2,11,22] are two stages of cross-modal hashing methods development. Shallow Cross-Modal Hashing uses hand-crafted features to learn the binary vector projection which is mapped from instances. However, most shallow cross-modal hashing retrieval methods just deal the feature with only a single layer and map data in a linear or nonlinear way. In recent years, the deep learning algorithm proposed in machine learning has been applied to cross modality retrieval. Deep cross-modal retrieval [18] also can be divided into unsupervised methods and supervised methods.

Supervised hashing methods [7,10,18,22] explore relative information, such as semantic information, or some other relative information by labels or tags, to enhance the ability of cross modality retrieval. Deep cross-modal hashing (DCMH) [7] is an end-to-end hashing method with deep neural networks, which can jointly learn hash codes and feature. In deep cross-modal hashing methods, generative adversarial network (GAN) is used to make adversarial learning. Self-Supervised Adversarial Hashing Networks (SSAH) [10] and Wang *et al.* [22] use image and text adversarial networks to generate hashing codes of both modalities, the learned features are used to keep the semantic relevance and preserve the semantic of different modalities.

Although some supervised methods perform well in practical applications, supervised information, such as label, is hard for us to collect, which is not suitable in reality.

Unsupervised hashing methods aim to learn hashing functions without supervised information, such as labeled data. For example, inter-media hashing (IMH) [20] considers the inter-media consistency and intra-media consistency with linear hash functions, and learns the hash function of image modality and text modality jointly. CVH [9] proposes a principled method to learn a hash function of different modality instances. Collective Matrix Factorization Hashing (CMFH) [4] learns the hash codes of an instance from two modalities and proposes the upper and lower bound. Latent Semantic Sparse Hashing (LSSH) [26] copes the instances of image and text with different methods and performs search by Sparse Coding and Matrix Factorization. Unsupervised Deep Cross-Modal Hashing (UDCMH) [24] makes a combination of deep learning and matrix factorization, considering the neighbour information and the weight assignment of optimization stage. Deep joint semantics reconstructing hashing (DJSRH) [21] considers the neighborhood information of different modalities.

Although the performance of these methods are remarkable, the features they focus on are not comprehensive. Moreover, they neglect the deep similarity information of two modalities and have bad performance at bridging the "heterogeneity gap".

3 Proposed Method

3.1 Problem Fomulation

Assume the training dataset of our methods is a collection of the pairwise image-text instances, written as $O = (X, Y)$. X is the instance of image modality and Y is the text modal instance. The number of instances of each modality is n. The goal of our method is to learn the modality-specific hash function for image modality and text modality which can generate hash codes with rich semantic information. For each modality, two branches (uniform branch and modality-specific auxiliary branch) are used to generate different features for each modality. MF_{*_i} are the i-th multi-scale features of image or text modality, which generate from the auxiliary branch with different dimensions. MH_{*_i} denotes the i-th multi-scale hash code of image or text which is generated from

MF_{*_i}. F_* and H_* are the feature and hash code gained from the uniform branch which is same as other unsupervised methods. The notations used in SAH are summarized in Table 1.

Table 1. Notations and their descriptions.

Notations	Descriptions	Modality
$S^F_{x,y}$	Similarity of uniform feature	$(x, y) \in \{(I, I), (T, T), (I, T)\}$
$S^H_{x,y}$	Similarity of uniform hash code	$(x, y) \in \{(I, I), (T, T), (I, T)\}$
$S^{CH}_{x,y}$	Similarity of complex hash code	$(x, y) \in \{(I, T)\}$
$S^{MH}_{x,y}$	Similarity of multi-scale hash code	$(x, y) \in \{(I, I), (T, T)\}$
F_*	Uniform feature of image or text	$* \in \{I, T\}$
MF_{*_i}	The multi-scale feature of image or text	$* \in \{I, T\}$
H_*	Uniform hash code of image or text	$* \in \{I, T\}$
H_{*_mix}	The mix hash code of image or text	$* \in \{img, txt\}$
MH_{*_i}	The i-th multi-scale hash code of image or text	$* \in \{I, T\}$
MH_{*_com}	The comprehensive hash code of image or text	$* \in \{I, T\}$
CH_*	The complex hash code of image or text	$* \in \{I, T\}$

3.2 Network Architecture

Figure 1 is a flowchart of our SAH. Our method is composed of two networks, image network and text network, both of them can be divided into the uniform branch and the modality-specific auxiliary branch. For image network, the uniform branch is composed of AlexNet [8]. Image auxiliary branch, shown by Fig. 2, deals the input image with a fully connected layer and gains the auxiliary data. For text network, the uniform branch consists of MLP. Text auxiliary branch, drawn in Fig. 3, tackles the input text data with a pooling layer first and gets the auxiliary data which is convenient for the later procession.

Feature Extraction. For image modality, we adopt the pre-trained AlexNet as the uniform feature extractor which is widely used in unsupervised methods. However, features gained from AlexNet are not comprehensive enough, which is the common disadvantage of previous works. Features obtained from a single scale often comes from the same measurement perspective, ignoring the details that may be obtained from other perspectives. Benefit from feature learning at multiple scales, multi-scale features can better represent the semantics of instances.

To gain multi-scale features, we process the input of image modality by fully connected layer and three pooling layers respectively. Then we get three sizes of image feature which we called the auxiliary data. And we resize them into the same size. These three auxiliary data are single-channel, we make them expand to three channels. To tackle with the auxiliary data of image modality, we deal

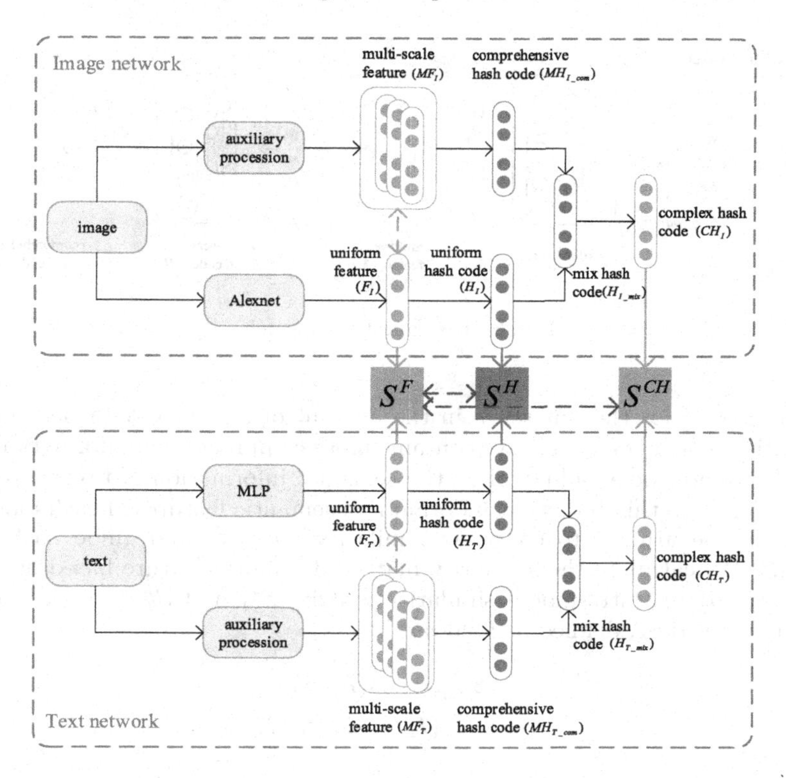

Fig. 1. The overview of our proposed SAH.

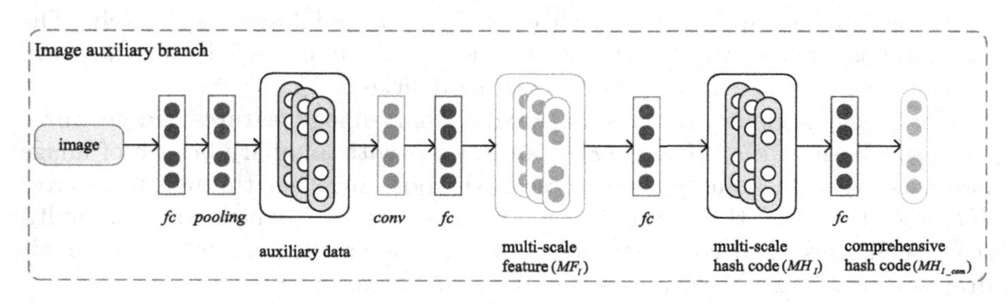

Fig. 2. Image auxiliary branch.

them with five convolution layer and three fully connected layer networks and obtain three multi-scale features MF_{I_1}, MF_{I_2} and MF_{I_3}. For text modality, we set four pooling layers to tackle with the input text data respectively and get four different size auxiliary data. Due to the character of text data is sparse, we deal these four data just with a fully connected layer and get four multi-scale features MF_{T_1}, MF_{T_2}, MF_{T_3} and MF_{T_4} of text modality.

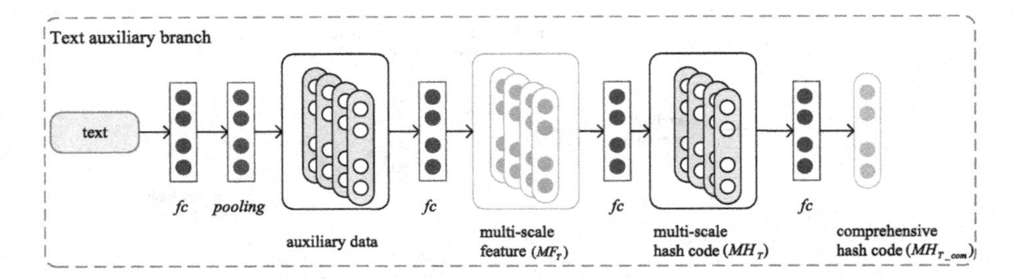

Fig. 3. Text auxiliary branch.

The reason for the difference in the amount of auxiliary data between two modalities is that, image always contains more comprehensive information than text. Therefore, we should explore the semantic information of text more comprehensively. To this end, we can obtain rich semantic features of each modality, which can be utilized to construct similarity matrices and guide hash codes learning. We calculate the similarity matrix of uniform feature based on cosine similarity. S_{IT}^{F} is intra-modal similarity matrix, S_{II}^{F} and S_{TT}^{F} are inter-modal similarity matrices, defined as follows:

$$\begin{aligned} S_{x,y}^{F} &= cos(F_x, F_y), \\ s.t.(x,y) &\in (I,I),(T,T),(I,T). \end{aligned} \tag{1}$$

Hash Code Generation. We will generate two kinds of hash code, uniform hash code and comprehensive hash code for two modalities respectively. The uniform hash codes (H_I and H_T) is obtained by the uniform feature in uniform branch with a simple hash layer for each modalities.

For image modality, we process three auxiliary image features and get three same size hash codes MH_{I_1}, MH_{I_2}, and MH_{I_3} with auxiliary branch of image modality. We concatenate these three hash codes together through hash layer $HILayer$ to obtain the comprehensive hash code H_{I_com} which contains multi-scale semantics. The concatenation will not change the semantic of each bit dramatically, it can be seen as a way of data enhancement.

$$H_{I\_com} = HILayer(ConCat(MH_{I_1}, MH_{I_2}, MH_{I_3})), \tag{2}$$

where $ConCat()$ denotes the concatenation of vectors.

For text modality, we have four auxiliary features, and we process them with four different hash layer and get four same size auxiliary hash codes, MH_{T_1}, MH_{T_2}, MH_{T_3}, MH_{T_4}. We also concatenate them four and make this hash code into a hash layer $HTLayer$ and get the comprehensive hash code H_{T_com}.

$$H_{T\_com} = HTLayer(ConCat(MH_{T_1}, MH_{T_2}, MH_{T_3}, MH_{T_4})). \tag{3}$$

Concatenation is a compression of semantic information which can preserve different scales of semantic information and similarity. Furthermore, we fuse the uniform hash code and the comprehensive hash code according to a certain proportion μ $(0 < \mu < 1)$ and get the mixed hash code. The mixed hash code can maintain more semantic information than the uniform hash code of each modality.

$$H_{img\_mix} = \mu M H_{I\_com} + (1 - \mu) H_I. \tag{4}$$

$$H_{txt\_mix} = \mu M H_{t\_com} + (1 - \mu) H_T. \tag{5}$$

The inter-similarity matrix of uniform hash codes can be calculated by similarity function:

$$\begin{gathered} S^H_{x,y} = cos(H_x, H_y), \\ s.t.(x, y) \in (I, I), (T, T), (I, T). \end{gathered} \tag{6}$$

Similarity Matrices Learning. Since dimension reduction during the procession from features to hash codes will cause some semantic lose, we aim to keep the semantic consistency of instance pairs. To this end, we introduce a loss function that can measure the semantic consistency between hash codes and features of intra-modality and inter-modality. The loss function L_1 can be written as follows:

$$L_1 = \sum_{i=1}^{n} \sum_{j=1}^{n} \| S^H_{x,y}(i,j) - S^F_{x,y}(i,j) \| . \tag{7}$$

For intra-modality, the multi-scale hash codes and the uniform hash codes are generated from features of different scale, they preserve richer semantic information of different view. Uniform feature similarity matrix S^F offers us the degree of similarity among different instances in a single modality. Loss function L_1 makes the multi-scale hash codes retains the semantic consistency, too. To ensure the accuracy of hash codes, the similarity matrix of hash codes should approximate to the feature similarity matrix. Therefore, we can minimize the distance between the similarity of the multi-scale hash codes of each modality and its intra-modality feature similarity. The loss function L_2 can be written as follows:

$$L_2 = \sum_{ni=1}^{3} \| S^{MH}_{I_{ni}} - S^F_{I,I} \| + \sum_{mi=1}^{4} \| S^{MH}_{T_{mi}} - S^F_{T,T} \| . \tag{8}$$

For inter-modality, the similarity matrix of features should also contains the inherent pair-wise information. The feature similarity of pair-wise instance of different modalities can be seen as converging to the maximum value in the cosine similarity. Apart from that, complex hash code will generate from the mix hash code to make sure the mixed hash codes still retain similarity consistency. To this end, loss function L_3 and L_4 can be written as:

$$L_3 = \sum_{i=1}^{n} | S^{CH}_{I,T} - E | + \sum_{i=1}^{n} | S^{CH}_{I,T} - S^F_{I,T} | . \tag{9}$$

$$L_4 = \sum_{i=1}^{n} \mid S_{I,T}^{F} - E \mid . \tag{10}$$

where E is an identity matrix.

3.3 Optimization

As mentioned above, the final loss function can be written as follows:

$$min\ \alpha L_1 + \beta L_2 + \gamma L_3 + \delta L_4.$$

The goal of our method is to generate hash codes, a kind of discrete data. The optimization of our objective function should satisfy the discrete condition. The sign function can map the input into -1 or 1. The gradient of this function is zero for all non-zero inputs, and may cause gradient explosion in backpropagation:

$$\lim_{\delta \to \infty} tanh(\eta x) = sgn(x). \tag{11}$$

where η is a hyper-parameter and will rise during network training.

With sign function as the activation function, the network will finally converge to our hash layer by changing the problem into a sequence of smoothed optimization problems.

4 Experiment

4.1 Datasets

MIRFlickr25k [6] contains 25,000 image-text pairs collected from the image website Flickr. The image-text pairs are labeled from 24 categories. All the images are denoted as SIFT feature. We use BoW vector to form the text tags with 1386 dimensions.

NUS-WIDE [3] consists of 269,648 pairs of images and texts. There are 81 label categories in the dataset, but we only used the top 10 most frequent categories, resulting in a total of 186,577 image-text pairs that can be used. The setup for this dataset is the same as the other methods. We use BoW vector to form the text tags with 500 dimensions.

4.2 Baselines and Evaluaton

We compare our SAH with 6 baseline methods, including CVH [9], IMF [20], CMFH [4], LSSH [25], UDCMH [24], and DJSRH [21].

Evaluation Criterion. Mean Average Precision (mAP) [14] and the top-K precision curves are used to evaluate the performance of the proposed SAH and baselines. Two instances of different modalities are considered semantically similar if they have the same label.

4.3 Implementation Details

The network of each modality is composed of the uniform branch and the auxiliary branch. For image modality, our uniform branch is composed of AlexNet which is same with UDCMH [24] for the sake of fairness. The auxiliary branch deal with the input image data and get three scale auxiliary data of image, 1024 $\times$ 1024, 512 $\times$ 512, 256 $\times$ 256, respectively. For text modality, MLP is uniform branch. In auxiliary branch, the lengths of text auxiliary data in four scale are 1024, 512, 256 and 128, respectively. For hyper-parameter, we set $\alpha = 1$, $\beta = 0.1$, $\gamma = 1$, $\delta = 1$, and $\mu = 0.1$ to achieve best performance. We implement our method by PyTorch on the NVIDIA RTX 1660Ti. We fix batch size as 32 and the learning rate for image network and text network is 0.005. During the optimization phase, we employ mini-batch optimizer to optimize our networks of two modalities.

4.4 Comparison with Existing Methods

Results on MIRFlickr25k. Table 2 shows the MAP@50 on MIRFlickr25K dataset of our proposed SAH and other previous methods. As can be seen, the proposed SAH significantly outperforms the baselines. We show the curve of 128 bits length hash code and can easily find that our SAH has the best performance. For the I$\rightarrow$T retrieval, we get more than 50% improvement in MAP in 128 bits compared with CVH. Compare with the latest method DJSRH, we get 3.1% enhancement in 128 bits. For the T$\rightarrow$I retrieval, we also achieve the superior performance compare with methods. The difference value of I$\rightarrow$T and T$\rightarrow$I has a shrink than any other works, which means that the auxiliary data bridges the heterogeneous gap (Fig. 4).

Results on NUS-WIDE. Table 3 also shows the MAP@50 on NUS-WIDE dataset of six methods, which shows that our SAH performs better than other methods. It can be seen that we get the best performance on four kinds of code length for two datasets, which means that our method is effective for cross modality retrieval. The results indicate that the auxiliary data of both modalities could mine more latent information in both modalities and remain the similarity consistency.

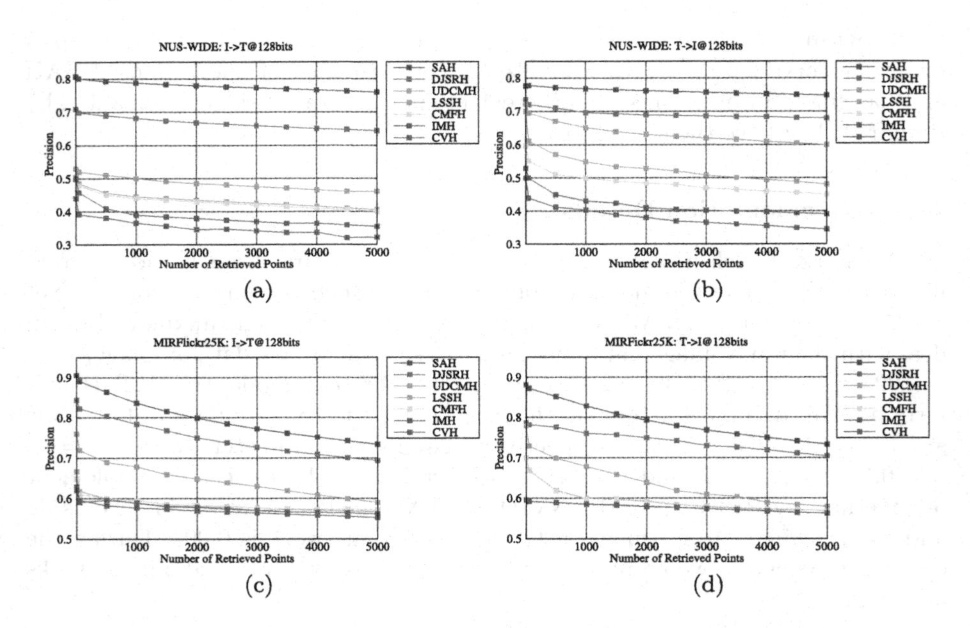

Fig. 4. Precision@top-K curves on two datasets at 128 bits.

Table 2. Mean average precision (MAP@50) comparison results.

Task	Method	MIRFlickr25K				NUS-WIDE			
		16 bit	32 bit	64 bit	128 bit	16 bit	32 bit	64 bit	128 bit
I→T	CVH	0.606	0.599	0.596	0.598	0.372	0.362	0.406	0.390
	IMH	0.612	0.601	0.592	0.579	0.470	0.473	0.476	0.459
	CMFH	0.642	0.662	0.676	0.685	0.529	0.577	0.614	0.645
	LSSH	0.584	0.599	0.602	0.614	0.481	0.489	0.507	0.507
	UDCMH	0.689	0.698	0.714	0.717	0.511	0.519	0.524	0.558
	DJRSH	0.810	0.843	0.862	0.876	0.724	0.773	0.798	0.817
	OURS	**0.852**	**0.879**	**0.889**	**0.903**	**0.753**	**0.779**	**0.804**	**0.818**
Task	Method	MIRFlickr25K				NUS-WIDE			
		16 bit	32 bit	64 bit	128 bit	16 bit	32 bit	64 bit	128 bit
T→I	CVH	0.591	0.583	0.576	0.576	0.401	0.384	0.442	0.432
	IMH	0.603	0.595	0.589	0.580	0.478	0.483	0.472	0.462
	CMFH	0.642	0.662	0.676	0.685	0.529	0.577	0.614	0.645
	LSSH	0.584	0.599	0.602	0.614	0.455	0.459	0.468	0.473
	UDCMH	0.692	0.704	0. 718	0.733	0.637	0.653	0.695	0.716
	DJRSH	0.786	0.822	0.835	0.847	0.712	0.744	0.771	0.789
	OURS	**0.852**	**0.864**	**0.878**	**0.885**	**0.765**	**0.772**	**0.786**	**0.791**

4.5 Ablation Study

We verify our method with 3 variants as diverse baselines of SAH:

(a) SAH-1 is built by removing the intra-modality multi-scale hash codes semantic enhancement;
(b) SAH-2 is built by removing the similarity matrices difference between uniform hash codes and uniform features;
(c) SAH-3 is built by removing the consistency between complex hash codes similarity and uniform features similarity.

Table 3 shows the results on MIRFlickr25K dataset with 64 bits and 128 bits. From the results, we can observe that each part is important to our method. Especially the part of similarity consistency between features and complex hash codes, which ensures the semantic consistency.

Table 3. The mAP@50 results for ablation analysis on MIRFlickr25k.

Method	64 bits		128 bits	
	I→T	T→I	I→T	T→I
SAH	0.889	0.878	0.903	0.885
SAH-1	0.881	0.869	0.885	0.883
SAH-2	0.877	0.863	0.898	0.880
SAH-3	0.855	0.849	0.863	0.834

5 Conclusion

In this paper, we propose a novel unsupervised deep hashing model named self-auxiliary hashing. We propose a two-branch network for each modality, mixing the uniform hash codes and the comprehensive hash codes, which can preserve richer semantic information and bridge the gap of different modalities. Moreover, we make a full use of inter-modality similarity matrices and the multi-scale intra-modality similarity matrices to learn the similarity information. Extensive experiments conducted on two datasets show that our SAH outperforms several baseline methods for cross modality retrieval.

References

1. Cao, Y., Long, M., Wang, J., Yang, Q., Yu, P.S.: Deep visual-semantic hashing for cross-modal retrieval. In: Proceedings of the 22nd ACM SIGKDD International Conference on Knowledge Discovery and Data Mining, San Francisco, CA, USA, 13–17 August 2016, pp. 1445–1454. ACM (2016). https://doi.org/10.1145/2939672.2939812

2. Cao, Y., Long, M., Wang, J., Zhu, H.: Correlation autoencoder hashing for supervised cross-modal search. In: Proceedings of the 2016 ACM on International Conference on Multimedia Retrieval, ICMR 2016, New York, New York, USA, 6–9 June 2016, pp. 197–204. ACM (2016). https://doi.org/10.1145/2911996.2912000
3. Chua, T., Tang, J., Hong, R., Li, H., Luo, Z., Zheng, Y.: NUS-WIDE: a real-world web image database from national university of Singapore. In: Proceedings of the 8th ACM International Conference on Image and Video Retrieval, CIVR 2009, Santorini Island, Greece, 8–10 July 2009. ACM (2009)
4. Ding, G., Guo, Y., Zhou, J.: Collective matrix factorization hashing for multimodal data. In: 2014 IEEE Conference on Computer Vision and Pattern Recognition, CVPR 2014, Columbus, OH, USA, 23–28 June 2014, pp. 2083–2090. IEEE Computer Society (2014)
5. Du, G., Zhou, L., Yang, Y., Lü, K., Wang, L.: Deep multiple auto-encoder-based multi-view clustering. Data Sci. Eng. **6**(3), 323–338 (2021)
6. Huiskes, M.J., Lew, M.S.: The MIR flickr retrieval evaluation. In: Proceedings of the 1st ACM SIGMM International Conference on Multimedia Information Retrieval, MIR 2008, Vancouver, British Columbia, Canada, 30–31 October 2008, pp. 39–43. ACM (2008)
7. Jiang, Q., Li, W.: Deep cross-modal hashing. In: 2017 IEEE Conference on Computer Vision and Pattern Recognition, CVPR 2017, Honolulu, HI, USA, 21–26 July 2017, pp. 3270–3278. IEEE Computer Society (2017)
8. Krizhevsky, A., Sutskever, I., Hinton, G.E.: ImageNet classification with deep convolutional neural networks. Commun. ACM **60**(6), 84–90 (2017)
9. Kumar, S., Udupa, R.: Learning hash functions for cross-view similarity search. In: Walsh, T. (ed.) IJCAI 2011, Proceedings of the 22nd International Joint Conference on Artificial Intelligence, Barcelona, Catalonia, Spain, 16–22 July 2011, pp. 1360–1365. IJCAI/AAAI (2011)
10. Li, C., Deng, C., Li, N., Liu, W., Gao, X., Tao, D.: Self-supervised adversarial hashing networks for cross-modal retrieval. In: 2018 IEEE Conference on Computer Vision and Pattern Recognition, CVPR 2018, Salt Lake City, UT, USA, 18–22 June 2018, pp. 4242–4251. IEEE Computer Society (2018)
11. Liong, V.E., Lu, J., Tan, Y., Zhou, J.: Cross-modal deep variational hashing. In: IEEE International Conference on Computer Vision, ICCV 2017, Venice, Italy, 22–29 October 2017, pp. 4097–4105. IEEE Computer Society (2017)
12. Liu, H., Ji, R., Wu, Y., Huang, F., Zhang, B.: Cross-modality binary code learning via fusion similarity hashing. In: 2017 IEEE Conference on Computer Vision and Pattern Recognition, CVPR 2017, Honolulu, HI, USA, 21–26 July 2017, pp. 6345–6353. IEEE Computer Society (2017)
13. Liu, W., Mu, C., Kumar, S., Chang, S.: Discrete graph hashing. In: Ghahramani, Z., Welling, M., Cortes, C., Lawrence, N.D., Weinberger, K.Q. (eds.) Advances in Neural Information Processing Systems 27: Annual Conference on Neural Information Processing Systems 2014, 8–13 December 2014, Montreal, Quebec, Canada, pp. 3419–3427 (2014). https://proceedings.neurips.cc/paper/2014/hash/f63f65b503e22cb970527f23c9ad7db1-Abstract.html
14. Liu, W., Mu, C., Kumar, S., Chang, S.F.: Discrete graph hashing (2014)
15. Liu, W., Wang, J., Ji, R., Jiang, Y., Chang, S.: Supervised hashing with Kernels. In: 2012 IEEE Conference on Computer Vision and Pattern Recognition, Providence, RI, USA, 16–21 June 2012, pp. 2074–2081. IEEE Computer Society (2012)

16. Liu, X., Nie, X., Zeng, W., Cui, C., Zhu, L., Yin, Y.: Fast discrete cross-modal hashing with regressing from semantic labels. In: 2018 ACM Multimedia Conference on Multimedia Conference, MM 2018, Seoul, Republic of Korea, 22–26 October 2018, pp. 1662–1669. ACM (2018)
17. Lu, J., Chen, M., Sun, Y., Wang, W., Wang, Y., Yang, X.: A smart adversarial attack on deep hashing based image retrieval. In: Proceedings of the 2021 International Conference on Multimedia Retrieval, ICMR 2021, pp. 227–235. Association for Computing Machinery, New York, NY, USA (2021). https://doi.org/10.1145/3460426.3463640
18. Nie, X., Wang, B., Li, J., Hao, F., Jian, M., Yin, Y.: Deep multiscale fusion hashing for cross-modal retrieval. IEEE Trans. Circ. Syst. Video Technol. **31**(1), 401–410 (2021). https://doi.org/10.1109/TCSVT.2020.2974877
19. Simonyan, K., Zisserman, A.: Very deep convolutional networks for large-scale image recognition. In: Bengio, Y., LeCun, Y. (eds.) 3rd International Conference on Learning Representations, ICLR 2015, San Diego, CA, USA, 7–9 May 2015, Conference Track Proceedings (2015)
20. Song, J., Yang, Y., Yang, Y., Huang, Z., Shen, H.T.: Inter-media hashing for large-scale retrieval from heterogeneous data sources. In: Proceedings of the ACM SIGMOD International Conference on Management of Data, SIGMOD 2013, New York, NY, USA, 22–27 June 2013, pp. 785–796. ACM (2013). https://doi.org/10.1145/2463676.2465274
21. Su, S., Zhong, Z., Zhang, C.: Deep joint-semantics reconstructing hashing for large-scale unsupervised cross-modal retrieval. In: 2019 IEEE/CVF International Conference on Computer Vision, ICCV 2019, Seoul, Korea (South), 27 October–2 November 2019, pp. 3027–3035. IEEE (2019)
22. Wang, B., Yang, Y., Xu, X., Hanjalic, A., Shen, H.T.: Adversarial cross-modal retrieval. In: Proceedings of the 2017 ACM on Multimedia Conference, MM 2017, Mountain View, CA, USA, 23–27 October 2017, pp. 154–162. ACM (2017)
23. Wawrzinek, J., Pinto, J., Wiehr, O., Balke, W.T.: Exploiting latent semantic subspaces to derive associations for specific pharmaceutical semantics. Data Sci. Eng. **5**, 333–345 (2020)
24. Wu, G., et al.: Unsupervised deep hashing via binary latent factor models for large-scale cross-modal retrieval. In: Lang, J. (ed.) Proceedings of the Twenty-Seventh International Joint Conference on Artificial Intelligence, IJCAI 2018, 13–19 July 2018, Stockholm, Sweden, pp. 2854–2860. ijcai.org (2018)
25. Zhou, J., Ding, G., Guo, Y.: Latent semantic sparse hashing for cross-modal similarity search. In: The 37th International ACM SIGIR Conference on Research and Development in Information Retrieval, SIGIR 2014, Gold Coast, QLD, Australia, 6–11 July 2014, pp. 415–424. ACM (2014)
26. Zhou, J., Ding, G., Guo, Y., Liu, Q., Dong, X.: Kernel-based supervised hashing for cross-view similarity search. In: IEEE International Conference on Multimedia and Expo, ICME 2014, Chengdu, China, 14–18 July 2014, pp. 1–6. IEEE Computer Society (2014)

RCBERT an Approach with Transfer Learning for App Reviews Classification

Shiqi Duan[1,2], Jianxun Liu[1,2](✉), and Zhenlian Peng[1,2]

[1] School of Computer Science and Engineering, Hunan University of Science and Technology, Xiangtan, China
ljx529@gmail.com, zlpeng@hnust.edu.cn

[2] Hunan Provincial Key Laboratory for Services Computing and Novel Software Technology, Hunan University of Science and Technology, Xiangtan, China

Abstract. Nowadays, extracting valuable information from user reviews of mobile applications has been a critical channel to obtaining user requirements in the process of mobile application development. However, it's difficult to extract accurately valuable information from massive reviews that are unstructured and uneven. Existing works classify the user reviews to assist the process of requirement elicitation. However, due to the differences in wording, sentence structure, and requirements of reviews, the generalization of existing methods is poor. Therefore, we propose a **R**eview **C**lassification using **BERT** model (RCBERT) for user reviews classification. RCBERT contains the great capability of generalization that can complete specific fine-tuning tasks only by providing less training data. We evaluate the performance on an English dataset with 4014 user reviews and a Chinese review dataset that contains 18331 user reviews. The experimental results show that the approach proposed in this paper improves the review classification accuracy effectively (F1-score of up to 88% on the English dataset and 91% on the Chinese dataset).

Keywords: Application · User reviews · Transfer learning · Reviews classification

1 Introduction

In the age of information, mobile applications (Apps) have spread across various domains. The number of apps on the mainland Chinese market was 3.87 million by the end of April 2021 [24]. After the Apps were released to the application store, users can download or update the previous version to newest version of the App. At the same time, users can rate the App and post reviews to feedback on program errors encountered during use, suggest some new features, or describe the advantages and disadvantages of the App. Related research [1] shows the interrelationship between App and user reviews. The quality of the App will determine the user's choice, and the reputation of users has a significant influence on the success of the App: Apps with better reviews and ratings will have a higher ranking in the application store, leading to higher downloads [1–3].

Y. Sun et al. (Eds.): ChineseCSCW 2021, CCIS 1492, pp. 444–457, 2022.
https://doi.org/10.1007/978-981-19-4549-6_34

User reviews and ratings are essential to the development of App and subsequent evolution and maintenance. Classifying user reviews effectively according to different types can help developers to satisfy the different requirements of users and provide a new version that proves more popular with users, which is conducive to the rapid development of the App. And we hold the opinion that analyzing requirements from the perspective of users will shorten the distance between software teams and final users. However, hundreds or even thousands of user reviews will be received every day after the new version of the App is released, which contains a large number of low-quality and meaningless reviews [2, 3]. For such a huge amount of user data, it is extremely difficult and tedious to analyze and filter useless reviews manually. Furthermore, relevant researches report that App user reviews have the following characteristics: 1): The text of user reviews is short in most Apps [2]. User reviews are usually several words to express the requirements of users, or a few short sentences summarizing their point of view. Therefore, the length of reviews is usually short and seldom is published in the form of paragraphs. 2): The proportion of valuable informative user reviews is low [2, 7–9]. Previous research has found that around 33% of the reviews contain messages that are beneficial to the development of the App. 3): Unstructured expression of review [2]. The user reviews are written in free form, without using predetermined fields, and users usually express their opinions freely with their own opinions. Therefore, it is necessary to classify user reviews through an automated method.

Existing researches [4–6, 10, 12, 26] have done a lot of work on App review classification. Manually rules to preprocess the dataset and feature engineering is common practice for review classification. In order to obtain more precise information from limited review texts, Maalej et al. [6] preprocessed the review text through stopword removal and lemmatization. In addition, predefined rules are also used for review classification. By examining 500 reviews from different types of Apps manually, Panichella et al. [5] identified 246 recurrent linguistic patterns to access wide information.

The sentence structure of reviews highly depends on the review corpus, which means additional user reviews will affect the content of the existing corpus and even the effect of the classification task. The existing approaches can't achieve a satisfying result in different review datasets. The results of Stanik et al. [4] show that traditional machine learning model algorithms outperform classification results than those using CNN models because they lack a training set. Therefore, it is urgent to propose a new automated classification approach with great generalization.

This paper focuses on how the performance of transfer learning in the review classification task. In fact, the application of transfer learning is widespread in natural language processing tasks. The underlying meanings and concepts of natural interlanguage are captured by training on a large textual dataset. Afterward, fine-tuned for each different specific task. These approaches are effective in improving the overall performance and generality with comparatively small training data.

In this paper, we proposed our approach RCBERT which means **R**eview **C**lassification using **BERT** based on Bidirectional Encoder Representations from Transformers (BERT) [17]. A large number of corpus has been used for the pre-training process of BERT and can complete fine-tuning tasks only by providing less training data. We present our approach that classifies user reviews in the App through the fine-tuning mechanism of BERT. The review dataset provided by Maalej et al. [6]. was applied to our RCBERT approach and achieve F1-score of up to 88%. In addition, we construct a Chinese dataset since there is not much Chinese user review data available. We conduct a lot of comparative experiments, the result shows that our approach outperforms the results of the deep learning approach by Stanik et al. [4] at 7% points (F1-score of 89% vs. 82%). This paper includes the following contributions:

1) We investigate whether and to what extent transfer learning improves review classification.
2) We create a new Chinese dataset which has been labeled as Inquiry, Problem Report, and Irrelevant.
3) We evaluate the performance of an approach based on transfer learning by applying different datasets.

The rest of the paper is organized as follows. Section 2 presents the related work. Section 3 explains how to fine-tune BERT for review classification tasks. Section 4 describes the study design including the research questions, data, and methodology. Then, we present the experiment in Sect. 5. Finally, Sect. 6 concludes the paper.

2 Related Work

With the evolution of mobile applications, review classification is discussed by researchers widely. User reviews contain information like requests for enhancements to current features, suggestions for new features, or information about App errors that are helpful for the development and evolution of the App. In addition, there is still a lot of information in user reviews that are irrelevant to the App. Existing researches have done a lot of work on review classification.

Pagano et al. [2] conduct an in-depth survey of large number of user reviews in the Apple App Store. They listed 17 topics related to user feedback, and the quality of those reviews varies considerably from another, in the form of useful opinions and innovative perspectives to insulting remarks. Guzman et al. [14] proposed a taxonomy based on the categories found in [2] which can detect App reviews that are related to software evolution. In the study of [2], there are nine categories related to software evolution. However, Guzman et al. [14] redefine some of those categories into more descriptive terms based on their research. Finally, seven categories related to the evolution of software were identified, including Feature shortcoming, Complaint, User request, Feature strength, Usage scenario Praise, and Bug report. Gu et al. [15] proposed SUR-Miner in order to classify review categories and measure user preferences, they obtained user review data from 17 popular Apps in Google play, and trained a separate model for each App to classify user reviews, and performed aspect opinion extraction and sentiment analysis from the corresponding categories. In the study of Panichella et al. [5] user reviews are classified according to whether they are related to the evolution of the software. They obtained the classification by analyzing the sentence categories of derived reviews and developer emails and used language rules and machine learning to classify the reviews. The study of Guzman et al. [14] is similar to theirs, but different from

Panichella et al. [5], they do not exploit predefined rules, but rather evaluate individual classifiers or integrated classifiers for classifying user reviews. Maalej et al. [6] designed an annotation guide in which each review type is precisely defined, in addition, to reduce manual annotation divergence, they also listed some examples to improve annotation quality. The classification methods and dataset they created have been cited by a large number of studies. Stanik et al. [4], similarly, wrote an annotation guide, which describes reviews labeled as Problem Report, Inquiry, and Irrelevant. In this paper, we refer to their coding guide.

The automated classification of reviews has been the focus of researchers, which resulting in many approaches. Guzman et al. [14] compared some machine learning algorithms and the performance of neural networks, and they integrated these different classifiers for predicting. In addition, to train each classifier they apply a series of manual preprocessing. To consider the different sentences in the reviews, Gu et al. [15] utilize the Stanford CoreNLP tool to split the original review text into each individual sentence, with the split sentences carrying a timestamp and a specified rating. Furthermore, typos and contractions in reviews were replaced by regular expressions. They considered kinds of aspects of the review text including sentiment and speech, which achieved promising results. However, this approach shows some limitations because of the need for manual preprocessing of parts specific to the dataset. The lexical and structural features created by Gu et al. [15] were applied and a Convolutional Neural Network model was constructed by Shah et al. [10], and the experimental results of Shah et al. showed that since they lack training dataset, traditional machine learning model algorithms were able to achieve better classification results than CNN models. Maalej et al. [6] created a complete dataset by removing stopwords, stemming, lemmatization, and tense detection. They also considered the review metadata and added the user rating and review length information into the classifier. However, their experimental results show that not more preprocessing operations lead to better classification results, lemmatization decreased the precision when predicting Feature Requests in their results. Their datasets were applied to our experiments for comparison. Stanik et al. [4] aim at investigating whether the application of deep learning algorithms can classify user reviews into Inquiries, Problem Reports, and Irrelevant performance improvements. They report classification results for both deep learning and machine learning models, and their results show that, despite the thousands of reviews they collected, comparable results to deep learning can still be obtained using traditional machine learning. Unfortunately, they did not publish the dataset. We created a Chinese dataset based on their method, which can be seen in Section IV. Furthermore, their classification approaches are also used to compare with transfer learning.

The approaches described above are prospective in certain contexts and demonstrate the ability of different techniques to review classification problems. However, in practice, many approaches are infeasible because they have to be preprocessed manually, or are highly data-dependent and can only target specific languages. Furthermore, the ability of these approaches to generalize to different datasets is not stated, one of the reasons for this is the lack of datasets available for training.

Transfer learning is the transfer of model parameters to a new model, which leads to a better trained model. Due to the large amount of relevant data and tasks, the parameters of the learned model can be shared with the new model, allowing the model to be learned with maximum efficiency. While BERT stands for Bidirectional Encoder Representations from Transformers [17] and applies pre-trained language representations to down-stream tasks through a fine-tuning approach to learn the language representations. Therefore, BERT is also a transfer learning model. Based on the approach of transfer learning, a large number of researchers have applied it to the task of natural language processing.

Because of the lack of available data on labeling hate speech, Mozafari et al. [11] improve the detection of hate speech by BERT which had learned on English Wikipedia and BookCorpus. And the evaluation results show that their approaches outperform the works of Waseem et al. [16] and Davidson et al. [28] adopted a fine-tuning strategy based on CNNs to combine syntactic information in a different encoder for the BERT model. Similarly, due to the lack of standard labeled datasets, Islam et al. [25] created two Bengali datasets for sentiment analysis, and they also reveal that the current work on sentiment classification could be improved by using transfer learning to train multilingual association extension models.

We apply transfer learning approaches to classify reviews to ensure better performance and generalization with less training data.

3 RCBERT: Review Classification Using BERT

Jacob Devlin et al. [17] proposed the BERT based on the Transformer architecture of [22], which aims to pre-train unlabeled text by conditioning the contextual environment in all layers so that it can achieve optimal performance on a large number of token-level and sentence-level tasks, better than many task-specific architectures.

BERT consists of two main components, i.e. pre-training and fine-tuning. The model is trained on unlabeled data using both the Masked Language Model (MLM) and Next Sentence Prediction (NSP) tasks during pre-training. For the MLM task, 15% of all tokens are randomly selected. Of the 15% selected tokens, 80% of them are processed for [mask], 10% of them are randomly replaced with another token, and the remaining 10% remain unchanged. For the NSP task, in each training sample, for a pair of sentences A and B, there is a 50% chance that B is a later sentence of A, i.e. the two sentences A and B are related (marked as IsNext), while there is a 50% chance that B and A are not related and are randomly selected from the corpus (marked as NotNext).

Since a large corpus of text has already been used for pre-training BERT and only a small amount of data is required for task-specific fine-tuning, we can fine-tune the network structure of the pre-trained BERT model for the review classification task.

We investigate the performance of transfer learning on the review classification task by fine-tuning the BERT. This decision is based on the expectation that the BERT model will generalize better even with a small amount of training data, in order to improve the classification of different datasets. For fine-tuning, as different language datasets, we use the corresponding pre-training models, the BERT-base Cased and BERT-base Chinese. The tokenizer of BERT can split the review sentences, so we don't perform preprocessing operations on the reviews.

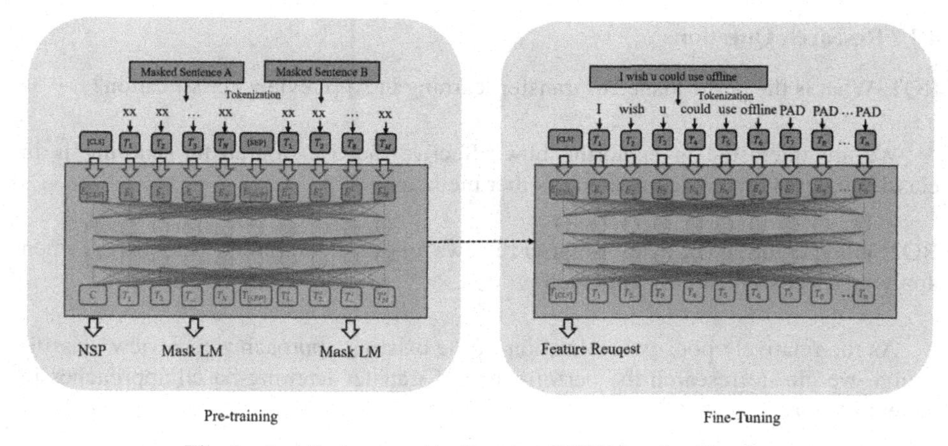

Fig. 1. Architecture used to fine-tune BERT for classification

Figure 1 shows how we fine-tune BERT for review classification. On the left is the step of pre-training, trained by a large corpus, the pre-trained model can be used. In this paper, we utilize the weight of the pre-trained model that has been trained to fine-tune the downstream tasks, and we define different output layers for the different review classification tasks based on the pre-trained models. The process of fine-tuning is trained in the way of supervised learning, as shown on the right side of Fig. 1. The sequence "*I wish u could use offline*" and the category was labeled are input into the model, and the text content is token to "*I*", "*wish*", "*U*", "*could*", "*use*", "*offline*". The input representation of each token is composed of corresponding token embeddings, segment embeddings, and position embeddings. This paper research mainly on text classification tasks, therefore the input in the fine-tuning process is a single sentence, and the self-attention mechanism of BERT based on Transformer allowed modeling of different downstream tasks for individual sentences or pairs of sentences. All parameters are fine-tuned during the fine-tuning process. The first symbol before each input sample is [CLS] for the classification token, and [PAD] in the sample is used for padding. For classification, all tokens are pooled and aggregated output to the first [CLS], which is the only output. This [CLS] is then fed into a single-layer feed-forward neural network, which calculates the probability of the category to which it belongs by means of a softmax function. The specific logic is as follows.

$$loss = log(softmax([C]_H [W]_{K \times H}^T)) \tag{1}$$

where $[C]_H$ represents the last layer of the vector corresponding to the first input token "[CLS]", and $[W]_{K \times H}$ is the hierarchical weight matrix, H is the hidden dimension of each token, K is the category label.

4 Research Design

In this section, we identify the questions we are trying to solve and explain the dataset and methods that we used.

4.1 Research Questions

RQ1: What is the performance of transfer learning in App review classification?

We are interested in exploring how effective the use of transfer learning is in classifying user reviews compared to other methods.

RQ2: Whether the generalization of user reviews for classification can be improved when applying a transfer learning approach?

As the relatively poor generalization of the existing approaches to review classification, we aim to research the performance of transfer learning-based approaches in generalization.

4.2 Existing Dataset

We made use of two different language datasets to answer RQ1 and RQ2. An existing data set is from Maalej et al. [6], they collected large amount of user review data, with approximately 1.1 million reviews in the App Store, and 146,057 reviews from 40 apps in the Google Play. In summary, they labeled 4,400 reviews as bug reports, feature requests, ratings, and user experience, and we obtained 4014 reviews from this data indeed. Table 1 shows the categories distribution for the dataset of Maalej. As can be observed from Table 1, the distribution of categories in the review data is uneven seriously, the number of each category varying greatly, with only 295 Feature Requests in the category with the lowest number of reviews, and 2605 Ratings in the category with the highest number of reviews.

Table 1. Categories distribution of Maalej's dataset

Category	Feature requests	Bug reports	User experiences	Ratings	Total
Quantity	295	379	735	2605	4014
Proportion	7.3%	9.4%	18.3%	64.9%	–

Furthermore, in this paper, 18331 reviews were crawled from App Store [29] in the Chinese mainland to increase the diversity of data. We created a Chinese dataset according to Stanik et al. [4]. First, we listed some typical examples based on the coding guide provided by Stanik et al. Table 2 shows part of our example. Then, all the user review data is assigned to three graduate students, who label the user review content as Inquiry, Problem Report, and Irrelevant according to the coding guide and our typical examples. After all the reviews are labeled, we discussed the inconsistency of the results of the three coders of the same review data, or even the inconsistency of the label of any two of the three, and determined the unique result, which is to ensure that each review is labeled as the correct category. Finally, the truth dataset was obtained as can be seen in Table 3. In our dataset, there is no significant difference in the quantity of

Inquiry and Problem Report, while the categories of Irrelevant account for the highest proportion. Useful information, i.e. Inquiry, and Problem Report reviews account for about one-third, which is close to the result of [2, 7–9].

Table 2. Examples of our dataset

Review	Category
Why can't the avatar be changed! ! !	Inquiry
This software does not have a look-back function. Sometimes after a lesson, there are places where I didn't hear clearly and I can't look back!	Inquiry
iOS13.3.1 freezes and restarts when using shared screen	Problem report
It's stuck to death, and the screen goes black, otherwise, there will be sound and the screen keeps stuck	Problem report
Thank you for this app for letting us experience the joy of class at home, thank you for enriching our empty vacation, and we must give you five stars (in installments!	Irrelevant
Five-star staging	Irrelevant

Table 3. Categories distribution of our dataset

Category	Inquiry	Problem report	Irrelevant	Total
Quantity	2504	2234	13593	18331
Proportion	13.7%	12.2%	74.2%	–

4.3 Research Methodology

We apply the approach of this paper to the task based on the above dataset:

Task1: Classification of four categories on Maalej's dataset.
Task2: Classification of three categories on our Chinese dataset.

In this paper, we evaluate the effectiveness of the classification task with Precision (P), Recall (R), and F1-score (F1) as evaluation metrics. Since our task of classification is multiclass classification, the true classification of reviews is indicated as $AC = \{AC_1, AC_2, AC_3\}$ and $PC = \{PC_1, PC_2, PC_3\}$ indicates the category predicted by the classifier, then the precision, recall, and F1-score are represented by the following equations:

$$precision(PC_i) = \frac{|AC_i \cap PC_i|}{|PC_i|} \tag{2}$$

$$recall(PC_i) = \frac{|AC_i \cap PC_i|}{|AC_i|} \quad (3)$$

$$F_1 - score(PC_i) = 2\frac{precision_{PC_i} \times recall_{PC_i}}{precision_{PC_i} + recall_{PC_i}} \quad (4)$$

where $|AC_i|$ is the number of categories, $|PC_i|$ is the number of reviews which classified into category PC_i, and $|AC_i \cap PC_i|$ presents the number of reviews is classified into AC_i correctly.

In this paper, we conduct a variety of settings for the task. The training and test sets are randomly divided by 7:3, i.e. 70% of the review dataset is allocated for training the model and 30% for testing the classification performance of the model. In addition, for reducing the coincidence arising from a single division of the training and validation sets, we also adopted ten-fold cross-validation by dividing all the data into ten parts, taking one data as the validation set and the other data as the training set for each of the ten experiments.

We attempt to reduce the influence of highly unbalanced data sets through undersampling (US) and oversampling (OS) strategies. For categories with a high number of reviews, we randomly sample from them by an undersampling strategy to equate their number to the category with a lower number. In the case of categories with a small number of reviews, the number of samples is increased from the training set through an oversampling strategy to extend the number to match that of the more numerous categories. Thus, the strategy of undersampling reduces the number of categories with a relatively large number of samples, while the categories with a relatively small number of samples are not affected, and the categories with a relatively small number of samples are expanded by the strategy of oversampling samples, while the categories with a relatively large number of samples remain unchanged. In both settings, we do not process the test set.

5 Experiment and Evaluation

In this paper, the same training set and test set were duplicated five times for the same experiment and the average was taken as the final result.

5.1 Task1. Classification of Four Categories on Maalej's Dataset

For the first task, we are interested in evaluating the transfer learning classification performance on a dataset labeled by Maalej et al. To investigate how well transfer learning performs in review classification which mentions in RQ1, we adopt different review classification approaches for comparison experiments. We compare our results to the approach by Maleej et al. [6]. The study of Stanik et al. [4] compares the effect of traditional machine learning with deep learning on review classification, where they classify reviews into Inquiry, Problem Reports, and Irrelevant, and we modify the output layer of their deep learning model as a baseline method. Table 4 reports the results of our approach compared to the best results reported by the Maalej and Stanik methods.

RCBERT achieves promising results with precision of 88%, recall of 89%, and 88% of F1-score on average, better than the best results of Maalej et al. [6] and Stanik et al. [4] by more than 10% points. On Ratings, RCBERT achieve better results than the approach of Maalej et al. [6] which relied on manual rules. In contrast, RCBERT does not require preprocessing of the dataset in advance. On User Experiences, Maleej et al. obtained better F1-score through a model with BoW features and metadata, which could be overfitted to the dataset and thus with poor generalization.

For the approaches of Stanik et al. [4], compared to deep learning, machine learning achieves better classification results. The probable reason for this situation is the small amount of data available for training, with only 4014 review data. Since the deep learning based approach is to learn feature representation between texts by feeding raw text to the neural network, so a large amount of data is required to train the model. We believe that deep learning may perform better when there is more training data. For machine learning, before the model is trained, the raw data is performed manual operations such as preprocessing and feature engineering. Thus, compared to deep learning, with feature engineering preprocessing, machine learning approach usually achieve better performance with smaller training sets. However, this approach may have the issue that it can only target specific data sets and does not provide good generalization capability.

Table 4. Classification results for the Maalej's dataset. ML = Machine Learning, DL = Deep Learning

Approaches	Feature requests			Bug reports			Ratings			User experiences			Average		
	P	R	F1	P	R	F1	P	R	F1	P	R	F1	P	R	F1
Maleej's	0.70	0.72	0.71	0.65	0.79	0.72	**0.90**	0.67	0.77	0.80	0.80	0.80	0.76	0.75	0.75
Stanik's ML	0.73	0.81	0.77	0.65	0.73	0.69	0.77	0.88	0.82	0.75	0.72	0.73	0.73	0.79	0.75
Stanik's DL	0.64	0.70	0.67	0.63	0.68	0.65	0.81	0.88	0.84	0.76	0.77	0.76	0.71	0.76	0.73
RCBERT	**0.85**	**0.89**	**0.87**	**0.87**	**0.88**	**0.87**	0.88	**0.90**	**0.89**	**0.90**	**0.89**	**0.89**	**0.88**	**0.89**	**0.88**

In contrast, the result of RCBERT reports higher precision, recall, and F1-score without performing any text preprocessing. Furthermore, the dataset size does not restrict the performance of RCBERT which is based on transfer learning approaches, using them for the specific task requires very little training data. This can address the challenge of not having a lot of available training data.

5.2 Task2. Classification of Three Categories on Our Chinese Dataset

In this evaluation, we evaluated the performance of transfer learning on the Chinese dataset to answer the RQ2. We created a new Chinese dataset according to Stanik et al. [4], and based on this dataset, we conducted several experiments to compare our approach with theirs. However, their original approaches are only applicable to English and Italian. Based on their work, we implemented their approaches, which can be applied to the Chinese dataset as a comparison.

Table 5 reports the results. For the case of dividing the training set and test set by 7:3, RCBERT achieves better effects than all but the highest scoring machine learning by

Stanik et al. [4] for the majority category, i.e. Irrelevant, which depends on manual rules to preprocess the dataset and feature engineering. On average, the F1-score of RCBERT exceeds 90% and outperforms all the approaches proposed by Stanik et al. [4]. Compared with the experiment on the dataset of Maalej et al. [6], the performance of the proposed approach didn't diminish when applied to different datasets as the results are similar. In addition, undersampling and oversampling are adopted to address data imbalance. The results show that the approaches proposed by Stanik et al. [4] using machine learning are more influenced by the dataset. However, on the original dataset, deep learning achieves a better result with F1-score of 84% on average which outperforms traditional machine learning by more than 16% points, which indicates the approaches proposed by Stanik et al. [4] lack of ability to generalize from different datasets.

The results of the 10-fold cross-validation are also reported in Table 5. RCBERT outperforms the results of the deep learning approach by Stanik et al. [4] at 7% points (F1-score of 89% vs. 82%) on original dataset. Even if undersampling or oversampling, the F1-score of transfer learning exceeds deep learning by 6% points (US F1-score of 89% vs. 83%) or 5% points (OS F1-score of 89% vs. 84%), which indicated that transfer learning provides better results than deep learning based on word embedding. Furthermore, the results of RCBERT exceeds all the results of traditional machine learning with manual preprocessing and feature selection by Stanik et al. [4], and the F1-score of RCBERT exceeds the traditional machine learning 26 (original), 19 (US), 24 (OS) percentage points separately. In addition, manual preprocessing and feature engineering result overfitting of traditional machine learning as all the performance decreased. Furthermore, compared with the .70-split, deep learning performs better than machine learning in 10-fold cross-validation.

Table 5. Classification results for our dataset ori = original US = Undersampling OS = Oversampling

	Approaches	Inquiry			Problem report			Irrelevant			Average		
		P	R	F1	P	R	F1	P	R	F1	P	R	F1
.70-split	$Stanik'sML_{ori}$	0.48	0.73	0.58	0.47	0.71	0.56	**0.97**	0.86	0.91	0.64	0.77	0.68
	$Stanik'sDL_{ori}$	0.80	0.78	0.79	0.83	0.80	0.81	0.95	0.90	0.92	0.86	0.83	0.84
	$RCBERT_{ori}$	0.88	0.88	0.88	**0.90**	0.91	**0.90**	0.93	**0.96**	**0.94**	0.90	**0.92**	**0.91**
	$Stanik'sML_{US}$	0.73	0.81	0.77	0.65	0.83	0.73	0.95	0.58	0.72	0.78	0.74	0.74
	$Stanik'sDL_{US}$	0.82	0.85	0.83	0.83	0.84	0.83	0.96	0.92	0.93	0.87	0.87	0.87
	$RCBERT_{US}$	0.88	**0.89**	**0.89**	0.89	**0.92**	**0.90**	0.96	0.92	**0.94**	0.91	0.91	**0.91**
	$Stanik'sML_{OS}$	0.68	0.77	0.72	0.61	0.65	0.63	0.80	0.63	0.71	0.78	0.74	0.74
	$Stanik'sDL_{OS}$	0.85	0.82	0.83	0.83	0.84	0.83	0.96	0.91	0.93	0.88	0.86	0.87
	$RCBERT_{OS}$	**0.89**	0.90	**0.89**	**0.90**	0.91	**0.90**	**0.97**	0.92	**0.94**	**0.92**	0.91	**0.91**
10-fold	$Stanik'sML_{ori}$	0.42	0.63	0.50	0.44	0.67	0.53	0.92	0.80	0.86	0.59	0.70	0.63
	$Stanik'sDL_{ori}$	0.78	0.76	0.77	0.78	0.77	0.77	0.94	0.91	0.92	0.83	0.81	0.82
	$RCBERT_{ori}$	**0.88**	0.86	**0.87**	0.85	**0.91**	0.88	**0.95**	**0.93**	**0.94**	0.89	**0.90**	**0.89**

(*continued*)

Table 5. (*continued*)

Approaches	Inquiry			Problem report			Irrelevant			Average		
	P	R	F1	P	R	F1	P	R	F1	P	R	F1
$Stanik'sML_{US}$	0.69	0.77	0.73	0.60	0.71	0.65	0.81	0.64	0.72	0.70	0.71	0.70
$Stanik'sDL_{US}$	0.79	0.83	0.81	0.80	0.81	0.80	0.88	0.86	0.87	0.82	0.83	0.83
$RCBERT_{US}$	0.85	**0.89**	**0.87**	**0.88**	0.87	0.87	0.93	0.90	0.91	0.89	0.89	**0.89**
$Stanik'sML_{OS}$	0.66	0.73	0.69	0.58	0.63	0.60	0.75	0.60	0.67	0.66	0.65	0.65
$Stanik'sML_{OS}$	0.80	0.84	0.92	0.81	0.78	0.79	0.90	**0.93**	0.91	0.84	0.85	0.84
$RCBERT_{OS}$	0.87	0.85	0.86	**0.88**	0.90	**0.89**	**0.95**	0.91	0.93	**0.90**	0.89	**0.89**

Overall, task 2 answered RQ2, RCBERT improve the generalization ability of user reviews classification. In contrast, the traditional machine learning approach requires domain experts to mine text features from raw data to improve performance, which needs a lot of manual effort. For the deep learning approach, it entails a pre-trained word embedding model for a specific language based on the training set, and the performance is limited by the dataset.

6 Conclusion

Users will seek the App they need from an application store. To get more downloads, developers will continue to satisfying user requirements. In the context of the rapid development of App, the acquisition of user requirements is no longer a single aspect of domain experts and developers, but interactive with users. Users can post reviews after downloading the App. However, the length and quality of the reviews are different. In addition to eagerly proposing new functional requirements or finding some errors, the reviews will also make a lot of complaints, including a lot of irrelevant information such as verbal abuse, which makes it impractical to classify effective feedback information manually.

In this paper, we propose RCBERT which is a review classification approach based on transfer learning. We have conducted extensive experiments on two different language datasets, and the results show that the proposed approach in this paper can effectively improve the classification results even with a small number of reviews. F1-score of up to 88% for English dataset and 91% for Chinese dataset. This research is a work in progress. We are also constantly collecting different types of Chinese App review data to expand the user review dataset. In addition, we will combine the App update information to further analyze the evolution process of the App and give some suggestions to developers.

Acknowledgments. This work was supported by China's National Key Research and Development Program project under grant No: 2018YFB1402804, Hunan Provincial Education Commission of China, under Grant No: 20B244, National Natural Science Foundation of China, under grant No: 61872139, 61873316 and 61702181.

References

1. Finkelstein, A., Harman, M., Jia, Y., et al.: App store analysis: mining app stores for relationships between customer, business and technical characteristics. RN **14**(10), 24 (2014)
2. Pagano, D., Maalej, W.: User feedback in the appstore: an empirical study. In: 2013 21st IEEE International Requirements Engineering Conference (RE), pp. 125–134. IEEE (2013)
3. Li, H., Zhang, L., Zhang, L., et al.: A user satisfaction analysis approach for software evolution. In: 2010 IEEE International Conference on Progress in Informatics and Computing, vol. 2, pp. 1093–1097. IEEE (2010)
4. Stanik, C., Haering, M., Maalej, W.: Classifying multilingual user feedback using traditional machine learning and deep learning. In: 2019 IEEE 27th International Requirements Engineering Conference Workshops (REW), pp. 220–226. IEEE (2019)
5. Panichella, S., Di Sorbo, A., Guzman, E., et al.: How can i improve my app? Classifying user reviews for software maintenance and evolution. In: 2015 IEEE International Conference on Software Maintenance and Evolution (ICSME), pp. 281–290. IEEE (2015)
6. Maalej, W., Nabil, H.: Bug report, feature request, or simply praise? On automatically classifying app reviews. In: 2015 IEEE 23rd International Requirements Engineering Conference (RE), pp. 116–125. IEEE (2015)
7. Hoon, L., Vasa, R., Schneider, J.G., et al.: An analysis of the mobile app review landscape: trends and implications. Faculty of Information and Communication Technologies, Swinburne University of Technology, Technical Report (2013)
8. Carreno, L.V.G., Winbladh, K.: Analysis of user comments: an approach for software requirements evolution. In: 2013 35th International Conference on Software Engineering (ICSE), pp. 582–591. IEEE (2013)
9. Chen, N., Lin, J., Hoi, S.C.H., et al.: AR-miner: mining informative reviews for developers from mobile app marketplace. In: Proceedings of the 36th International Conference on Software Engineering, pp. 767–778 (2014)
10. Shah, F.A., Sirts, K., Pfahl, D.: Simple app review classification with only lexical features. In: ICSOFT, 146–153 (2018)
11. Mozafari, M., Farahbakhsh, R., Crespi, N.: A BERT-based transfer learning approach for hate speech detection in online social media. In: Cherifi, H., Gaito, S., Mendes, J., Moro, E., Rocha, L. (eds.) COMPLEX NETWORKS 2019. SCI, vol. 881, pp. 928–940. Springer, Cham (2020). https://doi.org/10.1007/978-3-030-36687-2_77
12. Iacob, C., Harrison, R.: Retrieving and analyzing mobile apps feature requests from online reviews. In: 2013 10th Working Conference on Mining Software Repositories (MSR), pp. 41–44. IEEE (2013)
13. Dumitru, H., Gibiec, M., Hariri, N., et al.: On-demand feature recommendations derived from mining public product descriptions. In: Proceedings of the 33rd International Conference on Software Engineering, pp. 181–190 (2011)
14. Guzman, E., El-Haliby, M., Bruegge, B.: Ensemble methods for app review classification: an approach for software evolution (n). In: 2015 30th IEEE/ACM International Conference on Automated Software Engineering (ASE), pp. 771–776. IEEE (2015)
15. Gu, X., Kim, S.: What parts of your apps are loved by users? (T). In: 2015 30th IEEE/ACM International Conference on Automated Software Engineering (ASE), pp. 760–770. IEEE (2015)
16. Waseem, Z., Hovy, D.: Hateful symbols or hateful people? Predictive features for hate speech detection on Twitter. In: Proceedings of the NAACL Student Research Workshop, pp. 88–93 (2016)
17. Devlin, J., Chang, M.W., Lee, K., et al.: BERT: pre-training of deep bidirectional transformers for language understanding. arXiv preprint arXiv:1810.04805 (2018)

18. Gao, C., Zeng, J., Lyu, M.R., et al.: Online app review analysis for identifying emerging issues. In: Proceedings of the 40th International Conference on Software Engineering, pp. 48–58 (2018)
19. Villarroel, L., Bavota, G., Russo, B., et al.: Release planning of mobile apps based on user reviews. In: 2016 IEEE/ACM 38th International Conference on Software Engineering (ICSE), pp. 14–24. IEEE (2016)
20. Guzman, E., Maalej, W.: How do users like this feature? A fine grained sentiment analysis of app reviews. In: 2014 IEEE 22nd International Requirements Engineering Conference (RE), pp. 153–162. IEEE (2014)
21. Munikar, M., Shakya, S., Shrestha, A.: Fine-grained sentiment classification using BERT. In: 2019 Artificial Intelligence for Transforming Business and Society (AITB), vol. 1, pp. 1–5. IEEE (2019)
22. Vaswani, A., Shazeer, N., Parmar, N., et al.: Attention is all you need. arXiv preprint arXiv: 1706.03762 (2017)
23. Manning, C.D., Surdeanu, M., Bauer, J., Finkel, J., Bethard, S.J., McClosky, D.: The Stanford CoreNLP natural language processing toolkit. In: Proceedings of 52nd Annual Meeting of the Association for Computational Linguistics: System Demonstrations, pp. 55–60 (2014)
24. Ministry of Industry and Information Technology of the People's Republic of China. https://www.miit.gov.cn/gxsj/tjfx/rjy/index.html. Accessed 29 May 2021
25. Islam, K.I., Islam, M.S., Amin, M.R.: Sentiment analysis in Bengali via transfer learning using multi-lingual BERT. In: 2020 23rd International Conference on Computer and Information Technology (ICCIT), pp. 1–5. IEEE (2020)
26. Wang, C., Wang, T., Liang, P., et al.: Augmenting app review with app changelogs: an approach for app review classification. In: SEKE, pp. 398–512 (2019)
27. Messaoud, M.B., Jenhani, I., Jemaa, N.B., Mkaouer, M.W.: A multi-label active learning approach for mobile app user review classification. In: Douligeris, C., Karagiannis, D., Apostolou, D. (eds.) KSEM 2019. LNCS, vol. 11775, pp. 805–816. Springer, Cham (2019). https://doi.org/10.1007/978-3-030-29551-6_71
28. Davidson, T., Warmsley, D., Macy, M., et al.: Automated hate speech detection and the problem of offensive language. In: Proceedings of the International AAAI Conference on Web and Social Media, vol. 11, no. 1 (2017)
29. Apple App Store. https://www.apple.com.cn/app-store/. Accessed 29 May 2021

The Impact of COVID-19 on Online Games: Machine Learning and Difference-in-Difference

Shuangyan Wu[1], Haoran Hu[2], Yufan Zheng[2], Qiaoling Zhen[2], Shuntao Zhang[2], and Choujun Zhan[1](✉)

[1] South China Normal University, Guangzhou 510631, China
20175031@m.scnu.edu.cn, zchoujun2@gmail.com
[2] Nanfang College Guangzhou, Guangzhou 510000, China

Abstract. By intervening in people's behavior, governments in several nations have established a variety of strategies to slow down the spread of COVID-19 pandemic. At the same time, it has a different impact on everyone. Data from the Steam platform online games between January 2018 and February 2021 was used for this project's analysis. Through the difference-in-difference model in Synthetic Control Methods to quantify and analyze, crucial positive effect on Steam's online players during COVID-19 and the increase of the number of online players and the released games of the platform in 2020 had been found. The machine learning prediction model was created using the daily totals of the online gaming players of the most popular games on the site. The Ridge regression, whose R squared reached 0.805, had been demonstrated by the experimental results that it got the best performance. Simultaneously, this work found the features of the COVID-19 pandemic and the features of the human mobility, which helps to build a great majority of the predictive models.

Keywords: Steam platform · COVID-19 · Machine learning · Online game · Difference-in-difference

1 Introduction

Since the declaration of COVID-19 a world pandemic by the World Health Organization (WHO) on March 11, 2020 [5], COVID-19 has already spread to countries around the world. As of April 7, 2021, the cumulative number of infections in the world exceeded 100 million and the cumulative number of deaths exceeded 2.8 million [18]. In order to combat the spread of the disease, governments around the world have implemented a series of strict pandemic prevention and control policies, such as quarantine measures, travel bans, canceled social events, and closed public services [8]. These policies can effectively control the pandemic and reduce the risk of people being infected, but they have also brought varying

Y. Sun et al. (Eds.): ChineseCSCW 2021, CCIS 1492, pp. 458–470, 2022.
https://doi.org/10.1007/978-981-19-4549-6_35

degrees of impact to various industries. [11]: the global stock market had evaporated 6 trillion dollars in just one week [14], global airlines had lost more than half of the market share [7], the film industry lost about 5 billion dollars [13].

With the continuous spread of the COVID-19 pandemic and the implemented policy have unprecedented consequences for human life and entertainment. Under regular circumstances, people can gather for meals, travel, participate in exhibitions, and other entertainment activities for amusement. But it is difficult to achieve during the COVID-19 pandemic because of the restriction of human mobility. According to a finding by Ortiz Luz in Italy, playing online games is the fourth thing people most want to do when they are home in isolation. Online games commonly refer to electronic games in which multiple players engage in interactive entertainment through the computer Internet. More recently, researchers have started exploring how to do data mining online games. In 2009, Marcoux et al. addresses the issue of sales forecasting using a new approach based on connectionist and subspace decomposition methods [12]. In 2011, Debeauvais et al. analyze mechanisms of player retention and commitment in massively multiplayer online games. Measured the influence of gameplay, ingame sociality, and real-life status on player commitment [6]. In 2012, Bauckhage et al. introduced methods from random process theory into game data mining in order to draw inferences about player engagement [2]. In 2014, Sifa et al. conducted a large-scale analysis using player behavior data and identified four different patterns of game time frequency distribution [17]. Runge et al. focuses on predicting churn for high value players of casual social games and attempts to assess the business impact that can be derived from a predictive churn model [16]. Rezaei et al. examined the effects of values and behavioral intentions in virtual environments, and their results can help game developers understand what players expect from their games, thereby reducing or slowing churn [15]. In 2017 Liu et al. explored the motivations of gamers [10]. Ahn et al. using the Bass diffusion model identify success factors that will help drive future growth in the games industry [1]. In 2018, Baumann et al. reveal the specific behavioral categories of hardcore players using an unsupervised machine learning approach [3]. In 2019, Lin et al. studied the characteristics of game reviews over time [9]. Cheuque et al. tested the potential of the latest game recommendation model based on decomposing machine (FM), Deep Neural Network (DeepNN), and a new model derived from a mixture of the two (DeepFM) [4]. With the universal of the internet, people's gaming style is gradually changing from console games to online games. So how to predict daily online players accurately becomes a key question. An accurately and effectively predicted model can help game developers to adjust theirs sales strategies. And the game operator can reasonably assess the carrying capacity of the game server to safeguard the server effectively.

For now, there has not been a reliable and valid quantified result of the impact of COVID-19 on online games. The 2779 popular games data from Jan 2018 to Feb 2021 was used for this research to build a game research dataset. Based on the established research game dataset, the DID estimator is proposed to quantify the impact of COVID-19 on the number of game player on the

Steam platform. At the same time, in order to better aid game developers, we use machine learning to build a model and predict the number of online game player. The main contributions of this work are as follows:

- Build a Steam platform research dataset that includes 2779 popular games from Jan 2018 to Feb 2021.
- Quantified the impact of COVID-19 on daily online players by the difference-in-difference method.
- Use machine learning, ensemble learning, and deep learning to build a predicted model of daily online players. And explore the impact of pandemic and human mobility on the online game player predict model.

2 Data Exploratory Analysis

2.1 Data Description

For now, the Steam platform is known as the biggest digital distribution platform in the computer game field around the world, which was released as a game release platform by the US video game dealer Valve on 12, Sep 2003. This work was based on data of Steam platform (https://store.steampowered.com) to build a game research dataset from January 1, 2018 to February 14, 2021 containing 2,779 popular games, And also from Google's COVID-19 community mobility report website (https://www.google.com/covid19/mobility) and the Hopkins website (https://coronavirus.jhu.edu/map.html) to obtain the data to build human mobility dataset and the pandemic dataset, and finally merged and cleaned them. For each day t, there are 32 features in the research dataset as Table 1.

This paper proposes to build a predict model to predict the number of online game players based on machine learning.

$$\begin{aligned} x(t) = \{&C(t), R(t), D(t), \Delta C(t), \Delta R(t), \Delta D(t), E(t),\\ &N_T(t), N(t), N_C(t), N_A(t), N_R(t), G(t), G_S(t),\\ &G_I(t), G_M(t), G_D(t), G_L(t), G_A(t), W_K(t),\\ &W(t), N_W(t), T_D(t), T_M(t), T_N(t), T_Y(t)\} \end{aligned} \tag{1}$$

where $x(t) \in \mathbb{R}^{889\times 1}$. Here, we provide m-day forecasts for n consecutive days with quantified uncertainty based on machine learning models. In this work set $n = 7$, $m = 30$. The prediction model is:

$$\hat{y}(t+n) = f(x(t-m), x(t-m+1), \cdots, x(t-1)) \tag{2}$$

where $\hat{y}(t+n)$ represents the predicted value, $f(\cdot)$ stands for the machine learning model. Then, the prediction problem can be formulated as:

$$\begin{aligned} &\min_{f(\cdot)} \sum \|y(t+n) - \hat{y}(t+n)\|_2^2\\ &s.t. \quad \hat{y}(t+n) = f\left(x(t-m), x(t-m+1), \cdots, x(t-1)\right) \end{aligned} \tag{3}$$

where $y(t+n)$ is the true value.

Table 1. Original database feature table

Category	Feature	Symbol
COVID-19 (C)	Cumulative confirmed cases in each country	$C(t)$
	Cumulative recovered cases in each country	$R(t)$
	Cumulative deaths cases in each country	$D(t)$
	Daily confirmed cases in each country	$\Delta C(t)$
	Daily recovered cases in each country	$\Delta R(t)$
	Daily deaths cases in each country	$\Delta D(t)$
	Daily existing confirmed cases in each country	$E(t)$
Human mobility (H)	In and Out of Retail and Recreation Index	$H_R(t)$
	In and Out of Grocery Stores and Pharmacies Index	$H_G(t)$
	In and Out of Park Index	$H_P(t)$
	In and Out of Transit stations Index	$H_T(t)$
	In and Out of Workplaces Index	$H_W(t)$
	Residential Index	$H_h(t)$
Steam (S)	Daily online players in the top 20 games	$N_T(t)$
	Daily online players	$N(t)$
	Daily online players of each game type	$N_C(t)$
	Average daily online players	$N_A(t)$
	Daily games released	$N_R(t)$
	Daily games with players online	$G(t)$
	Variance daily online players	$G_S(t)$
	Minimum daily online players	$G_I(t)$
	Quartile daily online players	$G_M(t)$
		$G_D(t)$
		$G_L(t)$
	Maximum daily online players	$G_A(t)$
Time (T)	Weekend or not	$W_K(t)$
	Week	$W(t)$
	Day of the week	$N_W(t)$
	Day of the month	$T_D(t)$
	Month	$T_M(t)$
	Day of the year	$T_N(t)$
	Year	$T_Y(t)$

• The COVID-19 pandemic features include 7 countries: Brazil, Canada, China, Germany, Russia, the United Kingdom and the United States. $N_C(t)$ include 23 type of the game.
• The Human mobility features include 6 countries: Brazil, Canada, Germany, Russia, the United Kingdom and the United States.

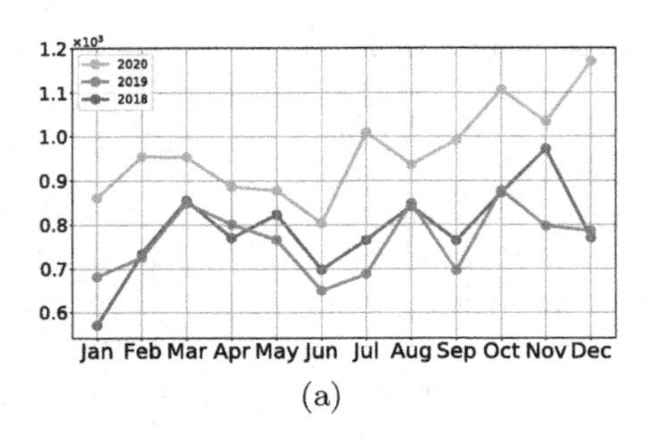

(a)

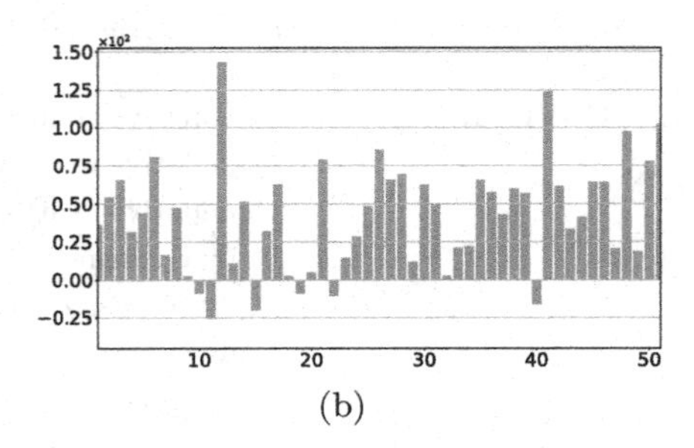

(b)

Fig. 1. (a) A line chart comparing the total monthly release of games from 2018 to 2020. The x-axis represents the month. The y-axis represents the number of games released; (b) A bar graph comparing the average number of released games in 2020 and 2018–2019. The x-axis represents the number of weeks, and the y-axis represents the difference in game released number between 2020 and the average number of 2018–2019.

2.2 Data Analysis

Daily Games Released. The number of games released means the number of new games released, which can reflect the investment intensity of game developers to the game development. With the COVID-19 continues to spread, most people have to stay at home or work online, thus increasing the demand for games, which in turn has boosted the distribution of online games. Figure 1(a) is the number of games released every month from 2018 to 2020. It can be found that there is not much difference between the monthly game released in 2018 and 2019 in general, only in January and November is the difference more obvious. The number of monthly releases in 2020 has increased significantly compared to 2018 and 2019, with an average increase of 190 games per month. Figure 1(b) is the comparison of the average number weekly games released between 2020 and 2018 to 2019. It shows that only a few weeks in 2020 have seen a decrease in the average number of game releases compared to the previous two years, and the majority of weeks have seen an increase. The number of games released in Jan 2020 increased by 75 games compared to Dec 2019, while the number of games released in the whole year of 2020 showed an upward trend. The number of games released achieved 1171 in Dec 2020. It means the number of online games released has a significant improvement during the COVID-19 pandemic. It reflects that many game developers intensify efforts to develop during the pandemic to meet the people's demand for the game.

Online Players. Daily online players can reflect the popularity of the online game. Figure 2(a) is monthly online players from 2018 to 2020. It shows online players per month trended downward in 2018 and 2019. But the daily online players has a growth trend from Feb 2020 to Apr 2020, which coincided with the COVID-19 pandemic. The Jan-Apr trends are similar between 2020 and 2018–2019. But after Apr 2020, the monthly online players has a certain growth

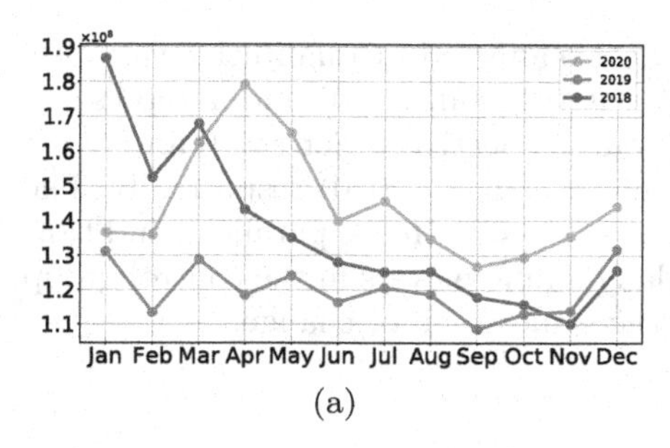

(a)

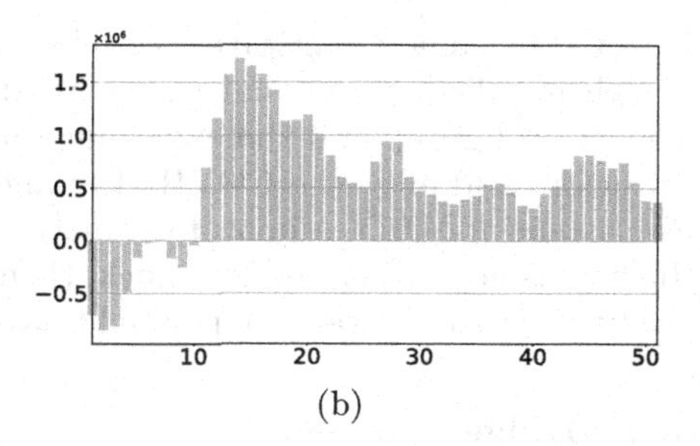

(b)

Fig. 2. (a) A line chart comparing the monthly online players of popular games from 2018 to 2020. The x-axis represents the month. The y-axis represents the online players per month; (b) A bar graph comparing the monthly total online players of popular games in 2020 and 2018–2019. The x-axis represents the number of weeks. The y-axis represents the difference in online quantity.

than 2019 and 2018, which increased by 25,070,967 and 20,536,477 people times respectively in July, increased by 21,591,836 and 25,291,587 people times respectively in November. Figure 2(b) is the comparison of the weekly online players between 2020 and 2018 to 2019. It shows starting in week 11, the total number of weekly popular games online in 2020 has increased compared to the previous two years, which shows the increase in demand for online games after the pandemic outbreak spread around the world. The most significant increase was in week 14, with an increase of 1,722,382 people times. The trend of growth has diminished after the 14th week of 2020, but it is still more than 2018 and 2019 on the whole.

3 Method

3.1 Difference-in-Difference Estimator

Difference-in-Difference (DID) estimator is a Synthetic Control Method used in econometrics for quantitative evaluation of public policy implementation. The DID estimator is a quantitative method to estimate the causal effect. Its basic idea is regarding public policy as a natural experiment. In order to evaluate the net impact of public policy implementation, it divides all sample data into two groups: the group affected by the policy is the treatment group and the group not affected by the policy is the control group. This study quantified the impact of COVID-19 on daily online players by the DID estimator:

$$Y_{it} = b_0 + b_1 \cdot T_{it} + b_2 \cdot A_{it} + b_3 \cdot T_{it} \cdot A_{it} + e_{it} \tag{4}$$

where Y is the explained variable which is daily online players, T is the time virtual variable, A is the group virtual variable which is the year virtual variable, $T \cdot A$ is the interaction term between the group virtual variable and the time virtual variable. In the regression analysis, the explained variable is not only affected by some quantitative variables but also affected by qualitative variables which is called the virtual variable. And e is residual, i = 0 and 1 represents the control group and the experimental group respectively. Set t = 0 and t = 1 to before and after the COVID-19 pandemic(March 11, 2020) respectively. The consistency of DID estimator estimate results depends on the parallel trend test. If there is no external policy shock, then daily online players in the control group and the experimental group should have the same trend of change.

3.2 Ridge Regression

Ridge Regression (RG) is a method of estimating the coefficients of a multiple regression model when the independent variables are highly correlated. For the data set $D = \{(X_1, y_1), (X_2, y_2), ..., (X_n, y_n)\}$, where $X_i = \{x_i^1, x_i^2, ..., x_i^m\}$. Initialization parameters to $\theta = (\theta_1, \theta_2, ..., \theta_m)$, where n is the number of samples, m is the number of features. Therefore the multiple linear regression is as follows:

$$h_\theta(x)_i = \theta_0 + \theta_1 x_i^1 + \theta_2 x_i^2 + \cdots + \theta_m x_i^m \tag{5}$$

where X_i is the ith samples, y_i is the ith true value, x_i^j is the jth feature of the ith samples. θ_j is the parameter of the model. Ridge regression adds a complexity penalty factor after the objective function $J(\theta)$, that is, a regular term to prevent overfitting. The cost function is:

$$J(\theta) = \frac{1}{2m}\sum_{i=1}^{m}(y_i - h_\theta(X_i))^2 + \lambda\sum_{j}^{n}|\theta_j| \tag{6}$$

Finally, the optimal model parameters are found by minimizing the cost function.

3.3 Pearson's Correlation Coefficient

Pearson's correlation coefficient is a method to help understand the linear relationship between two variables. Its value range is between −1 and +1, −1 means complete negative correlation, +1 means complete positive correlation, and 0 means no linear correlation. Pearson correlation coefficient of X and Y is defined as:

$$\rho(X, y) = \frac{\sum_{i=1}^{n}\left(X_i - \bar{X}\right)\left(y_i - \bar{y}\right)}{\sqrt{\sum_{i=1}^{n}\left(X_i - \bar{X}\right)^2}\sqrt{\sum_{i=1}^{n}\left(y_i - \bar{y}\right)^2}} \tag{7}$$

where X_i is the eigenvalue of sample i, y_i is the predicted value of sample i, $\bar{X}$ is the mean value of the feature value, $\bar{y}$ is the mean of the predicted value.

4 Experiment

4.1 Quantified Results

Because the WHO announced the COVID-19 pandemic on March 11, 2020, so set March 11, 2020 as the first day of the COVID-19 pandemic in the world. Set

Table 2. Quantified result of the impact of COVID-19 on daily online players

Variable	$T \cdot A$	Constant	R^2	Sample size
	791898*** (88,433)	3.942e+06*** (40,752)	0.393	669

• Robust standard errors in parentheses,
***, **, * indicate P < 0.01, P < 0.05 and P < 0.1 respectively.

11 March 2020 as the threshold point of time virtual variable before building DID estimator. Set the 2019 daily online players as the control group, and the 2020 one as the treatment group. Table 2 showed the result of building the DID estimator. The interaction coefficient of T and A is 791898, which shows an evident positive effect in daily online players ($P < 0.01$). It means the 2020 daily online players increased 791898 people times. Figure 3 is the result of the parallel trend test. It shows that there are no evident differences in daily online players between the control group and the treatment group before the COVID-19 pandemic. In the parallel trend test, the estimated coefficients of the time virtual variable have not reached 1% significance level ($P < 0.01$). It means the 2019 and the 2020 parallel trend assumption is true before the COVID-19 pandemic.

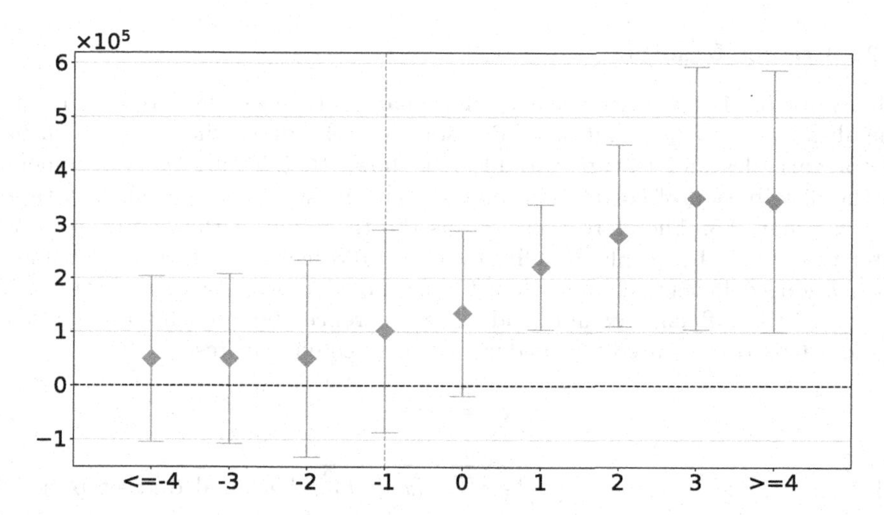

Fig. 3. The result of the parallel trend test. The x-axis represents the time dummy variables of the advance and lag of the outbreak. The 0 in the x-axis represents the first week of the pandemic. The y-axis is the estimated coefficient of the variable, and draws the estimated coefficient and its 95% confidence interval (error bars). The dotted orange line represents the week prior to the worldwide outbreak, from 4 to 11 March 2020. (Color figure online)

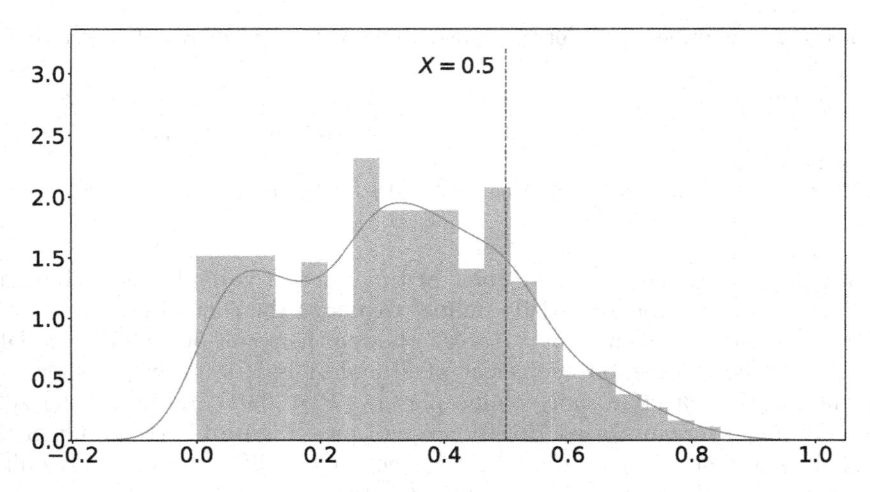

Fig. 4. The absolute value distribution of Pearson's correlation coefficient between true variables and feature variables. The x-axis is the correlation coefficient after taking the absolute value. The y-axis is the frequency. The red dashed line represents the selected threshold. (Color figure online)

4.2 Predict Results

This work used the Pearson correlation analysis to select features. Figure 4 is the absolute value distribution of Pearson's correlation coefficient between predictor variables and feature variables. It shows that a long tail phenomenon in the distribution of correlation analysis, and most absolute coefficient values are less than 0.5. Therefore chose 0.5 as the threshold of the feature selection after observing the graph. Finally, we chose 163 features which are Pearson's correlation coefficient of more than 0.5 as model input from 889 features. To eliminate the influence of unit and scale differences between features, Z-score standardization was used to standardize the selected features.

$$x_i^* = \frac{x_i - \bar{x}}{s} \tag{8}$$

where $\bar{x} = \frac{1}{n}\sum_{i=1}^{n} x_i$, $s = \sqrt{\frac{1}{n-1}\sum_{i=1}^{n}(x_i - \bar{x})^2}$. The final dataset period is from 29 Feb 2020 to 14 Feb 2021 and set 10 Dec 2020 as the split point between the training set and testing set. And used the machine learning based on grid search hyperparameter: Support Vector Regression (SVR), Ridge Regression (RG), Decision Tree (DT), K-Nearest Neighbors (KNN). Ensemble learning: Random Forest (RF), Gradient Boosting Decision Tree (GBDT), Extra-trees Random Forest (EXT), and deep learning: Multilayer Perceptron (MLP). The above 8 models perform data modeling predictions on daily online players.

And to evaluate the performance of the prediction model, regression evaluation indicators are used: Root Mean Squared Error (RMSE), Mean Absolute Error (MAE), Median Absolute Error (MAD), Mean Absolute Percentage Error

(MAPE), and Coefficient Of Determination (R^2). Set n is the sample size, and each evaluation index is calculated as follows:

- RMSE: It represents the sample standard error between the predicted value and the true value. It describes the degree of dispersion of the sample.

$$RMSE = \sqrt{\frac{1}{n}\sum_{i=1}^{n}(y_i - \hat{y}_i)^2} \tag{9}$$

- MAE: It represents a mean value of absolute error between the predictive value and true value

$$MAE = \frac{1}{n}\sum_{i=1}^{n}|y_i - \hat{y}_i| \tag{10}$$

- MAD: It is a robust metric. MAD is more flexible than MSE in dealing with outliers in data sets and can greatly reduce the impact of outliers on data sets.

$$MAD = \text{median}\left(|y_1 - \hat{y}_1|, \cdots, |y_n - \hat{y}_n|\right) \tag{11}$$

- MAPE: It expresses accuracy as a percentage of error. Since it is a percentage value, it may be easier to understand than other accuracy measurement statistics.

$$MAPE = \frac{100\%}{n}\sum_{i=1}^{n}\left|\frac{\hat{y}_i - y_i}{y_i}\right| \tag{12}$$

- R^2: It is an important statistic reflecting the degree of model fit, and its value reflects the relative degree of regression contribution.

$$R^2 = 1 - \frac{\sum_i(\hat{y}_i - y_i)^2}{\sum_i(y_i - \bar{y}i)^2} \tag{13}$$

Table 3. Predictive model evaluation results

Model	ST					HCST				
	RMSE	MAE	R^2	MAD	MAPE	RMSE	MAE	R^2	MAD	MAPE
SVR	352888	293797	0.133	264513	6.006	283091	234736	0.382	218617	4.727
RF	347314	283993	0.161	248243	5.827	281522	229086	0.389	196459	4.717
MLP	937397	858666	−5.114	859229	18.502	2344609	2003713	−41.39	2278619	40.255
RG	446734	362648	−0.389	337588	7.296	**159007**	**122909**	**0.805**	**100684**	**2.518**
KNN	361039	296071	0.093	256343	6.139	369801	292320	−0.055	208040	5.873
GBDT	**308194**	**250904**	**0.339**	**202034**	**5.258**	218000	169394	0.634	126804	3.415
EXT	344764	281602	0.173	244365	5.79	247919	207426	0.526	186008	4.283
DT	309003	259469	0.336	212610	5.308	274835	234179	0.418	202812	4.7

To explore the impact of COVID-19 features on modeling, this experiment had divided into two parts. One dataset is the HCST, which added the COVID-19 features and the human mobility features. The other is the ST dataset, which did not add the COVID-19 features and the human mobility features. The evaluation results of all predict models where in Table 3. It shows that: In the ST dataset, GBDT which R^2 and MAPE reached 0.339 and 5.258 separately, had the best predictive effect among the 5 evaluation index. And in the HCST dataset, RG had the best predictive effect among the 5 evaluation index, and R^2 increased from −0.389 to 0.805.

The results of model SVR, RF, GBDT, EXT, and DT, whose R2 were increased by 0.248, 0.228, 0.294, 0.353, and 0.081 respectively, had improved compared to the results that did not add the COVID-19 features and the human mobility. Whereas the models KNN and MLP, which simultaneously added the COVID-19 and human mobility features, had a decrease in R2. According to that, the two features were helpful in establishing models to predict the online players' numbers. In Fig. 5, it is the predictive and fitting result based on the HCST dataset of model RG. A good fit and robustness are demonstrated.

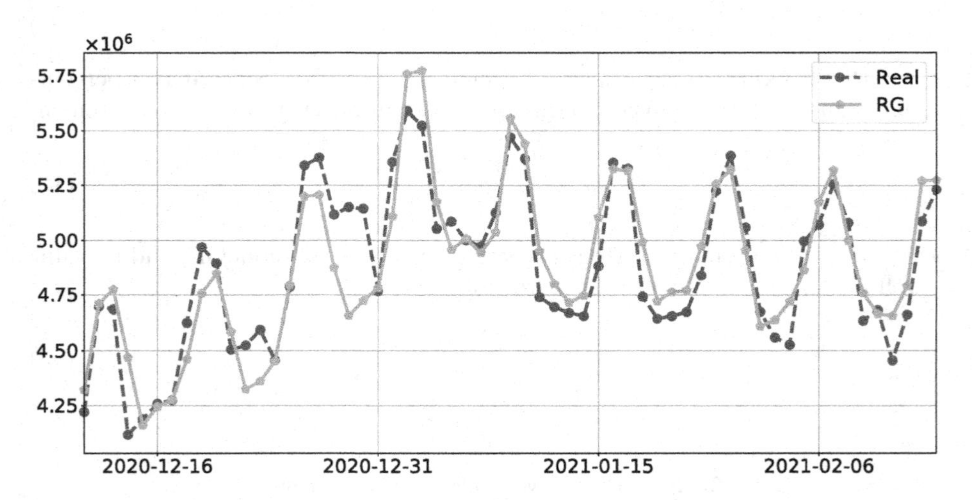

Fig. 5. Fitting graph of prediction results. The x-axis represents time. The y-axis represents daily online players. The blue dotted line represents the true value. The orange solid line represents the predicted value of RG based on the HCST data set. (Color figure online)

5 Conclusion

This work builds an online game research dataset that from January 2018 to February 2021 containing data from the Steam platform, the COVID-19 epidemic, and human mobility. Through analysis of daily game released and daily online players. Reveals that compared to 2020 the number of game releases and the number of online players in 2018–2019 has increased in varying degrees. Among them, the online games released in 2020 is average more than 190 per month than 2019-2018. And daily online players in 2020 begins more than 2018 and 2019 average in the 11th week. Finally, this work found that people's demand for the game has increased after the COVID-19 outbreak, promoting the development of the online game industry. Then quantified the impact of COVID-19 on daily online players by the difference-in-difference method. The result shows that the impact of the COVID-19 pandemic on daily online players has a significant positive effect. The weekly online players in 2020 is more than 791,898 people times than control group 2019. In order for the online game developers and the operators to better predict the demand for online games, this work used machine learning, ensemble learning, and deep learning to build a predicted model, which predicts daily online players, and explore the impact of epidemics and human mobility on the online game player predict model. The result shows the COVID-19 features and the human mobility features are helpful in most models when predicting daily online players. And the RG is the best model, its R^2 reach 0.805 and MAPE is 2.518.

Through the establishment of a predictive model, game developers and operators can adjust their investment in game development and marketing to obtain more benefits, thereby promoting the steady development of the online game industry. In the future, daily online players will be refined by game category, data modeling of online numbers of various types of games will be carried out, and the impact of the COVID-19 epidemic on each game type will be quantified. In addition, if to improve the accuracy of the prediction model, more features related to online games need to be included in future work.

Acknowledgment. This work was supported by Science and Technology Program of Guangzhou, China (201904010224), and Natural Science Foundation of Guangdong Province, China (2020A1515010761).

References

1. Ahn, S., Kang, J., Park, S.: What makes the difference between popular games and unpopular games? Analysis of online game reviews from steam platform using word2vec and bass model. ICIC Express Lett. **11**(12), 1729–1737 (2017)
2. Bauckhage, C., Kersting, K., Sifa, R., Thurau, C., Drachen, A., Canossa, A.: How players lose interest in playing a game: an empirical study based on distributions of total playing times. In: 2012 IEEE Conference on Computational Intelligence and Games (CIG), pp. 139–146. IEEE (2012)
3. Baumann, F., Emmert, D., Baumgartl, H., Buettner, R.: Hardcore gamer profiling: results from an unsupervised learning approach to playing behavior on the steam platform. Procedia Comput. Sci. **126**, 1289–1297 (2018)
4. Cheuque, G., Guzmán, J., Parra, D.: Recommender systems for online video game platforms: the case of STEAM. In: Companion Proceedings of the 2019 World Wide Web Conference, pp. 763–771 (2019)
5. Cucinotta, D., Vanelli, M.: WHO declares COVID-19 a pandemic. Acta Bio Medica: Atenei Parmensis **91**(1), 157 (2020)

6. Debeauvais, T., Nardi, B., Schiano, D.J., Ducheneaut, N., Yee, N.: If you build it they might stay: retention mechanisms in world of warcraft. In: Proceedings of the 6th International Conference on Foundations of Digital Games, pp. 180–187 (2011)
7. Gössling, S., Scott, D., Hall, C.M.: Pandemics, tourism and global change: a rapid assessment of COVID-19. J. Sustain. Tour. **29**(1), 1–20 (2020)
8. Hale, T., Petherick, A., Phillips, T., Webster, S.: Variation in government responses to COVID-19. Blavatnik School of Government Working Paper, vol. 31 (2020)
9. Lin, D., Bezemer, C.-P., Zou, Y., Hassan, A.E.: An empirical study of game reviews on the steam platform. Empir. Softw. Eng. **24**(1), 170–207 (2019)
10. Liu, X., Merrick, K., Abbass, H.: Toward electroencephalographic profiling of player motivation: a survey. IEEE Trans. Cogn. Dev. Syst. **10**(3), 499–513 (2017)
11. Mahar, I.: Impact of COVID-19 on global economy structure. Mod. Dipl. **22** (2020)
12. Marcoux, J., Selouani, S.-A.: A hybrid subspace-connectionist data mining approach for sales forecasting in the video game industry. In: 2009 WRI World Congress on Computer Science and Information Engineering, vol. 5, pp. 666–670. IEEE (2009)
13. Ozili, P.K., Arun, T.: Spillover of COVID-19: impact on the global economy, March 2020. Available at SSRN 3562570
14. Randewich, N.: Coronavirus, oil collapse erase $5 trillion from US stocks (2020)
15. Rezaei, S., Ghodsi, S.S.: Does value matters in playing online game? An empirical study among massively multiplayer online role-playing games (MMORPGs). Comput. Hum. Behav. **35**, 252–266 (2014)
16. Runge, J., Gao, P., Garcin, F., Faltings, B.: Churn prediction for high-value players in casual social games. In: 2014 IEEE conference on Computational Intelligence and Games, pp. 1–8. IEEE (2014)
17. Sifa, R., Bauckhage, C., Drachen, A.: The playtime principle: large-scale cross-games interest modeling. In: 2014 IEEE Conference on Computational Intelligence and Games, pp. 1–8. IEEE (2014)
18. Johns Hopkins University: COVID-19 dashboard by the center for systems science and engineering (CSSE) at Johns Hopkins University (JHU) (2021)

Uncertain Graph Publishing of Social Networks for Objective Weighting of Nodes

Chengye Liu[1,2], Jing Yang[1](✉), and Lianwei Qu[1]

[1] College of Computer Science and Technology, Harbin Engineering University, Harbin, Heilongjiang, China
yangjing@hrbeu.edu.cn
[2] North China Institute of Computing Technology, Beijing, China

Abstract. With the rapid rise of social network platforms, more users share their daily lives through social networks, which has led to a large amount of personal privacy being leaked or stolen. Therefore, secure social network data publishing has become the focus of research in the field of data publishing. Aiming at the problem of low data availability caused by node weights that are not considered in the existing privacy protection methods for publishing uncertain graphs, this paper proposes a privacy protection method for publishing uncertain graphs based on objective weighting. In this paper, the entropy weight method is used to objectively weight the nodes, and then the Laplacian mechanism is used to add noise to the edges and convert the noise into the probability of edges, and finally generate and publish an uncertain graph. Experiments have proved that the method proposed in this paper can effectively improve the usability of social network structure while ensuring privacy.

Keywords: Social network · Entropy method · Objective weighting · Probability · Uncertainty graph

1 Introduction

With the rapid popularization of mobile devices such as mobile phones, the mobile Internet has also developed rapidly, and social networking platforms have also taken the opportunity to rise. Nowadays, more and more users use social networking platforms to communicate, and at the same time conduct daily data release and sharing on social networking platforms. This has led to the retention of a large number of users' personal information on social networking platforms, which includes sensitive information such as social information, identity information, and location information. If this sensitive information is obtained by people with ulterior motives, it is extremely easy to cause infringements on the safety of users.

The structure of social networks is intricate and complex. Social networks include not only nodes and node attributes, but also the relationship between nodes and nodes. There are also many privacy protection methods for existing social networks, such as differential privacy, anonymization and uncertainty graphs in social network data. The

Y. Sun et al. (Eds.): ChineseCSCW 2021, CCIS 1492, pp. 471–483, 2022.
https://doi.org/10.1007/978-981-19-4549-6_36

published privacy protection has a wide range of applications. Wang Tingting [1] proposed a protection algorithm based on random projection and differential privacy in response to the problem of poor data availability when publishing data on social networks. Gu Zhen [2] and others proposed a differential privacy protection algorithm based on probabilistic principal component analysis for the problem of poor privacy protection when high-dimensional data is released. Wang Leixia and Meng Xiaofeng [3] proposed a hybrid privacy protection framework, ESA (encode-shuffle-analyze) framework, in view of the poor data availability characteristics of local differential privacy technology to improve the availability of data when data is released. Li [4] proposed a personalized differential privacy data publishing algorithm in order to reduce the waste of privacy budget when different users have different privacy requirements.

Esmerdag [5] found that the implementation of differential privacy is too cumbersome and requires additional additional protocols and changes to the database. In response to this problem, a platform for differential private data analysis is proposed for data release. Zhu Suxia [6], in order to further improve the accuracy of continuous numerical data mean estimation by differential privacy, proposed a classification transformation perturbation mechanism method that satisfies local differential privacy, which improves the data availability when data is released. Liu Shuangying [7] proposed a social network differential privacy model that focuses on the protection of edge weights in view of the weak privacy protection of edge weights in social networks. Zhang Xiaojian [8] proposed a privacy protection model for data release and analysis in response to the leakage of private data and sensitive data during data release and analysis. Jia Xu [9] proposed a histogram data publishing model based on differential privacy in view of the existing data publishing models mostly only simple aggregation. The histogram data publishing model greatly improves the privacy of social network data publishing. The users of social networks can be regarded as nodes, and the relationship between users can be represented by edges. Therefore, research on social networks can be transformed into research on graphs. Wu Zhenqiang [10] proposed a social network privacy protection method that combines differential privacy and uncertain graphs. This method has strong privacy protection, but its data availability needs to be improved. Song Yuning [11] proposed a differential privacy protection method oriented to random graph models. This method circumvents sensitivity analysis and cannot optimize the amount of noise, resulting in a waste of privacy budget to a certain extent. Uncertainty graphs are also widely used in the privacy protection of social networks. Boldi [12] proposed a new social network anonymization method (k,ε)-obfuscation algorithm, which adds uncertainty to the social network graph. So that the social network graph maintains a high data availability. Yan Jun [13] proposed a privacy protection method for uncertain graphs based on node characteristics in response to the network security problems caused by the leakage of privacy data in social networks. There are also many traditional privacy protection methods for uncertain graphs. For example, a wander-based method for publishing uncertain graph data is proposed to prevent chain attacks [14]. Nguyen [15] analyzed the deficiencies in the uncertainty graph and proposed a method based on the maximum variance method to measure the relationship between data utility and privacy. Based on the previous work, Nguyen [16] proposed a fuzzy model UMA based on the uncertain adjacency matrix based on the concept of uncertain graphs, and introduced

UAM into the previously proposed variance maximization algorithm. In recent years, the privacy protection of uncertain graphs has been widely used in social networks.

Although the uncertainty graph privacy protection method proposed in literature 10 has high privacy protection, the data availability is very low. Aiming at the problem of low data availability, this paper proposes an uncertain graph data release based on objective weighting. The privacy protection method improves data availability while ensuring privacy protection.

2 Related Basic Knowledge

2.1 Objective Weighting Based on Entropy Method

The entropy weight method has a strong objective mathematical basis, which can objectively analyze the amount of information of the node to measure the importance of the node, so this article adopts the objective weighting based on the entropy weight method. The information entropy formula is as follows:

$$E_i = -ln(n)^{-1} \sum\nolimits_{j=1}^{n} p_{ij} ln p_{ij} \quad (1)$$

where: $p_{ij} = y_{ij} / \sum_{j=1}^{n} y_{ij}$, it's the probability of a certain value. If $p_{ij} = 0$, then define $\lim_{p_{ij} \to 0} p_{ij} ln p_{ij} = 0$.

Then, the weight is determined according to the entropy value:

$$w_i = (1 - E_i)/(m - \sum\nolimits_{j=1}^{n} E_j) \quad (2)$$

where m is the number of nodes.

2.2 Differential Privacy and Related Concepts

Definition 1 (Differential privacy). There are two adjacent data sets D and D′, and algorithm M. M(D) represents the output set of algorithm M on data set D, and represents the set of all output values of algorithm M. If algorithm M Arbitrary output results on data sets D and D′ satisfy the following inequality (3):

$$Pr[M(D) \in S_M] \le e^{\varepsilon} \times Pr[M(D') \in S_M] \quad (3)$$

Definition 2 (Neighboring graph). Given graph $G_1 = (V_1, E_1)$ and $G_2 = (V_2, E_2)$, if satisfies $|V_1 \oplus V_2| + |E_1 \oplus E_2| = 1$, then it is called adjacent graph.

In this article, because it does not involve adding or deleting nodes, so $V_1 = V_2$, as long as $|E_1 \oplus E_2| = 1$, it is called a neighboring graph. As shown in Fig. 1, Fig. 1 (a) and Fig. 1 (b) are adjacent diagrams.

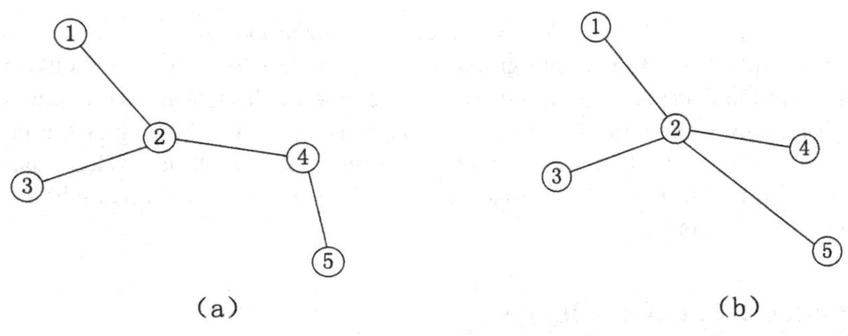

Fig. 1. The example of neighboring graphs

Definition 3 (Sensitivity). Given a function $f: G \rightarrow G''$, where G and G'' have the same set of vertices, the global sensitivity of function f is:

$$\Delta f = \max_{G_1, G_2} ||f(G_1) - f(G_2)|| \tag{4}$$

where G_1, G_2 is the adjacent graph, G'' is the graph output by the random algorithm, f represents the query function, for the edge e_i in G_1, G_2, query whether e_i exists in the same graph G_1, G_2, in Fig. 1 (a) and (b) Sensitivity $\Delta f = 2$.

Definition 4 (Laplace mechanism). For function $f: G \rightarrow G''$, if the output result of algorithm M satisfies the following equation, then algorithm M is said to satisfy ε-differential privacy.

$$M(G) = f(G) + Lap(\Delta f / \varepsilon) \tag{5}$$

where lap is the noise of the Laplacian distribution with the expected $\mu = 0$ position parameter $b = \Delta f / \varepsilon$ added to the query result, and the magnitude of the noise value is related to the sensitivity Δf and ε.

2.3 Uncertainty Graph

Definition 5 (Uncertain graph). Given a graph $G = (V, E)$, if the mapping $p: V_p \rightarrow [0, 1]$ is the probability function of each edge, then the graph $G' = (V, p)$ Is the uncertain graph of graph G, where V_p set contains all possible fixed-point pairs in set V, namely $V_p = \{(v_i, v_j)|1 \leq i < j \leq n\}$, $|V_p| = \frac{n(n-1)}{2}$.

3 Uncertain Graph Publishing Privacy Protection Algorithm for Objectively Empowered Nodes

3.1 Node Objective Weighting Based on Entropy Weight Method

First, transform the social relationship between users into a social network graph, and define the graph $G(V, E)$ to represent the social network, where $V = \{v_1, v_2, \ldots, v_n\}$

represents the set of nodes, and $E = \{e_1, e_2, \ldots, e_m\}$ represents the set of edges. Then the graph G is transformed into an adjacency matrix. The adjacency matrix of the graph is represented by $A_{n\times n} = (a_{ij})$, $a_{ij} = 1$ means that the nodes v_i and v_j are directly connected, otherwise $a_{ij} = 0$. Here *i, j* are the node numbers at both ends of the edge. After transforming the graph G into an adjacency matrix, use formulas (1) and (2) to weight each node into a graph $G'(V, W)$, where V represents the set of nodes and W represents the set of weights.

3.2 Privacy Protection Algorithm Based on the Release of Uncertain Graph Data

The UGDP algorithm proposed in literature 10 does not consider the weights of nodes in the social network graph, resulting in low availability of published uncertain graphs. In response to this problem, this article proposes a privacy protection algorithm (Privacy Preserving Algorithm for Uncertain Graph Data Publishing, PUGDP) based on the release of uncertain graph data. The algorithm model is as follows (Fig. 2):

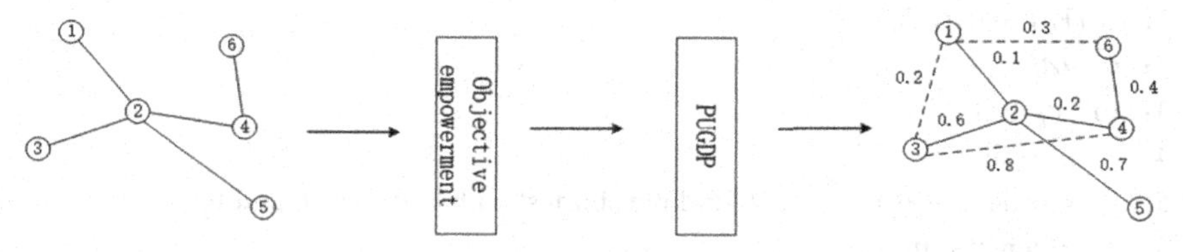

Fig. 2. Uncertain graph construction model of PUGDP algorithm

Where the PUGDP algorithm uses differential privacy to assign edge probabilities to the objectively weighted social network graph to generate an uncertain graph for publication. This algorithm mainly adds noise to the structure of social networks, so as to achieve the effect of privacy protection. It mainly consists of the following 4 steps.

1) First, judge whether the two nodes are associated or not, if they are associated, find the shortest path, and then calculate the sum of the weights of the path nodes to get the average value β. The calculation formula for β is:

$$\beta = \left(\sum\nolimits_{i=1}^{n} w_i\right)/n \tag{6}$$

A threshold α is set here. If β is greater than or equal to α, the two sides directly add noise.

2) In order to construct an uncertain graph, each noise value y_i corresponds to a probability value p_i, The probability value $p_i = Pr[y_i] = F(y_i)$.
3) The probability value p_i is added to the graph G_1 or G_2 to form an uncertain graph, and the probability value p_i is used as the probability that there is an edge between the vertex and the vertex.

$$F(y_i) = \int_{-\infty}^{y_i} g(X)dX \tag{7}$$

where, g(x) is the Laplace distribution that obeys the expected value μ and the position parameter b.

$$g(X) = \frac{1}{2b}exp(-\frac{|X - \mu|}{b}) \tag{8}$$

4) When publishing the constructed uncertain graph, in order to better protect the information in the original graph, use the adjacent uncertain graph G'' as the release graph.

The algorithm is described in detail as follows:

Algorithm 1.PUGDP .

Input: Original image:$G=(V,E)$, sensitivity:Δf;
Output : $G''=(V,p)$.
1: $G'(V,W) \leftarrow G(V,E)$;
2: $b \leftarrow (\Delta f/\varepsilon)$;
3: for v_i in V:
4: for $v_j(i<j)$ in V:
5: shortlest(v_i,v_j); //Find the shortest path between v_i and v_j.
5: if len(v_i ,v_j)==1 :
6: $y_i \leftarrow Lap(b), y_i \in Y$;
7: else:
8: $\beta = (\sum_{i=1}^{n} w_i)/n$; //Find the average value of the shortest path node-weight.
9: if $\beta \geq \alpha$:
10: $y_i \leftarrow Lap(b), y_i \in Y$;
11: end for;
12: end for;
13: $Pr[y_i] \leftarrow F(y_i)$;
14: $p_i \leftarrow Pr[y_i], p_i \in p$;
15: $e_i \leftarrow p_i, p_i \in p, e_i \in E$; //Add probability p_i to edge e_i.
16: Return $G''=(V,p)$;

Algorithm 1 is to transform the original social network graph into an uncertain graph for publication. The first line gives objective weights to the original graph, the second line calculates the differential privacy budget, and the fourth to 12th lines calculate the shortest path of any two nodes in the original graph. If the number of shortest path edges is 1, add noise directly, If the edge of the shortest path is greater than 1, calculate the average value β of the node weights on the path, if β is greater than the threshold α, add noise. Lines 14–16 convert the noise into probability and add it to the corresponding edge, and finally return to the uncertain graph G'.

3.3 Privacy Analysis

Theorem. The PUGDP algorithm satisfies differential privacy.

Proof: Let $f(\cdot)$ be the function $f: G \rightarrow G''$, G'' is the neighboring uncertain graph output by the algorithm, G_1, G_2 are the neighboring graphs, that is, the Hamming distance of the edge in the graph is 1. P_{G_1} means PUGDP (G_1, f, ε) probability density function, P_{G_2} represents the probability density function of PUGDP (G_2, f, ε). The relationship between probability and noise is $y_i \rightarrow p_i$. G_3 is the noise map generated by the Laplace function during the algorithm. Because The post-processing technology of the process from the noise map to the uncertain map meets the differential privacy. Therefore, in order to prove that the PUGDP algorithm meets the differential privacy, it is only necessary to prove that the original map is converted to the noise map to meet the differential privacy. The specific proof process is as follows:

$$\frac{P_{G_1}[G_3]}{P_{G_2}[G_3]} = \frac{P_{G_1}[PUGDP(G_1,f,\varepsilon)-f(G_1)]}{P_{G_2}[PUGDP(G_2,f,\varepsilon)-f(G_2)]}$$
$$= \prod_{i=1}^{j}\left(\frac{exp\left(-\frac{|f(G_1)_i-G_{3i}|}{\Delta f/\varepsilon}\right)}{exp\left(-\frac{|f(G_2)_i-G_{3i}|}{\Delta f/\varepsilon}\right)}\right)$$
$$= \prod_{i=1}^{j} exp\left(\frac{|f(G_2)_i-G_{3i}|-|f(G_1)_i-G_{3i}|}{\Delta f/\varepsilon}\right)$$
$$\leq \prod_{i=1}^{j} exp\left(\varepsilon \cdot \frac{|f(G_2)_i-f(G_1)_i|}{\Delta f}\right)$$
$$= exp\left(\varepsilon \cdot \frac{\|f(G_2)-f(G_1)\|_1}{\Delta f}\right) \leq exp(\varepsilon).$$

The PUGDP algorithm not only has the privacy protection effect of the uncertain graph, but also has the protection effect of differential privacy. Under the double privacy protection effect, the PUGDP algorithm has a strong privacy protection effect. In the privacy protection of uncertain graphs, we use α to control the number of edges. The larger the α, the fewer edges will be added, and the smaller the privacy protection effect on social networks, but the stronger the data availability, and vice versa. The more edges you add, the stronger the privacy protection effect on social networks, but the smaller the data availability, and the size of α can be controlled according to different data requirements.

4 Experimental Analysis and Comparison

In order to evaluate the privacy protection effect of different algorithms, according to the data characteristics of the uncertain graph, this paper introduces edge entropy to measure the privacy protection effect of the uncertain graph generated by the algorithm. In terms of data availability, a statistical indicator is introduced for measurement. This article uses Python language and the third-party function package NetworkX to conduct comparative experiments. The experimental data set is the gowalla data set publicly available on the Stanford University website. The data set has a total of 196591 nodes. 50, 100, 200, and 500 nodes were selected for experiments. In order to reduce randomness, the experiment was performed 10 times and averaged value.

4.1 Objective Weighting Based on Entropy Method

Data availability is measured according to some metrics proposed in the literature 17, 18. where *NE* represents the number of edges in the structure graph, *AD* represents the average degree of nodes in the structure graph, and *DV* represents the variance of the node degrees in the structure graph.

In the social network structure graph, use d_1, d_2,..., d_n to express the sequence of the node degree in the structure graph, but the edges in the uncertainty graph exist in the form of probability, and you cannot directly use the node degree in the original graph to express the The degree of the node in the missing graph, in the uncertain graph d_1, d_2,..., d_n is a random probability sequence, where the expected degree of the node is used to represent the degree of the node in the uncertain graph, that is, the probability of the edge connected to the node is used The sum to represent the degree of this node.

Such as the following equation:

$$d_v = \sum p(i, j) \tag{9}$$

In formula (10), $i = v$ or $j = v$ and $i \neq j$.

Formula (11) and formula (12) are the calculations of *NE* and *AD* in the original graph. For the calculation of NE' and AD' in the uncertain graph, please refer to formula (14) and formula (15), where *DV* is in the original graph and the uncertainty The calculation in the figure is consistent with formula (13).

$$NE = \frac{1}{2}\sum\nolimits_{v \in V} d_v \tag{10}$$

$$AD = \frac{1}{n}\sum\nolimits_{v \in V} d_v \tag{11}$$

$$DV = \frac{1}{n}\sum\nolimits_{v \in V} (d_v - AD)^2 \tag{12}$$

$$NE' = \frac{1}{2}\sum\nolimits_{v \in V}\sum\nolimits_{u \in V \backslash v} p(u, v) = \sum\nolimits_{e \in V_2} p(e) \tag{13}$$

$$AD' = \frac{1}{n}\sum\nolimits_{v \in V}\sum\nolimits_{u \in V \backslash v} p(u, v) = \frac{2}{n}\sum\nolimits_{e \in V_2} p(e) \tag{14}$$

4.2 Privacy Measurement

Because the edges of uncertain graphs are strongly random, this paper introduces edge entropy Ent_e to measure the privacy protection effect when measuring the privacy of uncertain graphs. Entropy is used to measure the degree of chaos in a system in information theory. The larger the information entropy, the more chaotic the system, the smaller the information entropy, the more orderly the system, and the side entropy reflects the degree of chaos in the graph structure. The definition of Ent_e is shown in the following formula:

$$I_{e_i} = -p(e_i) \times log_2 p(e_i) \tag{15}$$

where, $e_i \in G'$, $p(e_i)$ is the probability that the edge exists.

$$Ent_e = \sum_{e \in G'} I_{e_i} \tag{16}$$

The larger the Ent_e value reflects the more chaotic the structure of the uncertain graph, that is, the greater the degree of uncertainty, the better the privacy of the published uncertain graph.

4.3 Data Availability Analysis

While ensuring privacy protection, *NE, AD*, and *DV* are used to measure data availability. The closer the three metrics of *NE, AD*, and *DV* in the uncertain figure are to the original figure, the better the data availability. Table 1 and Table 2 list the data availability comparison between the original graph and the UGDP algorithm at $\varepsilon = 0.01, \varepsilon = 0.1, \varepsilon = 1$. Through a large number of experiments, when the PUGDP algorithm threshold α is around 0.02, the privacy protection does not decrease too much, but data availability is greatly improved. When α is 0.02, the algorithm effect is the best. so this article only lists three values 0.01, 0.02, and 0.05 for comparison experiments, so Tables 3, 4 and 5 list $\varepsilon = 0.01, \varepsilon = 0.1, \varepsilon = 1$ when $\alpha = 0.01, \alpha = 0.02, \alpha = 0.05$ Comparison of data availability.

Table 1. Data availability comparison between the original graph and the uncertain graph generated by the UGDP algorithm when $\varepsilon = 0.1$.

Metrics	Nodes					
	The original image			Uncertain graph $\varepsilon = 0.1$		
	NE	*AD*	*DV*	*NE'*	*AD'*	*DV'*
50	81	3.18	46.69	618.55	12.12	150.80
100	264	5.23	105.80	2522.17	24.97	630.83
200	731	7.27	217.46	9955.64	49.53	2469.53
500	3735	14.91	696.03	62568.19	124.89	15638.55

Table 2. Usability ratio of uncertain graph data generated by UGDP algorithm when $\varepsilon = 0.01$ and $\varepsilon = 1$.

Metrics	Nodes					
	Uncertain graph $\varepsilon = 0.01$			Uncertain graph $\varepsilon = 1$		
	NE'	*AD'*	*DV'*	*NE'*	*AD'*	*DV'*
50	647.59	12.70	164.12	631.76	12.39	157.51
100	2507.17	24.82	623.30	2537.78	25.13	623.13
200	10061.30	50.06	2520.88	10126.38	50.38	2555.28
500	62581.82	124.91	15641.60	62684.87	125.12	15692.18

Table 3. Comparison of availability of uncertain graph data generated by PUGDP algorithm when $\varepsilon = 0.01, \alpha = 0.01, \varepsilon = 0.01, \alpha = 0.0\,2$ and $\varepsilon = 0.01, \alpha = 0.0\,5$.

Metrics	Nodes								
	PUGDP $\varepsilon = 0.01$, $\alpha = 0.01$			PUGDP $\varepsilon = 0.01$, $\alpha = 0.02$			PUGDP $\varepsilon = 0.01$, $\alpha = 0.05$		
	NE'	AD'	DV'	NE'	AD'	DV'	NE'	AD'	DV'
50	606.51	11.89	146.67	344.79	6.76	51.56	175.51	3.44	18.80
100	1323.15	13.10	186.75	839.28	8.31	90.05	435.18	4.31	42.04
200	3171.28	15.78	292.96	1871.03	9.31	131.53	980.53	4.88	72.40
500	9513.41	18.99	490.63	5833.85	11.64	279.12	3477.14	6.94	207.02

Table 4. Comparison of the availability of uncertain graph data generated by the PUGDP algorithm when $\varepsilon = 0.1, \beta = 0.01, \varepsilon = 0.1, \alpha = 0.0\,2$ and $\varepsilon = 0.1, \alpha = 0.0\,5$.

Metrics	Nodes								
	PUGDP $\varepsilon = 0.1, \alpha = 0.01$			PUGDP $\varepsilon = 0.1, \alpha = 0.02$			PUGDP $\varepsilon = 0.1$, $\alpha = 0.05$		
	NE'	AD'	DV'	NE'	AD'	DV'	NE'	AD'	DV'
50	626.23	12.30	156.54	330.35	6.48	47.22	182.40	3.58	21.50
100	1319.36	13.06	184.57	828.11	8.20	86.32	419.50	4.15	38.15
200	3181.64	15.83	297.19	1881.16	9.36	132.22	1008.51	5.01	76.44
500	9442.04	18.84	474.04	5841.73	11.66	275.89	3468.14	6.92	212.69

Table 5. Comparison of the availability of uncertain graph data generated by the PUGDP algorithm when $\varepsilon = 1, \alpha = 0.01, \varepsilon = 1, \alpha = 0.02$ and $\varepsilon = 1, \alpha = 0.05$.

Metrics	Nodes								
	PUGDP $\varepsilon = 1, \alpha = 0.01$			PUGDP $\varepsilon = 1, \alpha = 0.02$			PUGDP $\varepsilon = 1, \alpha = 0.05$		
	NE'	AD'	DV'	NE'	AD'	DV'	NE'	AD'	DV'
50	614.00	12.04	150.66	334.54	6.55	49.63	178.52	3.50	20.73
100	1313.73	13.00	180.64	801.49	7.94	81.49	430.64	4.26	37.55
200	3186.73	15.85	296.17	1875.95	9.33	128.28	986.72	4.90	67.06
500	9484.90	18.93	493.72	5778.94	11.53	270.22	3476.72	6.94	216.13

By comparing the data obtained from the original map, UGDP and PUGDP, the PUGDP data is closer to the original map to a certain extent, so it can be concluded that PUGDP has higher data availability than the UGDP.

4.4 Privacy Analysis

This paper uses edge entropy to measure the privacy of existing algorithms and the proposed algorithms. The UGDP algorithm was verified by experiments at $\varepsilon = 0.01$, $\varepsilon = 0.1$ and $\varepsilon = 1$. When the PUGDP algorithm and UGDP algorithm ε are the same, the experiments are verified when α is 0.01, 0.02 and 0.05. Table 6 shows the UGDP algorithm. The edge entropy changes, Tables 7, 8, 9 are the edge entropy changes of the PUGDP algorithm.

Table 6. Changes in edge entropy in UGDP algorithm.

Metrics	Nodes		
	ε		
	$\varepsilon = 0.01$	$\varepsilon = 0.1$	$\varepsilon = 1$
50	471.47	455.59	453.60
100	1821.83	1824.89	1819.58
200	7242.42	7265.50	7231.89
500	45132.72	45123.00	45198.44

Table 7. PUGDP algorithm changes in edge entropy when $\varepsilon = 0.01$.

Metrics	Nodes		
	ε, α		
	$\varepsilon = 0.01, \alpha = 0.01$	$\varepsilon = 0.01, \alpha = 0.02$	$\varepsilon = 0.01, \alpha = 0.05$
50	463.37	238.71	131.53
100	934.63	603.38	315.86
200	2284.45	1344.04	714.65
500	6846.42	4161.70	2529.68

Through the comparison of Tables 6, 7, 8, 9, the side entropy of UGDP is always larger than the side entropy of PUGDP. According to Tables 7, 8, 9, when ε is equal, the side entropy gradually decreases with the increase of α. Therefore, the privacy protection of UGDP algorithm is better than that of PUGDP algorithm.For the PUGDP algorithm, privacy protection gradually becomes smaller as α increases, and the size of α can be adjusted according to different application scenarios.

Table 8. The change of edge entropy of PUGDP algorithm when $\varepsilon = 0.1$

Metrics	Nodes		
	ε, α		
	$\varepsilon = 0.1, \alpha =$ 0.01	$\varepsilon = 0.1, \alpha =$ 0.02	$\varepsilon = 0.1, \alpha =$ 0.05
50	458.50	243.53	135.09
100	949.96	611.78	322.65
200	2287.83	1327.69	730.36
500	6859.00	4191.55	2506.25

Table 9. The change of edge entropy of PUGDP algorithm when $\varepsilon = 1$

Metrics	Nodes		
	ε, α		
	$\varepsilon = 1, \alpha =$ 0.01	$\varepsilon = 1, \alpha =$ 0.02	$\varepsilon = 1, \alpha =$ 0.05
50	445.90	243.70	129.53
100	948.71	607.86	318.61
200	2277.78	1345.85	720.66
500	6846.50	4192.39	2502.28

5 Conclusion

Aiming at the poor data availability of traditional uncertain graph privacy protection algorithms, this paper proposes a privacy protection method for uncertain graph data release based on objective weighting. This method not only has the dual protection effects of differential privacy and uncertain graphs, but also corrects whether it is not. The algorithm for publishing the deterministic graph has been improved to objectively assign weights to nodes, making the published uncertain graph data more usable. Through a large number of experimental verifications, the PUGDP algorithm improves data availability while ensuring data privacy. This article adopts the entropy weight method for objective weighting. The entropy weight method has a strong mathematical basis, but the entropy weight method only analyzes the structure of the social network, and does not analyze the node attributes. Therefore, in the next step, we will explore the study of weighting methods that combine social network structure and node attributes.

Acknowledgments. This work was supported by The Natural Science Foundation of China.

References

1. Wang, T., Long, S., Ding, H.: Differential privacy protection algorithm for large social networks. Comput. Eng. Des. **41**(06), 1568–1574 (2020)
2. Gu, Z., Zhang, G., Ma, C., Song, L.: Differential privacy data publishing method based on probabilistic principal component analysis. J. Hal Univ. Technol., 1–8 (2021)
3. Wang, L., Meng, X.: ESA: a new privacy protection framework. Comput. Res. Dev., 1–27 (2021)
4. Li, Y., Liu, S., Li, D.: Release connection fingerprints in social networks using personalized differential privacy. Chin. J. Electron. **27**(5), 1104–1110 (2018)
5. Esmerdag, E., Gursoy, M.E., Inan, A.: Explode: an extensible platform for differentially private data analysis. In: Proceedings of the 16th IEEE International Conference on Data Mining Workshops, pp. 1300–1303. IEEE (2016)
6. Zhu, S., Wang, L., Sun, G.: Classification transformation perturbation mechanism satisfying local differential privacy. Comput. Res. Dev., 1–10 (2021)
7. Liu, S., Zhu, Y.: Differential privacy protection for social network edge weights. Comput. Eng. Des. **39**(01), 44–48 (2018)
8. Zhang, X., Meng, X.: Differential privacy protection for data publishing and analysis. Acta Comput. Sinica **37**(04), 927–949 (2014)
9. Xu, J., Zhang, Z., Xiao, X., Yang, Y., Yu, G., Winslett, M.: Differentially private histogram publication. VLDB J. **22**(6), 797–822 (2013)
10. Wu, Z., Hu, J., Tian, Y., Shi, W., Yan, J.: Privacy preserving algorithm for uncertain graphs in social networks. Acta Sinica Sinica Sinica **30**(4), 1106–1120 (2019)
11. Song, Y., Ding, L., Li, Y., Liu, X.: A differential privacy protection algorithm for graph model data. Autom. Technol. Appl. **39**(07), 86–90 (2020)
12. Boldi, P., Bonchi, F., Gionis, A.: Injecting uncertainty in graphs for identity obfuscation. Proc. VLDB Endow. **5**(11), 1376–1387 (2012)
13. Yan, J., Hu, J., Wen, G., Tian, Y.: Privacy protection method for uncertain graph social networks based on node characteristics. Inf. Secur. Res. **4**(06), 533–538 (2018)
14. Hu, J., Shi, W.C., Yan, J., Wu, Z.Q.: Research on privacy protection method based on uncertain graph. Comput. Technol. Dev. **28**(12), 116–121 (2018)
15. Nguyen, H.H., Imine, A., Rusinowitch, M.: A maximum variance approach for graph anonymization. In: Cuppens, F., Garcia-Alfaro, J., Zincir Heywood, N., Fong, P. (eds.) FPS 2014. LNCS, vol. 8930, pp. 49–64. Springer, Cham (2015). https://doi.org/10.1007/978-3-319-17040-4_4
16. Nguyen, H., Imine, A., Rusinowitch, M.: Anonymizing social graphs via uncertainty semantics. In: Proceedings of the 10th ACM Symposium on Information, Computer and Communications Security, Singapore, pp. 495–506 (2015)
17. Nguyen, H.H., Imine, A.: Anonymizing social graphs via uncertainty semantics. In: Proceedings of the 10th ACM Symposium on Information, Computer and Communications Security, New York, pp. 495–506 (2015)

Federated Multi-label Propagation Based on Neighbor Node Influence for Community Detection on Attributed Networks

Panpan Yang[1,2,3], Kun Guo[1,2,3(✉)], Ximeng Liu[1,2,3], and Yuzhong Chen[1,2,3]

[1] Fujian Provincial Key Laboratory of Network Computing and Intelligent Information Processing, Fuzhou University, Fuzhou 350108, China
{gukn,yzchen}@fzu.edu.cn

[2] College of Mathematics and Computer Science, Fuzhou University, Fuzhou 350108, China

[3] Key Laboratory of Spatial Data Mining and Information Sharing, Ministry of Education, Fuzhou 350108, China

Abstract. The research on community detection is usually based on the topological structure and attribute information of complex networks to improve computation precision. However, as more and more people pay attention to the disclosure of personal privacy, detecting communities without leaking sensitive information has become a hot topic in complex network analysis. In this paper, we first propose a distributed privacy-preserving graph learning model. Second, we develop a multi-label propagation algorithm (MLPA) based on the model to detect overlapping communities securely on the horizontally distributed networks with attributes. A novel perturbation strategy is combined with homomorphic encryption to achieve flexible privacy control and strict privacy protection. Moreover, a node similarity calculation method is proposed to consider the structural and attribute influences of each node's neighbors in label propagation no matter the attributes are numeric or categorical. The experiments on real-world and artificial networks demonstrate that our algorithm achieves identical results as the standalone MLPA and higher accuracy (200%) than the simple distributed MLPA without federated learning.

Keywords: Multi-label propagation · Neighbor node influence · Community detection · Federal learning · Attributed network

1 Introduction

Community detection aims to mine the community structure of a complex network to help us obtain an in-depth understanding of the functions and behaviors of its nodes and edges. Label propagation algorithms (LPAs) [2] are a type of typical community detection algorithm that have many merits. First, their time complexity is near-linear to network size, which makes them particularly suitable for large networks. Second, they rely on the iterative propagation of node

Y. Sun et al. (Eds.): ChineseCSCW 2021, CCIS 1492, pp. 484–498, 2022.
https://doi.org/10.1007/978-981-19-4549-6_37

labels to let the communities composed of nodes with the same labels emerge automatically. Therefore, we do not have to input the number of communities in advance. To extend LPAs to handle networks with overlapping nodes, a node is allowed to carry multiple labels, which means that it can belong to multiple communities. The representatives of such improved MLPAs include COPRA [3], SLPA [4] and so on. Furthermore, BMLPA [5], LPANNI [6] and NALPA [7] were proposed to eliminate the randomness of label propagation, which makes MLPAs run more steadily.

With people's increasing concern about personal privacy, detecting communities without leaking sensitive information has become a prerequisite in secure complex network analysis. The anonymization-based methods such as k-anonymity [8], l-diversity [9] and differential privacy (DP) [10], have been used for community detection. However, they all incur inevitable accuracy loss because the noises added to a network, such as added or removed nodes and edges severely interfere with the detection of the true communities. In contrast, cryptography-based methods such as homomorphic encryption [21], secret sharing [22] can protect the sensitive information of networks strictly. However, the intensive encryption and decryption operations are time-consuming, which makes them inapplicable to large networks.

The paradigm of federated learning [1] was first proposed by Google which provides a possible solution to the above problems. Federated learning is a new fundamental technology of artificial intelligence to solve the problem of data island. It establishes a machine learning model by combining the datasets distributed on multiple participants. All participants carry out collaborative training to share only the intermediate results instead of their private data.

In this paper, we first propose a federated graph learning model (FGLM) for privacy-preserving network data mining on distributed networks. Second, depending on the model, we propose a horizontal MLPA based on neighbor node influence (NNI) for community detection on attributed networks (HF-MLAPNNI-AN). A novel node perturbation strategy is designed to conceal network privacy in label propagation without losing any accuracy. Besides, a new node similarity calculation method is developed to consider both the structural and attribute influences of a node's neighbors. The theoretical analysis and the experiments on real-world and artificial networks demonstrate that our algorithm achieves identical results as the standalone MLPA while providing full protection on network privacy. They also gain higher accuracy over the simple distributed MLPA without federated learning. The main contributions of our study are as follows:

1. The proposed FGLM is suitable for horizontal federated learning paradigms. We can develop the horizontal federated versions of a standalone network data mining algorithm based on the same model, which is proved in this study.
2. The node perturbation strategy is combined with homomorphic encryption to achieve flexible control over privacy-level without accuracy lost and strict protection on the intermediate results in distributed label propagation.

3. The node similarity computation method not only considers the structural and attribute similarity between the nodes and the influence of their neighbors but also utilizes all numeric and categorical attributes of the nodes, therefore improving the computation precision.

2 Related Work

2.1 Multi-label Propagation Algorithm

The COPRA algorithm [3] proposed by Gregory et al. is the pioneering work that extends LPA to overlapping community detection by relaxing LPA's constraint on the number of labels. Wu et al. proposed a balanced pair label propagation algorithm (BMLPA) [5] which eliminated the randomness of label propagation according to a concept of balanced attribution coefficient. Lu et al. proposed an improved LPA overlapping community detection algorithm (LPANNI) [6] which uses a fixed label propagation sequence based on the ascending order of node importance scores and a label update strategy based on NNI to detect overlapping communities. In 2020, Zhang et al. proposed an LPA based on node capability (NALPA) [7] which adopts four-node capabilities (propagation capability, attraction capability, launching ability and sender capability) and label influence and a new label propagation mechanism to improve the stability and efficiency of LPA.

2.2 Privacy Protection Community Detection

At present, only the technique of anonymization has been applied to privacy-preserving community detection. Sarah et al. proposed a (k, x) isomorphism method [8] to protect personal privacy in weighted social networks. For some social networks where nodes and their labels have structural attributes, Tripathy et al. proposed an algorithm [9] that can be used to realize k-anonymity and l-diversity on the networks. Ji et al. proposed a DP community detection (DPCD) scheme [10] which detects communities in social networks through a probability generation model. The anonymization-based community detection algorithms run fast and provide flexible control over privacy-level. However, they have to sacrifice accuracy for privacy because the added noises make discovering communities more difficult. Although encryption is another popular technique for privacy-preserving graph data mining [23,24], it has not yet been applied to privacy-preserving community detection.

2.3 Federal Learning on Graphs

Wang et al. proposed the first federated learning framework (GraphFL) [11] for semi-supervised node classification of graphs. The framework is driven by a meta-learning method that solves the non-IID problem in federated graph learning and handles new labels well. Ialitha et al. proposed a distributed point-to-point federated learning algorithm [12] to train deep neural networks (DNNS).

Mei et al. proposed an adjacency matrix protection method [13] based on DP to protect a network's topology and an adjacency matrix aggregation method to improve a global model with multiple local models based on GNN (Graph Neural Network).

3 Preliminaries

3.1 Network Privacy

We assume that our framework is carried out under the semi-honest adversary model. Each participant does not disclose his information actively or attack other participants. A privacy protection algorithm on the graph is mainly to prevent the following two parts of information from being leaked out:

1. *The topology of a network, including nodes ID, nodes degree and edges existence.* On a complex network, the degree of most nodes may be different, so attackers can accurately locate the location of nodes through different node degrees. Moreover, the attacker can determine whether two nodes are connected through edges, although the two nodes' ID may not be known.
2. *The attributes of nodes and edges, including type, weight, profile, and so on.* In large complex networks, few nodes and edges have the same attributes. Therefore, if the attributes of sensitive nodes or edges are compromised, the attacker can accurately locate them.

3.2 Standalone MLPANNI-AN Algorithm

We present a general standalone MLPA based on LPANNI [6] for attributed networks (SMLPANNI-AN) which plays as a template to design the federated MLPA. SMLPANNI-AN contains the following three stages.

Stage 1: NNI computation. We first calculate node importance (NI) and the similarity between each pair of nodes (s). Then, the NNI is calculated based on NI and s. A new attribute similarity method is designed to calculate s.

STEP 1: *NI computation.* Calculate the NI value of each node according to Eq. (1).

$$NI(u) = \frac{k_u}{n} \tag{1}$$

where n indicates the number of nodes and k_u the sum of the attribute-based similarity between the node and its neighbors. k_u is calculated according to Eq. (2).

$$k(u) = \sum_{v \in N_b(u)} s_a(u, v) \tag{2}$$

where $N_b(u)$ denotes the set of neighbors of node u, $s_a\,(u, v)$ represents the attribute-based similarity between nodes u and v, which is calculated according to Eq. (3).

$$s_a(u, v) = \frac{1}{w} \sum_{k=1}^{w} \begin{cases} I(a_{uk}, a_{vk}), & categorical \\ |a_{uk} - a_{vk}|, & numeric \end{cases} \tag{3}$$

where a_{uk} is the value of the kth attribute of node u. $I(x, y)$ is the indicator function. When a categorical attribute values of node u and node v are identical, $I = 1$; otherwise, $I = 0$. Both categorical and numeric attributes are supported by Eq. (3).

STEP 2: *Node similarity computation.* Calculate the similarity between two nodes according to $s_a(u, v)$ and an α-hop distance between the nodes, as shown in Eq. (4).

$$s(u,v) = \frac{s_t(u,v)^2}{\sum_{x \in N_b(u)} s_a(u,x) + \sum_{y \in N_b(v)} s_a(v,y)} \tag{4}$$

$$s_t(u,v) = \sum_{|p|=1}^{\alpha} \frac{A_{uv}^{|p|}}{|p|} \tag{5}$$

where p represents the path from node u to node v directly or indirectly, $|p|$ indicates the path length of p and $A^{|p|}$ denotes the measure of p. For unweighted network, $A^{|p|} = 1$, α is a parameter of the path length. As suggested in [6], $\alpha = 2$ or 3 suffices to obtain acceptable results.

STEP 3: *NNI computation.* Calculate the influence of neighbor node v on node u according to Eq. (6).

$$NNI(u,v) = \sqrt{NI(u) \times \frac{s(u,v)}{\sum_{h \in N_b(u)} s(u,h)}} \tag{6}$$

Stage 2: Label propagation.

STEP 1: *Label initialization.* Each node initializes its label dictionary according to Eq. (7).

$$L_u = \{(c_1, b_1), (c_2, b_2), ..., (c_v, b_v)\}, v \in N_b(u) \tag{7}$$

where c denotes the label of node u, b denotes the belonging coefficient of label c.

STEP 2: *Label update.* The belonging coefficient of node u to label i is updated according to Eq. (8).

$$b_i = \frac{\sum_{v \in N_b(u)} L_u(c) \times NNI(u,v)}{\sum_{c' \in L_u.keys} \sum_{v \in N_b(u)} L_u(c') \times NNI(u,v)} \tag{8}$$

where $L_{u.keys}$ denotes a set of key values of the dictionary L_u.

STEP 3: *Label selection and normalization.* For each node, remove the labels whose belonging coefficient is less than $\frac{1}{L_{u'}}$ from its label set and normalize the belonging coefficients to the remaining labels according to Eq. (9).

$$b_i' = \frac{b_i}{\sum_{c' \in L_u.keys} L_u(c')} \tag{9}$$

Stage 3: Community generation. Merge the nodes with the same label into a community and output the communities.

4 HF-MLPANNI-AN Algorithm

4.1 FGLM

The scenarios of horizontal federated learning are often met in the real world. For example, the banks in different cities offer services to different customers with similar account features. We can easily extend the horizontal federated learning to graph or network data. Nodes play the role of customers and node attributes are similar to customer features. Given this, we design a federated graph learning model (FGLM) as shown in Fig. 1. The nodes belonging to different participant's local networks are rendered in different colors. Vectors $a_{A,i}$ and $a_{B,i}$ denote the attribute vectors of node v_i in participants A and B, respectively.

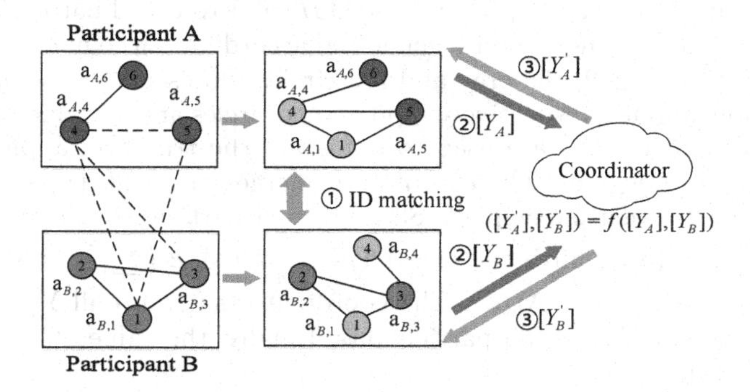

Fig. 1. Federated graph learning model

For the FGLM model, we first add the missing edges in the dotted line to each participant's network according to the vertex-cut-based graph partition rule [25] to ensure that there are no missing edges when the two subgraphs are combined. Second, we use the private set intersection (PSI) [19] protocol proposed in De et al. to match the overlapping of two networks without revealing the ID of other nodes. Third, the two participants send the encrypted information of the overlapping nodes Y_a and Y_b to the coordinator for intermediate information exchange, where $[\cdot]$ represents the encryption algorithm including hash, homomorphic encryption, etc. Finally, participants A and B can conduct data mining by exchanging intermediate information with the coordinator many times without disclosing any sensitive information. It seems clear that FGLM can be extended to the situation where more participants work.

4.2 Implementation of HF-MLPANNI-AN Algorithm

HF-MLPANNI-AN is composed of four stages, as shown follow.

Stage 1: Node ID Matching. We use the PSI protocol to match the overlapping node set X_h of each participant.

Stage 2: NNI computation. Stage 2 includes three steps to calculate the NNI values.

STEP 1: *NI computation.* First, each participant calculates s_a according to Eq. (3). Second, they calculate $Y_h = \left\{H(u), E\left(\sum_{i=1}^{|N_b(u)|} s_a(u,v)\right)\right\}$, where $u \in X_h$, $v \in N_b(u)$, $H()$ and $E()$ are a hash function such as SHA [20] and a homomorphic encryption function, respectively. After all Y_hs are sent to coordinator T, it sums up all $Y_h\,(H\,(u))$s to Y_T and sends back Y_T to each participant. Finally, each participant calculates the NI value of each node based on $D\left(\sum_{i=1}^{p} Y_T(H(u))\right)$ according to Eq. (1). $D()$ is a homomorphic decryption function and p is the number of participants.

STEP 2: *Node similarity computation.* For each node $u \in X_h$, a participant first sends a dictionary $Y_h = \{H(u), H(v) \cup H\,(u')\}$ to coordinator T, where $v \in N_b(u)$ and u' is a set of noise nodes generated according to the node perturbation strategy shown in Fig. 2, the grey and red circles represent the overlapping and non-overlapping nodes, respectively. The orange circles are the added noise nodes with random indices that are used to obfuscate the true overlapping nodes to conceal their true degrees. The coordinator merges all $Y_h\,(H\,(u))$s into Y_T and sends back Y_T to each participant. Second, each participant calculates $s_t(u,v)$ according to Eq. (5). Third, they send $Y_{h'} = \left\{H(u), E\left(\sum_{i=1}^{|N_b(u)|} s_t(u,v)\right)\right\}$ to coordinator T, where $v \in N_b(u)$. The coordinator sums up all $Y_{h'}(H(u))$s and sends back the sum $Y_{T'}$ to all participants. Finally, the value of s is calculated according to Eq. (4) and $D\left(\sum_{i=1}^{p} Y_{T'}\,(H\,(u))\right)$.

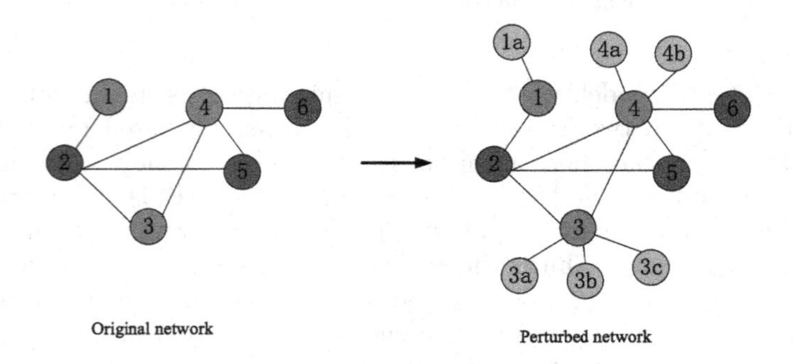

Fig. 2. Node perturbation strategy

STEP 3: *NNI computation.* First, for $Y_h = \left\{H(u), E\left(\sum_{i=1}^{|N_b(u)|} s(u,v)\right)\right\}$, all participant send it to coordinator T for each node $u \in X_h$. Second, the coordinator sums up all $Y_h\,(H\,(u))$s and sends back the sum Y_T to all participants. Finally, each participant calculates each node's NNI value based on $D\left(\sum_{i=1}^{p} Y_T\,(H\,(u))\right)$ according to Eq. (6).

Stage 3: Label propagation. The stage of label propagation contains three steps.

STEP 1: *Label update.* When this step is first to run, each participant initializes each node's label to its index and the belonging coefficient of the label to 1.0 to build an initial label set L_u, as shown in Eq. (7). Otherwise, each participant updates his label set according to equation (8).

STEP 2: *Label set length computation.* Each participant adopts a homomorphic encryption system [14] which supports mixed operations of integer addition and comparison in a dense state to encrypt the lengths of the updated label sets and sends them to coordinator T. Coordinator T sums up all lengths and returns the sum $L_u^{'}$ to all participants.

STEP 3: *Label selection and normalization.* Each participant removes the labels whose belonging coefficient is smaller than $\frac{1}{|L_u^{'}|}$ and normalizes the remaining labels according to Eq. (9).

Stage 4: Community generation. Merge the nodes with the same label into a community and output the communities.

4.3 Time Complexity Analysis

Let k be the average degree of nodes, T the maximum iteration number, p is the number of participants and w is the number of node attributes. The time complexity of our algorithms is analyzed as follows. In stage 1, the time for running the PSI protocol is $O(n \times p^2)$. Stage 2 contains 3 steps. In step 1, the time complexity of computing s_a and k_u are $O(w)$ and $O(k)$ respectively. The time for attribute similarity computation on the coordinator side is $O(n)$. Therefore, the time for computing NI is $O(w \times k \times n \times p)$. In step 2, computing $s_t(i, j)$ requires $O(k)$ time. The computation conducted by the coordinator requires $O(n)$ time. Therefore, the time for computing $s(u, v)$ is $O(k \times m \times p)$. In step 3, the time complexity of computing NNI is $O(m \times p)$. The total time complexity of stage 2 is $O\,(w \times k \times n \times p + (1 + k) \times m \times p)$. Stage 3 contains 3 steps. In step 1, the time complexity of label update is $O(n \times k)$. In step 2, the time to compute the length of the updated label sets is $O(p \times n + n)$. The time of step 3 is $O(n \times k)$. The total time complexity of stage 3 is $O(T \times n \times k \times p)$. In stage 4, the time for community generation is $O(n \times k)$. In summary, the time complexity of HF-MLPANNI-AN is $O(((p + (w \times T) \times k) \times p + k) \times n + ((1 + k) \times p) \times m) \approx O(h \times n + l \times m)$, which can be reduced to $O(n + m)$ because h, l are constants and $w, T, k, p, h, l \ll n, m$.

4.4 Correctness Analysis

Theorem 1. *The communities generated by HF-MLPANNI-AN are identical to those generated by the SMLPANNI-AN.*

Proof. We prove the theorem by analyzing the correctness of the four stages of HF-MLPANNI-AN.

In stage 1, the PSI protocol guarantees that the overlapping vertices between each pair of participants are correctly identified.

Stage 2 contains three steps to compute NNI. In step 1, the NI values of the non-overlapping nodes are computed locally because their neighbors are all in the local network. The neighbor nodes of the overlapping nodes may be distributed among different participants. Each participant sends the correct sum of s_a of the overlapping nodes to coordinator T. Coordinator T sums up the s_as in ciphertext and returns the correct Y_T. Therefore, each participant can calculate the NI values for the overlapping nodes correctly by decrypting the returned Y_T. The node perturbation strategy guarantees that the coordinator can not deduce the true number of neighbors of the overlapping nodes in step 2. The security of step 3 can be analyzed similarly to step 1. Therefore, stage 2 of HF-MLPANNI-AN obtains the same results as stage 1 of SMLPANNI-AN.

Stage 3 contains three steps of label propagation. Steps 1 and 3 are the same as steps 1 and 3 of SMLPANNI-AN. In step 2, each participant employs an improved homomorphic encryption function [14] to encrypt the length of their updated label set and send them to the coordinator for summation. The correctness of the sum is guaranteed by the addition property of homomorphic encryption. Therefore, step 2 of HF-MLPANNI-AN is also identical to step 2 of SMLPANNI-AN.

Stage 4 is equivalent to running stage 3 of the SMLPANNI-AN for each participant.

In summary, the results of each stage of HF-MLPANNI-AN are consistent with those of SMLPANNI-AN. Therefore, the communities detected by the two algorithms are identical, which proves the correctness of HF-MLPANNI-AN.

4.5 Privacy Protection Analysis

We analyze the possibility of privacy disclosure of the four stages of HF-MLPANNI-AN to prove its security.

Stage 1 is secure because the PSI protocol ensures that no participant can decrypt the local nodes of other participants from the encrypted messages he receives. The proof can be found in [19].

Stage 2 contains three steps. In step 1, the participants use homomorphic encryption to encrypt the attribute similarity of the overlapping nodes and send it to coordinator T. Since homomorphic encryption satisfies IND-CCA (indistinguishability under chosen-ciphertext attack), the coordinator T cannot get the node information of participants through the encrypted attribute similarity values. In step 2, the coordinator or other participants cannot deduce the true number of neighbors of a participant's overlapping nodes because of the node perturbation strategy. The security of step 3 can be analyzed similarly to step 1.

Stage 3 contains three steps. Steps 1 and 3 are conducted locally, therefore not causing any privacy leakage. In step 2, participants use an improved homomorphic encryption system [14] to compare two integers in a dense state. Therefore, the lengths of each participant's update label sets are not disclosed to the coordinator or other participants.

Stage 4 is conducted only on each participant's local network and causes no privacy leakage.

In summary, each stage of HF-MLPANNI-AN does not disclose any local network privacy of any participant. Therefore HF-MLPANNI-AN is secure to the semi-honest adversaries' attack.

5 Experiments

5.1 Datasets

Six real-world networks and two sets of artificial networks are used in the experiments. There are two non-overlapping real-world networks of Cora[1] and Washington[2] and four overlapping networks which are a set of Facebook ego networks and Twitter ego networks.[3] The detailed description of the networks is shown in Table 1. We use a method, which proposed by Huang et al. [15] to generated node attributes in artificial networks. The description of the artificial networks is shown in Table 2 and the meanings of LFR parameters are explained in Table 3.

Table 1. Real-world networks

Networks	n	m	w
Cora	2708	5278	1433
Washington	230	417	1704
FB-1684	792	14024	319
FB-1912	755	30025	480
T-356963	126	495	247
T-745823	242	4372	1174

5.2 Evaluation Metric

Normalized Mutual Information (NMI) [16] metric is an objective metric to evaluate the accuracy of a community division compared with the true one. However, it cannot apply directly to overlapping community detection. Aaron et al. proposed a new Overlapping Normalized Mutual Information (ONMI) [17] which overcame the problem. ONMI is defined as:

$$ONMI = \frac{\frac{1}{2}\left[H(X) - H(X/Y) + H(Y) - H(Y/X)\right]}{max(H(X), H(Y))} \tag{10}$$

where $H(X)$ and $H(Y)$ are the information entropy values of X and$Y$, respectively. $H(X|Y)$ denotes the entropy of X conditioned on Y.

[1] https://linqs.soe.ucsc.edu/data.
[2] http://www.cs.cmu.edu/afs/cs/project/theo-20/www/data/.
[3] http://snap.stanford.edu/data/.

Table 2. Artificial networks

Networks	n	k	$maxk$	mu	$minc$	$maxc$	om	on
D1	1000–5000	10	50	0.4	10	50	2	100
D2	3000	10	50	0.1–0.7	10	50	2	100

Table 3. Real description of parameters in artificial networks

Parameter	Description
n	Number of nodes
k	Average degree
$maxk$	Maximum degree
mu	Mixing parameter
$minc$	Minimum for the community sizes
$maxc$	Maximum for the community sizes
om	Number of memberships of the overlapping nodes
on	Number of overlapping nodes

5.3 Consistency Experiment

In Fig. 3, the number following the name of HF-MLPANNI-AN denotes the number of participants. From the figures, we can see that the ONMI values of SMLPANNI-AN and HF-MLPANNI-AN are identical. The experimental results shown in the figures demonstrate that the node perturbation strategy and homomorphic encryption adopted in HF-MLPANNI-AN do not affect the correctness of HF-MLPANNI-AN, which is consistent with Theorem 1.

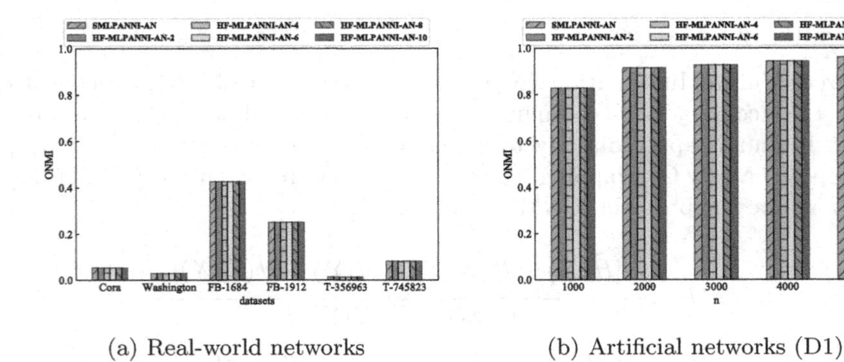

(a) Real-world networks

(b) Artificial networks (D1)

Fig. 3. Consistency experiment of HF-MLPANNI-AN and SMLPANNI-AN

5.4 Attribute Influence Experiment

We define the SMLPANNI-AN algorithm without attributes as SMLPANNI. As shown in Fig. 4, the accuracy of HF-MLPANNI-AN is higher than that of SMLPANNI in both real-world and artificial networks. Specifically, on the artificial networks, when $mu > 0.4$, the improvement is far more than 50%. The experimental results show that the new node similarity computation method can effectively improve the algorithm's accuracy by measuring the similarity between two nodes more finely according to structural and different types of attributes.

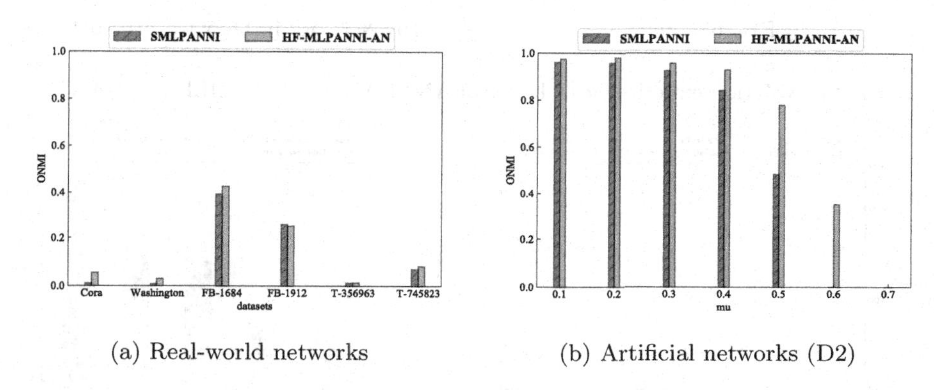

(a) Real-world networks (b) Artificial networks (D2)

Fig. 4. Accuracy of HF-MLPANNI-AN and SMLPANNI

5.5 Ablation Experiment

In the ablation experiments, we define a simplified version of HF-MLPANNI-AN as SHF-MLPANNI-AN by removing the mechanism of federated learning and the processing of overlapping nodes. That is, each participant discovers his local communities independently. The accuracy of the algorithms is computed by averaging the ONMI values calculated on the local communities found in each participant's network. For SHF-MLPANNI-AN, a two-step post-process is conducted. First, we compute the Jaccard coefficient [18] by comparing the local communities of each participant with that of others. Second, the two communities with the largest Jaccard coefficient greater than 0.5 are merged.

As shown in Fig. 5, the accuracy of HF-MLPANNI-AN is significantly higher than SHF-MLPANNI-AN (over 200%). Moreover, with the increase of the number of participants, the accuracy of SHF-MLPANNI-AN drops remarkably because it cannot utilize the overlapping nodes provided by other participants.

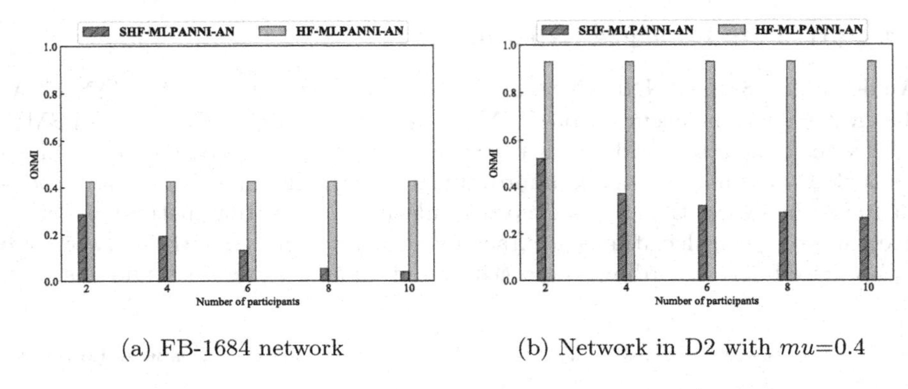

(a) FB-1684 network

(b) Network in D2 with mu=0.4

Fig. 5. Ablation experiment of HF-MLPANNI-AN and SHF-MLPANNI-AN

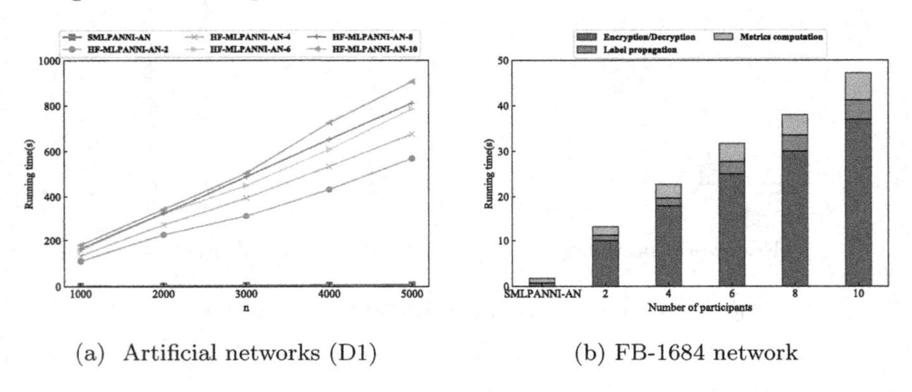

(a) Artificial networks (D1)

(b) FB-1684 network

Fig. 6. Results of the running time experiment

5.6 Runtime Experiment

As shown in Fig. 6a, the running time of HF-MLPANNI-AN increases linearly with the number of network sizes, which is consistent with the theoretical analysis in Sect. 4.3. Moreover, as shown in Fig. 6b, most of the running time is spent on the encryption/decryption operations (more than 70% of the total time) and the attribute similarity computation because the existing homomorphic encryption technique is still very time-consuming and there are many intermediate data to transfer under the distributed computing environment.

6 Conclusion

In this paper, we first propose a federated graph learning model for privacy-preserving data mining on distributed attributed networks. Second, an algorithm of HF-MLPANNI-AN is developed based on the model to execute privacy-protecting distributed community detection. The node perturbation strategy and homomorphic encryption are used to protect the sensitive network information of each participant's local network without sacrificing accuracy. The theoretical analysis and experimental study on real-world and artificial networks verify the correctness and efficiency of our algorithm. In the future, we plan to integrate the vertical federated learning and federated transfer learning paradigms into the FGLM model and develop more privacy-preserving distributed community detection algorithms on it. More complex networks such as dynamic and heterogeneous networks will also be tested in our experiments.

References

1. Bonawitz, K., et al.: Towards federated learning at scale: system design. arXiv preprint arXiv:1902.01046 (2019)
2. Raghavan, U.N., Albert, R., Kumara, S.: Near linear time algorithm to detect community structures in large-scale networks. Phys. Rev. E **76**(3), 036106 (2007)
3. Gregory, S.: Finding overlapping communities in networks by label propagation. New J. Phys. **12**(10), 103018 (2010)
4. Xie, J., Szymanski, B.K.: Towards linear time overlapping community detection in social networks. In: Tan, P.-N., Chawla, S., Ho, C.K., Bailey, J. (eds.) PAKDD 2012. LNCS (LNAI), vol. 7302, pp. 25–36. Springer, Heidelberg (2012). https://doi.org/10.1007/978-3-642-30220-6_3
5. Wu, Z.-H., Lin, Y.-F., Gregory, S., Wan, H.-Y., Tian, S.-F.: Balanced multi-label propagation for overlapping community detection in social networks. J. Comput. Sci. Technol. **27**(3), 468–479 (2012)
6. Lu, M., Zhang, Z., Qu, Z., Kang, Y.: LPANNI: overlapping community detection using label propagation in large-scale complex networks. IEEE Trans. Knowl. Data Eng. **31**(9), 1736–1749 (2018)
7. Zhang, Y., Liu, Y., Zhu, J., Yang, C., Yang, W., Zhai, S.: NALPA: a node ability based label propagation algorithm for community detection. IEEE Access **8**, 46642–46664 (2020)
8. Sarah, A.-K., Tian, Y., Al-Rodhaan, M.: A novel (k, x)-isomorphism method for protecting privacy in weighted social network. In: 2018 21st Saudi Computer Society National Computer Conference (NCC), pp. 1–6. IEEE (2018)
9. Tripathy, B., Mitra, A.: An algorithm to achieve k-anonymity and l-diversity anonymisation in social networks. In: 2012 Fourth International Conference on Computational Aspects of Social Networks (CASoN), pp. 126–131. IEEE (2012)
10. Ji, T., Luo, C., Guo, Y., Ji, J., Liao, W., Li, P.: Differentially private community detection in attributed social networks. In: Asian Conference on Machine Learning, pp. 16–31. PMLR (2019)
11. Wang, B., Li, A., Li, H., Chen, Y.: GraphFL: a federated learning framework for semi-supervised node classification on graphs. arXiv preprint arXiv:2012.04187 (2020)
12. Lalitha, A., Kilinc, O.C., Javidi, T., Koushanfar, F.: Peer-to-peer federated learning on graphs. arXiv preprint arXiv:1901.11173 (2019)
13. Mei, G., Guo, Z., Liu, S., Pan, L.: SGNN: a graph neural network based federated learning approach by hiding structure. In: 2019 IEEE International Conference on Big Data (Big Data), pp. 2560–2568. IEEE (2019)
14. Liu, X., Lu, R., Ma, J., Chen, L., Qin, B.: Privacy-preserving patient-centric clinical decision support system on Naive Bayesian classification. IEEE J. Biomed. Health Inform. **20**(2), 655–668 (2015)

15. Huang, B., Wang, C., Wang, B.: NMLPA: uncovering overlapping communities in attributed networks via a multi-label propagation approach. Sensors **19**(2), 260 (2019)
16. Danon, L., Diaz-Guilera, A., Duch, J., Arenas, A.: Comparing community structure identification. J. Stat. Mech. Theory Exp. **2005**(09), P09008 (2005)
17. McDaid, A.F., Greene, D., Hurley, N.: Normalized mutual information to evaluate overlapping community finding algorithms. arXiv preprint arXiv:1110.2515 (2011)
18. Lancichinetti, A., Fortunato, S., Radicchi, F.: Benchmark graphs for testing community detection algorithms. Phys. Rev. E **78**(4), 046110 (2008)
19. De Cristofaro, E., Tsudik, G.: Practical private set intersection protocols with linear complexity. In: Sion, R. (ed.) FC 2010. LNCS, vol. 6052, pp. 143–159. Springer, Heidelberg (2010). https://doi.org/10.1007/978-3-642-14577-3_13
20. Ki, Y., Ji, W.Y.: PD-FDS: purchase density based online credit card fraud detection system (2018)
21. Boemer, F., Lao, Y., Cammarota, R., Wierzynski, C.: nGraph-HE: a graph compiler for deep learning on homomorphically encrypted data. In: Proceedings of the 16th ACM International Conference on Computing Frontiers, pp. 3–13 (2019)
22. Ying, S., Xu, S.W., Chen, X.B., Niu, X., et al.: Expansible quantum secret sharing network. Quantum Inf. Process. **12**(8), 2877–2888 (2013). https://doi.org/10.1007/s11128-013-0570-4
23. Meng, X., Kamara, S., Nissim, K., Kollios, G.: GRECS: graph encryption for approximate shortest distance queries. In: Proceedings of the 22nd ACM SIGSAC Conference on Computer and Communications Security, pp. 504–517 (2015)
24. Zhang, C., Zhu, L., Xu, C., Sharif, K., Zhang, C., Liu, X.: PGAS: privacy-preserving graph encryption for accurate constrained shortest distance queries. Inf. Sci. **506**, 325–345 (2020)
25. Gonzalez, J.E., Low, Y., Gu, H., Bickson, D., Guestrin, C.: PowerGraph: distributed graph-parallel computation on natural graphs. In: 10th {USENIX} Symposium on Operating Systems Design and Implementation ({OSDI} 12), pp. 17–30 (2012)

AcaVis: A Visual Analytics Framework for Exploring Evolution of Dynamic Academic Networks

Qiang Lu[1,2,3,4(✉)], Dajiu Wen[1,2,3,4], Wenjiao Huang[2,3,4], Tianyue Lin[1,3], and Cheng Ma[1,3]

[1] Key Laboratory of Knowledge Engineering with Big Data, Hefei University of Technology, Ministry of Education, Hefei, Anhui, China
luqiang@hfut.edu.cn

[2] School of Computer Science and Information, Hefei University of Technology, Hefei, Anhui, China

[3] School of Foreign Studies, Hefei University of Technology, Hefei, Anhui, China

[4] Intelligent Interconnected Systems Laboratory of Anhui Province, Hefei University of Technology, Hefei, Anhui, China

Abstract. In today's academic society, the trend of complexity, internationalization, diversification and interdisciplinary of scientific research activities is becoming more and more obvious. Although people are increasingly interested in academic cooperation in different fields and disciplines, people still know little about cross-field and interdisciplinary cooperation mechanism in academic big data. In order to further understand the topic and structure change pattern of academic big data, and explore the collaborative relationship, we propose a visual analytics framework to explore the time evolution of the dynamic academic network in the context of its structure in an interactive way. The first step is to learn the potential representation of the structure and attribute characteristics of nodes in the network based on graph convolution neural network (GCN). The second step is to optimize the learning process based on the end-to-end network combinatorial optimization algorithm, and then find communities with similar structures and similar attributes. The third step is to design AcaVis, a visual analytics framework to enable users to explore the topic and structure change characteristics of nodes, communities and the whole network in the academic social network. We demonstrated the effectiveness of our approach through three case studies conducted with real-world datasets and positive feedback from users. Code for our method is available at https://github.com/VIMLab-hfut/AcaVis.

Keywords: Dynamic academic network · Graph convolution neural network · Combinatorial optimization · Visual analytics

1 Introduction

Scientific collaboration is becoming not only an important feature of scientific activities in modern academia, but an important research topic, and gradually developed into one of

Y. Sun et al. (Eds.): ChineseCSCW 2021, CCIS 1492, pp. 499–511, 2022.
https://doi.org/10.1007/978-981-19-4549-6_38

the important forms of knowledge production. Researchers and students tend to choose experts and scholars who are still active in a certain field. However, they often encounter some problems, such as the experts with whom they seek cooperation may have retired or changed their research interests, and some may only start to study in this field in recent years. In the end, they can't find a suitable collaborator. In addition, the seeker also needs to know more about the changes of experts' research interests, and compare with peers who have similar statistical data of publications and research fields.

Academic social network analysis methods traditionally focus on the static network representation of graphs. However, by representing dynamic academic data as static networks, the group structure may be difficult to identify or may not be fully displayed in a short time, resulting in the disappearance of some groups' evolutionary behavior. In view of this, we can explore the time pattern of researchers' cooperative relationship in the context of evolutionary structure through the combination of static and dynamic methods to reveal their career development pattern, and then provide the trajectory view of dynamic network evolution on the time axis.

Therefore, we aim to summarize the time evolution in the context of network structure and support exploration from node, community and network levels to promote comprehensive dynamic academic network analytics. For the first requirement, we apply graph convolution neural network (GCN) to the node embedding of academic network. We use GCN to express the link relations and attributes of nodes as dense vectors in high-dimensional vector space, so that we can retain the features of structure and attributes. Since the generated vectors can represent the characteristics of nodes, we can use k-means clustering nodes in the graph embedding space to solve the optimization problem, and finally extract the communities. From these two aspects, our method is novel and provides new insights for dynamic network analytics. According to the proposed method, we design AcaVis, an interactive visual analytics framework that enables users to effectively explore the time evolution of dynamic academic network. The interested reader can try the framework at https://driverlin.github.io/academicNetwork/#/Sankver2MOD.

In a word, this paper makes three main contributions:

(1) We use node embedding of GCN to realize the fusion of structure and attribute information in academic network;
(2) We use end-to-end combinatorial learning to realize community detection in academic network;
(3) We design AcaVis, a novel visualization framework that enables users to analyze and explore the evolution pattern of dynamic academic network.

2 Related Work

2.1 Community Discovery

The discovery of community structure on a certain time slice is inseparable from the community discovery algorithm. The importance of community discovery in social networks has been widely concerned, and Javed, M. A. summarized the classic community discovery algorithms [1]. In recent years, deep learning methods have been applied to graph related tasks, such as link prediction and node classification [2], and several methods have been proposed in community detection [3, 4, 7]. Particularly, graph neural network (GNN) shows more accurate and scalable potential in graph representation learning [5, 6].

In recent work, Wilder, B. proposed an end-to-end combinatorial learning model (ClusterNet) [8], which mainly includes three steps: firstly, node embedding based on GCN, secondly, node clustering based on k-means, thirdly, the combination of learning and optimization process in a network. The loss of the final optimization task can be backpropagated to the learning task, and the network parameters can be optimized together to improve the performance of the model in the optimization task. Based on this inspiration, we apply this end-to-end combinatorial learning model to community detection of academic network. In this model, the community is identified by node embedding based on GCN and node clustering based on *k-means*, and good results are achieved.

2.2 Learning and Optimization on Graph

There are many traditional network representation learning algorithms based on network structure [9–12]. In addition to connecting with other nodes, the nodes also have other rich attribute information, such as scholars' academic influence, gender, institutions and other personal information. Therefore, the network representation learning method combining with external information will be better represent scholars with vectorization. The graph convolution neural network (GCN), which appeared in recent years, can also generate useful feature representations of nodes in graph network, and has achieved good results [13].

The combination of learning and optimization can improve the performance of solving real-world problems [14–17], and many people are interested in training neural networks to solve combinatorial optimization problems related to graphs [18, 19]. A recent work pointed out that mainly focused on the combination of graphic learning and optimization, and its model solves an optimization problem that gives complete information input [8]. This work is not only to learn the potential feature representation of the network, but also to solve the optimization problem after graph embedding. Inspired by this end-to-end combinatorial optimization algorithm, we focus on the problem of finding highly combinatorial communities. We first use GCN to generate the feature representation of nodes in the network, and then use *k-means* in graph embedding space to solve the optimization problem. The results show that the learning optimization model based on graph learning can detect academic communities, which have good performance.

2.3 Visualization of Dynamic Academic Network

Researchers have proposed many visualization technologies to explore and analyze the dynamic evolution pattern of network structure. Dynamic network analytics and visualization methods are mainly divided into animation method and timeline method, as well as their comprehensive use.

The method based on animation enables users to browse snapshots at different times. Each snapshot is usually visualized based on the node graph to display its structural features [20, 21]. The methods based on the timeline can display the snapshot in any time step of the still image which allows users monitor network changes by comparing network snapshots with different time steps [22, 23].

In recent related research, some researchers also used the hybrid technologies of animation and timeline to visualize the network structure [24, 25]. We use graph convolution neural network to express the link relationship and attributes of nodes as dense vectors in high-dimensional vector space. With this method, we can retain the structure and attribute features, and record the network snapshot at each time on the time axis. This can not only provide users with a good time overview, but also allow users to explore the detailed network structure of a certain time slice.

3 End to End Network Model

In this section, we introduce the end-to-end network model, including node embedding and graph optimization.

3.1 Node Embedding Based on GCN

Dynamic networks usually contain activities with timestamps. We model dynamic networks as a series of snapshots, each of which is a directed graph or an undirected graph $G_{\mathrm{T}} = (V_{\mathrm{T}}, E_{\mathrm{T}}, T)$, T is a time slice. Snapshot G_{T} has a time window with a width of w, that is $[T_i - w/2, T_i - w/2]$.

The learning embedding of nodes has been widely used in network analytics, and we embed nodes based on graph convolution neural network (GCN). Traditional neural network models and deep learning models are mostly used for data mining and feature extraction on Euclidean datasets. However, with the gradual expansion of the field of practical problems to be solved, more and more practical application scenarios often have non-Euclidean characteristics of the objects to be processed, and the traditional convolution operation can not be implemented on non-Euclidean datasets, therefore, we need to introduce the graph convolution method to extract data features. At present, graph convolution can be divided into two types: spectral convolution and spatial convolution. The latter has been widely used because of its obvious advantage in algorithm complexity. The essence of graph convolution neural network (GCN) is to extract features. The network transforms the features of nodes in the hidden layer, then the features of layer $l + 1$ are as follows:

$$X^{(l+1)} = \sigma(AX^{(l)}W^{(l)} + b^{(l)}) \tag{1}$$

where $X^{(l)}$ is the feature of layer l nodes, σ is the activation function, $W^{(l)}$ is the weight matrix of the first layer, and $b^{(l)}$is the intercept of the first layer.

If the neighbor nodes are normalized by degree matrix D, then:

$$X^{(l+1)} = \sigma(D^{-\frac{1}{2}}AD^{-\frac{1}{2}}X^{(l)}W^{(l)} + b^{(l)}) \tag{2}$$

let $\hat{A} = A + I$, I is the identity matrix, that is, adding self-circulation to the original connection so that each node can start from itself and then point to itself. $\hat{D}$ is the degree matrix corresponding to $\hat{A}$. At the same time, in order to simplify the formula, let $b = 0$, then the node characteristics of layer $l + 1$ are as follows:

$$X^{(l+1)} = \sigma(\hat{D}^{-\frac{1}{2}}\hat{A}\hat{D}^{-\frac{1}{2}}X^{(l)}W^{(l)}) \tag{3}$$

The graph convolution operator is as follows:

$$x_i^{(l+1)} = \sigma(\sum_{j \in V_i} \hat{D}^{-\frac{1}{2}}\hat{A}\hat{D}^{-\frac{1}{2}}x_j^{(l)}W^{(l)}) \tag{4}$$

where V_i is all neighbor nodes of node v_i, $x_i^{(l+1)}$ is the feature of node v_i in layer $l + 1$, $x_j^{(l)}$ is the feature of v_j in layer l.

It can be seen from the above, GCN model is a first-order model, which can be used to deal with the feature information of the first-order neighbor centered on a node in the graph. For nodes in academic network, each node needs to be considered not only the features of neighbor nodes, but also the influence of others nodes around this node. Therefore, a two-layer GCN is used to improve the processing ability of the model. In order to alleviate the problem of overfitting and increase the sparsity of the network, *relu* is selected as the activation function σ of the first layer GCN, and *softma* is selected as the activation function σ of the second layer GCN. We adopt cross entropy function as the loss function of the evaluation model.

$$Loss = \sum_{j=1}^{c} p(j)\log_2 q(j) \tag{5}$$

where $p(j)$ represents the real value, $q(j)$ represents the output value of GCN model, c is the total label category, $j = 1, 2 \ldots\ldots, c$.

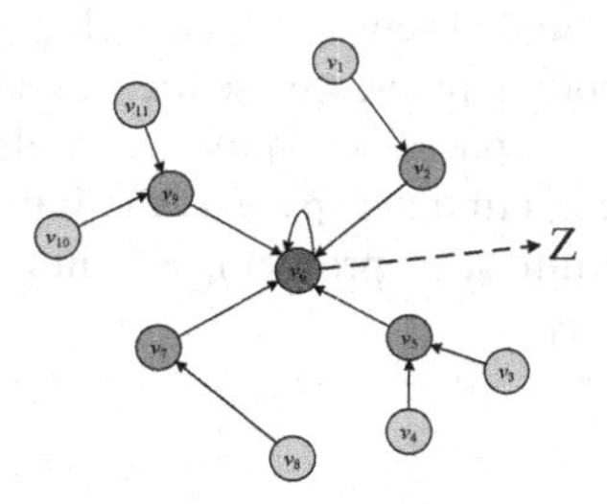

Fig. 1. Node embedding of GCN. The second-order neighbor nodes are aggregated, and Z is the feature of the aggregated nodes.

In GCN model, node features can not be extracted directly in one step. As shown in Fig. 1, GCN needs to calculate the loss value by using the above loss function and update the weight parameters in the model by comparing the predicted and reasoned features with the real eigenvalues in the network topology data. In the training process, the above process is repeated and iterated until the loss value does not change to make the model converge, and we can obtain the best characteristic Z of the output nodes of GCN model.

3.2 End to End Network Optimization

People are more and more interested in the flexible combination of learning and optimization process. The main idea is to combine learning and optimization to ensure the improvement of performance in the process of solving practical problems. These problems need to be made decisions based on machine learning prediction. Wilder, B., et al. proposed that the optimization problem be regarded as a differentiable layer, and then the initial problem can be mapped by learning a representation so that the interested problem can be transformed into an example of a simpler problem [8]. Based on this idea, we combine the node features in previous section with *k-means* clustering. The purpose of clustering is to let the graph neural network embed the nodes into a continuous space so that we can use optimization in the continuous embedding space to approximate the optimization on the discrete graph. To detect the communities, our optimization strategy is to use modularity.

The modularity of a community network is a measure to evaluate whether a community network is divided well or not. It means the difference between the number of connected edges of nodes in a community and the number of edges in a random situation. Its value range is (0, 1), and its definition is as follows:

$$Q = \frac{1}{2m}\sum_{i,j}\left[A_{i,j} - \frac{k_i k_j}{2m}\right]\delta(c_i, c_j) \tag{6}$$

$$\delta(u, v) = \begin{cases} 1 \; when \; u == v \\ 0 \qquad else \end{cases} \tag{7}$$

where A_{ij} represents the edge weight between node i and j (when the graph is not weighted, the edge weight can be regarded as 1), and k_i represents the edge weight sum of adjacent edges of node i, m is the edge weight sum of all edges in the graph, c_i is the community number of node i, and Q represents the value of modularity.

Firstly, in the graph embedding layer, we use the learned node features to embed the graph nodes into the continuous space. Secondly, we implement the *k-means* optimization layer, which embeds the continuous space as the input, and uses them to generate the solution x of the graph optimization problem, assigning the nodes to the community, and finally form the community.

4 Overview

4.1 Task Analysis

The proposed visual analytics methodology enables an integrated exploration of evolution from the node, community, and network levels to help users understand the structural and temporal properties of dynamic academic networks.

To explore the dynamic academic network effectively, we extract a task list.

T1: Evolution analysis of the whole academic network. The evolution pattern of the whole network over time can guide users to view the structure of the network globally so that they can focus on important periods.

T2: Evolution analysis of dynamic nodes. Finding dynamic nodes in each time step can facilitate the identification of changing nodes in each time step.

T3: Structural analysis of stable nodes. In the stable state of dynamic network, the stable node under a certain time slice is of great significance to understand the network structure.

T4: The evolution analysis of dynamic community. The community with time characteristics is helpful to understand the network pattern.

T5: Structural analysis of stable community. In the stable state of dynamic network, a stable community is very important to understand the network structure.

T6: Statistical analysis of key nodes and communities. In order to find the specific statistical information of nodes and communities, users should be able to explore a group of data with similar attributes.

4.2 Framework Architecture

Based on the above requirements, we designed AcaVis, which consists of two parts: the back end and the front end. The back-end pattern based on FSM is implemented with node.js, and the framework integrates the end-to-end combinatorial learning mining model. The front-end, written in JS and TypeScript, runs in the modern web browser, allows users to explore and analyze the complex propagation pattern interactively, through topology, propagation process and pattern instance.

5 Visual Design

5.1 Design Goals

Based on the above task analysis, we propose a set of design goals and design a visualization framework. In the next section, we will describe the framework with visualization and interaction to address these design goals.

G1: Time-oriented dynamic visual analytics. At the node level, users are allowed to explore the evolution pattern of nodes over time. We adopt the flow Sankey graph to show this dynamic evolution pattern, and users can pick out objects of interest and highlight their time-varying patterns (**T1**, **T2**). At the community level, users should be allowed to explore the evolution of the community over time. One of the strips in the Sankey graph represents a community, and users can dynamically study the dynamic patterns of when the community appears and when it dies based on the community they are interested in (**T4**). Flow graphs allow users to focus on important periods.

G2: Static visual analytics of key nodes and communities. At the node level, in the process of exploring the dynamic node evolution pattern in the time pattern, Sankey graph enables users to analyze the special node that they are interested in. Contour package graph and circular scatter graph enables users to explore the link pattern between this node and other nodes in detail, as well as its influence in the network (**T3**). At the

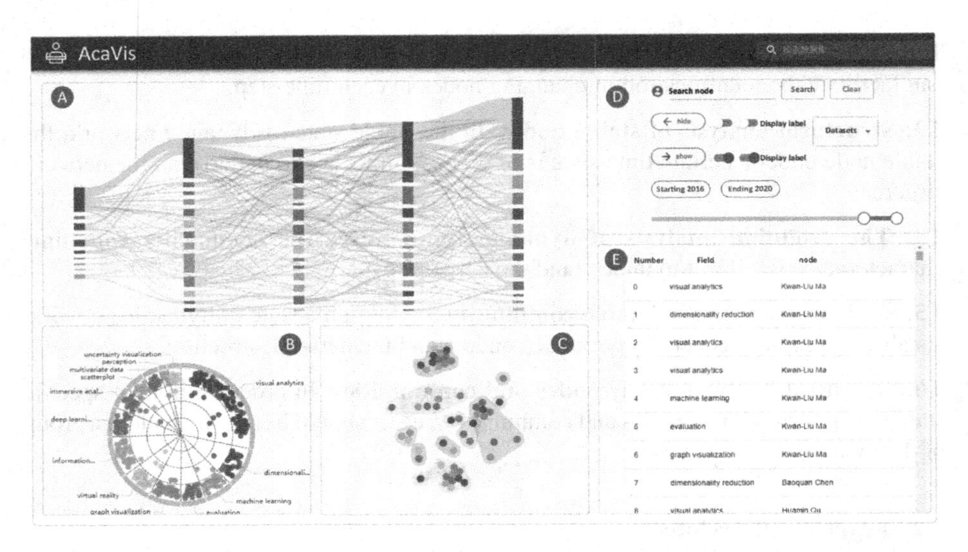

Fig. 2. Our visual analytics framework for dynamic academic network analysis.

community level, users are supported to analyze and explore larger communities in the Sankey graph. Wrapped Contour graph and circular scatter graph allow users to explore in detail the composition of nodes in the community, the distribution of topics and the size of the influence of the community in the network (**T5**).

G3: Multiple visual metaphors. A view will imply as many meanings as possible. For example, flow visualization based on Sankey graph, lines will represent nodes and bands will represent communities. An interface should integrate different visual metaphor views with a consistent style and allow the user to adjust the views according to the characteristics of the data at specific intervals, making it easier for the analyst to adjust the visual representation to the task at hand, at the same time, it also achieves more comprehensive coverage of the task **(T6)**.

5.2 Trend View

The trend view is shown in Fig. 2A. provides a temporal evolution pattern of network structure in context from the node and community levels. The evolution of nodes is a high-dimensional trajectory, which is difficult to understand and visualize. Therefore, we project these trajectories into a lower two-dimensional plane.

Design based on flow is often used to show evolution, which encodes time to a vertical or horizontal axis and we also adopt this idea. We propose a flow graph based trend view representation method, which maps the projection points to the vertical axis, and then locates them along the horizontal time axis. If two nodes have similar attributes, they will be coded closely in vertical position. The evolution of the community is represented by a band, that is, the band is composed of lines in the community.

Trend view can reveal some key events of the node and community evolution in time pattern, which correspond to different evolution patterns. We can define a set of covered rules to clarify these key events.

Birth: When a dynamic community appears and still exists after one time step, a new dynamic community appears in the network N.

Death: The disappearance of dynamic community occurs when at least d consecutive time steps are not observed (i.e. there is no corresponding step community).

Reappearance: The reappearance of dynamic community occurs after d time steps of community disappearance are observed again.

Splitting: If a dynamic community splits into two distinct communities at one time, new branches will no longer share the same time axis at the next time.

Merging: Merging occurs when two different dynamic groups are observed to overlap at a certain time. These communities share the same time axis from a certain point in time.

Expansion: If the corresponding community size of a dynamic community at a certain time is significantly larger than that of the previous time, the community will expand or grow.

Contraction: If the corresponding community size of a dynamic community at a certain time is significantly smaller than that of the previous time, then the community shrinks or decreases.

5.3 Topic Distribution View

As shown in Fig. 2B, we use the deformed circular scatter graph to show the statistical information of a certain time step. We use the sector size to encode the activity of a certain field and the nodes are mapped to the 2D plane and evenly distributed in the view. A single node may have multiple field features, therefore, a node in the topic distribution view is mapped to multiple nodes according to the number of research fields it belongs to. The distance between the node and the center indicates the level of activity in the field. If the node is close to the center of the circle, it means that the node is very active in the field. Otherwise, if the node is far from the center of the circle, it means that the node is less active in the field.

5.4 Structure View

Contrary to the trend view of time characteristics, structure view supports further exploration of view structure of specific time slice and visualizes academic network structure based on structure node clustering (including the stable community of stable nodes and the evolution of dynamic nodes in a stable state). To depict the relationship between more than two communities, we project them into 2D space in the structure view, as shown in Fig. 2C.

Each point is a metaphor for an egocentric. The distance between points represents the closeness between individuals, the distance between communities metaphors the closeness between communities. The emergence of the community and its change pattern in the life cycle depends on the link structure relationship and topic similarity, that is to say, the community is mainly formed by the link relationship and topic similarity within the community. If the link relationship dominates, we will show the community formed by this relationship in the form of a bubble, for the community formed by this kind of relationship, we use ribbon as a metaphor, ribbon community and bubble community are coded with different colors. A large node indicates that it is active, whereas a small node indicates that it is inactive.

6 Evaluation

To demonstrate the effectiveness and practicability of the end-to-end analysis method based on graph convolution neural network (GCN), we apply the visual analytics framework to real-world academic social network datasets. There are two types of datasets, one is the author co-authored network dataset, the other is the paper citation network dataset. IEEE VIS is a subset of the DBLP dataset [26], which includes all papers from three major visualization conferences (IEEE SciVis, IEEE InfoVis and IEEE VAST). We have prepared the publication dataset from 1990 to 2020, which includes two sub datasets: the author co-authored dataset of IEEE Vis publications from 1990 to 2020, and the paper citation dataset of IEEE Vis publications from 1990 to 2020. By analyzing and exploring these datasets, our method enables users to better understand the structural and temporal properties of the evolution from the node, community, and network levels.

In order to prove the wide applicability of our framework, we also apply our framework to an email dataset, which contains 1.2 million lines of communication records of employees in Hacking Team company. We try to classify employees and explore the development process of Hacking Team company through our visual analytics framework. For the case analysis of the email dataset and the paper citation dataset, please see our attachment (https://github.com/VIMLab-hfut/AcaVis).

In this case study, we demonstrate how our framework can effectively insight into the subtle changes of dynamic academic network from the node, community and network levels. We collected the author co-authored dataset of IEEE Vis publications from 1990 to 2020. Each paper contains the title, author, year of publication, abstract and keyword information. The dataset contains 6885 nodes and 24356 edges from 1990 to 2020.

First of all, we use the trend graph to explore the time evolution pattern of nodes and communities. Firstly, at the node level, each line in the flow Sankey graph represents the evolution track of a node over time. We can see that each line in the graph develops over time. We search for interested nodes and code them with highlighted colors. For example, we search Wei Chen, and Fig. 3a below shows the distribution of his personal research fields. We can see that Wei Chen's main research topic was visual analytics in 2014–2016, and in 2017, Wei Chen focused on high-dimensional data visualization. After 2017, Wei Chen's research scope is more extensive. In addition to visual analytics,

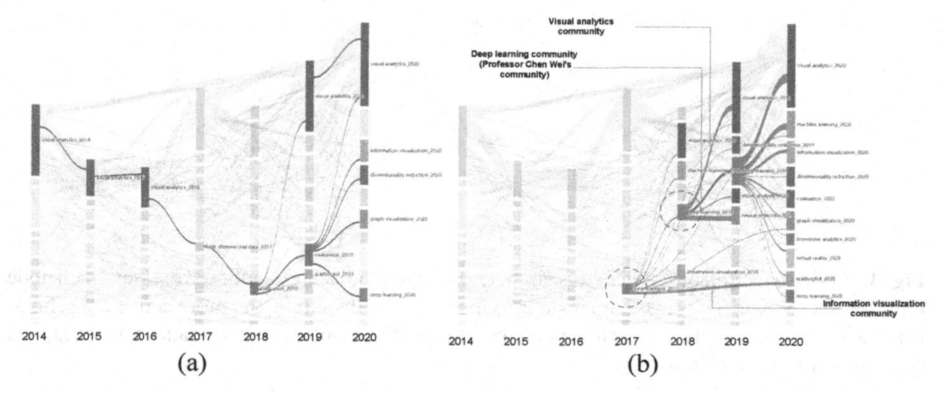

Fig. 3. (a) Distribution of research topics of Wei Chen, (b) evolution pattern of a scholar community from 2014 to 2020.

he has also interested in visual evaluation and scatterplot. In 2020, Wei Chen's research scope is more extensive, which can be said to be a year of fruitful scientific research output. He not only continues to pay attention to visual analytics and high dimensional data visualization, also has research achievement in the research topics of information visualization, graph visualization and deep learning.

Secondly, at the community level, we use highlight color coding to represent the evolution pattern of the community. As mentioned above, a community contains multiple nodes, and nodes are metaphorized by lines, therefore lines come together to form a "band", which means the community. Figure 3b shows the evolution pattern of a scholar community from 2014 to 2020. We can see that the community is mainly composed of "deep learning" enthusiasts in 2017. As time goes on, the scholars in the community begin to split, and the community also begins to split into multiple "sub communities". By 2018, the members of the community gradually flow to machine learning, information visualization and other fields. From 2016 to 2020, there are scientific research achievements on deep learning, and we think that visualization research method based on deep learning is a hot research direction at present and in the future. It is worth noting that the purple line in Fig. 3b indicates the change of Wei Chen's deep learning research community. It can be seen that during 2017–2020, no matter whether the field of deep learning is hot or not, Wei Chen has been investing in this field.

In addition, our framework can further explore the detailed structure information of a certain time slice node and community. Figure 4a below shows the distribution of scholars' research topics in 2020. The nodes represent scholars. We can see that the main research topics in 2020 include visual analytics, information visualization, deep learning and dimensional reduction, etc., and the scholars are the most active in the field of visual analytics, the nodes represent scholars. The connections between scholars depend on whether they have co-authored papers. We can see that there are many communities, and there is overlap between communities in Fig. 4b. For example, Wei Chen cooperated with many scholars in 2020, and the research topics include visual analysis, deep learning, dimensional reduction and other fields, therefore, Professor Chen Wei belongs to multiple communities.

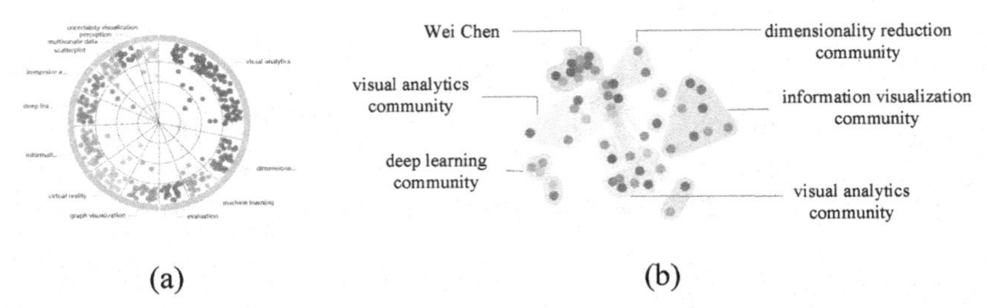

Fig. 4. (a) Distribution of scholars' research fields in the scholars' co-authored dataset. Each node represents a scholar. The closer the node is to the center of the circle, the more active the scholar is in the field, (b) the outline of scholar collaboration in the scholar coauthor dataset. the wrapped contour represents a community.

7 Conclusion and Future Work

In this study, we propose a new visual analytics method for the analytics of nodes and communities in dynamic academic networks. In our work, we extract communities from academic networks based on graph convolution neural network and combinatorial learning optimization algorithm, which are used to aggregate nodes with similar attributes and connection patterns. We designed AcaVis that integrates novel visual views. These new views present the evolution trajectory of dynamic nodes and communities on the time axis and helps users explore the attributes and structural patterns at different levels and scales of nodes, communities and networks. Three examples and user interviews verify the effectiveness of the framework (please refer to the attachment for email contact network cases and user interview records). For future work, we plan to further improve our dynamic community extraction method. In addition, we also consider applying our research analytics method to more other entity networks (such as schoolfellow network, disease transmission network) for analytics.

Acknowledgments. This work was supported in part by Provincial Key Research and Development Program of Anhui of China under grant number 201904d07020010 and the Science and Technology Program of Huangshan of China under grant number 2019KN-05.

References

1. Chunaev, P.: Community detection in node-attributed social networks: a survey. Comput. Sci. Rev. **37**, 1–24 (2020)
2. Cai, H., Zheng, V.W., Chang, C.C.: A comprehensive survey of graph embedding: problems, techniques and applications. IEEE Trans. Knowl. Data Eng. **30**(9), 1616–1637 (2018)
3. Ma, T., Liu, Q., Cao, J., Tian, Y.: LGIEM: global and local node influence based community detection. Future. Gener. Comput. Syst. **105**(20), 533–546 (2020)
4. Liu, F., Xue, S., Wu, J.: Deep learning for community detection: progress, challenges and opportunities. In: 29th International Joint Conference on Artificial Intelligence (IJCAI 20), Yokohama, pp. 4981–4987. Morgan Kaufmann (2020)

5. Chen, Z., Li, L., Bruna, J.: Supervised community detection with line graph neural networks. In: International Conference on Learning Representations, New Orleans, pp. 1–24. Microtome (2019)
6. Shchur, O., Günnemann, S.: Overlapping community detection with graph neural networks. In: KDD Workshop DLG 2019, Anchorage, pp. 1–7. ACM (2019)
7. Luo, X., Wu, J., Zhou, C.: Deep semantic network representation. In: 20th IEEE International Conference on Data Mining (ICDM), Sorrento, pp. 1154–1159. IEEE (2020)
8. Wilder, B., Ewing, E., Dilkina, B.: End to end learning and optimization on graphs. In: Advances in Neural Information Processing Systems, Vancouver, pp. 4674–4685. MIT Press (2019)
9. Roweis, S.T., Saul, L.K.: Nonlinear dimensionality reduction by locally linear embedding. Science **290**(5500), 2323–2326 (2000)
10. Perozzi, B., Al-Rfou, R., Skiena, S.: DeepWalk: online learning of social representations. In: Proceedings of the 20th ACM SIGKDD International Conference on Knowledge Discovery and Data Mining, pp. 701–710. ACM, New York (2014)
11. Cao, S.S., Lu, W., Xu, Q.K.: Grarep: learning graph representations with global structural information. In: Proceedings of the 24th ACM International on Conference on Information and Knowledge Management, pp. 891–900. ACM, New York (2015)
12. Yang, J., Leskovec, J.: Overlapping community detection at scale: a nonnegative matrix factorization approach. In: Proceedings of the sixth ACM International Conference on Web Search and Data Mining, pp. 587–596. ACM, New York (2013)
13. Jie, Z.A., Gc, A.: Graph neural networks: a review of methods and applications. AI Open **1**, 57–81 (2020)
14. Burgess, M., Adar, E., Cafarella, M.: Link-prediction enhanced consensus clustering for complex networks. PLoS ONE **11**(5), 1–23 (2016)
15. Xue, F., He, X., Wang, X.: Deep item-based collaborative filtering for top-N recommendation. ACM Trans. Inf. Syst. (TOIS) **37**(3), 1–25 (2019)
16. Balghiti, O.E., Elmachtoub, A.N., Grigas, P.: Tewari, A.: Generalization bounds in the predict-then-optimize framework. In: Advances in Neural Information Processing Systems, Vancouver, pp. 14412–14421. MIT Press (2019)
17. Elmachtoub, A.N., Grigas, P.: Smart "predict, then optimize". Manag. Sci. **16**, 1–19 (2021)
18. Khalil, E., Dai, H., Zhang, Y., Dilkina, B., Song, L.: Learning combinatorial optimization algorithms over graphs. In: Advances in Neural Information Processing Systems, Long Beach, pp. 6351–6361. MIT Press (2017)
19. Kool, W., Van Hoof, H.: Attention, learn to solve routing problems! In: International Conference on Learning Representations, New Orleans, pp. 1–25. Microtome (2019)
20. Eades, P., Huang, M.: Navigating clustered graphs using force-directed methods. J. Graph Algorithms Appl. **4**(3), 157–181 (2004)
21. Stef, V., Holten, D., Blaas, J.: Reducing snapshots to points: a visual analytics approach to dynamic network exploration. IEEE Trans. Vis. Comput. Graph. **22**(1), 1–10 (2015)
22. He, Q., Zhu, M., Xie, Z., Liu, H.: Revoler: visual analysis of relation evolution in temporal data. J. Comput. Inf. Syst. **11**(18), 6555–6569 (2015)
23. Lu, Q., Huang, J., Ge, Y.F: EgoVis: visual analysis for social network based on egocentric research. Int. J. Coop. Inf. Syst. **29**(2), 1–20 (2019)
24. Beck, F., Burch, M., Diehl, S., Weiskopf, D.: A taxonomy and survey of dynamic graph visualization. Comput. Graph. Forum **36**(1), 133–159 (2017)
25. Col, A.D., Valdivia, P., Petronetto, F.: Wavelet-based visual analysis of dynamic networks. IEEE Trans. Vis. Comput. Graph. **24**(8), 2456–2469 (2018)
26. Isenberg, P., Heimerl, F., Koch, S., Isenberg, T.: Vispubdata.org: a metadata collection about IEEE visualization (VIS) publications. IEEE Trans. Vis. Comput. Graph. **23**(9), 2199–2206 (2016)

Dynamic Information Diffusion Model Based on Weighted Information Entropy

Zekun Liu, Jianyong Yu(✉), Linlin Gu, and Xue Han

School of Computer Science and Engineering, Hunan University of Science and Technology, Xiangtan 411201, China
yujyong@hnust.edu.cn

Abstract. Today's social networks present multiplex and heterogeneous attributes, and they are becoming more and more difficult to grasp the unique attributes and structures of the network from the whole network structures. The attribute factors such as Degree Centrality and Closeness Centrality are inadequate in the study of social networks if they are not modified and taken fully into account the real characteristics. Although the evolution of a social network is a macroscopic process, the state of each node at different moments determines the trend of the overall state on the network. In order to better identify the key nodes in the social network and more really simulate information diffusion process, this paper proposes the method using WIE (Weighted Information Entropy) to measure node's influent force and importance degree based on the concept of information entropy in information theory, which carries higher-order local information of each node rather than only contains first-order local information like degree does. Different from the conventional SIR model, this paper realizes the dynamic calculation of the diffusion probability β in the improved SIR model according to node's local influence and weighted value. Finally, four kinds of real social networks are selected and analyzed. The experiment shows that the WIE measurement and the dynamically calculated diffusion probability can more truly reflect the information transmission in real social networks, and the node sets selected by WIE values have at least 10 percent improvement in diffusion efficiency.

Keywords: Social networks · Local influence · Global selection · SIR model

1 Introduction

In the past when person-to-person communication was still underdeveloped, the government and news media were the main information disseminators, and the information interaction between people was mainly concentrated in a small range of limited areas. Nowadays, with the increasing diversification of communication means, social networks are becoming more and more complex. Everyone is the producer and transmitter of information, and the distance of communication is no longer a limitation. Because of this, the study of complex networks has gradually shifted from the macroscopic universal law among different networks to the study of community structure, group structure

Y. Sun et al. (Eds.): ChineseCSCW 2021, CCIS 1492, pp. 512–524, 2022.
https://doi.org/10.1007/978-981-19-4549-6_39

and microscopic node properties themselves, and explored the different attributes among different networks through a more in-depth detailed study. Many studies [1–3] found that macroscopic research would "average" the characteristics displayed by individuals, and such "average" led to that we could only give qualitative explanations in statistical sense, but it was difficult to give quantitative prediction results. Due to the transformation of the research in the whole field of network science from the macro statistical characteristics to the community order and then to the node and link order, the important node search focusing on the overall characteristics of the network and individual attributes has become the focus of research in recent years [4–8].

With the development of related disciplines, information dissemination has expanded from its own specific information to public opinion [9], disease and other aspects with a common nature. Research on information transmission under different backgrounds can help us to quickly know the law of information transmission [10–14], predict the transmission process [15], and make corresponding interventions according to different situations [16] to provide a theoretical basis for the realization of public opinion control.

2 Related Works

2.1 Diffusion Model

In the past decade, due to a large number of available real data and the emergence of complex network theories, network research based on infectious disease transmission model has attracted extensive attention [17]. In the early 1760s, Daniel Bernoulli studied the transmission of smallpox virus through ordinary differential equations and proposed the Bernoulli equation [18], which became one of the earliest transmission dynamics models. In 1926, McKendrick and Kermack developed a simple, deterministic model by studying the transmission patterns of the Black Death around 1665 and the plague in 1906, known as the SIR chamber model in modern mathematical epidemiology, in which individuals infected with infectious diseases in a given area, they are divided into several species according to their physical health, and each species is called a chamber. It has successfully predicted the outbreak behavior of many recorded epidemics [19]. Infectious disease model is a relatively mature model in the field of information transmission [20]. This model believes that when the disease transmission rate of infected people is greater than a certain critical value, infected people will spread the disease to susceptible people, and this process will continue until the whole network of infected people is in a stable state [21]. In addition to the classic SIR model, there are also SI model and SIS model [22], which have evolved into several variants in the process of development, such as cascading model similar to SI [23] model and SIRS model in heterogeneous network. Xiong et al. [24] proposed a SCIR model to divide the population in the network into four categories. That is, the susceptible group S (susceptible), the susceptible group C (Contacted), the infected group I (Infected) and the refractory group R (Reached), The model considers that I and R are the final stable states of message propagation. Li et al. [25] proposed an improved SIQRS model, in which Quarantined Individuals Q (Quarantined Individuals) were added to study the propagation dynamics model in scale-free networks. Xiong et al. [26] introduced the role of lurker and proposed the SIR (Susceptible-Infected- latent-refractory) model, etc. The propagation process of this type of model is contact. Virus or rumor propagation occurs between two nodes. Except for the time cost, there is generally no transmission cost.

2.2 SIR Model and Information Entropy

Infectious disease model of differential equation is used to establish mathematical model, which can accurately describe the individual information in the process of communication with the state change of time, so as to find out the rule of information dissemination, found that influence factors in the process of information transmission, and then to effectively control the spread of information, enhance the effectiveness of information communication, in addition to the scope and the trend of information communication of users to predict in advance. In the study of modern information transmission model, researchers liken the transmission process of information to the transmission process of infectious diseases. The user nodes in social networks are divided into three categories: Nodes that have never heard of the message (S), that have received the current information and are willing to tell other neighboring nodes (I), and that have received the current information but are not willing to tell other neighboring nodes (R). The communication rules are defined as follows:

(1) If an infected node contacts its neighboring uninfected nodes, the healthy node will become infected node with probability β;
(2) In the process of transmission, a propagation node loses interest in transmission with probability γ and becomes an immune node. After becoming an immune state, the node will no longer carry out information transmission and its state will remain unchanged forever.
(3) The dynamic equation of SIR information transmission model is:

$$\frac{dS(t)}{dt} = -\beta S(t)I(t) \tag{1}$$

$$\frac{dI(t)}{dt} = \beta S(t)I(t) - \gamma I(t) \tag{2}$$

$$\frac{dR(t)}{dt} = \gamma I(t) \tag{3}$$

Shannon borrowed the concept of thermodynamics and pointed out that there is redundancy in any information, and the degree of redundancy is related to the probability or uncertainty of each symbol (number, letter or word) in the information. In general, there is uncertainty about what symbols are sent from a source, and the degree of uncertainty can be measured by the probability of symbols appearing. Large probability, more opportunities, less uncertainty, and vice versa. In information theory, information entropy is defined as follows. Given a discrete random variable X, it has possible values of x_i, and the probability of each value occurring is p_i, i = 1, 2, …, n, among them, $p_i \leq 1$, $\sum_{i=1}^{n} p_i = 1$, so entropy H of X is defined as [27]:

$$H(X) = -\sum_{i=1}^{n} p_i log_b p_i \tag{4}$$

generally b is 2, 10 or *e*. Dehmer [28] recommends introducing an array of non-negative integers $(t_1, t_2, \ldots, t_n)$, and that gives you a probability distribution $p = (p_1, p_2, \ldots, p_n)$, the calculation formula of p_i is:

$$p_i = \frac{t_i}{\sum_{j=1}^{n} t_j}, i = 1, 2, \ldots, n \tag{5}$$

3 WIE and Improved Diffusion Model

3.1 Weighted Information Entropy

The object of this paper is directed graph, G (V, E) (shorthand for G) stands for the network, a collection of nodes in a network is denoted by $V = \{v_1, v_2, \ldots, v_n\}$, $|V| = n$, the set of edges in the network is denoted by$E = \{e_1, e_2, \ldots, e_m\}$, $|E| = m$. e_{ij} is the directed edge from node i to node j, the weight of this edge is how close node i is to node j. $\Gamma_i = \{v_k | v_k \in V, (v_i, v_k) \in E\}$ represents the set of all nodes that have a connecting edge to node i, that is the first order neighbors of node i.

In graph theory, the concept of degree is the number of edges that a node connects to other neighbor nodes. Many node mining algorithms will use the local information of the node, so the degree of the node has become the basic index for most algorithms to measure the influence of the node itself, and further algorithm design is based on this index. However, this index only contains the first-order local information of a node, that is, the degree of a node cannot distinguish the importance of its neighbors. As is shown in the Fig. 1.

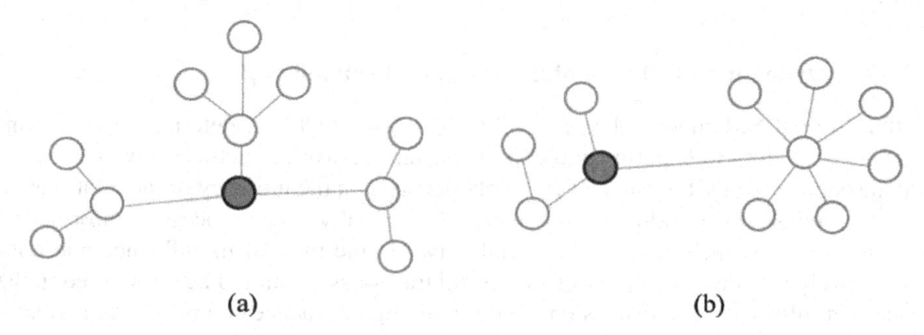

Fig. 1. (a) The degree of the colored node is 5; (b) The degree of the colored node is 5. However, the distributions of the two nodes' neighbors are obviously uneven.

The colored nodes in image (a) and image (b) have the same degree, since the degree only contains the local information of the first order of the node, it is impossible to distinguish the importance of the node's neighbors in the measurement process of the node. As can be seen from the second-order neighbors of the nodes in the figure, the colored node in image (a) has more important position than the colored node in image (b). But how to prove it? We introduce the concept of information entropy and modify it according to the actual situation of social network, the improvement is due to the following two problems:

(1) In the social network graph, node with high degree only has more opportunity to transmit information to more nodes, but it is not guaranteed that the success rate of transmission is proportional to the number of edges of the node itself.
(2) Even a node with a small degree, if its connection weight with its neighbors is large, it should have a high probability to realize the success of information transmission.

According to the summarized problems, the improved weighted information entropy is defined as follows:

$$E_v = \sum_{u\in\Gamma_v} H_{vu} = \sum_{u\in\Gamma_v} -P_{vu}log_b P_{vu} \tag{6}$$

$$P_{vu} = \frac{\frac{k_u}{\sum_{q\in\Gamma_v} k_q}\sum_{i\in\Gamma_u} W_{ui}}{\sum_{u\in\Gamma_v}\frac{k_u}{\sum_{q\in\Gamma_v} k_q}\sum_{i\in\Gamma_u} W_{ui}} \tag{7}$$

In the original formula (4), P is the probability of information i. In the improved formula (6), P is the probability of node v sending information to node u. In the formula (7), $\sum_{u\in\Gamma_v} P_{vu} = 1$, k represents the outdegree of the node, Γ represents the set of adjacent nodes of the node, W represents the weight of the edge from one node to another.

Now go back to the situation shown in Fig. 1, we use Formulas (6) and (7) to calculate and get that the WIE of the two nodes are 0.45 and 0.02 respectively. Thus, it is proved that the node in image (a) has more important network location than the node in image (b), it can be seen that it is more accurate to measure the importance of nodes through WIE.

3.2 The Dynamic Calculation of the Infection Probability β

In the original SIR model, the probability of infection β was determined by a constant set according to experimental experience and network structure. However, in the actual social network, the value of β largely depends on the intimacy of the relationship between nodes, so this value should change dynamically in the process of information transmission. For each node in the social network, the ratio of its influence and connection weight to the probability of successful transmission should be considered in the process of information transmission. Nodes with high influence certainly have a greater chance to contact other nodes, but for nodes with low influence, if it has close neighbors, it should also have a greater probability to realize the success of information transmission. In these two cases, weight and influence have different effects on the β value. Therefore, this paper proposes the following formula for the calculation of β value:

$$\beta_{vu} = \alpha Z_v + (1-\alpha)C_{vu} \tag{8}$$

In the formula (8), β represents the propagation probability of one node to another, Z represents the probability distribution of weighted information entropy of nodes, and C represents the probability distribution of weights between nodes.

4 Proposed Algorithm

In this section, we mainly introduce the pseudocode implementation of WIE and β values. WIE comes from the concept of information entropy proposed by Shannon in Informatics, the previous section illustrates by example that WIE is a better measure of the importance of nodes in the social network structure than the degree of nodes in it. The β value is also dynamically calculated according to the attributes of the nodes themselves and the connection weights between nodes. Through the improvement of these two parts, the simulation experiment of information transmission can be carried out more realistically in the information transmission model.

4.1 WIE Function Based on the Information Entropy

This paper proposes a new algorithm called Weighted Information Entropy to measure the importance of the nodes in social networks. The algorithm comes from the idea, which is that not only the node with high degree having more opportunities to contact other nodes should be considered, but also the node with low degree and high weight with neighboring nodes should also be paid attention to.

The WIE algorithm is described by pseudocode in Algorithm 1.

Algorithm 1 Weighted Information Entropy(G)

Input:

Graph G = (V, E)

Output:

G with WIE values

```
 1: for each node v in V do
 2:     neighbors = list(all the neighbors of node v)
 3:     ks = the sum of all neighbors' outdegrees
 4:     for each node u in neighbors do
 5:         Wu = the sum of the weights of all the edges starting from node u
 6:         if u's outdegree = 0 then
 7:             p = 0
 8:         else
 9:             p = Wu *(u's outdegree / ks)
10:             sw += p
11:         end if
12:     for each node u in neighbors do
13:         if sw = 0 or p = 0 then
14:             continue
15:         else
16:             Pvu = p/sw
17:             Hvu = -Pvu* log(Pvu)
18:             node's influnce += Hvu
19:         end if
20:     v's WIE = round(node's influnce,2)
```

In Algorithm 1, all the neighbors of node v are obtained through traversal, and the outdegree and the sum of the weights of all the edges starting from the node u are successively calculated. The method similar to mathematical expectation is used to calculate and represent the important situation of node u in the neighbors of node v.

4.2 Improved Dynamic Calculation Function of β

Based on the better node importance measurement function WIE, this paper proposes a comprehensive calculation of β value which uses the α value to determine the ratio of the WIE and the connection weight between two nodes to it.

The improved β function is described by pseudocode in Algorithm 2.

Algorithm 2 Function Of β

```
Input:
    Graph G = (V, E), node v
Output:
    β of node v
1: P = {}
2: for edge in E
3:     if edge's weight in P then
4:         times[weight] += 1
5:     else
6:         times[weight] = 1
7:     end if
8:     sort(P)
9:     Q = {}
10:    for weight, times in P then
11:        Q[weight] = times / |E|
12:        Q[probability of weight] += Q[weight]
13: The steps for calculating the probability of WIE are similar to do 1~12, the
results saved in W.
14: β = alpha*W[v's WIE]+(1-alpha)*Q[edge(v, u) 's weight]
15: return β
```

In Algorithm 2, the probability distribution of weights can be calculated by accumulating the number of weights of each value in the data set. And the probability distribution of WIE is also computed by a similar calculation. For each node in the state of information received, the β value will be calculated through algorithm 2 before the information is transmitted to a neighboring node that has not received the information. Through the determination of α, the proportion of weight and WIE value can be adjusted in the calculation process.

5 Experiments

In order to prove the validity of the above functions through experiments, this paper selects four real social network datasets with directed and weighted that are available online (http://www-personal.umich.edu/~mejn/netdata/). The selected network size varies from several thousand to tens of thousands of nodes, and their network structures are also different from each other. Experiments and analysis on these different data are the best way to assess the true effectiveness and general applicability of the WIE and the improved β value functions. The detailed data sets' contents are shown in Table 1.

(1) netscience. Co-authorship network of scientists working on network theory and experiment, as compiled by M. Newman.
(2) hep-th. Weighted network of co-authorships between scientists posting preprints on the High-Energy Theory E-Print Archive.
(3) cond-mat-2005. Weighted network of co-authorships between scientists posting preprints on the Condensed Matter E-Print Archive.
(4) astro-ph. Weighted network of co-authorships between scientists posting preprints on the Astrophysics E-Print Archive.

Table 1. Data sets description.

ID	Data sets	Nodes	Edges	Average degree
1	netscience	1589	2742	1.726
2	hep-th	8361	15751	1.884
3	cond-mat-2005	40421	175692	4.347
4	astro-ph	16706	121251	7.258

5.1 Comparison Method Description

There are many explanations for the importance of nodes, and the measures to determine centrality are different under different explanations. But the most major measurements at present are Degree Centrality, Closeness Centrality, Betweenness Centrality and Eigenvector Centrality.

Degree Centrality measures the degree to which a node in the network is associated with all other nodes, with values ranging from 0.0 to 1.0, where 0.0 is associated with no nodes and 1.0 is associated with every node. Closeness Centrality is reflected in the degree of proximity between a node and other nodes in the network. In other words, for a node, the closer it is to other nodes, the more centrality it has. Betweenness Centrality refers to the number of times that one node acts as the shortest paths' bridge between two other nodes. The more times a node acts as a "mediator," the more centrality it becomes. The basic idea of Eigenvector Centrality is that the centrality of one node is a function of the centrality of adjacent nodes. In other words, the more important the people you connect with, the more important you are. Eigenvector Centrality is different from Degree Centrality. A node with high Degree Centrality that has many connections may not have high Eigenvector Centrality, because all connectors may have low Eigenvector Centrality. Similarly, high Eigenvector Centrality does not mean that it has high Degree Centrality. It can also have high Eigenvector Centrality if it has few but very important connectors.

5.2 Performance Evaluation

By setting the values of k for different initial node group sizes, the effect of the algorithm is measured under different initial diffusion scales. Set the information immunity rate γ of the propagation model, $\gamma = 1/\langle k \rangle$. The information propagation probability β is calculated by using the improved function (8) in this paper. Measure the propagation range of different node sizes in a specific same time step. According to the node sizes of the network, the variation range of the size of the initial node group with different data is also different.

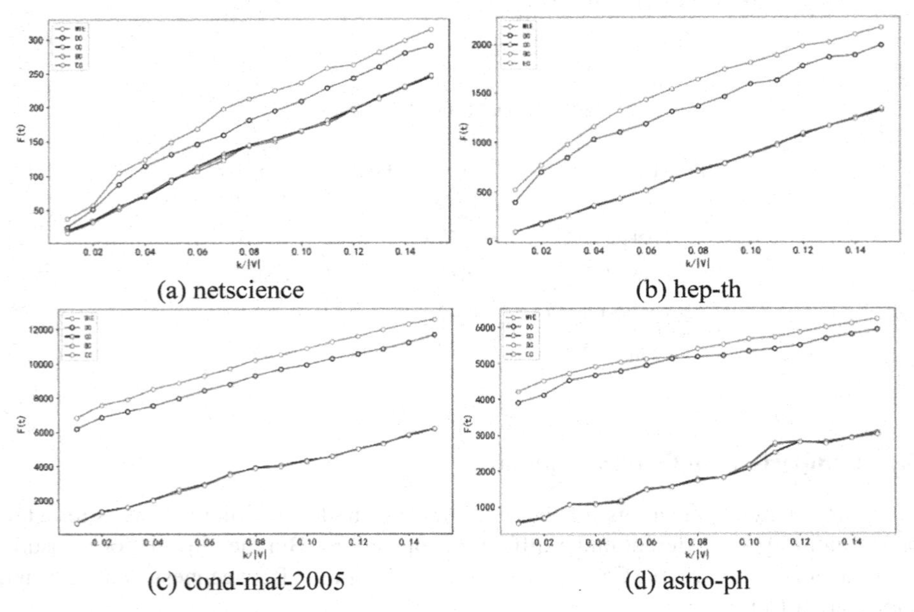

(a) netscience (b) hep-th

(c) cond-mat-2005 (d) astro-ph

Fig. 2. An influence contrast of the different proposition of top k on diffusion range at time t, when the proposed method is applied in four types of datasets. Here, seed node sets with different sizes depend on k/|v|.

As shown in Fig. 2, the same proportion of seed set size is used for experimental research on different data sets. In every subgraph, with the increase of the size of seed node set, the corresponding dispersal range is also obviously on the rise. In this experiment, the β value of WIE in function (8) was set to 0.5 in advance, other methods' probability of information diffusion. Nevertheless, the first k nodes picked out using WIE still have the largest propagation range over the same time period. For group (a) experiments, the sets of nodes selected by different methods are not far apart in the final propagation range, in the remaining experimental groups, the diffusion ranges of the node sets selected by WIE and DC were significantly larger than that of the node sets selected by other methods. It can be seen from the figure that the seed node sets found by CC, EC and BC are not far apart, a significant part of this is due to the inherent nature of the datasets themselves. From what has been discussed above, It can be concluded that using WIE to measure node influence can more accurately calculate and distinguish different nodes according to their actual status in the network. In addition, by querying the WIE of each node to select the node with a large value, the information can be diffusing in a shorter time.

5.3 The Optimal Parameter

As shown in function (8), α can change the importance ratio of the influence of the node itself and weight between nodes to β of information diffusion. But what should α be for different social networks? If the value tends to 1.0, then for β, the main factor determining the value of β is the influence of the node itself, on the contrary, if the value of α approaches 0.0, then the weight of the connection between the two points becomes the main factor affecting the value of β. Therefore, it is necessary to determine the optimal α values for different data sets. The details are shown in the Fig. 3.

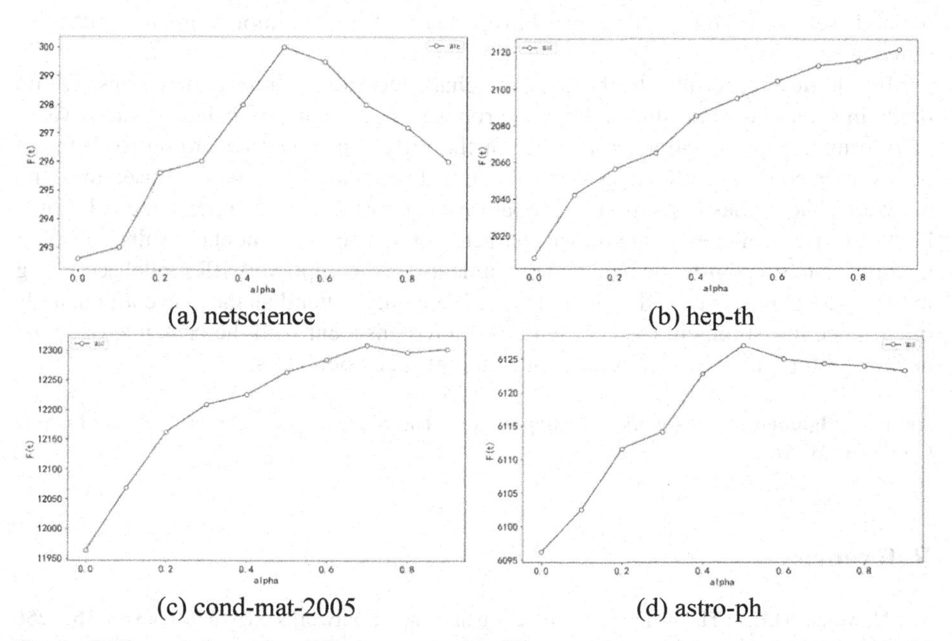

(a) netscience (b) hep-th

(c) cond-mat-2005 (d) astro-ph

Fig. 3. Different information diffusing ranges for different values of α in sets of seed nodes with the same size of top k at time t. Each subplot denotes the influence of diffusion ranges on different α in four kinds of datasets, respectively.

In Fig. 3, what we can see is that the curves of experiments (b) and (c) show a similar upward trend, their optimal α values are 0.9 and 0.7, respectively. This indicates that in these two data sets, the influence of the nodes themselves can determine the success of information transmission to a greater extent than the weight of the connection between nodes. In other words, nodes with high WIE values result in more efficient diffusion of information on both networks. In contrast, experiments (a) and (d) concluded that the best α value is 0.5, and that leads to the conclusion, which is that the influence of the nodes themselves are as important as the weights of the connections between nodes in both datasets. Therefore, for these two networks, to realize efficient information diffusion, the influence of the nodes themselves and the weights of the connecting edges between nodes should be considered comprehensively.

6 Conclusion

In order to better distinguish the key nodes in the social network and more really simulate information diffusion process, this paper proposes a new attribute, which can be calculated by improved information entropy formula considering both the node's attributes and characteristics of the information transmission in society, to measure the importance of nodes. Unlike most measurements, which only contain first-order local information and are inadequate in the study of social networks, WIE applies to directed weighted graphs. The local importance of the current node can be calculated more effectively by considering second level neighbors' information of the node. Experiments show that the node sets selected by WIE values have at least 10 percent improvement in diffusion efficiency.

In addition, according to the intrinsic characteristics of information transmission mode in social network, this paper also proposes to dynamically calculate the β value of information propagation probability in the diffusion model according to different nodes' own conditions. It solves the problem that when information is diffused in social networks, the actual transmission probability cannot accurately reflect the relevancy between specific nodes and their own characteristics. The experimental results show that the diffusion probability β calculated dynamically in the improved SIR model according to node's local influence and weighted value is not only better than the conventional node importance measurement method, but also more consistent with the principle of social network information dissemination for different social networks.

Acknowledgements. This work was supported by The National Social Science Fund of China (No. 20BXW096).

References

1. Newman, M.E.J.: The structure and function of complex networks. SIAM Rev. **45**(2), 167–256 (2003)
2. Pastor-Satorras, R., Vespignani, A.: Epidemic spreading in scale-free networks. Phys. Rev. Lett. **86**(14), 3200–3203 (2001)

3. Castellano, C., Pastor-Satorras, R.: Thresholds for epidemic spreading in networks. Phys. Rev. Lett. **105**(21), 218701 (2010)
4. Zhong, J., Zhang, F., Li, Z.: Identification of vital nodes in complex network via belief propagation and node reinsertion. IEEE Access **6**, 29200–29210 (2018)
5. Wang, Y., Yan, G., Ma, Q., et al.: Identifying influential nodes based on vital communities. In: 2018 IEEE 16th International Conference on Dependable, Autonomic and Secure Computing, 16th International Conference on Pervasive Intelligence and Computing, 4th International Conference on Big Data Intelligence and Computing and Cyber Science and Technology Congress, pp. 314–317. IEEE (2018)
6. Lei, M., Wei, D.: Identifying influence for community in complex networks. In: 2018 Chinese Control and Decision Conference (CCDC), pp. 5346–5349. IEEE (2018)
7. Dai, J., Wang, B., Sheng, J., et al.: Identifying influential nodes in complex networks based on local neighbor contribution. IEEE Access **7**, 131719–131731 (2019)
8. Basaras, P., Iosifidis, G., Katsaros, D., et al.: Identifying influential spreaders in complex multilayer networks: a centrality perspective. IEEE Trans. Netw. Sci. Eng. **6**(1), 31–45 (2017)
9. Valerio, A., Marco, C.,Massimiliano, L.G., et al.: Ego network structure in online social networks and its impact on information diffusion. Comput. Commun. **76**(57), 26–41 (2016)
10. He, X.S., Zhou, M.Y., et al.: Predicting online ratings based on the opinion spreading process. Physica A **436**, 658–664 (2015)
11. Fu, R., Gutfraind, A., Brandeau, M.L.: Modeling a dynamic bi-layer contact network of injection drug users and the spread of blood-borne infections. Math. Biosci. **273**, 102 (2016)
12. Kang, H.Y., Fu, X.C.: Epidemic spreading and global stability of an SIS model with an infective vector on complex networks. Commun. Nonlinear Sci. Numer. Simul. **5**(4), 27–30 (2016)
13. Wang, Q.Y., Lin, Z., Jin, Y.H., et al.: ESIS: emotion-based spreader-ignorant-stifler model for information diffusion. Knowl. Based Syst. **81**, 46 (2015)
14. Ai, J.,Su, Z.,Li, Y., et al.: Link prediction based on a spatial distribution model with fuzzy link importance. Physica A **527**, 121155 (2019)
15. Ai, J., Liu, Y.Y.,Su, Z., et al.: Link prediction in recommender systems based on multi-factor network modeling and community detection. Europhys. Lett. **126**, 12003–12011 (2019)
16. Fine, P.E.M.: Herd immunity: history, theory. Pract. Epidemiol. Rev. **15**(265), 302 (1993)
17. Dietz, K., Heesterbeek, J.A.P.: Daniel Bernoulli's epidemiological model revisited. Math. Biosci. **180**(1–2), 1–21 (2002)
18. Anderson, R.M., Anderson, B., May, R.M.: Infectious Diseases of Humans: Dynamics and Control. Oxford University Press, Oxford (1992)
19. Romualdo, P.S., Claudio, C., Piet, V.M., et al.: Epidemic processes in complex networks. Rev. Mod. Phys. **87**(3), 925–979 (2015)
20. Liu, Q.M., Deng, C.S., Sun, M.C.: The analysis of an epidemic model with time delay on scale-free networks. Physica A **410**, 79–84 (2014)
21. Kempe, D., Jon, K., Éva, T.: Maximizing the spread of influence through a social network authors. In: Proceedings of the Ninth ACMSIGKDD International Conference on Knowledge Discovery and Data Mining, p. 137 (2003)
22. Li, C.H., Tsai, C.C., Yang, S.Y.: Analysis of epidemic spreading of an SIRS model in complex heterogeneous networks. Commun. Nonlinear Sci. Numer. Simul. **19**, 1042–1054 (2014)
23. Chen, L.J., Sun, J.T.: Global stability and optimal control of an SIRS epidemic model on heterogeneous networks. Physica A **410**, 196–204 (2014)
24. Xiong, F.,Liu, Y.,Zhang, Z.J., et al.: An information diffusion model based on retweeting mechanism for online social media. Phys. Lett. A **376**, 2103–2108 (2012)
25. Li, T., Wang, Y.M.,Guan, Z.H.: Spreading dynamics of a SIQRS epidemic model on scale-free networks. Commun. Nonlinear Sci. Num. Simul. **19**, 686–692 (2014)

26. Xiong, F., Wang, X.M., Cheng, J.J.: Subtle role of latency for information diffusion in online social networks. Chin. Phys. B **25**(10), 108904 (2016)
27. Shannon, C.E.: A mathematical theory of communication. Bell Syst. Tech. J. **27**(3), 379–423 (1948)
28. Dehmer, M.: Information processing in complex networks: graph entropy and information functionals. Appl. Math. Comput. **201**(1–2), 82–94 (2008)

Author Index

Printed in the United States
by Baker & Taylor Publisher Services